Springer Proceedings in Mathematics & Statistics

Volume 174

Springer Proceedings in Mathematics & Statistics

This book series features volumes composed of selected contributions from workshops and conferences in all areas of current research in mathematics and statistics, including operation research and optimization. In addition to an overall evaluation of the interest, scientific quality, and timeliness of each proposal at the hands of the publisher, individual contributions are all refereed to the high quality standards of leading journals in the field. Thus, this series provides the research community with well-edited, authoritative reports on developments in the most exciting areas of mathematical and statistical research today.

More information about this series at http://www.springer.com/series/10533

Syed Tariq Rizvi · Asma Ali
Vincenzo De Filippis
Editors

Algebra and its Applications

ICAA, Aligarh, India, December 2014

Springer

Editors
Syed Tariq Rizvi
Department of Mathematics
The Ohio State University
Lima, OH
USA

Asma Ali
Department of Mathematics
Aligarh Muslim University
Aligarh, Uttar Pradesh
India

Vincenzo De Filippis
Department of Mathematics and Computer Science
University of Messina
Messina
Italy

ISSN 2194-1009 ISSN 2194-1017 (electronic)
Springer Proceedings in Mathematics & Statistics
ISBN 978-981-10-1650-9 ISBN 978-981-10-1651-6 (eBook)
DOI 10.1007/978-981-10-1651-6

Library of Congress Control Number: 2016943788

Mathematics Subject Classification (2010): 08-xx, 13-xx, 16-xx, 17-xx, 20-xx

Printed on acid-free paper

This Springer imprint is published by Springer Nature
The registered company is Springer Science+Business Media Singapore Pte Ltd.

Preface

The international conference on “Algebra and its Applications” was organized by the Department of Mathematics, Aligarh Muslim University, Aligarh, India, and was held during December 15–17, 2014, under the UGC-DRS (SAP-II) programme. The conference was sponsored by Aligarh Muslim University (AMU), Science and Engineering Research Board (SERB), Department of Science and Technology (DST), New Delhi and the National Board of Higher Mathematics (NBHM), Mumbai.

The purpose of the conference was to bring together algebraists from all over the world working in Algebra and related areas to present their recent research works, exchange new ideas, discuss challenging issues and foster future collaborations in Algebra and is applications. An important aim of the conference was to expose young researchers to new research developments and ideas in Algebra via the talks presented and the research interactions the conference provided.

This research volume based on the proceedings of the conference consists of research literature on latest developments in various branches of algebra. It is the outcome of the invited lectures and research papers presented at the conference. It also includes some articles by invited algebraists who could not attend the conference. This includes Professors Jae Keol Park, Busan National University, Busan, South Korea; Akihiro Yamamura, Akita University, Japan; Shuliang Huang, Chuzhou University, China; Shervin Sahebi and V. Rahmani, Islamic Azad University, Tehran, Iran; Shreedevi K. Masuti and Parangama Sarkar, IIT Bombay, Mumbai; C. Selvaraj, Periyar University, Salem, Tamil Nadu; T. Tamizh Chelvam, Sundarnar University, Tamil Nadu; S. Tariq Rizvi, The Ohio State University, Ohio, Lima, USA; N.K. Thakare, Pune University, Pune; A. Tamilselvi, Ramanujan Institute for Advanced Study in Mathematics, Chennai; and V.S. Kapil, Himachal Pradesh University, Shimla.

To maintain the quality of the work, all papers of the research volume are peer-reviewed by global subject experts. As Algebra continues to experience tremendous growth and diversification, these articles highlight the cross-fertilization of ideas between various branches of algebra and exhibit the latest methods and techniques needed in solving a number of existing research problems while provide new open questions for further research investigations. These will cover a broad range of topics and variety of methodologies. It is expected that this research volume will be a valuable resource for young as well as experienced researchers in Algebra.

Professor Asma Ali was the convener of the conference and Professor M. Mursaleen, the coordinator of the DRS programme. The conference had Professor Patrick W. Keef, Whitman College, Walla, Walla, USA, the chief guest and Professor Ashish K. Srivastava, Saint Louis University, St. Louis, USA, the guest of honor. The enriching programme contained a keynote address on computer-aided linear algebra by no less a mathematician than Professor Vasudevan Srinivas of a premier research centre of India, the Tata Institute of Fundamental Research (TIFR). Professor Vasudevan is also the recipient of Indian National Science Academy (INSA) Medal for Young Scientists, elected Fellow of Indian Academy of Science (IAS) and has received the B.M. Birla Science Award, Swarnajayanthi Fellowship Award, Bhatnagar Prize, J.C. Bose Fellowship, and TWAS Mathematics Prize.

A total of 13 plenary talks and 20 invited talks on current topics of algebra and its applications were delivered by distinguished algebraists. The speakers included, Professors Luisa Carini; Vincenzo De Filippis, Italy; Nanqing Ding, China; Sudhir R. Ghorpade, IIT Bombay; Jugal K. Verma, Tony Joseph and Ananthanarayan, IIT Bombay; Manoj Kummini, Chennai Mathematical Society, Chennai; Sarang Sane, Indian Institute of Science, Bangalore; Kapil Hari Paranjape, IISER Mohali, Chandigarh; M.K. Sen, University of Calcutta; B.N. Waphare, University of Pune, Pune; A.R. Rajan, University of Kerala, Kerala; B.M. Pandeya, Banaras Hindu University; P.G. Romeo, Cochin University of Sciences and Technology (CUSAT), Kerala; Manoj Kumar Yadav, Harish-Chandra Research Institute (HRI), Allahabad; R.P. Sharma, Himachal Pradesh University, Summerhill, Shimla, and others. Another 44 research papers were presented by young researchers in algebra. Overall, the conference was greatly successful in its aims and objectives.

The organizing committee, for the first time in mathematics conferences held at A.M.U. Aligarh, initiated the best paper presentation award. The award was initiated to motivate and inspire the young talents below the age of 32 years, and carried a participation certificate and a modest prize in cash.

We thank all our colleagues who contributed papers to this research volume and those who graciously accepted to serve as referees of the submitted papers. We also thank the organizing team, the members of the department and the student workers who actively helped in making the conference a success. The conference could not be so successful without their active help and participation. The financial support from all agencies listed is gratefully acknowledged. We also express our thanks to Springer for bringing out this volume in a nice form. The professional help and cooperation provided by Mr. Shamim Ahmad, Editor (Math. Sciences) is thankfully appreciated and acknowledged.

Lima, USA — Syed Tariq Rizvi
Aligarh, India — Asma Ali
Messina, Italy — Vincenzo De Filippis

Contents

Editors and Contributors

About the Editors

Syed Tariq Rizvi is Professor and the Coordinator of Department of Mathematics, the Ohio State University (OSU), Lima, USA. After receiving his Ph.D. from McMaster University (Canada) in 1981 and teaching at the University of Waterloo (Canada), he joined OSU in 1982. A recipient of the Outstanding Scholar Award from OSU and a Fulbright Fellow Award from the Fulbright Board, Washington, DC, Professor Rizvi is world renowned for his research contributions to the theory of rings and modules. He has numerous research papers to his credit published in several international journals of high repute, published a monograph by Birkhäuser and coedited five research volumes with other reputed publishers. He also serves on editorial boards of ten international journals, including two as a managing or executive editor. Professor Rizvi has given over 120 invited talks at several national and international conferences, many as a plenary speaker. He has regularly organized sessions at a number of conferences including at the American Mathematical Society conferences and the Ohio State-Denison Math Conference series.

Asma Ali is Professor of Mathematics at the Department of Mathematics, Aligarh Muslim University (AMU), Aligarh, India. She has made valuable contributions to mathematical research with publishing a number of papers in international journals of repute. She has been awarded the honor of recognition for the achievements as a successful professional at the International Women's Day at AMU in 2015. She is also life member of many organizations including the American Mathematical Society, the Indian Mathematical Society and the Indian Science Congress.

Vincenzo De Filippis is Associate Professor of Mathematics at the University of Messina, Italy. He received the M.S. (1993) and Ph.D. (1999) degrees from the University of Messina. He is an author and coauthor of numerous professional publications and has been on research visits to several institutions in Europe, India, China and Turkey. His research interests include ring theory, theory of associative algebras, linear and multilinear algebra. His theoretical work is primarily aimed at studying the structure of algebras satisfying functional identities.

Contributors

R. Akhila Department of Mathematics, Cochin University of Science and Technology, Kochi, Kerala, India

Farhat Ali Department of Mathematics, Aligarh Muslim University, Aligarh, India

T. Asir Department of Mathematics (DDE), Madurai Kamaraj University, Madurai, Tamil Nadu, India

P.A. Azeef Muhammed Department of Mathematics, University of Kerala, Trivandrum, Kerala, India

S. Chattopadhyay Sovarani Memorial College, Jagathballavpur, India

A.K. Chaturvedi Department of Mathematics, University of Allahabad, Allahabad, India

Basudeb Dhara Department of Mathematics, Belda College, Belda Paschim Medinipur, West Bengal, India

Basudeb Dhara Department of Mathematics, Belda College, Paschim Medinipur, India

R.D. Giri RTM Nagpur University, Nagpur, India

Shuliang Huang School of Mathematics and Finance, Chuzhou University, Chuzhou, People's Republic of China

Selby Jose School of Mathematics, Tata Institute of Fundamental Research, Mumbai, India; Department of Mathematics, Institute of Science, Mumbai, India

V.S. Kapil Department of Mathematics, Himachal Pradesh University, Shimla, India

B. Kethesan Ramanujan Institute for Advanced Study in Mathematics, University of Madras, Chennai, India

Anil Khairnar Department of Mathematics, Abasaheb Garware College, Pune, India

Mohammad Aasim Khan Department of Mathematics, Aligarh Muslim University, Aligarh, India

Noor Mohammad Khan Department of Mathematics, Aligarh Muslim University, Aligarh, India

Shahoor Khan Department of Mathematics, Aligarh Muslim University, Aligarh, India

Shreedevi K. Masuti Chennai Mathematical Institute, Chennai, Tamil Nadu, India

Jae Keol Park Department of Mathematics, Busan National University, Busan, South Korea

Rajni Parmar Department of Mathematics, Himachal Pradesh University, Shimla, India

Manoj Kumar Patel Department of Mathematics, NIT Nagaland, Dimapur, India

Avinash Patil Garware College of Commerce, Pune, India

Tony J. Puthenpurakal Department of Mathematics, Indian Institute of Technology Bombay, Powai, Mumbai, India

Venus Rahmani Department Of Mathematics, Islamic Azad University, Tehran, Iran

A.R. Rajan Department of Mathematics, University of Kerala, Kariavattom, India

Ravi A. Rao School of Mathematics, Tata Institute of Fundamental Research, Mumbai, India; Department of Mathematics, Institute of Science, Mumbai, India

P.G. Romeo Department of Mathematics, Cochin University of Science and Technology, Kochi, Kerala, India

Shervin Sahebi Department of Mathematics, Islamic Azad University, Tehran, Iran

R. Saravanan Department of Mathematics, Periyar University, Salem, Tamil Nadu, India

Parangama Sarkar Department of Mathematics, Indian Institute of Technology Bombay, Mumbai, India

K. Selvakumar Department of Mathematics, Manonmaniam Sundaranar University, Tirunelveli, Tamil Nadu, India

C. Selvaraj Department of Mathematics, Periyar University, Salem, Tamil Nadu, India

M.K. Sen Department of Pure Mathematics, University of Calcutta, Kolkata, India

Ram Parkash Sharma Department of Mathematics, Himachal Pradesh University, Shimla, India

Ashish K. Srivastava Department of Mathematics and Computer Science, St. Louis University, St. Louis, MO, USA

A. Tamilselvi Ramanujan Institute for Advanced Study in Mathematics, University of Madras, Chennai, India

T. Tamizh Chelvam Department of Mathematics, Manonmaniam Sundaranar University, Tirunelveli, Tamil Nadu, India

N.K. Thakare Dhule, India

J.K. Verma Department of Mathematics, Indian Institute of Technology Bombay, Mumbai, India

A. Vidhya Ramanujan Institute for Advanced Study in Mathematics, University of Madras, Chennai, India

B.N. Waphare Department of Mathematics, Center for Advanced Studies in Mathematics, Savitribai Phule Pune University, Pune, India

Manoj K. Yadav School of Mathematics, Harish-Chandra Research Institute, Jhunsi, Allahabad, India

Akihiro Yamamura Department of Mathematical Science and Electrical-Electronic-Computer Engineering, Akita University, Akita, Japan

On Some Classes of Module Hulls

Jae Keol Park and S. Tariq Rizvi

Abstract The study of various types of hulls of a module has been of interest for a long time. Our focus in this paper is to present results on some classes of these hulls of modules, their examples, counter examples, constructions and their applications. Since the notion of hulls and its study were motivated by that of an injective hull, we begin with a detailed discussion on classes of module hulls which satisfy certain properties generalizing the notion of injectivity. Closely linked to these generalizations of injectivity, are the notions of a Baer ring and a Baer module. The study of Baer ring hulls or Baer module hulls has remained elusive in view of the underlying difficulties involved. Our main focus is to exhibit the latest results on existence, constructions, examples and applications of Baer module hulls obtained by Park and Rizvi. In particular, we show the existence and explicit description of the Baer module hull of a module N over a Dedekind domain R such that $N/t(N)$ is finitely generated and $\mathrm{Ann}_R(t(N)) \neq 0$, where $t(N)$ is the torsion submodule of N. When $N/t(N)$ is not finitely generated, it is shown that N may not have a Baer module hull. Among applications, our results yield that a finitely generated module N over a Dedekind domain is Baer if and only if N is semisimple or torsion-free. We explicitly describe the Baer module hull of the direct sum of $\mathbb{Z}$ with $\mathbb{Z}_p$ (p a prime integer) and extend this to a more general construction of Baer module hulls over any commutative PID. We show that the Baer hull of a direct sum of two modules is not necessarily isomorphic to the direct sum of the Baer hulls of the modules, even if each relevant Baer module hull exists. A number of examples and applications of various classes of hulls are included.

Dedication: Dedicated to the memory of Professor Bruno J. Müller

---

J.K. Park (✉)
Department of Mathematics, Busan National University,
Busan 609-735, South Korea
e-mail: jkp1128@yahoo.com

S.T. Rizvi
Department of Mathematics, The Ohio State University,
Lima, OH 45804-3576, USA
e-mail: rizvi.1@osu.edu

S.T. Rizvi et al. (eds.), *Algebra and its Applications*, Springer Proceedings
in Mathematics & Statistics 174, DOI 10.1007/978-981-10-1651-6_1

Keywords Hull · Quasi-injective · Continuous · Quasi-continuous · (FI-) Extending · Baer module · Baer hull · Baer ring · Quasi-Baer ring · Dedekind domain · Quasi-retractable · Fractional ideal

Classifications 16D10 · 16D50 · 16D25 · 16D40 · 16D80 · 16E60 · 16P40

1 Introduction

Since the discovery of the existence of the injective hull of an arbitrary module independently in 1952 by Shoda [49] and in 1953 by Eckmann and Schopf [14], there have been numerous papers dedicated to the study and description of various types of hulls. These hulls are basically smallest extensions of rings and modules satisfying some generalizations of injectivity (for example, quasi-injective, continuous, quasi-continuous hulls, etc.) or satisfying properties related to such generalizations of injectivity. For a given module M (or a given ring R), the investigations include in general, to construct the smallest essential extension of M (or of R) which belongs to a particular class of modules (or of rings) within a fixed injective hull of M (or a fixed maximal quotient ring of R). We call this a hull of M (or of R) belonging to that particular class. One benefit of these hulls is that such hulls generally lie closer to the module M (or to the ring R) than its injective hull. This closeness may allow for a better transfer of information between M (or R) and that particular hull of M (or of R) from these classes than between M (or R) and its injective hull. These hulls have also proved to be useful tools for the study of the structure of M (or of R). So an important focus of investigations has been to obtain results on the existence and explicit descriptions of various types of module hulls. This is the topic of this survey paper.

We recall that a module M is said to be *quasi-injective* if, for each $N \leq M$, any $f \in \text{Hom}(N, M)$ can be extended to an endomorphism of M. Among other well-known generalizations of injectivity, the study of the continuous, quasi-continuous, extending, and the FI-extending properties has been extensive in the literature (see for example [4, 8, 13, 34–36, 43]). A module M is said to be *extending* if, for each $V \leq M$, there exists a direct summand $W \leq^{\oplus} M$ such that $V \leq^{\text{ess}} W$. And an extending module M is called *quasi-continuous* if for all direct summands M_1 and M_2 of M with $M_1 \cap M_2 = 0$, $M_1 \oplus M_2$ is also a direct summand of M. Furthermore, an extending module M is said to be *continuous* if every submodule N of M which is isomorphic to a direct summand is also a direct summand of M. A module M is called *FI-extending* if every fully invariant submodule is essential in a direct summand of M. For more details on FI-extending modules, see [4, 8], and [10, Sect. 2.3]. The following implications hold true for modules:

$$\text{injective} \Rightarrow \text{quasi-injective} \Rightarrow \text{continuous}$$
$$\Rightarrow \text{quasi-continuous} \Rightarrow \text{extending} \Rightarrow \text{FI-extending}$$

while each of reverse implications does not hold true, in general.

Since the injective module hull of a module always exists [14, 49], the study of module hulls with certain properties inside the injective hull of the module is more natural in contrast to the study of ring hulls of a ring (the injective hull of a ring may not even be a ring in general–and even if it is, for it to have a compatible ring structure with the ring is another hurdle).

Section 1 of the paper is devoted to results and examples (of either existence or non-existence) of various hulls which generalize injective hulls. This includes the consideration of quasi-injective, continuous, quasi-continuous and (FI-)extending module hulls. For a given module M, let $H = \mathrm{End}_R(E(M))$ denote the endomorphism ring of its injective hull $E(M)$. By Johnson and Wong [23], the unique quasi-injective hull of the module M is precisely given by HM. Goel and Jain [16] showed that there always exists a unique quasi-continuous hull of every module. The quasi-continuous hull of M is given by ΩM, where Ω is the subring generated by all idempotents of $H = \mathrm{End}(E(M))$. In contrast to this, it was shown by Müller and Rizvi in [35] that continuous module hulls do not always exist. However, they did show the existence of continuous hulls of certain classes of modules over a commutative ring (such as nonsingular cyclic ones) and provided a description of these continuous hulls (see [35, Theorem 8]). Similar to the case of continuous module hulls, it is also known that extending module hulls do not always exist (for example, see [10, Example 8.4.13, p. 319]). For the case of FI-extending module hulls, it was proved in [8, Theorem 6] that every finitely generated projective module over a semiprime ring has an FI-extending hull.

Closely linked to these notions, are the notions of a Baer ring and a Baer module. A ring R in which the left (right) annihilator of every nonempty subset of R is generated by an idempotent is called a Baer ring. It is well-known that this is a left-right symmetric notion for rings. Kaplansky introduced the notion of Baer rings in [26] (also see [27]). Having their roots in Functional Analysis, the class of Baer rings and the more general class of quasi-Baer rings (discussed ahead) were studied extensively by Kaplansky and many others who obtained a number of interesting results on these classes of rings (see [1, 3, 6–12, 18, 19, 21, 22, 31–33, 37, 38, 41]).

More recently, the notion of a Baer ring was extended to an analogous module theoretic notion using the endomorphism ring of the module by Rizvi and Roman in [44]. According to [44], a module M is called a *Baer module* if, for any $N_R \leq M_R$, there exists $e^2 = e \in S$ such that $\ell_S(N) = Se$, where $\ell_S(N) = \{f \in S \mid f(N) = 0\}$ and $S = \mathrm{End}(M_R)$. Equivalently, a module M is Baer if and only if for any left ideal I of S, $r_M(I) = fM$ with $f^2 = f \in S$, where $r_M(I) = \{m \in M \mid Im = 0\}$. Examples of Baer modules include any nonsingular injective module. In particular, it is known that every ($\mathcal{K}$-)nonsingular extending module is a Baer module while the converse holds under a certain dual condition. To study Baer module hulls, we provide relevant results and properties of Baer modules and related notions in Sect. 3 of the paper. These results will also be used in Sect. 4 of the paper.

In the main section, Sect. 4 of this expository paper, we introduce and discuss Baer module hulls of certain classes modules over a Dedekind domain from our recent work in [40]. We exhibit explicit constructions and examples of Baer module hulls and provide their applications in this section. Properties of Baer module hulls will also be discussed.

Extending the notion of a Baer ring, a quasi-Baer ring was introduced by Clark in [12]. A ring for which the left annihilator of every ideal is generated by an idempotent, as a left ideal is called a quasi-Baer ring. It was initially defined by Clark to help characterize a finite dimensional algebra over an algebraically closed field F to be a *twisted semigroup algebra* of a matrix units semigroup over F. Historically, it is of interest to note that the Hamilton quaternion division algebra over the real numbers field $\mathbb{R}$ is a *twisted group algebra* of the Klein four group V_4 over $\mathbb{R}$. It was also shown in [12] that any finite distributive lattice is isomorphic to a certain sublattice of the lattice of all ideals of an artinian quasi-Baer ring. It is clear that every Baer ring is quasi-Baer while the converse is not true in general. It is also obvious that the two notions coincide for a commutative ring and for a reduced ring. In [41], a number of interesting properties of quasi-Baer rings are obtained. See [10] for more details on quasi-Baer rings.

Quasi-Baer modules were defined and investigated by Rizvi and Roman [44] in the module theoretic setting. Recall from [44] that a module M_R is called a *quasi-Baer module* if for each $N \trianglelefteq M$, $\ell_S(N) = Se$ for some $e^2 = e \in S$, where $S = \text{End}(M_R)$. Thus M_R is quasi-Baer if and only if for any ideal J of S, $r_M(J) = fM$ for some $f^2 = f \in S$. In [44] and [47], it is shown that the endomorphism ring of a (quasi-)Baer module is a (quasi-)Baer ring. It is proved that there exist close connections between quasi-Baer modules and FI-extending modules. A number of interesting properties of quasi-Baer modules and applications have also been presented.

As mentioned earlier, the notion of a "hull" with a certain property allows us to work with an overmodule or overring which has better properties than the original module or ring. It is worth mentioning that very little is known even about Baer ring hulls. Recall from [10, Chap. 8] that the *Baer* (resp., *quasi-Baer*) *ring hull* of a ring R is the smallest Baer (resp., quasi-Baer) right essential overring of R in $E(R_R)$. To the best of our knowledge, the only explicit results about *Baer ring hulls* in earlier existing literature have been due to Mewborn [33] for commutative semiprime rings, Oshiro [37] and [38] for commutative von Neumann regular rings, and Hirano, Hongan and Ohori [19] for reduced right Utumi rings. All these results were recently extended and a unified result was obtained for the case of an arbitrary semiprime ring using *quasi-Baer ring hulls* by Birkenemier, Park, and Rizvi [7, Theorem 3.3]. The focus of the present paper is on module hulls, more specifically on results and study of Baer module hulls.

For a given module M, the smallest Baer overmodule of M in $E(M)$ is called the *Baer module hull* of M. In short, we will often call it the *Baer hull* of M and denote it by $\mathfrak{B}(M)$.

Park and Rizvi in [40] recently initiated the study of the Baer module hulls. We introduce and discuss the results obtained in [40] on the Baer module hulls in Sect. 4. We show that the Baer module hull exists for a module N over a Dedekind domain R such that $N/t(N)$ is finitely generated and $\text{Ann}_R(t(N)) \neq 0$, where $t(N)$ is the torsion submodule of N. An explicit description of this Baer module hull has been provided. In contrast, an example exhibits a module N for which $N/t(N)$ is not finitely generated and which does not have a Baer module hull.

Among applications presented, we show that a finitely generated module N over a Dedekind domain is Baer if and only if N is semisimple or torsion-free. We explicitly describe the Baer module hull of $N = \mathbb{Z}_p \oplus \mathbb{Z}$, where p is a prime integer, as $V = \mathbb{Z}_p \oplus \mathbb{Z}[1/p]$ and extend this to a more general construction of Baer module hulls over any commutative PID. It is shown that unlike the case of (quasi-)injective hulls, the Baer hull of the direct sum of two modules is not necessarily isomorphic to the direct sum of the Baer hulls of the modules, even if all relevant Baer module hulls exist. Several interesting examples and applications of various types of module hulls are included throughout the paper.

All rings are assumed to have identity and all modules are assumed to be unitary. For right R-modules M_R and N_R, we use $\text{Hom}(M_R, N_R)$, $\text{Hom}_R(M, N)$, or $\text{Hom}(M, N)$ to denote the set of all R-module homomorphisms from M_R to N_R. Likewise, $\text{End}(M_R)$, $\text{End}_R(M)$, or $\text{End}(M)$ denote the endomorphism ring of an R-module M. For a given R-homomorphism (or R-module homomorphism) $f \in \text{Hom}_R(M, N)$, $\text{Ker}(f)$ denotes the kernel of f. A submodule U of a module V is said to be *fully invariant* in V if $f(U) \subseteq U$ for all $f \in \text{End}(V)$.

We use $E(M_R)$ or $E(M)$ for an injective hull of a module M_R. For a module M, we use $K \leq M$, $L \trianglelefteq M$, $N \leq^{\text{ess}} M$, and $U \leq^{\oplus} M$ to denote that K is a submodule of M, L is a fully invariant submodule of M, N is an essential submodule of M, and U is a direct summand of M, respectively.

If M is an R-module, $\text{Ann}_R(M)$ stands for the annihilator of M in R. For a module M and a set Λ, let $M^{(\Lambda)}$ be the direct sum of $|\Lambda|$ copies of M, where $|\Lambda|$ is the cardinality of Λ. When Λ is finite with $|\Lambda| = n$, then $M^{(n)}$ is used for $M^{(\Lambda)}$. For a ring R and a positive integer n, $\text{Mat}_n(R)$ and $T_n(R)$ denote the $n \times n$ matrix ring and the $n \times n$ upper triangular matrix ring over R, respectively.

For a ring R, $Q(R)$ denotes the maximal right ring of quotients of R. The symbols $\mathbb{Q}$, $\mathbb{Z}$, and $\mathbb{Z}_n (n > 1)$ stand for the field of rational numbers, the ring of integers, and the ring of integers modulo n, respectively. Ideals of a ring without the adjective "left" or "right" mean two-sided ideals.

As mentioned, we will use the term *Baer hull* for Baer module hull in this paper.

2 Quasi-Injective, Continuous, Quasi-Continuous, Extending, and FI-Extending Hulls

We begin this section with a discussion on some useful generalizations of injectivity which are related to the topics of study in this paper. In particular, we discuss the notions of quasi-injective, continuous, quasi-continuous, extending, and FI-extending modules. Relationships between these notions, their examples, characterizations, and other relevant properties are presented.

For a given module M, its injective hull $E(M)$ is the minimal injective overmodule of M (equivalently, its maximal essential extension) and is unique up to isomorphism over M (see [14] and [49]). We discuss module hulls satisfying some generalizations of injectivity. One may expect that such minimal overmodules H of a module M will allow for a rich transfer of information between M and H. This, because each of these hulls, with more general properties than injectivity, sits in between a module M and a fixed injective hull $E(M)$ of M. Therefore, that specific hull of the module M usually lies closer to the module M that $E(M)$.

A module M is said to be *quasi-injective* if for every submodule N of M, each $\varphi \in \text{Hom}(N, M)$ extends to an R-endomorphism of M. The following is a well-known result.

Theorem 2.1 *A module M is quasi-injective if and only if M is fully invariant in $E(M)$.*

Quasi-injectivity is an important generalization of injectivity. All quasi-injective modules satisfy the (C_1), (C_2), (C_3), and (FI) conditions given next.

Proposition 2.2 *Let M be a quasi-injective module. Then it satisfies the following conditions.*

(C_1) *Every submodule of M is essential in a direct summand of M.*
(C_2) *If $V \leq M$ and $V \cong N \leq^{\oplus} M$, then $V \leq^{\oplus} M$.*
(C_3) *If M_1 and M_2 are direct summands of M such that $M_1 \cap M_2 = 0$, then $M_1 \oplus M_2$ is a direct summand of M.*
(FI) *Any fully invariant submodule of M is essential in a direct summand of M.*

It is easy to see the relationship between the condition (C_2) and the condition (C_3) as follows.

Proposition 2.3 *If a module M satisfies (C_2), then it satisfies (C_3).*

Conditions (C_1), (C_2), (C_3), and (FI) help define the following notions.

Definition 2.4 Let M be a module.

(i) M is called *continuous* if it satisfies the (C_1) and (C_2) conditions.
(ii) M is said to be *quasi-continuous* if it has the (C_1) and (C_3) conditions.
(iii) M is called *extending* (or *CS*) if it satisfies the (C_1) condition.

(iv) M is called *FI-extending* if it satisfies the (FI) condition.

From the preceding, the following implications hold true for modules. However, the reverse implications do not hold as illustrated in Example 2.5.

$$\text{injective} \Rightarrow \text{quasi-injective} \Rightarrow \text{continuous}$$
$$\Rightarrow \text{quasi-continuous} \Rightarrow \text{extending} \Rightarrow \text{FI-extending}.$$

Example 2.5 (i) Every injective module and every semisimple module are quasi-injective. There exist simple modules which are not injective (e.g., $\mathbb{Z}_p$ for any prime integer p as a $\mathbb{Z}$-module). Further, there is a quasi-injective module which is neither injective nor semisimple. Let $R = \mathbb{Z}$ and $M = \mathbb{Z}_{p^n}$, with p a prime integer and n an integer such that $n > 1$. Then $E(M) = \mathbb{Z}_{p^\infty}$, the Prüfer p-group, and thus M is neither injective nor semisimple. But $f(M) \subseteq M$ for any $f \in \text{End}(E(M))$. So M is quasi-injective by Theorem 2.1 (see [15, Example, p. 22]).

(ii) Let K be a field and F be a proper subfield of K. Set $K_n = K$ for all $n = 1, 2 \ldots$. We take.

$$R = \left\{ (a_n)_{n=1}^{\infty} \in \prod_{n=1}^{\infty} K_n \mid a_n \in F \text{ eventually} \right\},$$

which is a subring of $\prod_{n=1}^{\infty} K_n$. Say $I \trianglelefteq R$. Then we can verify that $r_R(I) = eR$ with $e^2 = e \in R$. Therefore $I_R \leq^{\text{ess}} r_R(\ell_R(I)) = (1 - e)R_R$ as R is semiprime. So R_R is extending. Further, since R is von Nuemann regular, R_R also satisfies (C_2) condition. Thus R_R is continuous. As $E(R_R) = \prod_{n=1}^{\infty} K_n$, R_R is not injective, so R_R is not quasi-injective.

(iii) Let R be a right Ore domain which is not a division ring (e.g., the ring $\mathbb{Z}$ of integers). Then R_R is quasi-continuous. Take $0 \neq x \in R$ such that $xR \neq R$. Then $xR_R \cong R_R$, but xR_R is not a direct summand of R_R. Thus R_R is not continuous.

(iv) Let F be a field and $R = T_2(F)$, the 2×2 upper triangular matrix ring over F. Then we see that R_R is extending. Let $e_{ij} \in R$ be the matrix with 1 in the (i, j)-position and 0 elsewhere. Put $e = e_{12} + e_{22}$ and $f = e_{22}$. Then $e^2 = e$ and $f^2 = f$. Note that $eR \cap fR = 0$. But $eR_R \oplus fR_R$ is not a direct summand of R_R. Thus R_R is not quasi-continuous.

(v) Let $R = \text{Mat}_n(\mathbb{Z}[x])$ (n is an integer such that $n > 1$). Then R_R is FI-extending, but R_R is not extending. Further, the module $M = \oplus_{n=1}^{\infty} \mathbb{Z}$ is an FI-extending $\mathbb{Z}$-module which is not extending.

The next theorem allows us to transfer any given decomposition of the injective hull $E(M)$ of a quasi-continuous module M to a similar decomposition for M (the converse always holds). This fact is also helpful in transference of properties between between a quasi-continuous module M and its injective hull $E(M)$ or a module in between.

Theorem 2.6 ([16], [20], and [39]) *The following are equivalent for a module M.*

(i) M is quasi-continuous.
(ii) $M = X \oplus Y$ for any two submodules X and Y which are complements of each other.
(iii) $fM \subseteq M$ for every $f^2 = f \in End_R(E(M))$.
(iv) $E(M) = \oplus_{i\in\Lambda} E_i$ implies $M = \oplus_{i\in\Lambda}(M \cap E_i)$.
(v) Any essential extension V of M with a decomposition $V = \oplus_{\alpha\in\Gamma} V_\alpha$ implies that $M = \oplus_{\alpha\in\Gamma}(M \cap V_\alpha)$.

Remark 2.7 The equivalence of the conditions (i), (ii), (iii), and (iv) of Theorem 2.6 are comprised by results obtained in [16] and [20], while the condition (v) of Theorem 2.6 is obtained in [39].

Definition 2.8 Let $\mathfrak{M}$ be a class of modules and M be any module. We call, when it exists, a module H the $\mathfrak{M}$ *hull* of M if H is the smallest essential extension of M in a fixed injective hull $E(M)$ that belongs to $\mathfrak{M}$.

It is clear from the preceding definition that an $\mathfrak{M}$ hull of a module is unique within a fixed injective hull $E(M)$ of M. It may be worth to note that in [42, Definitions 4.7, 4.8, and 4.9, pp. 36–37], three types of continuous hulls of a module, Type I, Type II, and Type III are introduced (see also [35, Definitions]). The authors of [42] and [35] chose the Type III continuous hull of a module to be called as the continuous hull of an arbitrary module for several reasons provided in [42] and [35]. Our Definition 2.8 follows the definition of continuous hull of Type III.

The next result due to Johnson and Wong [23] describes precisely how the quasi-injective hull of a module can be constructed and that the quasi-injective hull of any module always exists.

Theorem 2.9 *Assume that M is a right R-module and let $S = End(E(M))$. Then $SM = \{\sum f_i(m_i) \mid f_i \in S \text{ and } m_i \in M\}$ is the quasi-injective hull of M.*

The following result for the existence of the quasi-continuous hull of a module is obtained by Goel and Jain [16].

Theorem 2.10 *Assume that M is a right R-module and $S = End(E(M))$. Let Ω be the subring of S generated by the set of all idempotents of S. Then $\Omega M = \{\sum f_i(m_i) \mid f_i \in \Omega \text{ and } m_i \in M\}$ is the quasi-continuous hull of M.*

Recall that a module is called *uniform* if the intersection of any two nonzero submodule is nonzero (i.e., the module $\mathbb{Z}_\mathbb{Z}$). If M is a uniform module, then $E(M)$ is also uniform. Thus $S = \text{End}(E(M))$ has only trivial idempotents, so $\Omega M = M$. Therefore the quasi-continuous module hull of M is M itself.

A module is said to be *directly finite* if it is not isomorphic to a proper direct summand of itself. A module is called *purely infinite* if it is isomorphic to the direct sum of two copies of itself. Recall that a ring R is called directly finite if $xy = 1$

implies $yx = 1$ for $x, y \in R$. We remark that a module M is directly finite if and only if $\mathrm{End}(M)$ is directly finite.

The following result was obtained by Goodearl [17] in a categorical way. In [36], Müller and Rizvi gave an algebraic proof of the result and extended it. They also proved a strong "uniqueness" of the decomposition. The result was further extended by them to a similar decomposition of a quasi-continuous module as provided in Theorem 2.13 ahead.

Theorem 2.11 ([36, Theorem 1]) *Every injective module E has a direct sum decomposition, $E = U \oplus V$, where U is directly finite, V is purely infinite, and U and V have no nonzero isomorphic direct summands (or submodules). If $E = U_1 \oplus V_1 = U_2 \oplus V_2$ are two such decompositions, then $E = U_1 \oplus V_2$ holds too, and consequently $U_1 \cong U_2$ and $V_1 \cong V_2$.*

Given a quasi-continuous module M and a submodule A of M, it is easy to find the direct summand of M in which A is essential (just consider $M \cap E(A)$). This summand was called an *internal quasi-continuous hull* of A in M by Müller and Rizvi [36].

Another interesting property of a quasi-continuous module M obtained is that if A and B are two isomorphic submodules of M then the direct summands of M which are essential over A and B respectively, are unique up to isomorphism as follows.

Theorem 2.12 ([36, Theorem 4]) *Assume that M is a quasi-continuous module and $A_i \leq^{ess} P_i \leq^{\oplus} M$ $(i = 1, 2)$. If $A_1 \cong A_2$, then $P_1 \cong P_2$.*

By using Theorem 2.12, the decomposition theorem of injective modules (Theorem 2.11) can be extended to the case of quasi-continuous modules as follows.

Theorem 2.13 ([36, Proposition 6]) *Every quasi-continuous module M has a direct sum decomposition, $M = U \oplus V$, where U is directly finite, V is purely infinite, and U and V have no nonzero isomorphic direct summands (or submodules). If $M = U_1 \oplus V_1 = U_2 \oplus V_2$ are two such decompositions, then $M = U_1 \oplus V_2$ holds too, and consequently $U_1 \cong U_2$ and $V_1 \cong V_2$.*

The existence and description of continuous hulls of certain modules have been investigated in [42] (and [35]). In contrast to Theorems 2.9 and 2.10, Müller and Rizvi [35, Example 3] construct the example of a nonsingular uniform cyclic module over a *noncommutative ring* which cannot not have a continuous hull as follows.

Example 2.14 Let V be a vector space over a field F with basis elements v_m, w_k $(m, k = 0, 1, 2, \ldots)$. We denote by V_n the subspace of V generated by the v_m $(m \geq n)$ and all the w_k. Also we denote by W_n the subspace generated by the w_k $(k \geq n)$. We write $\mathcal{S}$ for the shifting operator such that $\mathcal{S}(w_k) = w_{k+1}$ and $\mathcal{S}(v_i) = 0$ for all k, i.

Let R be the set of all $\rho \in \mathrm{End}_F(V)$ such that $\rho(v_m) \in V_m$, $\rho(w_0) \in W_0$ and $\rho(w_k) = \mathcal{S}^k \rho(w_0)$, for $m, k = 0, 1, 2, \ldots$. Note that $\tau\rho(w_k) = \mathcal{S}^k \tau\rho(w_0)$, for $\rho, \tau \in R$, and so $\tau\rho \in R$. Thus it is routine to check that R is a subring of $\mathrm{End}_F(V)$. Further, we see that $V_n = Rv_n$, $W_n = Rw_n$, and $V_{n+1} \subseteq V_n$ for all n. (When $f \in R$ and $v \in V$, we also use fv for the image $f(v)$ of v under f.)

Consider the left R-module $M = W_0$. First, we show that $M = Rw_0$ is uniform. For this, take $fw_0 \neq 0$, $gw_0 \neq 0$ in M, where $f, g \in R$. We need to find $h_1, h_2 \in R$ such that $h_1 f w_0 = h_2 g w_0 \neq 0$. Let

$$fw_0 = b_0 w_0 + b_1 w_1 + \cdots + b_m w_m \in Rw_0$$

and

$$gw_0 = c_0 w_0 + c_1 w_1 + \cdots + c_m w_m \in Rw_0,$$

where $b_i, c_j \in F, i, j = 0, 1, \ldots, m$, and some terms of b_i and c_j may be zero.

Put $h_1 w_0 = x_0 w_0 + x_1 w_1 + \cdots + x_\ell w_\ell$ and $h_2 w_0 = y_0 w_0 + y_1 w_1 + \cdots + y_\ell w_\ell$, where $x_i, y_i \in F$, $i = 0, 1, \ldots, \ell$ (also some terms of x_i and y_j may be zero). Since $h_1(w_k) = \mathcal{S}^k h_1(w_0)$ and $h_2(w_k) = \mathcal{S}^k h_2(w_0)$ for $k = 0, 1, 2 \ldots$, we need to find such $x_i, y_i \in F, 0 \leq i \leq \ell$ so that $h_1 f w_0 = h_2 g w_0 \neq 0$ from the following equations:

$$b_0 x_0 = c_0 y_0, \quad b_0 x_1 + b_1 x_0 = c_0 y_1 + c_1 y_0,$$

$$b_0 x_2 + b_1 x_1 + b_2 x_0 = c_0 y_2 + c_1 y_1 + c_2 y_0,$$

$$b_0 x_3 + b_1 x_2 + b_2 x_1 + b_3 x_0 = c_0 y_3 + c_2 y_1 + c_2 y_1 + c_3 y_0,$$

and so on.

Say $\alpha(t) = b_0 + \cdots + b_m t^m \neq 0$ and $\beta(t) = c_0 + \cdots + c_m t^m \neq 0$ in the polynomial ring $F[t]$. Then $\alpha(t)F[t] \cap \beta(t)F[t] \neq 0$.

We may note that finding such $x_0, x_1 \ldots, x_\ell, y_0, y_1 \ldots, y_\ell$ in F above is the same as the job of finding $x_0, x_1 \ldots, x_\ell, y_0, y_1, \ldots, y_\ell$ such that

$$\alpha(t)(x_0 + x_1 t + \cdots + x_\ell t^\ell) = \beta(t)(y_0 + y_1 t + \cdots + y_\ell t^\ell) \neq 0$$

in the polynomial ring $F[t]$. Observing that $0 \neq \alpha(t)\beta(t) \in \alpha(t)F[t] \cap \beta(t)F[t]$, take $h_1 w_0 = c_0 w_0 + c_1 w_1 + \cdots + c_m w_m$ by putting $\ell = m$, $x_i = c_i$ for $0 \leq i \leq m$, and $h_2 w_0 = b_0 w_0 + b_1 w_1 + \cdots + b_m w_m$ by putting $\ell = m$, $y_i = b_i$ for $0 \leq i \leq m$. Since $\alpha(t)\beta(t) \neq 0$, we see that $0 \neq h_1 f w_0 = h_2 g w_0 \in Rf w_0 \cap Rg w_0$. So M is uniform.

Next, we show that each V_n is an essential extension of M (hence each V_i is uniform). Indeed, let $0 \neq \mu v_n \in Rv_n = V_n$, where $\mu \in R$. Say

$$\mu v_n = a_{n+k} v_{n+k} + \cdots + a_{n+k+\ell} v_{n+k+\ell} + b_s w_s + \cdots + b_{s+m} w_{k+m}.$$

If $a_{n+k} = \cdots = a_{n+k+\ell} = 0$, then $\mu v_n \in W_0$. Otherwise, we may assume that $a_{n+k} \neq 0$. Let $\omega \in R$ such that $\omega(v_{n+k}) = w_0$ and $\omega(v_i) = 0$ for $i \neq n+k$ and $\omega(w_j) = 0$ for all j. Then $0 \neq \omega\mu v_n = a_{n+k} w_0 \in W_0$. Thus $M = W_0$ is essential in V_n. Since M is uniform, V_n is also uniform for all n.

We prove that ${}_R M$ is nonsingular. For this, assume that $u \in Z({}_R M)$ (where $Z({}_R M)$ is the singular submodule of ${}_R M$) and let $K = \{\alpha \in R \mid \alpha u = 0\}$. Then K is an essential left ideal of R. So $K \cap RS^2 \neq 0$. Thus there is $\rho \in R$ with $\rho S^2 \neq 0$ and $\rho S^2(u) = 0$. Say

$$u = a_k w_k + a_{k+1} w_{k+1} + \cdots + a_n w_n$$

with $a_k, a_{k+1}, \ldots, a_n \in F$. Assume on the contrary that $u \neq 0$. Then we may suppose that $a_k \neq 0$. Because $\rho(w_n) = S^n \rho(w_0)$ for $n = 0, 1, 2, \ldots$,

$$\begin{aligned} 0 = \rho S^2(u) &= a_k \rho S^2(w_k) + a_{k+1} \rho S^2(w_{k+1}) + \cdots + a_n \rho S^2(w_n) \\ &= a_k S^{k+2} \rho(w_0) + a_{k+1} S^{k+3} \rho(w_0) + \cdots + a_n S^{n+2} \rho(w_0).f \end{aligned}$$

Here we put $\rho(w_0) = b_\ell w_\ell + b_{\ell+1} w_{\ell+1} + \cdots + b_t w_t$.

If $\rho(w_0) = 0$, then $\rho S^2(w_0) = \rho(w_2) = S^2 \rho(w_0) = 0$. Also, $\rho S^2(w_m) = 0$ for all $m = 1, 2, \ldots$, and $\rho S^2(v_i) = 0$ for all $i = 0, 1, \ldots$. So $\rho S^2 = 0$, a contradiction. Hence $\rho(w_0) \neq 0$, and thus we may assume that $b_\ell \neq 0$. We note that

$$S^{k+2} \rho(w_0) = b_\ell w_{\ell+k+2} + b_{\ell+1} w_{\ell+k+3} + \cdots + b_t w_{t+k+2},$$

$$S^{k+3} \rho(w_0) = b_\ell w_{\ell+k+3} + b_{\ell+1} w_{\ell+k+4} + \cdots + b_t w_{t+k+3},$$

and so on. Thus

$$0 = \rho S^2(u) = a_k b_\ell w_{\ell+k+2} + (a_k b_{\ell+1} + a_{k+1} b_\ell) w_{\ell+k+3} + \cdots,$$

and hence $a_k b_\ell = 0$, which is a contradiction because $a_k \neq 0$ and $b_\ell \neq 0$. Therefore $u = 0$, and so M is nonsingular.

We show now that V_n is continuous. Note that V_n is uniform. So clearly, V_n has the (C_1) condition. Thus, to show that V_n is continuous, it suffices to prove that every R-monomorphism of V_n is onto for V_n to satisfy the (C_2) condition.

Let $\varphi : V_n \to V_n$ be an R-monomorphism. We put

$$\varphi(v_n) = \rho v_n \in Rv_n = V_n, \quad \text{where} \quad \rho \in R.$$

We claim that $\rho v_n \notin V_{n+1}$. For this, assume on the contrary that $\rho v_n \in V_{n+1}$. Let $\lambda \in R$ such that $\lambda v_n = v_n$, $\lambda v_k = 0$ for $k \neq n$, and $\lambda w_m = 0$ for all m. Then $\varphi(\lambda v_n) = \lambda(\rho v_n) = 0$ since $\rho(v_n) \in V_{n+1}$. But $\lambda v_n = v_n \neq 0$. Thus φ is not one-to-one, a contradiction. Therefore $\rho v_n \notin V_{n+1}$.

As $\rho v_n \in V_n$, write

$$\rho v_n = a_n v_n + a_{n+1} v_{n+1} + \cdots + a_{n+\ell} v_{n+\ell} + b_0 w_0 + \cdots + b_h w_h,$$

where $a_n, a_{n+1}, \ldots, a_{n+\ell}, b_0, b_1, \ldots, b_h \in F$, and $a_n \neq 0$.

Take $\nu \in R$ such that $\nu v_n = a_n^{-1} v_n$, $\nu v_k = 0$ for $k \neq n$ and $\nu w_m = 0$ for all m. Then we see that $v_n = \nu \rho v_n \in R\rho v_n$. Therefore $Rv_n \subseteq R\rho v_n$, and hence $V_n = Rv_n =$

$R\rho v_n$. Thus $\varphi(Rv_n) = R\varphi(v_n) = R\rho v_n = V_n$, so φ is onto. Therefore V_n is continuous for all n.

Finally, note that the uniform nonsingular module $M = Rw_0$ is not continuous, since the shifting operator $\mathcal{S}$ provides an R-monomorphism which is not onto. Hence, M does not have a continuous hull (in $E(M) = E(V)$), because such a hull would have to be contained in each V_n, and hence in $M = \bigcap_n V_n$.

Despite Example 2.14, continuous hulls do exist for certain classes of modules over a *commutative ring*. For the class of cyclic modules, the next result and Theorem 2.17 due to Müller and Rizvi [35] show the existence of continuous hulls over commutative rings.

Theorem 2.15 ([35, Theorem 8]) *Every cyclic module over a commutative ring whose singular submodule is uniform, has a continuous hull.*

The next example, due to Müller and Rizvi [35], shows that in general, the quasi-continuous hull of a module is distinct from the continuous hull, which in turn is distinct from the (quasi-)injective hull of the module.

Example 2.16 ([35, Example 2]) Let $F_n = \mathbb{R}$ for $n = 1, 2, \ldots$, and put $A = \prod_{n=1}^{\infty} F_n$, where $\mathbb{R}$ is the field of real numbers. Let R be the subring of A generated by $\oplus_{n=1}^{\infty} F_n$ and 1_A. Then $E(R_R) = Q(R) = A$. In this case, we see that

$$V = \left\{(a_n)_{n=1}^{\infty} \in A \mid a_n \in \mathbb{Z} \text{ eventually}\right\}$$

is the quasi-continuous hull of R_R, while

$$W = \left\{(a_n)_{n=1}^{\infty} \in A \mid a_n \in \mathbb{Q} \text{ eventually}\right\}$$

is the continuous hull of R_R because W is the smallest continuous von Neumann regular ring between R and $Q(R)$ (so W is the intersection of all intermediate continuous von Neumann regular rings between R and $Q(R)$). We note that A_W is an injective hull of W_W, and also A_W is a quasi-injective hull of W_W.

When M is a uniform cyclic module over a commutative ring, the following theorem shows that M has a continuous hull (see [42]). Furthermore, it explicitly describes the continuous hull of M. Recall that when M_R is a right R-module, an element $c \in R$ is said to *act regularly on* M if $mc = 0$ with $m \in M$ implies that $m = 0$. Let C be the multiplicative set of elements of R which act regularly on M, and let $MC^{-1} = \{mc^{-1} \mid m \in M, c \in C\}$.

Theorem 2.17 ([42, Theorem 4.15] and [10, Theorem 8.4.11, p. 319]) *Let R be a commutative ring, and M a uniform cyclic R-module. Then MC^{-1} is a continuous hull of M.*

In view of the existence of quasi-injective and quasi-continuous hulls for all modules and from the existence of continuous hulls for some classes of modules in

Theorems 2.15 and 2.17, it is natural to consider the existence of extending hulls of modules. However, the following example exhibits that there exists a free module of finite rank over a commutative domain which has no extending hull.

Example 2.18 (cf. [40, Example 2.19] and [10, Example 8.4.13, p. 319]) We let $R = \mathbb{Z}[x]$, the polynomial ring over $\mathbb{Z}$. Then $(R \oplus R)_R$ has no extending hull.

We recall that a module M satisfying the (FI) condition is called *FI-extending*. Thus a module M is FI-extending if and only if every fully invariant submodule of M is essential in a direct summand of M. A ring R is called *right FI-extending* if R_R is FI-extending. Similarly left FI-extending ring is defined. For more details on FI-extending modules and rings, see [4, 8, 10].

The notion of an FI-extending module generalizes that of an extending module by requiring that only *every fully invariant* submodule is essential in a direct summand rather than *every* submodule. Many well-known submodule of a given module are fully invariant. For example, the socle of a module, and the Jacobson radical of a module, and the singular submodule of a module, are fully invariant. For a ring R, all its fully invariant submodules are precisely the ideals of R. It was shown in [4, Theorem 1.3] that any direct sum of FI-extending modules is FI-extending without any additional requirements. Thus while a direct sum of extending modules may not be extending, it does satisfy the extending property for all its fully invariant submodules.

There are close connections between the FI-extending property and the quasi-Baer property. For example, assume that R is a semiprime ring. Then R is right FI-extending if and only if R is left extending if and only if R is a quasi-Baer ring from [4, Theorem 4.7]. Further, every nonsingular FI-extending module is a quasi-Baer module (in fact, this also holds true under much weaker nonsingularity conditions).

A commutative domain R is called *Prüfer* if R is semihereditary. Thus a commutative domain is Prüfer if and only if every finitely generated ideal is projective. Note that every extending module is FI-extending. If R is a commutative domain which is not Prüfer (e.g., $R = \mathbb{Z}[x]$) and n is an integer such that $n > 1$, then $R_R^{(n)}$ is FI-extending, but $R_R^{(n)}$ is not extending (cf. Example 2.5(v)).

For a ring R, recall that $Q(R)$ denotes the maximal right ring of quotients of R. Let $\mathcal{B}(Q(R))$ be the set of all central idempotents of $Q(R)$. By [2], the subring $R\mathcal{B}(Q(R))$ of $Q(R)$ generated by R and $\mathcal{B}(Q(R))$ is called *the idempotent closure* of R.

Between R and $R\mathcal{B}(Q(R))$, LO (Lying Over), GU (Going Up), and INC (Incomparable) hold. Thereby, $\text{kdim}(R) = \text{kdim}(R\mathcal{B}(Q(R))$, where $\text{kdim}(-)$ is the classical Krull dimension of a ring, i.e., the supremum of all length of chains of prime ideals. For prime radicals and Jacobson radicals of R and $R\mathcal{B}(Q(R))$, we have that $P(R\mathcal{B}(Q(R)) \cap R = P(R)$ and $J(R\mathcal{B}(Q(R)) \cap R = J(R)$, where $P(-)$ and $J(-)$ denote the prime radical and the Jacobson radical of a ring, respectively. Also, R is strongly π-regular if and only if $R\mathcal{B}(Q(R))$ is strongly π-regular (recall that a ring A is called *strongly π-regular* if for each $a \in A$ there exist $x \in A$ and a positive integer n, depending on a, such that $a^n = a^{n+1}x$. (See [10, Lemma 8.3.26 and Theorem 8.3.28, pp. 296–297].)

Further, by [10, Corollary 8.3.30, p. 298], R is von Neumann regular if and only if $R\mathcal{B}(Q(R))$ is von Neumann regular. When R is a semiprime ring with exactly n (n a positive integer) minimal prime ideals $P_1, P_2, \ldots, P_n$, we have the following structure theorem

$$R\mathcal{B}(Q(R)) \cong R/P_1 \oplus R/P_2 \oplus \cdots \oplus R/P_n$$

as rings from [10, Theorem 10.1.20, p. 370].

By using the above structure theorem for $R\mathcal{B}(Q(R))$, it was shown in [7, Corollary 4.17] that if A is a unital C^*-algebra and n is a positive integer, then A has exactly n minimal prime ideals if and only if $A\mathcal{B}(Q(A))$ is a direct sum of n prime C^*-algebras if and only if the extended centroid $\text{Cen}(Q(A))$ of A is $\mathbb{C}^n$, where $\mathbb{C}$ is the field of complex numbers.

An overring T of a ring R is called a *right ring of quotients* of R if R_R is a dense submodule of T_R. Assume that R is a semiprime ring. Then from [7, Theorem 3.3], the ring $R\mathcal{B}(Q(R))$ is the smallest right FI-extending right ring of quotients of R. For more details on $R\mathcal{B}(Q(R))$, see [10, Sects. 8.3 and 10.1].

In the following definition, for a ring R, we fix a maximal right ring of quotients $Q(R)$ of R. Thus a right ring of quotients T of R is a subring of $Q(R)$.

Definition 2.19 (*see* [6, Definition 2.1]) The smallest right FI-extending right ring of quotients of a ring R is called the right FI-extending ring hull of R (when it exists). Such hull is denoted by $\widehat{Q}_{\mathbf{FI}}(R)$.

The existence of the right FI-extending ring hull $\widehat{Q}_{\mathbf{FI}}(R)$ of a semiprime ring R was obtained and explicitly described by Birkenmeier, Park, and Rizvi in the following interesting result.

Theorem 2.20 ([7, Theorem 3.3]) *Assume that R is a semiprime ring. Then $\widehat{Q}_{\mathbf{FI}}(R)$ exists and $\widehat{Q}_{\mathbf{FI}}(R) = R\mathcal{B}(Q(R))$.*

Let R be a commutative semiprime ring. Then $R\mathcal{B}(Q(R))$ is the smallest extending ring of quotients of R by Theorem 2.20.

In contrast to Theorem 2.20, there exists a semiprime ring for which the right extending ring hull does not exist. For this, we need the the next result.

Theorem 2.21 ([10, Theorem 6.1.4, p. 191]) *Let R be a commutative domain. Then the following are equivalent.*

(i) R is a Prüfer domain.
(ii) $Mat_n(R)$ is a (right) extending ring for every positive integer n.
(iii) $Mat_k(R)$ is a (right) extending ring for some integer $k > 1$.
(iv) $Mat_2(R)$ is a (right) extending ring.

The smallest right extending right ring of quotients of a ring R is called the *right extending ring hull* of R (when it exists). Such hull is denoted by $\widehat{Q}_{\mathbf{E}}(R)$. By using Theorem 2.21, we obtain the following example which exhibits that the right extending ring hull of a semiprime ring does not exist, in general.

Example 2.22 (see [10, Example 8.3.34, p. 300]) Let $R = \text{Mat}_k(F[x, y])$, where F is a field and k is an integer such that $k \geq 2$. Then the right extending ring hull $\widehat{Q}_{\mathbf{E}}(R)$ of R does not exist.

Assume on the contrary that $\widehat{Q}_{\mathbf{E}}(R)$ exists. Note that $F(x)[y]$ and $F(y)[x]$ are Prüfer domains, where $F(x)$ (resp., $F(y)$) is the field of fractions of $F[x]$ (resp., $F[y]$). So $\text{Mat}_k(F(x)[y])$ and $\text{Mat}_k(F(y)[x])$ are right extending rings by Theorem 2.21. Note $Q(R) = \text{Mat}_k(F(x, y))$, where $F(x, y)$ is the field of fractions of $F[x, y]$. Hence

$$\widehat{Q}_{\mathbf{E}}(R) \subseteq \text{Mat}_k(F(x)[y]) \cap \text{Mat}_k(F(y)[x]) = \text{Mat}_k(F(x)[y] \cap F(y)[x]).$$

To see that $F(x)[y] \cap F(y)[x] = F[x, y]$, let

$$\begin{aligned}\gamma(x, y) &= f_0(x)/g_0(x) + (f_1(x)/g_1(x))y + \cdots + (f_m(x)/g_m(x))y^m \\ &= h_0(y)/k_0(y) + (h_1(y)/k_1(y))x + \cdots + (h_n(y)/k_n(y))x^n\end{aligned}$$

be in $F(x)[y] \cap F(y)[x]$, where $f_i(x),\ g_i(x) \in F[x],\ h_j(y),\ k_j(y) \in F[y]$, and $g_i(x) \neq 0,\ k_j(y) \neq 0$ for $i = 0, 1, \ldots, m, j = 0, 1, \ldots, n$. Let $\overline{F}$ be the algebraic closure of F. If $\deg(g_0(x)) \geq 1$, then there exists $\alpha \in \overline{F}$ such that $g_0(\alpha) = 0$. Thus $\gamma(\alpha, y)$ cannot be defined. On the other hand, we note that

$$\gamma(\alpha, y) = h_0(y)/k_0(y) + (h_1(y)/k_1(y))\alpha + \cdots + (h_n(y)/k_n(y))\alpha^n,$$

which is a contradiction. Thus $g_0(x) \in F$. Similarly, $g_1(x), \ldots, g_m(x) \in F$. Hence $\gamma(x, y) \in F[x, y]$. Therefore $F(x)[y] \cap F(y)[x] = F[x, y]$, and so

$$\widehat{Q}_{\mathbf{E}}(R) = \text{Mat}_k(F(x)[y] \cap F(y)[x]) = \text{Mat}_k(F[x, y]).$$

Thus $\text{Mat}_k(F[x, y])$ is a right extending ring, a contradiction from Theorem 2.21 because the commutative domain $F[x, y]$ is not Prüfer. Therefore $R = \text{Mat}_k(F[x, y])$ has no right extending ring hull.

In contrast to Theorem 2.20, the existence of the right FI-extending ring hull of a ring is not always guaranteed, even in the presence of nonsingularity, as the next example shows.

Example 2.23 (see [5, Example 2.10(ii)], [6, Example 3.16], and [10, Example 8.2.9, p. 278]) Let F be a field and put

$$R = \left\{ \begin{bmatrix} a & 0 & x \\ 0 & a & y \\ 0 & 0 & c \end{bmatrix} \mid a, c, x, y \in F \right\} \cong \begin{bmatrix} F & F \oplus F \\ 0 & F \end{bmatrix}.$$

Then R is right nonsingular and $Q(R) = \text{Mat}_3(F)$.

Let

$$H_1 = \left\{ \begin{bmatrix} a & 0 & x \\ 0 & b & y \\ 0 & 0 & c \end{bmatrix} \mid a, b, c, x, y \in F \right\} \cong \begin{bmatrix} F \oplus F & F \oplus F \\ 0 & F \end{bmatrix},$$

and let

$$H_2 = \left\{ \begin{bmatrix} a+b & a & x \\ 0 & b & y \\ 0 & 0 & c \end{bmatrix} \mid a, b, c, x, y \in F \right\}.$$

Note that R, H_1, and H_2 are subrings of $\mathrm{Mat}_3(F)$. Define $\phi : H_1 \to H_2$ by

$$\phi \begin{bmatrix} a & 0 & x \\ 0 & b & y \\ 0 & 0 & c \end{bmatrix} = \begin{bmatrix} a & a-b & x-y \\ 0 & b & y \\ 0 & 0 & c \end{bmatrix}.$$

Then ϕ is a ring isomorphism. It is routine to check that the ring R is not right FI-extending. But, we can verify that H_1 is a right FI-extending ring. Therefore H_2 is also right FI-extending because $H_1 \cong H_2$ (ring isomorphic).

Let $F = \mathbb{Z}_2$. Then there is no proper intermediate ring between R and H_1, also between R and H_2. If $\widehat{Q}_{\mathbf{FI}}(R)$ exists, then $\widehat{Q}_{\mathbf{FI}}(R) \subseteq H_1 \cap H_2 = R$, so $\widehat{Q}_{\mathbf{FI}}(R) = R$. Hence R is a right FI-extending ring, which is a contradiction.

In contrast to Example 2.18 where the extending hull of a finitely generated free module of rank 2 does not exist, it was shown that the FI-extending hulls of every finitely generated projective module over a semiprime ring does exist in [8]. Also such an FI-extending hulls is described explicitly using Theorem 2.20 as in the next theorem. For a module M, let $\mathfrak{F}\mathfrak{I}(M)$ denote the FI-extending hull of M, when it exists.

Theorem 2.24 ([8, Theorem 6]) *Any finitely generated projective module P_R over a semiprime ring R has the FI-extending hull $\mathfrak{F}\mathfrak{I}(P_R)$. Indeed, $\mathfrak{F}\mathfrak{I}(P_R) \cong e(\oplus^n \widehat{Q}_{\mathbf{FI}}(R)_R)$ where $P \cong e(\oplus^n R_R)$, for some $e^2 = e \in End(\oplus^n R_R)$ and some positive integer n.*

From Theorems 2.20 and 2.24, the following result is obtained.

Corollary 2.25 ([8, Corollary 7]) *Assume that R is a semiprime ring and P_R is a finitely generated projective module. Then $\widehat{Q}_{\mathbf{FI}}(End(P_R)) \cong End(\mathfrak{F}\mathfrak{I}(P_R))$.*

An application of Theorem 2.24 yields the following consequences.

Corollary 2.26 ([8, Corollary 13]) *Let R be a semiprime ring. Then:*

(i) If P_R is a progenerator of the category Mod-R of right R-modules, then $\mathfrak{F}\mathfrak{I}(P_R)_{\widehat{Q}_{\mathbf{FI}}(R)}$ is a progenerator of the category Mod-$\widehat{Q}_{\mathbf{FI}}(R)$ of right $\widehat{Q}_{\mathbf{FI}}(R)$-modules.

(ii) If R and a ring S are Morita equivalent, then $\widehat{Q}_{\mathbf{FI}}(R)$ and $\widehat{Q}_{\mathbf{FI}}(S)$ are Morita equivalent.

3 Baer Modules

We introduce the definition of a Baer module M_R via its endomorphism ring $S = \text{End}(M_R)$ in contrast to defining this notion in terms of the base ring R. The use of the endomorphism ring instead of the base ring R appears to offer a more natural generalization of a Baer ring in the general module theoretic setting (see Definition 3.1 and the comments after Example 3.2).

Properties of Baer modules are included and examples are provided. Similar to the ring theoretic concepts of nonsingularity and cononsingularity, $\mathcal{K}$-nonsingularity and $\mathcal{K}$-cononsingularity, respectively are discussed for modules. Using these concepts, strong connections between extending modules and Baer modules are provided, which generalizes the Chatters-Khuri theorem to the module theoretic setting. We include a characterization of rings R for which every projective right R-module is Baer. Properties of Baer modules from this section will also be used in Sect. 4. For more details on Baer modules and their properties, see [44–47], and [10, Chap. 4].

We start with the following definition.

Definition 3.1 ([44, Definition 2.2]) A right R-module M is called a *Baer module* if, for any $N_R \leq M_R$, there exists $e^2 = e \in S$ such that $\ell_S(N) = Se$, where $S = \text{End}(M_R)$ and $\ell_S(N) = \{f \in S \mid f(N) = 0\}$. A right R-module M is Baer if and only if for any left ideal I of S, $r_M(I) = fM$ with $f^2 = f \in S$, where $r_M(I) = \{m \in M \mid Im = 0\}$.

A ring R is said to be a *Baer ring* if the right annihilator of any nonempty subset of R is generated, as a right ideal, by an idempotent of R. Thus a ring R is a Baer ring if and only if R_R is a Baer module. Further, we can verify that a ring R is Baer if and only if the left annihilator of any nonempty subset of R is generated, as a left ideal, by an idempotent of R (see [27, Theorem 3, p. 2]).

Example 3.2 (i) Every semisimple module is a Baer module.

(ii) If R is a Baer ring and $e^2 = e \in R$, then eR_R is a Baer module (see Theorem 3.12).

(iii) ([44, Proposition 2.19]) A finitely generated Baer abelian group M is a Baer $\mathbb{Z}$-module if and only if M is semisimple or torsion-free.

(iv) ([10, Corollary 4.3.6, p. 112]) Any finitely generated right Hilbert A-module over an AW^*-algebra A is a Baer module.

(v) ([44, Theorem 2.23]) A module M is an indecomposable Baer module if and only if any nonzero endomorphism of M is a monomorphism.

(vi) Any nonsingular extending module is a Baer module (see [44, Theorem 2.14]).

(vii) For a commutative domain R and an integer $n > 1$, $R_R^{(n)}$ is a Baer module if and only if $R_R^{(n)}$ is an extending module if and only if R is a Prüfer domain.

(viii) ([47, Theorem 3.16]) Let R be an n-fir (n a positive integer). Then $R_R^{(n)}$ is a Baer module (recall that a ring R is said to be an n-fir if any right ideal of R generated by at most n elements is free of unique rank).

In [30, Definition 3.1], Lee and Zhou also called a module M_R *Baer* if, for any nonempty subset X of M, $r_R(X) = eR$ with $e^2 = e \in R$. But Definition 3.1 is distinct from their definition. In fact, any semisimple module is a Baer module by Definition 3.1 (see Example 3.2(i)), but it may not be a Baer module in the sense of Lee and Zhou [30] (for example $\mathbb{Z}_p$ as a $\mathbb{Z}$-module, where p is a prime integer, is a Baer module in our sense).

Definition 3.3 ([44, Theorem 2.5]) Let M be a module. Then M is called $\mathcal{K}$-*nonsingular* if, for $\phi \in \text{End}_R(M)$, $\text{Ker}(\phi) \leq^{\text{ess}} M$ implies $\phi = 0$.

Example 3.4 (i) Any semisimple module is $\mathcal{K}$-nonsingular.

(ii) ([44, Proposition 2.10]) Every nonsingular module is $\mathcal{K}$-nonsingular.

(iii) ([44, Example 2.11]) The $\mathbb{Z}$-module $\mathbb{Z}_p$, where p is a prime integer, is $\mathcal{K}$-nonsingular, but it is not nonsingular.

(iv) Any polyform module is $\mathcal{K}$-nonsingular. Recall that a module M is said to be *polyform* if every essential submodule of M is a dense submodule. A polyform module M is also called *non-M-singular*.

(v) For a ring R, R_R is $\mathcal{K}$-nonsingular if and only if R_R is nonsingular if and only if R_R is polyform.

(vi) [46, Example 2.5]) Let $M = \mathbb{Q} \oplus \mathbb{Z}_2$ as a $\mathbb{Z}$-module. Then M is $\mathcal{K}$-nonsingular. But M is neither nonsingular nor polyform.

(vii) ([44, Lemma 2.15]) Every Baer module is $\mathcal{K}$-nonsingular.

(viii) ([44, Lemma 2.6]) A module M is $\mathcal{K}$-nonsingular if and only if, for any left ideal I of S, $r_M(I) \leq^{\text{ess}} M$ implies $I = 0$, where $S = \text{End}(M)$.

While the nonsingularity of a module M provides the uniqueness of essential closures in M (i.e., M is a UC-module), the $\mathcal{K}$-nonsingularity provides the uniqueness of closures which happen to be direct summands of M.

Theorem 3.5 ([46, Proposition 2.8]) *Assume that M is a $\mathcal{K}$-nonsingular module, and let $N \leq M$. If $N \leq^{\text{ess}} N_i \leq^{\oplus} M$, for $i = 1, 2$, then $N_1 = N_2$.*

We recall that a ring R is said to be *right cononsingular* if for $I_R \leq R_R$, $\ell_R(I) = 0$ implies $I_R \leq^{\text{ess}} R_R$. Dual to the notion in Definition 3.3, the following is a module theoretic version of cononsingularity introduced in [44].

Definition 3.6 ([44, Definition 2.7]) A module M_R is called $\mathcal{K}$-*cononsingular* if for all $N_R \leq M_R$, $\ell_S(N) = 0$ implies $N_R \leq^{\text{ess}} M_R$, where $S = \text{End}(M_R)$.

Example 3.7 (i) For a ring R, R_R is $\mathcal{K}$-cononsingular if and only if R is right cononsingular.

(ii) ([44, Lemma 2.13]) Every extending module is $\mathcal{K}$-cononsingular.

(iii) For a commutative semiprime ring R, $R_R^{(n)}$ is $\mathcal{K}$-cononsingular for every positive integer n.

(iv) Let $R = \mathbb{Z}[x]$. Then $(R \oplus R)_R$ is $\mathcal{K}$-cononsingular by part (iii). But $(R \oplus R)_R$ is not extending by Theorem 2.21. Hence the converse of part (ii) is not true.

Proposition 3.8 ([44, Proposition 2.8(ii)]) *Assume that M is a right R-module. Then M is $\mathcal{K}$-cononsingular if and only if, for $N \leq M$, $r_M(\ell_S(N)) \leq^{\oplus} M$ implies $N \leq^{ess} r_M\ell_S(N)$, where $S = End_R(M)$.*

It is shown by Chatters and Khuri [11, Theorem 2.1] that a ring R is right extending right nonsingular if and only if R is a Baer ring and right cononsingular. This result is extended to an arbitrary module in the next theorem which exhibits strong connections between a Baer module and an extending module.

Theorem 3.9 ([44, Theorem 2.12]) *A module M is extending and $\mathcal{K}$-nonsingular if and only if M is Baer and $\mathcal{K}$-cononsigular.*

Definition 3.10 ([47, Definition 2.3]) Let M_R be an R-module and $S = \mathrm{End}_R(M)$. Then M_R is called *quasi-retractable* if $\mathrm{Hom}_R(M, r_M(I)) \neq 0$ for every left ideal I of S with $r_M(I) \neq 0$ (or, equivalently, if $r_S(I) \neq 0$ for every left ideal I with $r_M(I) \neq 0$).

Recall from [29] that a module M is said to be *retractable* if any $0 \neq N \leq M$, $\mathrm{Hom}(M, N) \neq 0$. Examples of retractable modules include free modules, generators, and semisimple modules. Obviously retractable modules are quasi-retractable. But there exists a quasi-retractable module which is not retractable. For example, let F be a field. Put

$$R = \begin{bmatrix} F & F \\ 0 & F \end{bmatrix} \text{ and } e = \begin{bmatrix} 1 & 0 \\ 0 & 0 \end{bmatrix} \in R.$$

Consider the module $M = eR$. Note that $S := \mathrm{End}(M_R) \cong eRe \cong F$, which is a field. Let I be a left ideal of S such that $r_M(I) \neq 0$. Then $I = 0$ and so $r_M(I) = M$. Hence, $\mathrm{Hom}(M_R, r_M(I)) = \mathrm{End}(M_R) \cong F \neq 0$. Thus, M_R is quasi-retractable. But M_R is not retractable, since the endomorphism ring S of M_R, which is isomorphic to F, consists of isomorphisms and the zero endomorphism. On the other hand, as M_R is not simple, retractability of M_R implies that there exist nonzero endomorphisms of M_R which are not onto (see [10, Example 4.2.4, p. 101]).

By [44, Theorem 4.1], the endomorphism ring of a Baer module is a Baer ring. But the converse does not hold by [44, Example 4.3]. Indeed, let $M = \mathbb{Z}_{p^\infty}$, the Prüfer p-group (p a prime integer), as $\mathbb{Z}$-module. Then $S := \mathrm{End}_{\mathbb{Z}}(M)$ is the ring of p-adic integers, so S is a commutative domain. Hence S is a Baer ring. But M is not a Baer $\mathbb{Z}$-module.

In spite of the above example, the following result shows a connection between the Baer property of a module and its endomorphism ring via its quasi-retractability.

Theorem 3.11 ([47, Theorem 2.5]) *A module M_R is Baer if and only if $\mathrm{End}_R(M)$ is a Baer ring and M_R is quasi-retractable.*

Theorem 3.12 ([44, Theorem 2.17]) *Any direct summand of a Baer module is a Baer module.*

We noted before, $\mathbb{Z}_p \oplus \mathbb{Z}$ (p a prime integer) is not Baer as a $\mathbb{Z}$-module, while both $\mathbb{Z}_p$ and $\mathbb{Z}$ are Baer $\mathbb{Z}$-modules. For the Baer property of a finite direct sum of Baer modules, we need the following. Let M and N be R-modules. Then M is said to be N-*injective* if, for any $W \leq N$ and $f \in \mathrm{Hom}(W, M)$, there exists $\varphi \in \mathrm{Hom}(N, M)$ such that $\varphi|_W = f$. Recall from [47, Definition 1.3] that two modules M and N are said to be *relatively Rickart* if, for every $f \in \mathrm{Hom}(M, N)$, $\mathrm{Ker}(f) \leq^{\oplus} M$ and for every $g \in \mathrm{Hom}(N, M)$, $\mathrm{Ker}(g) \leq^{\oplus} N$.

Theorem 3.13 ([47, Theorem 3.19] also see [10, Theorem 4.2.17, p. 105]) *Assume that $\{M_i \mid 1 \leq i \leq n\}$ be a finite set of Baer modules. Let M_i and M_j be relatively Rickart for $i \neq j$, and M_i be M_j-injective for $i < j$. Then $\oplus_{i=1}^{n} M_i$ is a Baer module.*

The study of rings R for which a certain class of R-modules is Baer is of natural interest. In the following, R is semisimple artinian if and only if every injective R-module is Baer.

Theorem 3.14 ([46, Theorem 2.20]) *The following are equivalent for a ring R.*

(i) Every injective (right) R-module is Baer.
(ii) Every (right) R-module is Baer.
(iii) R is semisimple artinian.

A ring R is said to be *semiprimary* if $R/J(R)$ is artinian and $J(R)$ is nilpotent. Recall that a ring R is *right* (resp., *left*) *hereditary* if every right (resp., left) ideal of R is projective. It is well-known that if a ring R is semisprimary, then R is right hereditary if and only if R is left hereditary.

The following result provides a characterization of rings R for which every projective right R-module is Baer. Also see Theorem 4.11.

Theorem 3.15 ([47, Theorem 3.3]) *The following are equivalent for a ring R.*

(i) Every projective right R-module is a Baer module.
(ii) Every free right R-module is a Baer module.
(iii) R is a semiprimary, hereditary (Baer) ring.

Since condition (iii) is left-right symmetric, the left-handed versions of (i) and (ii) also hold.

A module M_R is called *torsionless* if it can be embedded in a direct product of copies of R_R. The following result characterizes a ring R for which every finitely generated right R-module is a Baer module.

Recall that an R-module M is said to be *finitely presented* if there exists a short exact sequence of R-modules $0 \to K \to R^{(n)} \to M \to 0$, where n is a positive integer and K is a finitely generated R-module.

A ring R is called *right Π-coherent* if every finitely generated torsionless right R-module is finitely presented. Left Π-coherent ring is defined similarly. Recall that a ring R is said to be *right semiheditary* if every finitely generated right ideal of R is projective. A left semihereditary ring is denied similarly.

Theorem 3.16 ([47, Theorem 3.5]) *The following are equivalent for a ring R.*

(i) Every finitely generated free right R-module is a Baer module.
(ii) Every finitely generated projective right R-module is a Baer module.
(iii) Every finitely generated torsionless right R-module is projective.
(iv) Every finitely generated torsionless left R-module is projective.
(v) R is left semihereditary and right Π-coherent.

(vi) *R is right semihereditary and left* Π*-coherent.*
(vii) $Mat_n(R)$ *is a Baer ring for every positive integer n.*

For a positive integer n, we recall that an *n-generated module* means a module which is generated by n elements. A ring R is said to be *right n-hereditary* if every n-generated right ideal of R is projective. Thus, a ring R is right semihereditary if and only if it is right n-hereditary for all positive integers n. Given a fixed positive integer n, we introduce the following characterization for every n-generated free R-module to be Baer.

Theorem 3.17 ([47, Theorem 3.12]) *Let R be a ring and n a positive integer. Then the following are equivalent.*

(i) *Every n-generated free right R-module is a Baer module.*
(ii) *Every n-generated projective right R-module is a Baer module.*
(iii) *Every n-generated torsionless right R-module is projective (therefore R is right n-hereditary).*
(iv) $Mat_n(R)$ *is a Baer ring.*

Corollary 3.18 *Let R be a ring. Then R is a Baer ring if and only if every cyclic torsionless right R-module is projective.*

4 Baer Module Hulls

We present recent results and examples on Baer hulls in this section. As mentioned in the introduction, the study of even Baer ring hulls has been rather limited. And the only results on Baer ring hulls that exist in earlier literature are from [19, 33, 37, 38], respectively for the classes of commutative semiprime rings, commutative von Neumann regular rings, and reduced right Utumi rings. Some newer developments on ring hulls were presented in [5–7, 9, 10]. The question about the existence of Baer module hulls and their existence has not been addressed till now and is quite challenging. The results presented here are the latest developments on Baer module hulls of finitely generated modules over a commutative domain.

From [44] it is known that $N = \mathbb{Z}_p \oplus \mathbb{Z}$ (p a prime integer) is not a Baer $\mathbb{Z}$-module, while $\mathbb{Z}_p$ and $\mathbb{Z}$ are. We construct the Baer hull of the module N in a more general setting. Let R be a commutative noetherian domain. We first introduce a result from [40] for intermediate modules between an analogous direct sum as an R-module N and its injective hull *E(N)* to be Baer (Theorem 4.1). Then we use this result to construct and characterize the Baer hull of a module N over a Dedekind domain R, when $\mathrm{Ann}_R(t(N)) \neq 0$ and $N/t(N)$ is finitely generated, where $t(N)$ denotes the torsion submodule of N (Theorems 4.4, 4.5, and 4.8). As a consequence, every finitely generated module over a Dedekind domain, has a unique Baer hull precisely when its torsion submodule is semisimple. For a module N such that $N/t(N)$ is not finitely generated, an example shows that N does not have a Baer hull (Example 4.12).

Among applications presented, we show that a finitely generated module N over a Dedekind domain is Baer if and only if N is semisimple or torsion-free (Corollary 4.6). This extends a result on finitely generated abelian groups. The isomorphism problem between modules and their Baer hulls is discussed (Proposition 4.13 and Example 4.14). It is also shown that the Baer hull of a direct sum of two modules is not necessarily isomorphic to the direct sum of the Baer hulls of the modules, even if all Baer hulls exist (Example 4.16). The Baer hull of $N = \mathbb{Z}_p \oplus \mathbb{Z}$ (p a prime integer) as a $\mathbb{Z}$-module, is shown to be precisely $V = \mathbb{Z}_p \oplus \mathbb{Z}[1/p]$. The disparity of the Baer hull and the extending hull of $\mathbb{Z}_p \oplus \mathbb{Z}$ is discussed (Example 4.17). A number of other examples which illustrate the results are provided.

Let R be a commutative noetherian domain and F be its field of fractions. Assume that $N = M_R \oplus (\oplus_{i\in\Lambda} K_i)$, where M is semisimple with a finite number of homogeneous components, and $\{K_i\}_{i\in\Lambda}$ is a set of nonzero submodules of F_R.

By using the preceding results, we obtain the following which identifies intermediate modules between N and $E(N)$ which happen to be Baer modules.

Theorem 4.1 ([40, Theorem 2.6]) *Let R be a commutative noetherian domain, which is not a field. Assume that M is a nonzero semisimple R-module with only a finite number of homogeneous components, and $\{K_i \mid i \in \Lambda\}$ is a nonempty set of nonzero submodules of F_R, where F is the field of fractions of R. Let V_R be an essential extension of $M_R \oplus (\oplus_{i\in\Lambda} K_i)_R$. Then the following are equivalent.*

(i) V is a Baer module.
(ii) (1) $V = M \oplus W$ for some Baer essential extension W of $(\oplus_{i\in\Lambda} K_i)_R$.
(2) $Hom_R(W, M) = 0$.

Let R be a commutative domain with the field of fractions F. A submodule K of F_R is called a *fractional ideal* of R if $rK \subseteq R$ for some $0 \neq r \in R$. Thus $K_R \cong (rK)_R$ and rK is an ideal of R. We note that any ideal of R is a fractional ideal.

For a fractional ideal K of R, we put $K^{-1} = \{q \in F \mid qK \subseteq R\}$, which is called the *inverse* of K. We say that a fractional ideal K is *invertible* if $KK^{-1} = R$. It is well-known that for a nonzero ideal I of a commutative domain R, I_R is projective if and only if $II^{-1} = R$. In this case, I_R is finitely generated and I^{-1} is a fractional ideal of R.

Recall that a commutative domain R is a *Dedekind domain* if and only if R is hereditary. Thus for each nonzero ideal I of a Dedekind domain R, it follows that $II^{-1} = R$ because I_R is projective. Furthermore, every nonzero fractional ideal of a Dedekind domain is invertible. We note that a Dedekind domain is noetherian because every ideal is projective (hence every ideal is finitely generated). See [28, p. 37]and [48, Chap. 6] for more details on Dedekind domains.

Assume that I is an invertible ideal of a commutative domain R. Then we let

$$I^{-2} = I^{-1}I^{-1}, \quad I^{-3} = I^{-1}I^{-1}I^{-1}, \quad \text{and so on.}$$

For convenience, we put $I^0 = R$.

Assume that R is a Dedekind domain. Then for nonzero ideals $I_1, I_2, \ldots, I_n$ of R, it can be checked that $(I_1 I_2 \cdots I_n)^{-1} = I_n^{-1} \cdots I_2^{-1} I_1^{-1}$ (see [40, Lemma 2.8]).

Proposition 4.2 ([40, Lemma 2.9]) *Assume that R is a Dedekind domain and I is a nonzero ideal of R. We let $A = \sum_{\ell \geq 0} I^{-\ell}$. Then:*

(i) $A = R[q_1, q_2, \ldots, q_n]$, where $1 = \sum_{u=1}^{n} r_u q_u$ for some $r_u \in I$ and $q_u \in I^{-1}$ with $1 \leq u \leq n$.
(ii) A is a Dedekind domain.

In Proposition 4.2, since R is a Dedekind domain, R is a Prüfer domain. Because $A = R[q_1, q_2, \ldots, q_n]$ is an intermediate domain between R and its field of fractions, A is a Prüfer domain (note that any intermediate domain between a Prüfer domain and its field of fractions is a Prüfer domain). Since A is a noetherian domain, A is a Dedekind domain.

Theorem 4.3 ([48, Theorem 6.11, p. 171]) *Let R be a Dedekind domain and M an R-module with nonzero annihilator in R. Then there exists a unique family $\{P_i,\ n_i\}_{i \in \Gamma}$ such that:*

(i) The P_i are maximal ideals of R and there are only finitely many distinct ones.
(ii) $\{n_i \mid i \in \Gamma\}$ is a bounded family of positive integers.
(iii) $M \cong \oplus_{i \in \Gamma} (R/P_i^{n_i})$ as R-modules.

Let R be a Dedekind domain and N an R-module. Say $t(N)$ is the torsion submodule of N. Suppose that $N/t(N)$ is finitely generated as an R-module. Since $N/t(N)$ is torsion-free, $N/t(N) \cong (\oplus_{j=1}^{m} K_j)$ (as R-modules) for some fractional ideals K_j, $1 \leq j \leq m$, of R from [48, Theorem 6.16, p. 177] (see also Corollary 4.6). So $N/t(N)$ is projective, and hence

$$N \cong t(N) \oplus N/t(N) \cong t(N) \oplus (\oplus_{j=1}^{m} K_j)$$

as R-modules.

Our next result is a complete characterization for the existence of the Baer hull of a module N when $N/t(N)$ is finitely generated and $\mathrm{Ann}_R(t(N)) \neq 0$ (also see Theorem 4.5). Furthermore, we describe the Baer hull of N explicitly in this case.

We denote the Baer hull of a module M by $\mathfrak{B}(M)$ when it exists.

Theorem 4.4 ([40, Theorem 2.13]) *Let R be a Dedekind domain. Assume that M is an R-module with nonzero annihilator in R, and $\{K_1, K_2, \ldots, K_m\}$ is a finite set of nonzero fractional ideals of R. Then the following are equivalent.*

(i) $M_R \oplus (\oplus_{j=1}^{m} K_j)_R$ has a Baer hull.
(ii) M_R is semisimple.
(iii) $M_R \oplus (\oplus_{j=1}^{m} K_j)_R$ has a Baer essential extension.

In this case, $\mathfrak{B}(M_R \oplus (\oplus_{j=1}^m K_j)_R) = M_R \oplus (\oplus_{j=1}^m K_j A)_R$, *where* $A = \sum_{\ell \geq 0} I^{-\ell}$ *with* $I = \mathrm{Ann}_R(M)$. *Furthermore,* $A = R[q_1, q_2, \ldots, q_n]$, *where* $1 = \sum_{u=1}^n r_u q_u$ *with* $r_u \in I$ *and* $q_u \in I^{-1}$, $1 \leq u \leq n$.

The following is a restatement of Theorem 4.4 for characterization of the Baer hull of a module N over a Dedekind domain for the case when $N/t(N)$ is finitely generated and $\mathrm{Ann}_R(t(N)) \neq 0$.

Theorem 4.5 ([40, Theorem 2.15]) *Let R be a Dedekind domain. Assume that N is an R-module with $N/t(N)$ finitely generated and $\mathrm{Ann}_R(t(N)) \neq 0$. Then the following are equivalent.*

(i) N has a Baer hull.
(ii) $t(N)$ is semisimple.
(iii) N has a Baer essential extension.

By [44, Proposition 2.19 and Remark 2.20], a finitely generated module N over a commutative PID is a Baer module if and only if N is semisimple or torsion-free. This result is extended to the case when the base ring is a Dedekind domain as follows by applying Theorems 4.4 and 4.5.

Corollary 4.6 ([40, Corollary 2.17]) *Let R be a Dedekind domain and N be a finitely generated R-module. Then the following are equivalent.*

(i) N is a Baer module.
(ii) N is semisimple or torsion-free.

The next theorem details the structure of finitely generated modules over a Dedekind domain.

Theorem 4.7 ([48, Theorem 6.16, p. 177]) *Let R be a Dedekind domain and N a finitely generated R-module. Then there exist positive integers $n_1, n_2, \ldots, n_k$ (k is a nonnegative integer), nonzero maximal ideals $P_1, P_2, \ldots, P_k$, and nonzero fractional ideals $K_1, K_2, \ldots, K_m$ (m is a nonnegative integer) of R such that $N \cong (\oplus_{i=1}^k R/P_i^{n_i}) \oplus (\oplus_{j=1}^m K_j)$ as R-modules.*

Assume that N is a finitely generated module over a Dedekind domain. From Theorem 4.7, $N \cong (\oplus_{i=1}^k R/P_i^{n_i}) \oplus (\oplus_{j=1}^m K_j)$, where P_i are nonzero maximal ideals of R and K_j are nonzero fractional ideals of R (k and m are nonnegative integers). In the following theorem, we characterize the existence of the Baer hull of such N and describe the Baer hull of N explicitly.

Theorem 4.8 ([40, Theorem 2.18]) *Let R be a Dedekind domain, and let N be a finitely generated R-module. Then the following are equivalent.*

(i) N has a Baer hull.
(ii) $t(N)$ is semisimple.
(iii) N has a Baer essential extension.

In this case, $\mathfrak{B}(N_R) \cong (\oplus_{i=1}^k (R/P_i))_R \oplus (\oplus_{j=1}^m K_j A)_R$*, where* $A = \sum_{\ell \geq 0} I^{-\ell}$ *with* $I = \mathrm{Ann}_R(t(N))$*. Further,* $A = R[q_1, q_2, \ldots, q_n]$*, where* $1 = \sum_{u=1}^n r_u q_u$ *with* $r_u \in I$ *and* $q_u \in I^{-1}$*,* $1 \leq u \leq n$*.*

The following remark exhibits an explicit description of $A = \sum_{\ell \geq 0} I^{-\ell}$ in Theorem 4.4.

Remark 4.9 We have the following (see [40, Remark 3.1]).

(i) In Theorem 4.3, we put $A = \sum_{\ell \geq 0} I^{-\ell}$, where $I = \mathrm{Ann}_R(M)$. By Theorems 4.3 and 4.4, $M \cong \oplus_{i \in \Gamma} R/P_i$ and $\{P_i \mid i \in \Gamma\}$ is a finite set of maximal ideals P_i. Let $P_1, P_2, \ldots, P_s$ is all the distinct maximal ideals of $\{P_i \mid i \in \Gamma\}$.
We can verify that $A = \sum P_1^{-\ell_1} P_2^{-\ell_2} \cdots P_s^{-\ell_s}$, where $\ell_1, \ell_2, \ldots, \ell_s$ run through all nonnegative integers. In fact, $I \subseteq P_i$ for all i since $I = P_1 P_2 \cdots P_s$. For $i, 1 \leq i \leq s, P_i^{-1} \subseteq I^{-1}$ and therefore $P_i^{-\ell} \subseteq I^{-\ell}$ for every nonnegative integer ℓ. Hence,

$$P_1^{-\ell_1} P_2^{-\ell_2} \cdots P_s^{-\ell_s} \subseteq I^{-\ell_1} I^{-\ell_2} \cdots I^{-\ell_s} = I^{-(\ell_1+\ell_2+\cdots+\ell_s)} \subseteq A.$$

Thus $\sum P_1^{-\ell_1} P_2^{-\ell_2} \cdots P_s^{-\ell_s} \subseteq A$, where $\ell_1, \ell_2, \ldots, \ell_s$ run through all nonnegative integers. Conversely, $I^{-1} = (P_1 P_2 \cdots P_s)^{-1} = P_1^{-1} P_2^{-1} \cdots P_s^{-1}$. Therefore it follows that $I^{-\ell} = P_1^{-\ell} P_2^{-\ell} \cdots P_s^{-\ell}$ for any nonnegative integer ℓ. Hence we obtain that $A \subseteq \sum P_1^{-\ell_1} P_2^{-\ell_2} \cdots P_s^{-\ell_s}$, where $\ell_1, \ell_2, \ldots, \ell_s$ run through all nonnegative integers.
Consequently, $A = \sum P_1^{-\ell_1} P_2^{-\ell_2} \cdots P_s^{-\ell_s}$, where $\ell_1, \ell_2, \ldots, \ell_s$ run through all nonnegative integers.

(ii) Let R be a commutative PID. Assume that M is a nonzero semisimple R-module with nonzero annihilator in R. Then from Theorem 4.3, M has only a finite number of homogeneous components. Let $\{H_k \mid 1 \leq k \leq s\}$ be the set of all homogeneous components of M. For k, $1 \leq k \leq s$, we put $H_k = \oplus_\alpha M_{(k,\alpha)}$ with each $M_{(k,\alpha)}$ simple. So $M_{(k,\alpha)} \cong R/p_k R$ for $k, 1 \leq k \leq s$, with p_k a nonzero prime.
We put $P_k = \mathrm{Ann}_R(H_k)$ for k, $1 \leq k \leq s$. Then $P_k = p_k R$. For a nonnegative integer ℓ, we can routinely verify that $P_k^{-\ell} = (1/p_k^\ell)R$ for k, $1 \leq k \leq s$. Therefore,

$$P_1^{-\ell_1} P_2^{-\ell_2} \cdots P_s^{-\ell_s} = (1/p_1^{\ell_1})(1/p_2^{\ell_2}) \cdots (1/p_s^{\ell_s})R$$

for nonnegative integers $\ell_1, \ell_2, \ldots, \ell_s$.

Let $A = \sum_{\ell \geq 0} I^{-\ell}$, where $I = \mathrm{Ann}_R(M) = P_1 P_2 \cdots P_s = p_1 p_2 \cdots p_s R$. By the preceding argument, $A = R[1/p_1, 1/p_2, \ldots, 1/p_s]$. Put $a = p_1 p_2 \cdots p_s$. Then it follows that $A = R[1/a]$ because $I^{-\ell} = (1/a^\ell)R$. Also note that $\mathrm{Ann}_R(M) = aR$.

Example 4.10 ([40, Example 3.2]) Let Γ_i, $i = 1, 2, 3$ be nonempty sets and m be a positive integer. Then by Remark 4.9(ii), we have

$$\mathfrak{B}(\mathbb{Z}_2^{(\Gamma_1)} \oplus \mathbb{Z}_3^{(\Gamma_2)} \oplus \mathbb{Z}_5^{(\Gamma_3)} \oplus \mathbb{Z}^{(m)}) = \mathbb{Z}_2^{(\Gamma_1)} \oplus \mathbb{Z}_3^{(\Gamma_2)} \oplus \mathbb{Z}_5^{(\Gamma_3)} \oplus \mathbb{Z}[1/30]^{(m)}$$

as $\mathbb{Z}$-modules because $\mathrm{Ann}_{\mathbb{Z}}(\mathbb{Z}_2^{(\Gamma_1)} \oplus \mathbb{Z}_3^{(\Gamma_2)} \oplus \mathbb{Z}_5^{(\Gamma_3)}) = 30\mathbb{Z}$.

For a ring R and a nonempty set Λ, we use $\mathrm{CFM}_\Lambda(R)$ to denote the $\Lambda \times \Lambda$ column finite matrix ring over the ring R.

Theorem 4.11 ([50, Theorem 2] and [47, Theorem 3.3]) *Let R be a ring. Then the following are equivalent.*

(i) R is a semiprimary right (and left) hereditary ring.
(ii) $CFM_\Lambda(R)$ is a Baer ring for any nonempty set Λ.

Example 4.12 in the following shows that the hypothesis "$\{K_1, K_2, \ldots, K_m\}$ is a finite set" in Theorem 4.4 and the hypothesis "$N/t(N)$ is finitely generated" in Theorem 4.5 are not superfluous conditions for the existence of the Baer hull of N.

Example 4.12 (see [40, Example 3.6]) Let $\Gamma_i, i = 1, 2, 3$ be nonempty sets as in Example 4.10. Since $\mathbb{Z}[1/30]$ is not a field, $\mathbb{Z}[1/30]$ is not semiprimary because $\mathbb{Z}[1/30]$ is a domain. By Theorem 4.11, there exists a nonempty set Λ such that $\mathrm{CFM}_\Lambda(\mathbb{Z}[1/30])$ is not a Baer ring. Note that the set Λ is necessarily infinite. In fact, if Λ is finite with the cardinality n, then $\mathrm{CFM}_\Lambda(\mathbb{Z}[1/30]) = \mathrm{Mat}_n(\mathbb{Z}[1/30])$ is a Baer ring as $\mathbb{Z}[1/30]$ is a Prüfer domain (see [10, Theorem 6.1.4, p. 191]), a contradiction. Let

$$N = \mathbb{Z}_2^{(\Gamma_1)} \oplus \mathbb{Z}_3^{(\Gamma_2)} \oplus \mathbb{Z}_5^{(\Gamma_3)} \oplus \mathbb{Z}^{(\Lambda)}.$$

Then we have the following.

(i) $V := \mathbb{Z}_2^{(\Gamma_1)} \oplus \mathbb{Z}_3^{(\Gamma_2)} \oplus \mathbb{Z}_5^{(\Gamma_3)} \oplus \mathbb{Z}[1/30]^{(\Lambda)}$ is not a Baer $\mathbb{Z}$-module. In fact, if V is a Baer module, then $\mathbb{Z}[1/30]^{(\Lambda)}$ is Baer as a $\mathbb{Z}$-module by Theorem 3.12. We show that

$$\mathrm{End}_{\mathbb{Z}}(\mathbb{Z}[1/30]^{(\Lambda)}) = \mathrm{End}_{\mathbb{Z}[1/30]}(\mathbb{Z}[1/30]^{(\Lambda)}).$$

For this, first we note that $\mathrm{End}_{\mathbb{Z}[1/30]}(\mathbb{Z}[1/30]^{(\Lambda)}) \subseteq \mathrm{End}_{\mathbb{Z}}(\mathbb{Z}[1/30]^{(\Lambda)})$. Next, let $f \in \mathrm{End}_{\mathbb{Z}}(\mathbb{Z}[1/30]^{(\Lambda)})$. Assume on the contrary that $f \notin \mathrm{End}_{\mathbb{Z}[1/30]}(\mathbb{Z}[1/30]^{(\Lambda)})$. Then there exist $y \in \mathbb{Z}[1/30]^{(\Lambda)}$ and $q \in \mathbb{Z}[1/30]$ such that $f(yq) - f(y)q \neq 0$. Put $q = ab^{-1}$, where $a, b \in R$ and $b \neq 0$. So

$$0 \neq (f(yq) - f(y)q)b = f(yq)b - f(y)a = f(yqb) - f(ya) = f(ya) - f(ya) = 0,$$

which is a contradiction. Therefore, $f \in \mathrm{End}_{\mathbb{Z}}(\mathbb{Z}[1/30]^{(\Lambda)})$. Consequently, we have that $\mathrm{End}_{\mathbb{Z}}(\mathbb{Z}[1/30]^{(\Lambda)}) = \mathrm{End}_{\mathbb{Z}[1/30]}(\mathbb{Z}[1/30]^{(\Lambda)})$. From Theorem 3.11,

$$\mathrm{End}_{\mathbb{Z}}(\mathbb{Z}[1/30]^{(\Lambda)}) = \mathrm{End}_{\mathbb{Z}[1/30]}(\mathbb{Z}[1/30]^{(\Lambda)}) = \mathrm{CFM}_\Lambda(\mathbb{Z}[1/30])$$

is a Baer ring. So we get a contradiction.

(ii) $N/t(N)(\cong \mathbb{Z}^{(\Lambda)})$ is not finitely generated as a $\mathbb{Z}$-module because Λ is infinite.

(iii) *N has no Baer module hull* as a $\mathbb{Z}$-module.

In Proposition 4.13 and Example 4.14, we consider the isomorphism problem for Baer hulls as follows: Let N_1 and N_2 be modules with Baer hulls $\mathfrak{B}(N_1)$ and $\mathfrak{B}(N_2)$, respectively. Is it true that $N_1 \cong N_2$ if and only if $\mathfrak{B}(N_1) \cong \mathfrak{B}(N_2)$ in this case?

Proposition 4.13 ([40, Proposition 3.8]) *Let N_1 and N_2 are isomorphic modules. If N_1 has a Baer hull $\mathfrak{B}(N_1)$, then N_2 has a Baer hull $\mathfrak{B}(N_2)$, and $\mathfrak{B}(N_1) \cong \mathfrak{B}(N_2)$ as modules.*

The next example shows that the converse of Proposition 4.13 does not hold true. In other words, there exist modules N_1 and N_2 such that $\mathfrak{B}(N_1) = \mathfrak{B}(N_2)$ (hence $\mathfrak{B}(N_1) \cong \mathfrak{B}(N_2)$ as modules), but $N_1 \ncong N_2$. Thus the isomorphism problem does not hold for the case of Baer hulls.

Example 4.14 ([40, Example 3.9]) Let $N_1 = \mathbb{Z}_2 \oplus \mathbb{Z}_3 \oplus \mathbb{Z}$. Then by Theorem 4.4 or Example 4.10, $\mathbb{Z}_2 \oplus \mathbb{Z}_3 \oplus \mathbb{Z}[1/6]$ is the Baer hull of N_1 as $\mathbb{Z}$-modules.

Next, let $N_2 = \mathbb{Z}_2 \oplus \mathbb{Z}_3 \oplus \mathbb{Z}[1/3]$. Say V is a Baer module with $N_2 \le V \le E(N_2)$. From Theorem 4.1, $V = \mathbb{Z}_2 \oplus \mathbb{Z}_3 \oplus W$ for some Baer module W such that

$$\mathbb{Z}[1/3] \le W \le \mathbb{Q} \quad \text{and} \quad \mathrm{Hom}_{\mathbb{Z}}(W, \mathbb{Z}_2 \oplus \mathbb{Z}_3) = 0.$$

Thus $\mathrm{Hom}_{\mathbb{Z}}(W, \mathbb{Z}_2) = 0$, and so $2^k W = W$ for any nonnegative integer k (see the proof of [40, Theorem 2.13]). Therefore $1/2^k \in W$ for any positive integer k, and thus $\mathbb{Z}[1/2, 1/3] \le W$. Hence we have

$$\mathbb{Z}_2 \oplus \mathbb{Z}_3 \oplus \mathbb{Z}[1/2, 1/3] = \mathbb{Z}_2 \oplus \mathbb{Z}_3 \oplus \mathbb{Z}[1/6] \le V.$$

Because $\mathbb{Z}_2 \oplus \mathbb{Z}_3 \oplus \mathbb{Z}[1/6]$ is Baer as a $\mathbb{Z}$-module, $\mathbb{Z}_2 \oplus \mathbb{Z}_3 \oplus \mathbb{Z}[1/6]$ is the Baer hull of N_2. However, $N_1 \ncong N_2$ because $\mathbb{Z} \ncong \mathbb{Z}[1/3]$ as $\mathbb{Z}$-modules.

In the next examples, we compare the direct sum of Baer hulls with the Baer hull of a direct sum of modules.

Example 4.15 ([40, Example 2.19]) There exist two modules W_1 and W_2 such that both W_1 and W_2 have Baer module hulls, but $W_1 \oplus W_2$ has no Baer hull.

Let $R = \mathbb{Z}[x]$, the polynomial ring over $\mathbb{Z}$. Put $N = (R \oplus R)_R$. Then $t(N) = 0$, so $t(N)$ is semisimple. However, N has no Baer hull. For this, note that if N is a Baer module, then $\mathrm{End}_R(N) = \mathrm{Mat}_2(R)$ is a Baer ring from Theorem 3.12. So [10, Theorem 6.1.4, p. 191] yields that the ring $R = \mathbb{Z}[x]$ must be Prüfer, which is a contradiction.

Say B is the Baer hull of N. Put $F = \mathbb{Q}(x)$, the field of fractions of R. Note that $E(N) = F \oplus F$. Put $U = F \oplus R$. Then by [10, Theorem 4.2.18, p. 107], U_R is a Baer module. Similarly, $V_R := (R \oplus F)_R$ is a Baer module. Thus $B \subseteq U \cap V = N$, so $B = N$. Hence N is Baer, a contradiction. Therefore N has no Baer hull.

Example 4.16 ([40, Example 3.10]) There exist two modules M and N such that M, N, and $M \oplus N$ have Baer hulls $\mathfrak{B}(M)$, $\mathfrak{B}(N)$, and $\mathfrak{B}(M \oplus N)$, respectively. But

$$\mathfrak{B}(M \oplus N) \not\cong \mathfrak{B}(M) \oplus \mathfrak{B}(N).$$

Let $M = \mathbb{Z}_p$ (p a prime integer) and $N = \mathbb{Z}$ as $\mathbb{Z}$-modules. Then $\mathfrak{B}(M) = \mathbb{Z}_p$ and $\mathfrak{B}(N) = \mathbb{Z}$ since $\mathbb{Z}_p$ is a semisimple $\mathbb{Z}$-module and $\mathbb{Z}$ is a Baer ring. Therefore we have that $\mathfrak{B}(M) \oplus \mathfrak{B}(N) = \mathbb{Z}_p \oplus \mathbb{Z}$.

On the other hand, $\mathfrak{B}(M \oplus N) = \mathfrak{B}(\mathbb{Z}_p \oplus \mathbb{Z}) = \mathbb{Z}_p \oplus \mathbb{Z}[1/p]$ (see Theorem 4.8 and Remark 4.9(ii)). Hence $\mathfrak{B}(M \oplus N) \not\cong \mathfrak{B}(M) \oplus \mathfrak{B}(N)$ because $\mathbb{Z} \not\cong \mathbb{Z}[1/p]$ as $\mathbb{Z}$-modules.

The following example exhibits the disparity of the Baer hull and the extending hull of $\mathbb{Z}_p \oplus \mathbb{Z}$ (p a prime integer).

Example 4.17 [40, Example 3.7]) (i) Let $V = \mathbb{Z}_p \oplus \mathbb{Z}[1/p]$, where p is a prime integer. Then by Remark 4.9(ii), V is the Baer hull of $\mathbb{Z}_p \oplus \mathbb{Z}$ as a $\mathbb{Z}$-module. Hence in view of Theorem 3.9, one might expect that V is also the extending hull of $\mathbb{Z}_p \oplus \mathbb{Z}$ as a $\mathbb{Z}$-module. But this is not true. Further, V is not even extending from [25, Corollary 2]. In fact, the extending hull of $\mathbb{Z}_p \oplus \mathbb{Z}$ is $\mathbb{Z}_{p^\infty} \oplus \mathbb{Z}$, where $\mathbb{Z}_{p^\infty}$ is the Prüfer p-group.

(ii) In the chain of $\mathbb{Z}$-submodules $\mathbb{Z}_p \leq \mathbb{Z}_{p^2} \leq \cdots \leq \mathbb{Z}_{p^\infty}$ of $\mathbb{Z}_{p^\infty}$ (p a prime integer), $\mathbb{Z}_p$ is the Baer hull (also quasi-injective hull) of itself, and $\mathbb{Z}_{p^\infty}$ is the injective hull of each of the modules in the chain. However, $\mathbb{Z}_{p^n}$ ($n > 1$) has no Baer hull by Theorem 4.8. Also note that $\mathbb{Z}_{p^\infty}$ has no Baer hull.

Acknowledgments The authors are grateful for the partial research grant support received from The Ohio State University at Lima, Mathematics Research Institute, and the OSU College of Arts and Sciences. The authors also thank each others' institutions for the hospitality received during the research work on this paper.

References

1. Armendariz, E.P.: A note on extensions of Baer and P.P. rings. J. Aust. Math. Soc. **18**, 470–473 (1974)
2. Beidar, K., Wisbauer, R.: Strongly and properly semiprime modules and rings. In: Jain, S.K., Rizvi, S.T. (eds.) Ring Theory. Proceedings of Ohio State-Denison Conference, pp. 58–94. World Scientific, Singapore (1993)
3. Berberian, S.K.: Baer $*$-Baer Rings. Springer, Berlin (1972)
4. Birkenmeier, G.F., Müller, B.J., Rizvi, S.T.: Modules in which every fully invariant submodule is essential in a direct summand. Commun. Algebr. **30**, 1395–1415 (2002)
5. Birkenmeier, G.F., Park, J.K., Rizvi, S.T.: Ring hulls of generalized triangular matrix rings. Aligarh Bull. Math. **25**, 65–77 (2006)
6. Birkenmeier, G.F., Park, J.K., Rizvi, S.T.: Ring hulls and applications. J. Algebr. **304**, 633–665 (2006)
7. Birkenmeier, G.F., Park, J.K., Rizvi, S.T.: Hulls of semiprime rings with applications to C^*-algebras. J. Algebr. **322**, 327–352 (2009)

8. Birkenmeier, G.F., Park, J.K., Rizvi, S.T.: Modules with FI-extending hulls. Glasgow Math. J.**51**, 347–357 (2009)
9. Birkenmeier, G.F., Park, J.K., Rizvi, S.T.: Hulls of ring extensions. Canad. Math. Bull. **53**, 587–601 (2010)
10. Birkenmeier, G.F., Park, J.K., Rizvi, S.T.: Extensions of Rings and Modules Research Monograph. Birkhäuser/Springer, New York (2013)
11. Chatters, A.W., Khuri, S.M.: Endomorphism rings of modules over nonsingular CS-rings. J. London Math. Soc. **21**, 434–444 (1980)
12. Clark, W.E.: Twisted matrix units semigroup algebras. Duke Math. J. **34**, 417–424 (1967)
13. Dung, N.V., Huynh, D.V., Smith, P.F., Wisbauer, R.: Extending Modules. Longman, Harlow (1994)
14. Eckmann, B., Schopf, A.: Über injektive Moduln. Arch. Math. **4**, 75–78 (1953)
15. Faith, C.: Lectures on injective modules and quotient rings. Lecture Notes in Maths, vol. 49, Springer, Berlin-Heidelberg-New York (1967)
16. Goel, V.K., Jain, S.K.: π-injective modules and rings whose cyclics are π-injective. Commun. Algebr. **6**, 59–73 (1978)
17. Goodearl, K.R.: Direct sum properties of quasi-injective modules. Bull. Am. Math. Soc. **82**, 108–110 (1976)
18. Hattori, A.: A foundation of torsion theory for modules over general rings. Nagoya Math. J. **17**, 147–158 (1960)
19. Hirano, Y., Hongan, M., Ohori, M.: On right P.P. rings. Math. J. Okayama Univ. **24**, 99–109 (1982)
20. Jeremy, L.: Modules et anneaux quasi-continus. Can. Math. Bull. **17**, 217–228 (1974)
21. Jin, H.L., Doh, J., Park, J.K.: Quasi-Baer rings with essential prime radicals. Commun. Algebr. **34**, 3537–3541 (2006)
22. Jin, H.L., Doh, J., Park, J.K.: Group actions on quasi-Baer rings. Can. Math. Bull. **52**, 564–582 (2009)
23. Johnson, R.E., Wong, E.T.: Quasi-injective modules and irreducible rings. J. London Math. Soc. **36**, 260–268 (1961)
24. Jøndrup, S.: p.p. rings and finitely generated flat ideals. Proc. Am. Math. Soc. **28**, 431–435 (1971)
25. Kamal, M.A., Müller, B.J.: Extending modules over commutative domains. Osaka J. Math. **25**, 531–538 (1988)
26. Kaplansky, I.: Rings of operators. In: Berberian S.K., Blattner R. (eds.) University of Chicago Mimeographed Lecture Notes. University of Chicago (1955)
27. Kaplansky, I.: Rings of Operators. Benjamin, New York (1968)
28. Kaplansky, I.: Commutative Rings. University of Chicago Press, Chicago (1974)
29. Khuri, S.M.: Nonsingular retractable modules and their endomorphism rings. Bull. Aust. Math. Soc. **43**, 63–71 (1991)
30. Lee, T.K., Zhou, Y.: Reduced modules. In: Facchini A., Houston E., Salce L. (eds.) Rings, Modules, Algebras, and Abelian Groups. Lecture Notes in Pure and Applied Maths, vol. 236, pp.365–377, Marcel Dekker, New York (2004)
31. Lenzing, H.: Halberbliche endomorphismenringe. Math. Z. **118**, 219–240 (1970)
32. Maeda, S.: On a ring whose principal right ideals generated by idempotents form a lattice. J. Sci. Hiroshima Univ. Ser. A **24**, 509–525 (1960)
33. Mewborn, A.C.: Regular rings and Baer rings. Math. Z. **121**, 211–219 (1971)
34. Mohamed, S.H., Müller, B.J.: Continuous and Discrete Modules. Cambridge University Press, Cambridge (1990)
35. Müller, B.J., Rizvi, S.T.: On the existence of continuous hulls. Commun. Algebr. **10**, 1819–1838 (1982)
36. Müller, B.J., Rizvi, S.T.: On injective and quasi-continuous modules. J. Pure Appl. Algebr. **28**, 197–210 (1983)
37. Oshiro, K.: On torsion free modules over regular rings. Math. J. Okayama Univ. **16**, 107–114 (1973)

38. Oshiro, K.: On torsion free modules over regular rings III. Math. J. Okayama Univ. **18**, 43–56 (1975-1976)
39. Oshiro, K., Rizvi, S.T.: The exchange property of quasi-continuous modules with finite exchange property. Osaka J. Math. **33**, 217–234 (1996)
40. Park, J.K., Rizvi, S.T.: Baer module hulls of certain modules over a Dedekind domain. J. Algebr. Appl. **15**(8), 1650141, (24 pages) (2016)
41. Pollingher, A., Zaks, A.: On Baer and quasi-Baer rings. Duke Math. J. **37**, 127–138 (1970)
42. Rizvi, S.T.: Contributions to the theory of continuous modules. Dissertation, McMaster University (1981)
43. Rizvi, S.T.: Commutative rings for which every continuous module is quasi-injective. Arch. Math. **50**, 435–442 (1988)
44. Rizvi, S.T., Roman, C.S.: Baer and quasi-Baer modules. Commun. Algebr. **32**, 103–123 (2004)
45. Rizvi, S.T., Roman, C.S.: Baer property of modules and applications. In: Chen J.L., Ding N.Q., Marubayashi H. (eds.). Advances in Ring Theory. Proceedings of 4th China–Japan–Korea International Conference, pp.225–241. World Scientific, New Jersey (2005)
46. Rizvi, S.T., Roman, C.S.: On $\mathcal{K}$-nonsingular modules and applications. Commun. Algebr. **35**, 2960–2982 (2007)
47. Rizvi, S.T., Roman, C.S.: On direct sums of Baer modules. J. Algebr. **321**, 682–696 (2009)
48. Sharpe, D.W., Vámos, P. V.: Injective Modules. Cambridge University Press, Cambridge (1972)
49. Shoda, K.: Zur Theorie der algebraischen Erweiterungen. Osaka Math. J. **4**, 133–143 (1952)
50. Stephenson, W., Tsukerman, G.M.: Rings of endomorphisms of projective modules. Siberian Math. J. **11**, 181–184 (1970) (English translation)

Spined Product Decompositions of Orthocryptogroups

Akihiro Yamamura

Abstract A semigroup is said to be an internal spined product of its subsemigroups if it is naturally isomorphic to an external spined product of the subsemigroups. We shall show that internal spined products can be identified with external spined products in the class of orthocryptogroups. On the other hand, two concepts are not equivalent in general as we give examples of external spined products that admit no internal spined product decomposition. Further, we examine internal spined product of orthocryptogroups. Using a lattice theoretic method, we obtain a unique decomposition theorem similar to the Krull–Schmidt theorem in group theory. We also study completely reducible orthocryptogroups in which any normal sub-orthocryptogroup is a spined factor. We show that such an orthocryptogroup is an internal spined product of simple sub-orthocryptogroups.

Keywords Orthocryptogroups · Spined products · Krull–Schmidt theorem · Ore theorem

1 Introduction

An external spined product gives a convenient way to construct a new semigroup from old ones. It plays an important role in the structure theory of regular semigroups. Suppose S_1 and S_2 are semigroups. Let $\phi_1 : S_1 \to Q$ and $\phi_2 : S_2 \to Q$ be epimorphisms. The *external spined product* of S_1 and S_2 over Q with respect to ϕ_1 and ϕ_2 is defined to be the set of pairs (s_1, s_2) satisfying $\phi_1(s_1) = \phi_2(s_2)$. Obviously, an external spined product forms a subsemigroup of the external direct product $S_1 \times S_2$. We denote the external spined product by $S_1 \bowtie_Q S_2$. An external spined product is called just a *spined product* in the literature of semigroup theory. An external spined product of more than two factors is defined similarly. If Γ is the largest semilattice homomorphic image of S_1 and S_2, respectively, and both ϕ_1 and ϕ_2 are the natural

A. Yamamura (✉)
Department of Mathematical Science and Electrical-Electric-Computer Engineering,
Akita University, 1-1, Tegata-Gakuenmachi, Akita, Japan
e-mail: yamamura@ie.akita-u.ac.jp

S.T. Rizvi et al. (eds.), *Algebra and its Applications*, Springer Proceedings
in Mathematics & Statistics 174, DOI 10.1007/978-981-10-1651-6_2

homomorphisms, then the external spined product can be formed over Γ. Similarly, if Green's $\mathcal{H}$-relation of S_1 and S_2 are congruences and $B = S_1/\mathcal{H} = S_2/\mathcal{H}$, then the external spined product with respect to $\mathcal{H}$ can be formed over B.

A semigroup is called *cryptic* if Green's $\mathcal{H}$-relation is a congruence. A cryptic completely regular semigroup is called a *cryptogroup*. An *orthodox semigroup* is a regular semigroup in which the set of idempotents forms a subsemigroup and an *orthocryptogroup* is an orthodox cryptogroup. It was first studied by Yamada and called a *strictly inversive semigroup* in [7]. He showed that an external spined product of a Clifford semigroup and a band with respect to the structure decomposition is an orthocryptogroup, and conversely, every orthocryptogroup S is isomorphic to an external spined product $C \bowtie_\Gamma E(S)$ of the largest Clifford semigroup homomorphic image C and the band $E(S)$ of idempotents of S over the structure semilattice Γ.

The group inverse of an element a in a completely regular semigroup is denoted by a^{-1}. The identity element of the subgroup of a completely regular semigroup containing an element a is denoted by a^0, that is, $a^0 = aa^{-1} = a^{-1}a$. It is known (see [5]) that a completely regular semigroup satisfies the equation

$$(xy)^{-1} = (xy)^0 y^{-1} (yx)^0 x^{-1} (xy)^0 \tag{1.1}$$

and an orthocryptogroup satisfies the equation

$$(xy)^0 = x^0 y^0. \tag{1.2}$$

Therefore an orthocryptogroup satisfies the equation

$$(xy)^{-1} = x^0 y^{-1} x^{-1} y^0. \tag{1.3}$$

The equational class of completely regular semigroups defined by (1.3) includes the variety of orthocryptogroups but does not coincide [9].

The least band congruence of an orthocryptogroup S is Green's $\mathcal{H}$-relation and so S has the $\mathcal{H}$-decomposition $\bigcup_{e \in E(S)} S(e)$, where $S(e)$ is the maximal subgroup containing the idempotent e and $E(S)$ is the band of idempotents of S. Note that $E(S)$ is isomorphic to the largest band image of S and $E(S) \cong S/\mathcal{H}$.

A nonempty subset of an orthocryptogroup S is called a *sub-orthocryptogroup* if it forms an orthocryptogroup under the multiplication of S, that is, a nonempty subset is a sub-orthocryptogroup if and only if it is closed under taking an inverse and multiplication.

Suppose S is an orthocryptogroup and ϕ is the natural homomorphism of S onto the largest band image B, that is, $B \cong S/\mathcal{H}$. A sub-orthocryptogroup H of S is called *full* if $E(H) = E(S)$. If S has the $\mathcal{H}$-decomposition $\bigcup_{e \in E(S)} S(e)$, then H has the $\mathcal{H}$-decomposition $\bigcup_{e \in E(S)} H(e)$. The following lemma is obvious because an orthocryptogroup is isomorphic to an external spined product of a Clifford semigroup and a band, however, we give a direct proof.

Lemma 1.1 *Let $H_1, H_2, \ldots, H_n$ be full sub-orthocryptogroups of S. Suppose $s_1 \in H_1(e_1)$, $s_2 \in H_2(e_2)$, ..., $s_n \in H_n(e_n)$ for $e_1, e_2, \ldots, e_n \in E(S)$. Then there exists $s_1' \in H_1(e)$, $s_2' \in H_2(e)$, ..., $s_n' \in H_n(e)$ such that $s_1 s_2 \ldots s_n = s_1' s_2' \ldots s_n'$, where $e = e_1 e_2 \ldots e_n$.*

Proof Note that $s_1 = s_1 e_1$ because $s_1 \in S(e_1)$ and e_1 is the identity element of $S(e_1)$. Likewise, $s_2 \ldots s_n = e_2 \ldots e_n s_2 \ldots s_n$ because $s_2 \ldots s_n \in S(e_2 \ldots e_n)$ and $e_2 \ldots e_n$ is the identity element of $S(e_2 \ldots e_n)$. Then $s_1 s_2 \ldots s_n = s_1 e_1 e_2 \ldots e_n s_2 \ldots s_n = s_1(e_1 e_2 \ldots e_n)(e_1 e_2 \ldots e_n)s_2 \ldots s_n = s_1 e_2 \ldots e_n e_1 s_2 \ldots s_n$. Likewise we have $s_1 e_2 \ldots e_n e_1 s_2 s_3 \ldots s_n = s_1 e_2 \ldots e_n e_1 s_2 e_3 \ldots e_n e_1 e_2 s_3 \ldots s_n$ and similarly $s_1 s_2 \ldots s_n = (s_1 e_2 \ldots e_n)(e_1 s_2 e_3 \ldots e_n)(e_1 e_2 s_3 e_4 \ldots e_n) \ldots (e_1 \ldots e_{n-1} s_n)$. Now we set $s_i' = e_1 e_2 \ldots e_{i-1} s_i e_{i+1} \ldots e_n$. Then $s_1 s_2 \ldots s_n = s_1' s_2' \ldots s_n'$ and $s_i' \in H_i(e)$ for every $i = 1, 2, \ldots, n$. □

2 Internal Spined Products

Let S be a semigroup and ϕ a homomorphism of S onto Q. Suppose H_1 and H_2 are subsemigroups of S such that $\phi(H_1) = \phi(H_2) = Q$. If the external spined product $H_1 \bowtie_Q H_2$ over Q with respect to $\phi|_{H_1}$ and $\phi|_{H_2}$ is isomorphic to S under the mapping $(h_1, h_2) \mapsto h_1 h_2$ where $(h_1, h_2) \in H_1 \bowtie_Q H_2$, then S is said to be the *internal spined product* of H_1 and H_2 over Q. In such a case we denote $S = H_1 \bowtie_Q H_2$. Similarly, we can define an internal spined product $H_1 \bowtie_Q H_2 \bowtie_Q \ldots \bowtie_Q H_n$ of finitely many subsemigroups.

By the definition, every internal spined product is always isomorphic to an external spined product of its subsemigroups. On the other hand, an external spined product does not always admit an internal spined product decomposition as we shall see next. We note that an external direct product of groups always admits an internal direct product decomposition of its subgroups.

Example 1 A band is said to be *normal* (*left normal*, *right normal*, resp.) if it satisfies the equation $xyzx = xzyx$ ($xyz = xzy$, $yzx = zyx$, resp.). Let B be a band defined on the set $\{e, f, a, b, c, d\}$ with the following multiplication Table 1.

Table 1 Multiplication table of B

	e	f	a	b	c	d
e	e	b	a	b	a	b
f	c	f	c	d	c	d
a	a	b	a	b	a	b
b	a	b	a	b	a	b
c	c	d	c	d	c	d
d	c	d	c	d	c	d

Table 2 Multiplication table of L

	l_1	l_2	l_3	l_4
l_1	l_1	l_3	l_3	l_3
l_2	l_4	l_2	l_4	l_4
l_3	l_3	l_3	l_3	l_3
l_4	l_4	l_4	l_4	l_4

Table 3 Multiplication table of R

	r_1	r_2	r_3	r_4
r_1	r_1	r_4	r_3	r_4
r_2	r_3	r_2	r_3	r_4
r_3	r_3	r_4	r_3	r_4
r_4	r_3	r_4	r_3	r_4

Clearly, $\{e\}$, $\{f\}$, and $\{a, b, c, d\}$ are $\mathcal{D}$-classes of B, which are rectangular bands, and B is a strong semilattice of them. Note that a band is normal if it is a strong semilattice of rectangular bands [6]. Therefore, B is normal and has the structure decomposition $B = \{e\} \cup \{f\} \cup \{a, b, c, d\}$. It is well known that a normal band is an external spined product of a left normal band and a right normal band [6]. It is easy to see B is isomorphic to an external spined product of four element left normal band $L = \{l_1, l_2, l_3, l_4\}$ and a four element right normal band $R = \{r_1, r_2, r_3, r_4\}$ over the structure semilattice $\Gamma = \{\alpha, \beta, 0\}$ that is the three element non-chain semilattice (Tables 2 and 3).

Note that L and R have the structure decomposition $L = \{l_1\} \cup \{l_2\} \cup \{l_3, l_4\}$ and $R = \{r_1\} \cup \{r_2\} \cup \{r_3, r_4\}$, respectively. Then $L \bowtie_\Gamma R$ is isomorphic to B under the mapping; $(l_1, r_1) \mapsto e$, $(l_2, r_2) \mapsto f$, $(l_3, r_3) \mapsto a$, $(l_3, r_4) \mapsto b$, $(l_4, r_3) \mapsto c$, $(l_4, r_4) \mapsto d$.

On the other hand, it is easy to verify that there is no proper subsemigroup of B whose largest semilattice homomorphic image is Γ. It follows that there is no subsemigroups B_1 and B_2 of B so that B is the internal spined product $B_1 \bowtie_\Gamma B_2$. Therefore, B admits no internal spined product with respect to the structure decomposition even though B is an external spined product.

Example 2 Next we consider a spined product of completely simple semigroups. Let S_1 be the two element right zero semigroup. Note that S_1 can be considered as the Rees matrix semigroup $\mathcal{M}(G_1; I, \Lambda; P)$, where G_1 is the trivial group, $I = \{1\}$, $\Lambda = \{1, 2\}$, and the sandwich matrix P is defined by $\begin{pmatrix} p_{11} \\ p_{21} \end{pmatrix} = \begin{pmatrix} 1 \\ 1 \end{pmatrix}$.

Next let S_2 be a Rees matrix semigroup $\mathcal{M}(G_2; I, \Lambda; Q)$, where $G_2 = \{1, g, g^2\}$ is the cyclic group of order three, $I = \{1\}$, $\Lambda = \{1, 2\}$, and the sandwich matrix Q is defined by $\begin{pmatrix} q_{11} \\ q_{21} \end{pmatrix} = \begin{pmatrix} g \\ 1 \end{pmatrix}$. Note that S_2 is not a rectangular group because the set of idempotents does not form a subsemigroup. Clearly both $S_1/\mathcal{H}$ and $S_2/\mathcal{H}$ are the two element right zero semigroup B. Then the external spined product $S_1 \bowtie_B S_2$ over B can be considered as the Rees matrix semigroup $\mathcal{M}(G_1 \times G_2; I, \Lambda; R)$, where the sandwich matrix R is defined by $\begin{pmatrix} r_{11} \\ r_{21} \end{pmatrix} = \begin{pmatrix} (1, g) \\ (1, 1) \end{pmatrix}$. It is easy to see that S is isomorphic to S_2. On the other hand, there exists no subsemigroup isomorphic to S_1 because $\mathcal{M}(G_1 \times G_2; I, \Lambda; R)$ does not contain the two element right zero semigroup. Therefore, $S_1 \bowtie_B S_2$ admits no internal spined product with respect to $\mathcal{H}$.

3 Internal Spined Products of Orthocryptogroups

Suppose S is an orthocryptogroup and ϕ is the natural homomorphism of S onto the largest band image B, that is, $B \cong S/\mathcal{H}$. Let $H_1, H_2, \ldots, H_n$ be full sub-orthocryptogroups of S. They have the same largest band image B and so we can consider the external spined product over B. Recall that if S is isomorphic to the external spined product $H_1 \bowtie_B H_2 \bowtie_B \ldots \bowtie_B H_n$ under the mapping $(s_1, s_2, \ldots, s_n) \mapsto s_1 s_2 \ldots s_n$, then S is said to be the internal spined product of $H_1, H_2, \ldots, H_n$.

Lemma 3.1 *If S is the internal spined product of full sub-orthocryptogroups $H_1, H_2, \ldots, H_n$, then the following conditions hold.*

(A1) Elements of $H_i(e)$ and $H_j(e)$ $(i \neq j)$ commute for every $e \in E(S)$.
(A2) For $e \in E(S)$ every element s of $S(e)$ is expressed uniquely as $s = s_1 s_2 \ldots s_n$, where $s_i \in H_i(e)$.

Conversely, if full sub-orthocryptogroups $H_1, H_2, \ldots, H_n$ satisfy (A1) and (A2), then S is the internal spined product $H_1 \bowtie_B H_2 \bowtie_B \ldots \bowtie_B H_n$.

Proof If $S = H_1 \bowtie_B H_2 \bowtie_B \ldots \bowtie_B H_n$, then clearly (A1) and (A2) are satisfied. We now suppose (A1) and (A2) hold for full sub-orthocryptogroups $H_1, H_2, \ldots, H_n$. Define a mapping ψ of the external spined product $H_1 \bowtie_B H_2 \bowtie_B \ldots \bowtie_B H_n$ into S by $\psi(s_1, s_2, \ldots, s_n) = s_1 s_2 \ldots s_n$. We shall show that ψ is an isomorphism onto S. Take two elements $(s_1, s_2, \ldots, s_n)$ and $(t_1, t_2, \ldots, t_n)$ of $H_1 \bowtie_B H_2 \bowtie_B \ldots \bowtie_B H_n$. Suppose that $s_1 \in H_1(e), s_2 \in H_2(e), \ldots, s_n \in H_n(e)$ and $t_1 \in H_1(f), t_2 \in H_2(f), \ldots, t_n \in H_n(f)$. Let $h = ef$. Then we have $\psi((s_1, s_2, \ldots, s_n)(t_1, t_2, \ldots, t_n)) = \psi(s_1 t_1, s_2 t_2, \ldots, s_n t_n) = s_1 t_1 s_2 t_2 s_3 t_3 \ldots s_n t_n = s_1 h t_1 s_2 h t_2 s_3 t_3 \ldots s_n t_n = s_1 s_2 h h t_1 t_2 s_3 t_3 \ldots s_n t_n = s_1 s_2 t_1 t_2 s_3 t_3 \ldots s_n t_n$ since elements of $H_i(h)$ and $H_j(h)$ commute by (A1). Similarly we can show $s_1 s_2 t_1 t_2 s_3 t_3 \ldots s_n t_n = s_1 s_2 \ldots s_n t_1 t_2 \ldots t_n = \psi(s_1, s_2, \ldots, s_n)\psi(t_1, t_2, \ldots, t_n)$. Next, suppose that $\psi(s_1, s_2, \ldots, s_n) = \psi(t_1, t_2, \ldots, t_n)$, where $s_1 \in H_1(e), s_2 \in H_2(e), \ldots, s_n \in H_n(e)$ and $t_1 \in H_1(f)$,

$t_2 \in H_2(f), \ldots, t_n \in H_n(f)$. We have $s_1 s_2 \ldots s_n = t_1 t_2 \ldots t_n$. Then we have $e = f$ and (A2) implies $(s_1, s_2, \ldots, s_n) = (t_1, t_2, \ldots, t_n)$. Clearly ψ is surjective by (A2). Therefore ψ is an isomorphism. □

A sub-orthocryptogroup N of S is called *normal* if N is full and $s^{-1}Ns \subset N$ for every s in S (see [8]). For any $s \in S$ and $e \in E(S)$ we have $(s^{-1}es)(s^{-1}es) = s^{-1}(ess^{-1})(ess^{-1})s = s^{-1}ess^{-1}s = s^{-1}es$. Hence, $s^{-1}es \in E(S)$ and so $E(S)$ is normal. Obviously S itself is normal.

For a normal sub-orthocryptogroup N we define a relation ρ_N of S by $s\,\rho_N\,t$ if and only if $s\,\mathcal{H}\,t$ and $st^{-1} \in N$. It is easy to see that ρ_N is an idempotent-separating congruence of S and N coincides with its kernel $\mathrm{Ker}(\rho_N) = \{s \mid s\,\rho_N\,e$ for some $e \in E(S)\}$. Conversely, for every idempotent-separating congruence ρ the kernel $\mathrm{Ker}(\rho) = \{s \mid s\,\rho\,e$ for some $e \in E(S)\}$ is a normal sub-orthocryptogroup of S, and furthermore we have $\rho_{\mathrm{Ker}(\rho)} = \rho$.

Lemma 3.2 *If S is the internal spined product of full sub-orthocryptogroups $H_1, H_2, \ldots, H_n$, then the following conditions hold.*

(B1) Every H_i is normal.
(B2) $S = H_1 H_2 \ldots H_n$.
(B3) $H_i \cap (H_1 \ldots H_{i-1} H_{i+1} \ldots H_n) = E(S)$ for every $i = 1, 2, \ldots n$.

Conversely, if sub-orthocryptogroups $H_1, H_2, \ldots, H_n$ satisfy (B1), (B2), and (B3), then S is the internal spined product $H_1 \bowtie_B H_2 \bowtie_B \ldots \bowtie_B H_n$.

Proof First we suppose S is the internal spined product $H_1 \bowtie_B H_2 \bowtie_B \ldots \bowtie_B H_n$. Then $H_1, H_2, \ldots, H_n$ satisfy (A1) and (A2) by Lemma 3.1. Take elements h in H_1 and s in S. Suppose $h \in H_1(f)$ and $s \in S(e)$. By (A2), there exists an element s_i in $H_i(e)$ for $i = 1, 2, \ldots, n$ such that $s = s_1 s_2 \ldots s_n$. Since all s_i belong to the subgroup $S(e)$, we have $s^{-1} = s_n^{-1} \ldots s_2^{-1} s_1^{-1}$. Note that $s_1^{-1} h s_1 \in H_1 \cap S(efe) = H_1(efe)$ because $s_1, h \in H_1$. Then we have $s_2^{-1}(s_1^{-1} h s_1) s_2 = s_2^{-1}(efe)(s_1^{-1} h s_1)(efe) s_2 = (s_1^{-1} h s_1) s_2^{-1}(efe)(efe) s_2 = (s_1^{-1} h s_1) s_2^{-1} s_2$ because elements of $H_1(efe)$ and $H_2(efe)$ commute by (A1). On the other hand, $(s_1^{-1} h s_1) s_2^{-1} s_2 = (s_1^{-1} h s_1) e = s_1^{-1} h s_1$. Inductively we can show $s_n^{-1} \ldots s_2^{-1}(s_1^{-1} h s_1) s_2 \ldots s_n = s_1^{-1} h s_1$. It follows that $s^{-1} h s = s_1^{-1} h s_1 \in H_1$. Thus H_1 is normal. Similarly we can show H_i is normal for $i = 2, \ldots, n$. Obviously, (A2) implies (B2). Now, take an element s in $H_1 \cap (H_2 \ldots H_n)$. Then $s = s_2 \ldots s_n$ for some $s_2 \in H_2, \ldots, s_n \in H_n$. Suppose $s \in H_1(e)$ for some $e \in E(S)$. By Lemma 1.1 we may assume that $s_i \in H_i(e)$. Then we have $e = s^{-1}s = s^{-1}s_2 \ldots s_n$. By (A2) we have $e = s^{-1} = s_2 = \ldots = s_n$ and $s = e$. Hence, $E(S) = H_1 \cap (H_2 \ldots H_n)$. Similarly we can show $H_i \cap (H_1 \ldots H_{i-1} H_{i+1} \ldots H_n) = E(S)$ for every $i = 2, \ldots, n$. Therefore, (B1), (B2), and (B3) hold.

Conversely, we suppose (B1), (B2), and (B3). Take elements s in $H_i(e)$ and t in $H_j(e)$ $(i \neq j)$. Since $sts^{-1} \in H_j$ and $ts^{-1}t^{-1} \in H_i$, we have $sts^{-1}t^{-1} \in H_i \cap H_j$. By (B3) we have $sts^{-1}t^{-1} \in E(S) \cap S(e)$. Hence, $sts^{-1}t^{-1} = e$. On the other hand, $s^{-1}t^{-1}ts = s^{-1}es = s^{-1}s = e$. Then $st = ste = sts^{-1}t^{-1}ts = ets = ts$. Hence, (A1) holds. Next, take an element s in $S(e)$. By (B2) $s = s_1 s_2 \ldots s_n$

for some $s_i \in H_i$ $(i = 1, 2, \ldots, n)$. Moreover, we may take $s_i \in H_i(e)$ for $i = 1, 2, \ldots, n$ by Lemma 1.1. Suppose $s_1 s_2 \ldots s_n = t_1 t_2 \ldots t_n$, where $s_i, t_i \in H_i(e)$ for every $i = 1, 2, \ldots, n$. Then, $t_1^{-1} s_1 = t_2 \ldots t_n s_n^{-1} \ldots s_2^{-1}$. Note that $t_1^{-1} s_1 \in H_1(e)$. Since we have already shown (A1) holds, $t_2 \ldots t_n s_n^{-1} \ldots s_2^{-1} = t_2 s_2^{-1} t_3 s_3^{-1} \ldots t_n s_n^{-1}$. Thus $t_2 \ldots t_n s_n^{-1} \ldots s_2^{-1} \in H_2(e) H_3(e) \ldots H_n(e)$. By (B3) we have $t_1^{-1} s_1 = t_2 \ldots t_n s_n^{-1} \ldots s_2^{-1} = e$. Therefore, $s_1 = t_1$. Similarly we can show $s_i = t_i$ for every $i = 2, \ldots, n$. Consequently we obtained (A2). □

In group theory, the external direct product $G = G_1 \times G_2$ always admits an internal direct decomposition of its subgroups isomorphic to G_1 and G_2. Let H_1 be $\{(g_1, 1) \mid g_1 \in G_1\}$ and H_2 be $\{(1, g_2) \mid g_2 \in G_2\}$, respectively. Then G is the *internal direct product* of H_1 and H_2. Thus, the concept of external and internal direct products are equivalent. This is not the case with wider classes of semigroups as we have seen in the preceding section. Fortunately spined products of orthocryptogroups over the largest band image are similar to direct products of groups.

Theorem 3.3 *Every external spined product of orthocryptogroups over the largest band image admits an internal spined product decomposition.*

Proof Suppose S is the external spined product of S_1 and S_2 over the band B, where $S_1/\mathcal{H} \cong B \cong S_2/\mathcal{H}$. We define subsemigroups H_1 and H_2 of S to be $H_1 = \{(s, e) \mid s \in S_1(e), e \in E(S_2)\}$ and $H_2 = \{(e, t) \mid t \in S_2(e), e \in E(S_1)\}$, respectively. It is routine to check H_1 and H_2 satisfy (B1), (B2), and (B3). Hence, S is the internal spined product of H_1 and H_2. It can be similarly shown for $S_1 \bowtie_B S_2 \bowtie_B \ldots \bowtie_B S_n$ for $n \geq 3$. □

4 Spined Product Decompositions

Decomposing an algebraic system into indecomposable ones is an essential problem in mathematics. In group theory, the Krull–Schmidt theorem guarantees the uniqueness of direct product decompositions of groups satisfying certain finiteness conditions into indecomposable factors (see [2]). Ore [4] proved the Krull–Schmidt theorem using a lattice theoretic method. We shall prove the uniqueness of internal spined product decompositions of orthocryptogroups into indecomposable factors using a lattice theoretic method.

A lattice is said to be *of finite length* if there is a bound on the length of its chains. Two elements a and b in a lattice with the least element 0 and the greatest element 1 are said to be *complementary* if $a \vee b = 1$ and $a \wedge b = 0$ hold. In such a case, b is said to be *complement* of a and vice versa. Two elements in a lattice that have a common complement c are said to be *c-related*. A lattice L is called *modular* if it satisfies the modular law

$$a \leq b \Rightarrow (c \vee a) \wedge b = (c \wedge b) \vee a \quad (a, b, c \in L) \tag{4.1}$$

Let L be a modular lattice with the least element 0. A subset $\{a_1, a_2, \ldots, a_n\}$ of finitely many elements of L is said to be *independent* if $a_i \neq 0$ $(i = 1, 2, \ldots, n)$ and

$$a_i \wedge (a_1 \vee \ldots \vee a_{i-1} \vee a_{i+1} \vee \ldots \vee a_n) = 0 \tag{4.2}$$

for every $i = 1, 2, \ldots, n$. If an element $a \in L$ is represented as the join of an independent set, that is, $a = a_1 \vee \ldots \vee a_n$ where $\{a_1, a_2, \ldots, a_n\}$ is independent, then a is said to be the *direct join* of the elements $a_1, a_2, \ldots, a_n$ and we write $a = a_1 \times \ldots \times a_n$. An element a in a lattice L is said to be *indecomposable* if $a \neq 0$ and it admits no direct join $a = b \times c$ with $b \neq a$ and $c \neq a$. If a is written as a direct join of indecomposable elements, then it is called a *complete decomposition* of a. The following theorem is due to Ore (see [1, 3] for a proof).

Proposition 4.1 *In a modular lattice L of finite length, if*

$$1 = a_1 \times \ldots \times a_m$$

and

$$1 = b_1 \times \ldots \times b_n$$

are two complete decompositions of 1*, then each* a_i *is* a_i'*-related to some* b_j *for* $i = 1, 2, \ldots, m$*, where* $a_i' = a_1 \times \ldots \times a_{i-1} \times a_{i+1} \times \ldots \times a_m$. □

We shall show that the set of normal sub-orthocryptogroups of an orthocryptogroup S forms a lattice. Suppose M and N are normal sub-orthocryptogroups of S. Take $m \in M$ and $n \in N$. Note that S satisfies the Eqs. (1.2) and (1.3). Then we have $mn = mn(mn)^0 = mnm^{-1}mn^0 \in NM$ since $mnm^{-1} \in N$ and $n^0 \in M$. Thus, $MN \subset NM$ and vice versa. Hence, $MN = NM$. Then $(MN)(MN) = MMNN = MN$ and so MN is closed under multiplication. Next we take $m \in M$ and $n \in N$. We have $(mn)^{-1} = m^0n^{-1}m^{-1}n^0 \in MNMN = MN$. Hence, MN is closed under taking inverse. Since M and N are full, $M \subset ME(S) \subset MN$ and $N \subset E(S)N \subset MN$. Thus MN include both M and N. Next take $s \in MN$ and $h \in S$. Suppose $s \in S(e)$. By Lemma 1.1 we can write $s = mn$ for some $m \in M(e)$ and $n \in N(e)$. Then we have $h^{-1}sh = h^{-1}mnh = h^{-1}m(h^{-1}m)^0nh = h^{-1}mhh^{-1}m^0nh \in MN$ since $h^{-1}mh \in M$ and $h^{-1}m^0nh \in N$. Thus MN is normal. Therefore, MN is the smallest normal sub-orthocryptogroup including both M and N. On the other hand, $M \cap N$ is the largest normal sub-orthocryptogroup contained in both M and N. Consequently, the set of normal sub-orthocryptogroups forms a lattice with the join MN and the meet $M \cap N$. This lattice has the greatest element S and the least element $E(S)$. Moreover, we have the following.

Lemma 4.2 *The lattice of normal sub-orthocryptogroups of an orthocryptogroup is modular.*

Proof Let S be an orthocryptogroup. Suppose that A, B, C are normal sub-ortho cryptogroups of S satisfying $A \subset B$. It is enough to show $(CA) \cap B \subset (C \cap B)A$.

Take an arbitrary element s from $(CA) \cap B$. Then $s = ca$, where $c \in C$, $a \in A$ and $s \in B$. Note that $sa^{-1} = caa^{-1}$. Since $A \subset B$, we have $sa^{-1} \in B$. On the other hand, $caa^{-1} \in C$ because C is full. Therefore, $s = ca = caa^{-1}a = (sa^{-1})a \in (C \cap B)A$ and so $(CA) \cap B \subset (C \cap B)A$. □

It is easy to see that the lattice of normal sub-orthocryptogroups is isomorphic to the lattice of idempotent-separating congruences under the correspondence $N \leftrightarrow \rho_N$. Therefore the lattice of idempotent-separating congruences of an orthocryptogroup is modular by Lemma 4.2.

We say that an orthocryptogroup S is *spined indecomposable* if $S \neq E(S)$ and the internal spined product decomposition $S = S_1 \bowtie_B S_2$, where $B = S/\mathcal{H}$, implies either $S_1 = S$ or $S_2 = S$. Note that an orthocryptogroup S always admits the internal spined product decomposition $S = S \bowtie_B E(S)$.

We shall next give a sufficient condition for an orthocryptogroup to admit a spined product decomposition into spined indecomposable factors. An orthocryptogroup S is said to satisfy the *ascending chain condition* if $N_1 \subset N_2 \subset N_3 \subset \ldots$ is a chain of normal sub-orthocryptogroups, then there exists t for which $N_t = N_{t+1} = N_{t+2} = \ldots$, and S is said to satisfy the *descending chain condition* if $K_1 \supset K_2 \supset K_3 \supset \ldots$ is a chain of normal sub-orthocryptogroups then there exists t for which $K_t = K_{t+1} = K_{t+2} = \ldots$.

Theorem 4.3 *Suppose S satisfies either the ascending or descending chain condition. Then S is an internal spined product of a finitely many spined indecomposable factors.*

Proof Suppose the conclusion does not hold. Then S is not spined indecomposable and so it is decomposed as $H_0 \bowtie_B K_0$, where H_0 and K_0 are proper sub-orthocryptogroups. By the assumption, either H_0 or K_0 is not spined indecomposable, say H_0. By induction, there is a sequence of sub-orthocryptogroups $H_0, H_1, H_2, \ldots$, where every H_i is a proper spined factor of H_{i-1}. Then we have a descending chain $S \supsetneq H_0 \supsetneq H_1 \supsetneq H_2 \supsetneq \ldots$. It is easy to see H_i is normal in S. If S satisfies the descending chain condition, this is a contradiction. Now we suppose S satisfies the ascending chain condition. Since each H_i is a spined factor of H_{i-1}, there is a normal sub-orthocryptogroup K_i in H_{i-1} satisfying $H_{i-1} = H_i \bowtie_B K_i$. Since each K_i is normal in S, we have an ascending chain $K_0 \subsetneq K_0 \bowtie_B K_1 \subsetneq K_0 \bowtie_B K_1 \bowtie_B K_2 \subsetneq \ldots$, which is a contradiction. □

We note that a modular lattice is of finite length if and only if it satisfies both the chain conditions (see [3]). Therefore, if an orthocryptogroup S satisfies both the chain conditions, then the lattice of normal sub-orthocryptogroups is of finite length and vice versa.

Theorem 4.4 *Let S be an orthocryptogroup satisfying both the chain conditions and $B = S/\mathcal{H}$. If S has two spined product decompositions $H_1 \bowtie_B H_2 \bowtie_B \ldots \bowtie_B H_m$ and $K_1 \bowtie_B K_2 \bowtie_B \ldots \bowtie_B K_n$, where $H_i\ (i = 1, 2, \ldots, m)$ and $K_j\ (j = 1, 2, \ldots, n)$ are spined indecomposable, then $m = n$ and there exists a bijection Ψ of the family*

$\{H_i \mid i = 1, 2, \ldots, m\}$ *onto the family* $\{K_i \mid j = 1, 2, \ldots, n\}$ *such that* H_i *is isomorphic and* H_i'*-related to* $\Psi(H_i)$.

Proof Note that the lattice of normal sub-orthocryptogroups of S is modular by Lemma 4.2, and it is of finite length because S satisfies both the chain conditions. By Lemma 3.2, $\{H_1, H_2, \ldots, H_m\}$ and $\{K_1, K_2, \ldots, K_n\}$ are independent, respectively. Therefore both $H_1 \bowtie_B H_2 \bowtie_B \ldots \bowtie_B H_m$ and $K_1 \bowtie_B K_2 \bowtie_B \ldots \bowtie_B K_n$ are complete decompositions of S.

Suppose $n \leq m$. By Proposition 4.1, H_1 is H_1'-related to some K_j (say K_1). Recall that $H_1' = H_2 \bowtie_B \ldots \bowtie_B H_m$. We have $S = K_1 \bowtie_B H_1' = K_1 \bowtie_B H_2 \bowtie_B \ldots \bowtie_B H_m$. By induction, we obtain $S = K_1 \bowtie_B K_2 \bowtie_B \ldots \bowtie_B K_n \bowtie_B H_{n+1} \bowtie_B \ldots \bowtie_B H_m$ after renumbering K_j. On the other hand, we have $S = K_1 \bowtie_B K_2 \bowtie_B \ldots \bowtie_B K_n$. Therefore, we have $m = n$. Moreover, each H_i is H_i'-related to K_j for some j by Proposition 4.1.

Next we shall show that if H_1 and K_j (say K_1) is H_1'-related, then H_1 and K_1 are isomorphic. Suppose that $S = H_1 \bowtie_B H_1' = K_1 \bowtie_B H_1'$. Define a mapping $\psi : H_1 \to K_1$ as follows. For $h \in H_1(e)$ $(e \in E(S))$ we set $\psi(h) = k$ where k is an element of $K_1(e)$ satisfying $h = ka$ for some $a \in H_1'(e)$. Such an element is uniquely determined by Lemma 3.1 and so ψ is well defined.

Suppose that $\psi(h_1) = \psi(h_2) = k$ for $h_1, h_2 \in H_1(e)$ $(e \in E(S))$. Then $h_1 = ka_1$ and $h_2 = ka_2$ for some $a_1, a_2 \in H_1'(e)$. We have $h_1^{-1}h_2 = (ka_1)^{-1} ka_2 = a_1^{-1}k^{-1}ka_2 = a_1^{-1}a_2$ as $k, a_1 \in S(e)$. Thus $a_1^{-1}a_2 = h_1^{-1}h_2 \in H_1(e)$. On the other hand, $a_1^{-1}a_2 \in H_1'(e)$. Since $H_1(e) \cap H_1'(e) = \{e\}$, we have $a_1^{-1}a_2 = e$. Therefore, $a_1 = a_1e = a_1a_1^{-1}a_2 = ea_2 = a_2$. It follows that $h_1 = ka_1 = ka_2 = h_2$ and ψ is injective. It is easy to see ψ is surjective.

Next we shall show that ψ is a homomorphism. Take arbitrary elements $h_1 \in H_1(e)$ and $h_2 \in H_1(f)$, where $e, f \in E(S)$. Suppose $\psi(h_1) = k_1$ and $\psi(h_2) = k_2$. Then $h_1 = k_1a_1$ and $h_2 = k_2a_2$ for some $a_1 \in H_1'(e)$ and $a_2 \in H_1'(f)$. Then we have $h_1h_2 = k_1a_1k_2a_2 = k_1a_1efk_2a_2 = k_1a_1efefk_2a_2 = k_1efk_2a_1efa_2 = k_1k_2a_1a_2$ since S is orthodox and $a_1ef (\in H_1'(ef))$ and $efk_2 (\in K_1(ef))$ commute by Lemma 3.1. Note that $k_1k_2 \in K_1(ef)$ and $a_1a_2 \in H_1(ef)$. Therefore, $\psi(h_1h_2) = k_1k_2 = \psi(h_1)\psi(h_2)$. Consequently, ψ is an isomorphism of H_1 onto K_1. □

5 Completely Reducible Orthocryptogroups

In the preceding sections, we have considered internal spined products of finitely many sub-orthocryptogroups. We now consider internal spined product of an arbitrary family of sub-orthocryptogroups and examine orthocryptogroups in which any normal sub-orthocryptogroup is an internal spined product factor.

Let B be a band and $\{S_\lambda \mid \lambda \in \Lambda\}$ a nonempty family of orthocryptogroups such that $E(S_\lambda) \cong B$ for every λ in Λ. Note that each S_λ has the same largest homomorphic band image B. Consider the set P of functions defined on Λ for which there exists e_f in B satisfying the following.

1. $f(\lambda) \in S_\lambda(e_f)$.
2. $f(\lambda) = e_f$ for all but finitely many $\lambda \in \Lambda$.

For $f, g \in P$, we define a multiplication fg by $(fg)(\lambda) = f(\lambda)g(\lambda)$. It is easy to see that fg belongs to P and P forms an orthocryptogroup under this multiplication. Then the set of the idempotents is isomorphic to B and we identify it with B. We say that P is the *external spined product* of the family $\{S_\lambda \mid \lambda \in \Lambda\}$ and denote it by $\bowtie_{\lambda\in\Lambda} S_\lambda$. Note that if Λ is finite, then $\bowtie_{\lambda\in\Lambda} S_\lambda$ is exactly the external spined product defined in Sect. 1.

Suppose S is an orthocryptogroup and $\{H_\lambda \mid \lambda \in \Lambda\}$ is a family of full sub-orthocryptogroups of S. If the external spined product $\bowtie_{\lambda\in\Lambda} H_\lambda$ is isomorphic to S under the mapping $f \mapsto f(\lambda_1)f(\lambda_2)\dots f(\lambda_n)$, where $f(\tau) = e_f(\in E(S))$ for $\tau \in \Lambda \setminus \{\lambda_1, \lambda_2, \dots, \lambda_n\}$, then S is said to be the *internal spined product* of $\{H_\lambda \mid \lambda \in \Lambda\}$ and denoted by $S = \bowtie_{\lambda\in\Lambda} S_\lambda$.

A family of sub-orthocryptogroups $\{H_\lambda \mid \lambda \in \Lambda\}$ of S is called *independent* if any finite subset is independent in the lattice of sub-orthocryptogroups, that is, any finite subset satisfies (4.2). A proof of the following lemma is similar to the one for Lemma 3.2 and so we omit it.

Lemma 5.1 *If an orthocryptogroup S is the internal spined product of a family of full sub-orthocryptogroups $\{H_\lambda \mid \lambda \in \Lambda\}$, then the following conditions hold.*

(C1) Every H_λ is normal.
(C2) S is generated by $\bigcup_{\lambda\in\Lambda} H_\lambda$.
(C3) $\{H_\lambda \mid \lambda \in \Lambda\}$ is independent.

Conversely, if the family $\{H_\lambda \mid \lambda \in \Lambda\}$ of full sub-orthocryptogroups satisfy (C1), (C2), and (C3), then S is the internal spined product $\bowtie_{\lambda\in\Lambda} H_\lambda$. □

A full sub-orthocryptogroup H of S is said to be a *spined factor* if there exists a full sub-orthocryptogroup K such that $S = H \bowtie_B K$. For example, both $E(S)$ and S are spined factors. An orthocryptogroup S is called *simple* if $S \neq E(S)$ and there exists no proper normal sub-orthocryptogroup other than $E(S)$. We say that S is *completely reducible* if there exists a family $\{H_\lambda \mid \lambda \in \Lambda\}$ of simple full sub-orthocryptogroups such that $S = \bowtie_{\lambda\in\Lambda} H_\lambda$.

Theorem 5.2 *Let S be an orthocryptogroup. Then the following conditions are equivalent.*

(1) S is completely reducible.
(2) There is a family of simple normal sub-orthocryptogroups $\{H_\lambda \mid \lambda \in \Lambda\}$ such that S is generated by $\bigcup_{\lambda\in\Lambda} H_\lambda$.
(3) Every normal sub-orthocryptogroup H is a spined factor.

Proof (1) implies (2). Suppose $S = \bowtie_{\lambda\in\Lambda} H_\lambda$ where H_λ is simple. By Lemma 5.1, H_λ is normal and S is generated by $\bigcup_{\lambda\in\Lambda} H_\lambda$.

(2) implies (3). Let H be a normal sub-orhtocryptogroup of S. If $H = S$, then we can take $K = E(S)$. So we may assume $H \neq S$. Let $\mathcal{A}$ be the set of all subsets A

of Λ such that the family $\{H\} \cup \{H_\lambda \mid \lambda \in A\}$ is independent. Since $H \neq S$ and S is generated by $\bigcup_{\lambda\in\Lambda} H_\lambda$, there exists $\lambda \in \Lambda$ such that $H_\lambda \not\subset H$. Then $H_\lambda \cap H = E(S)$ because H_λ is simple. Therefore $\mathcal{A}$ is not empty. By Zorn's lemma, there exists a maximal element M in $\mathcal{A}$. Let L be the sub-orthocryptogroup generated by $H \cup (\bigcup_{\lambda\in M} H_\lambda)$. Using Lemma 1.1, we can show L is also normal as H and H_λ are normal. If $L \neq S$, then there exists $\rho \in \Lambda$ such that $H_\rho \not\subset L$. Since H_ρ is simple and $H_\rho \cap L$ is normal, we have $H_\rho \cap L = E(S)$. Then the family $\{H, H_\rho\} \cup \{H_\lambda \mid \lambda \in M\}$ is independent, which contradicts to the maximality of M. It follows that $L = S$. Let K be the sub-orthocryptogroup generated by $\bigcup_{\lambda\in M} H_\lambda$. By Lemma 3.2, $S = H \bowtie_B K$ and so H is a spined factor.

(3) implies (1). We may assume that $S \neq E(S)$. First, we shall show that for any proper normal sub-orthocryptogroup H, there exists a normal simple sub-orthocryptogroup T such that $H \cap T = E(S)$. Choose an element $u \in S$ such that $u \notin H$. Let $\mathcal{B}$ be the family of the normal sub-orthocryptogroups of S containing H but not u. By Zorn's lemma, there exists a maximal element M in the family. Next we shall show that M is a maximal normal sub-orthocryptogroup. Suppose M is not. Then $M \subsetneq L$ for some proper normal sub-orthocryptogroup L. By our assumption, there exists a proper sub-orthocryptogroup V such that $S = L \bowtie_B V$. If $MV \subset M$ then $V \subset E(S)V \subset MV \subset M$. This implies $L \bowtie_B V \subset LM = L$, which is a contradiction. Therefore, $M \subsetneq MV$. By the maximality of M we have $u \in MV \cap L$. By Lemma 4.2, $MV \cap L = M(V \cap L) = ME(S) = M$. This contradicts to the fact that $u \notin M$. Therefore, M is a maximal normal sub-orthocryptogroup. By our assumption, $S = M \bowtie_B T$ for some T. If T is not simple, then there exists a nontrivial proper normal sub-orthocryptogroup $D \subsetneq T$. Then $M \bowtie_B D$ is normal by Lemmas 1.1 and 3.2, but $M \subsetneq M \bowtie_B D \subsetneq S$, which is a contradiction. Therefore, T is simple. Since $H \subset M$ and $M \cap T = E(S)$, we have $H \cap T = E(S)$.

By the preceding argument, there exists a simple normal sub-orthocryptogroup $T \neq E(S)$ since we are assuming $S \neq E(S)$. We consider the family of independent sets of simple normal sub-orthocryptogroups of S. By Zorn's lemma, there exists a maximal set $\{H_\lambda \mid \lambda \in \Lambda\}$. Let H_0 be the sub-orthocryptogroup generated by $\bigcup_{\lambda\in\Lambda} H_\lambda$. Note that H_0 is normal. If $H_0 \subsetneq S$, there exists a normal simple sub-orthocryptogroup C such that $H_0 \cap C = E(S)$ by the preceding argument. Then the family $\{H_\lambda \mid \lambda \in \Lambda\} \cup \{C\}$ is independent, which contradicts to the maximality of the set $\{H_\lambda \mid \lambda \in \Lambda\}$. Hence, $H_0 = S$ and so $S = \bowtie_{\lambda\in\Lambda} H_\lambda$ by Lemma 5.1. Consequently, S is completely reducible. □

Finally we characterize simple orthocryptogroups. Recall that a completely regular semigroup S can be decomposed into a semilattice Γ of completely simple semigroups R_γ ($\gamma \in \Gamma$). Each R_γ is a $\mathcal{J}$ class of S. Such a completely simple semigroup is called *completely simple component* of S. In particular, every completely simple component is a rectangular group if S is an orthocryptogroup.

Theorem 5.3 *Let S be a simple orthocryptogroup. If S is a semilattice Γ of rectangular groups R_γ ($\gamma \in \Gamma$), then there exists an element δ in Γ such that $R_\delta \cong G \times B_\delta$, where G is a simple group and B_δ is a rectangular band, and R_γ is a rectangular band for every $\gamma \in \Gamma \setminus \{\delta\}$.*

Proof Since S is simple, $S \neq E(S)$ and so at least one completely simple component is not a rectangular band. We shall show that there exists exactly one completely simple component that is not a rectangular band. Suppose that $R_\delta \neq E(R_\delta)$ and $R_\tau \neq E(R_\tau)$ for $\delta, \tau \in \Gamma$ ($\delta \neq \tau$). We may assume $\tau \not\leq \delta$. Let H be a set defined by

$$\left(\bigcup_{\tau \leq \rho} E(R_\rho)\right) \cup \left(\bigcup_{\tau \not\leq \rho} R_\rho\right).$$

We shall show H is normal. Take $h \in H$ and $s \in S$. If either h or s belongs to $\bigcup_{\tau \not\leq \rho} R_\rho$ then so does $s^{-1}hs$. We now suppose that h and s belong to $\bigcup_{\tau \leq \rho} E(R_\rho)$. In this case, h and s are idempotents and so is $s^{-1}hs$. Thus, $s^{-1}hs$ belongs to $\bigcup_{\tau \leq \rho} E(R_\rho)$. It follows that H is a proper normal sub-orthocryptogroup of S. This contradicts to the assumption that S is simple. Hence, there exists exactly one completely simple component R_δ that is not a rectangular band.

Suppose $R_\delta = G \times B_\delta$ for some nontrivial group G and a rectangular band B_δ and that the other completely simple components are rectangular bands. Suppose that G is not simple. There exists a proper normal subgroup N of G. Let $R'_\delta = N \times B_\delta$. Let J be a set defined by

$$\left(\bigcup_{\gamma \in \Gamma \setminus \{\delta\}} R_\gamma\right) \cup R'_\delta.$$

It is easy to see that J is a proper normal sub-orthocryptogroup of S. This is a contradiction. Therefore, G must be simple. □

Acknowledgments The author would like to express his gratitude to Miyuki Yamada for his kind guidance to the theory of semigroups. The author also thanks Yefim Katsov for pointing out some references. This work is supported by Grants-in-Aid for Scientific-Research Program 20540055.

References

1. Cohn, P.M.: Universal Algebra. Harper & Row, New York (1965)
2. Kurosh, A.G.: The Theory of Groups. Chelsea, New York (1960)
3. McKenzie, R.N., McNulty, G.E., Taylor, W.F.: Algebras, Lattices, Valieties. Wadsworth and Brooks/Cole, Monterey (1987)
4. Ore, O.: On the foundations of abstract algebra, II. Ann. Math. **37**, 265–292 (1936)
5. Petrich, M., Reilly, N.R.: Completely Regular Semigroups. Wiley-Interscience Publication, New York (1999)

6. Yamada, M., Kimura, N.: Note on idempotent semigroups II. Proc. Jpn. Acad. **34**, 110–112 (1958)
7. Yamada, M.: Strictly inversive semigroups. Bull. Shimane Univ. (Natl. Sci.) **13**, 128–138 (1964)
8. Yamada, M.: Normal systems in an orthogroup. Mem. Fac. Sci. Shimane Univ. **9**, 1–6 (1975)
9. Yamamura, A.: Completely regular semigroups satisfying formular for inverse of products (in preparation)

Generalized Skew Derivations and g-Lie Derivations of Prime Rings

Vincenzo De Filippis

Abstract Let R be a prime ring of characteristic different from 2, Q_r its right Martindale quotient ring and C its extended centroid. Suppose that F is a nonzero generalized skew derivation of R, with the associated automorphism α, and $p(x_1, \ldots, x_n)$ a noncentral polynomial over C, such that

$$F\Big([x, y]\Big) = [F(x), \alpha(y)] + [\alpha(x), F(y)]$$

for all $x, y \in \{p(r_1, \ldots, r_n) : r_1, \ldots, r_n \in R\}$. Then α is the identity map on R and F is an ordinary derivation of R.

Keywords Generalized skew derivation · Prime ring

Classifications 16W25 · 16N60

1 Introduction

Let R be a prime ring of characteristic different from 2. Throughout this paper $Z(R)$ always denotes the center of R, Q_r the right Martindale quotient ring of R and $C = Z(Q_r)$ the center of Q_r (C is usually called the extended centroid of R). Let $S \subseteq R$ be a subset of R.

An additive map $d : R \to R$ is called derivation of S if $d(xy) = d(x)y + xd(y)$, for all $x, y \in S$. An additive map $G : R \to R$ is called generalized derivation of S if there exists a derivation d of R such that $G(xy) = G(x)y + xd(y)$, for all $x, y \in S$. The additive map $F : R \to R$ is called Lie derivation of S if $F([x, y]) = [F(x), y] + [x, F(y)]$, for any $x, y \in S$. Of course any derivation is a Lie derivation.

V. De Filippis (✉)
Department of Mathematics and Computer Science, University of Messina, 98166 Messina, Italy
e-mail: defilippis@unime.it

S.T. Rizvi et al. (eds.), *Algebra and its Applications*, Springer Proceedings in Mathematics & Statistics 174, DOI 10.1007/978-981-10-1651-6_3

The problem of whether a Lie derivation is a derivation has been studied by several authors (see for example [3, 21] and the references therein).

Motivated by this, here we introduce the definition of g-Lie derivations. More precisely, let $f, g : R \to R$ be two additive maps and $S \subseteq R$ a subset of R. If $f([x, y]) = [f(x), g(y)] + [g(x), f(y)]$, for any $x, y \in S$, then f is called g-Lie derivation of S. It is clear that any Lie derivation is a 1-Lie derivation. The simplest example of g-Lie derivation is the following:

Example 1.1 Let $n \geq 2$ be an integer and C a field of characteristic $2n - 1$. Let R be a prime C-algebra, $0 \neq \lambda \in C$, $f(x) = \lambda x$ and $g(x) = nx$, for any $x \in R$. Then f is a g-Lie derivation of R in the sense of the above definition. Moreover f is not a Lie derivation of R.

One natural question could be whether a g-Lie derivation of $S \subseteq R$ is a Lie derivation of S. Here we consider a first step of this problem. To be more specific, in this paper we study the form of a generalized skew derivation f acting as a g-Lie derivation on the subset $\{p(r_1, \ldots, r_n) : r_1, \ldots, r_n \in R\}$, where g is the associated automorphism with f and $p(x_1, \ldots, x_n)$ is a noncentral polynomial in n non-commuting variables.

More precisely, let α be an automorphism of R. An additive mapping $d : R \longrightarrow R$ is called a *skew derivation* of R if

$$d(xy) = d(x)y + \alpha(x)d(y)$$

for all $x, y \in R$ and α is called an *associated automorphism* of d. An additive mapping $G : R \longrightarrow R$ is said to be a *generalized skew derivation* of R if there exists a skew derivation d of R with associated automorphism α such that

$$G(xy) = G(x)y + \alpha(x)d(y)$$

for all $x, y \in R$; d is said to be an *associated skew derivation* of G and α is called an *associated automorphism* of G. Any mapping of R with form $G(x) = ax + \alpha(x)b$ for some $a, b \in R$ and $\alpha \in Aut(R)$, is called *inner generalized skew derivation*. In particular, if $a = -b$, then G is called *inner skew derivation*. If a generalized skew derivation (respectively, a skew derivation) is not inner, then it is usually called *outer*.

In light of above definitions, one can see that the concept of generalized skew derivation unifies the notions of skew derivation and generalized derivation.

It is well known that automorphisms, derivations, and skew derivations of R can be extended to Q_r. In [4] Chang extends the definition of generalized skew derivation to the right Martindale quotient ring Q_r of R as follows: by a (right) generalized skew derivation we mean an additive mapping $G : Q_r \longrightarrow Q_r$ such that $G(xy) = G(x)y + \alpha(x)d(y)$ for all $x, y \in Q$, where d is a skew derivation of R and α is an automorphism of R. Moreover, there exists $G(1) = a \in Q_r$ such that $G(x) = ax + d(x)$ for all $x \in R$.

The main result of this article is

Theorem 1 *Let R be a prime ring of characteristic different from 2, Q_r its right Martindale quotient ring and C its extended centroid. Suppose that F is a nonzero generalized skew derivation of R, with the associated automorphism α, and $p(x_1, \ldots, x_n)$ a noncentral polynomial over C, such that*

$$F\Big([x, y]\Big) = [F(x), \alpha(y)] + [\alpha(x), F(y)]$$

for all $x, y \in \{p(r_1, \ldots, r_n) : r_1, \ldots, r_n \in R\}$. Then α is the identity map on R and F is an ordinary derivation of R.

2 Preliminaries

We now collect some Facts which follow from results in [6–9] and will be used in the sequel.

Fact 2.1 In [11] Chuang and Lee investigate polynomial identities with skew derivations. They prove that if $\Phi(x_i, D(x_i))$ is a generalized polynomial identity for R, where R is a prime ring and D is an outer skew derivation of R, then R also satisfies the generalized polynomial identity $\Phi(x_i, y_i)$, where x_i and y_i are distinct indeterminates.

Fact 2.2 Let R be a prime ring and I be a two-sided ideal of R. Then I, R, and Q_r satisfy the same generalized polynomial identities with coefficients in Q_r (see [6]). Furthermore, I, R, and Q_r satisfy the same generalized polynomial identities with automorphisms (see [8, Theorem 1]).

Remark 2.3 We would like to point out that in [18] Lee proves that every generalized derivation can be uniquely extended to a generalized derivation of U and thus all generalized derivations of R will be implicitly assumed to be defined on the whole U. In particular Lee proves the following result:

Theorem 3 in [18] *Every generalized derivation g on a dense right ideal of R can be uniquely extended to U and assumes the form $g(x) = ax + d(x)$, for some $a \in U$ and a derivation d on U.*

We also need the following:

Remark 2.4 Let R be a non-commutative prime ring of characteristic different from 2, D_1 and D_2 be derivations of R such that $D_1(x)D_2(x) = 0$ for all $X \in R$. Then either $D_1 = 0$ or $D_2 = 0$.

Proof It is a reduced version of Theorem 3 in [22]. □

Remark 2.5 Let R be a prime ring and $0 \neq a \in R$. If $[x_1, x_2]a \in Z(R)$, for any $x_1, x_2 \in R$, then either R is commutative or $a = 0$.

Proof Since for all $x_1, x_2 \in R$ we have $[[x_1, x_2]a, [x_1, x_2]] = 0$, then $[x_1, x_2][a, [x_1, x_2]] = 0$. As a consequence of [19, Theorem 2], either R is commutative or $a \in Z(R)$. Moreover, in this last case and by our hypothesis, it follows that either $a = 0$ or $a \neq 0$ and $[x_1, x_2] \in Z(R)$, for all $x_1, x_2 \in R$, that is R is commutative. □

Remark 2.6 Let R be a non-commutative prime ring and $a \in R$ be such that

$$[x_1, x_2]a[y_1, y_2] - [y_1, y_2]a[x_1, x_2] \tag{2.1}$$

is a generalized polynomial identity for R. Then $a = 0$.

Proof In relation (2.1) we replace y_2 with y_2t, for any $t \in R$. Using again (2.1), we have that R satisfies

$$[y_1, y_2]\Big[a[x_1, x_2], t\Big] + \Big[[x_1, x_2]a, y_2\Big][y_1, t]. \tag{2.2}$$

For $t = a[x_1, x_2]$ in (2.2), it follows that

$$\Big[[x_1, x_2]a, y_2\Big]\Big[y_1, [x_1, x_2]a\Big] \tag{2.3}$$

is a generalized polynomial identity for R. Let $x_1, x_2 \in R$ and D_1 and D_2 be the inner derivations of R induced respectively by $[x_1, x_2]a$ and $a[x_1, x_2]$. By (2.3) we get $D_1(y_1)D_2(y_2) = 0$, for any $y_1, y_2 \in R$. By Remark 2.4 we have that either $D_1 = 0$ or $D_2 = 0$. This means that, for any $x_1, x_2 \in R$, either $[x_1, x_2]a \in Z(R)$ or $a[x_1, x_2] \in Z(R)$.

Let $x_1, x_2 \in R$ be such that $a[x_1, x_2] \in Z(R)$.

Thus, by (2.3) it follows $\Big[[x_1, x_2]a, y_2\Big][y_1, t] = 0$, for any $y_1, y_2, t \in R$ and, using again Remark 2.4, we have that $[x_1, x_2]a \in Z(R)$, for any $x_1, x_2 \in R$. Therefore, by Remark 2.5 and since R is not commutative, we get $a = 0$. □

Remark 2.7 Let R be a non-commutative prime ring and $F : R \to R$ a generalized derivation of R. If F acts as a Lie derivation of $[R, R]$, then F is an usual derivation of R.

Proof Since F acts as a Lie derivation, we have that $[R, R]$ satisfies $F([u, v]) - [F(u), v] - [u, F(v)]$. Using Remark 2.3, one has that there exist $a \in U$ and a derivation d on U such that $F(x) = ax + d(x)$, for any $x \in R$.

By easy calculations it follows that $[R, R]$ satisfies the generalized identity $uav - vau$, that is R satisfies

$$[x_1, x_2]a[y_1, y_2] - [y_1, y_2]a[x_1, x_2].$$

Hence, by Remark 2.6, we get $a = 0$ and $F = d$ is an ordinary derivation of R. □

As an easy consequence we also have that

Remark 2.8 Let R be a non-commutative prime ring, $a \in R$ and $F : R \to R$ be such that $F(x) = ax$, for any $x \in R$. If F acts as a Lie derivation of $[R, R]$, then $a = 0$, that is $F = 0$.

3 The Case of Inner Generalized Skew Derivations

In the first part of this section we will prove the following:

Proposition 3.1 *Let R be a non-commutative prime ring of characteristic different from 2, Q_r be its right Martindale quotient ring and C be its extended centroid. Suppose that α is an inner automorphism of R induced by the invertible element $q \in Q_r$ and F is an inner generalized skew derivation of R defined as follows: $F(x) = ax + qxq^{-1}b$, for all $x \in R$ and suitable fixed $a, b \in Q_r$. If*

$$F\Big([x, y]\Big) = [F(x), \alpha(y)] + [\alpha(x), F(y)]$$

for all $x, y \in [R, R]$, then $a + b = 0$ and either $q \in C$ or $q^{-1}b \in C$.

We assume that R satisfies the following generalized polynomial identity

$$\begin{aligned}&\Psi(x_1, x_2, y_1, y_2) = \\ &\Big[a[x_1, x_2] + q[x_1, x_2]q^{-1}b, q[y_1, y_2]q^{-1}\Big] + \Big[q[x_1, x_2]q^{-1}, a[y_1, y_2] + q[y_1, y_2]q^{-1}b\Big] \\ &\quad - a\Big[[x_1, x_2], [y_1, y_2]\Big] - q\Big[[x_1, x_2], [y_1, y_2]\Big]q^{-1}b. \end{aligned} \tag{3.1}$$

Lemma 3.2 *If $q^{-1}a \in C$, then $q^{-1}b \in C$ and $a + b = 0$.*

Proof Left multiplying (3.1) by q^{-1} and since $q^{-1}a \in C$, one has that R satisfies

$$\Big([x_1, x_2](a + q^{-1}bq)[y_1, y_2] - [y_1, y_2](a + q^{-1}bq)[x_1, x_2]\Big)q^{-1} \tag{3.2}$$

so that, right multiplying by q, it follows that

$$[x_1, x_2](a + q^{-1}bq)[y_1, y_2] - [y_1, y_2](a + q^{-1}bq)[x_1, x_2] \tag{3.3}$$

is a generalized polynomial identity for R. Therefore, by Remark 2.6 we get $a = -q^{-1}bq$. Moreover, since $q^{-1}a \in C$, we notice that $0 = [q^{-1}a, q] = q^{-1}aq - a$, that is $q^{-1}aq = a = -q^{-1}bq$. Therefore $q^{-1}b = -q^{-1}a \in C$ and $a + b = 0$, as required. □

Remark 3.3 Notice that, in case $q \in C$ then F is a generalized derivation of R and the conclusion of Proposition 3.1 follows directly from Remark 2.7. Moreover, in light of Lemma 3.2, we are also done in the case $q^{-1}a \in C$.

We begin with the following

Fact 3.4 (Lemma 1.5 in [12]) Let H be an infinite field and $n \geq 2$. If $A_1, \ldots, A_k$ are not scalar matrices in $M_m(H)$ then there exists some invertible matrix $P \in M_m(H)$ such that each matrix $PA_1P^{-1}, \ldots, PA_kP^{-1}$ has all nonzero entries.

Lemma 3.5 *Let $R = M_m(C)$, $m \geq 2$ and let C be infinite, $Z(R)$ the center of R, a, b, q elements of R and q is invertible. If R satisfies $\Psi(x_1, x_2, y_1, y_2)$ then $a + b = 0$ and one of the following holds:*

(a) $q \in Z(R)$;
(b) $q^{-1}b \in Z(R)$.

Proof If either $q \in Z(R)$ or $q^{-1}a \in Z(R)$, then the conclusion follows from Remark 3.3.

We assume that $q^{-1}a \notin Z(R)$ and $q \notin Z(R)$, that is both $q^{-1}a$ and q are not scalar matrices, and prove that a contradiction follows.

By Fact 3.4, there exists some invertible matrix $P \in M_m(C)$ such that each matrix $P(q^{-1}a)P^{-1}$, PqP^{-1} has all nonzero entries. Denote by $\varphi(x) = PxP^{-1}$ the inner automorphism induced by P. Without loss of generality we may replace q and $q^{-1}a$ with $\varphi(q)$ and $\varphi(q^{-1}a)$, respectively, and denote $q = \sum q_{lm}e_{lm}$ and $q^{-1}a = a_{lm}e_{lm}$, for some $q_{lm}, a_{lm} \in C$.

Let e_{ij} be the usual matrix unit, with 1 in the (i, j)-entry and zero elsewhere. For any $i \neq j$ and $[x_1, x_2] = [e_{ii}, e_{ij}] = e_{ij}$, $[y_1, y_2] = [e_{ji}, e_{ij}] = e_{jj} - e_{ii}$ in (3.1), then

$$\begin{aligned}&\left[ae_{ij} + qe_{ij}q^{-1}b, q(e_{jj} - e_{ii})q^{-1}\right]\\&\quad + \left[qe_{ij}q^{-1}, a(e_{jj} - e_{ii}) + q(e_{jj} - e_{ii})q^{-1}b\right]\\&\quad - 2ae_{ij} - 2qe_{ij}q^{-1}b = 0.\end{aligned} \tag{3.4}$$

Left multiplying (3.4) by $e_{ij}q^{-1}$ and right multiplying by qe_{ij}, we get $4e_{ij}q^{-1}ae_{ij}qe_{ij} = 0$, that is $a_{ji}q_{ji} = 0$, which is a contradiction. □

Lemma 3.6 *Let $R = M_m(C)(m \geq 2)$. Then Proposition 3.1 holds.*

Proof If one assumes that C is infinite, the conclusion follows from Lemma 3.5.

Now, let E be an infinite field which is an extension of the field C and let $\overline{R} = M_t(E) \cong R \otimes_C E$. Consider the generalized polynomial $\Psi(x_1, x_2, y_1, y_2)$, which is a multilinear generalized polynomial identity for R. Clearly, $\Psi(x_1, x_2, y_1, y_2)$ is a generalized polynomial identity for $\overline{R}$ too, and the conclusion follows from Lemma 3.5.

Lemma 3.7 *Either $\Psi(x_1, x_2, y_1, y_2)$ is a nontrivial generalized polynomial identity for R or $a + b = 0$ and one of the following holds:*

(a) $q \in Z(R)$;
(b) $q^{-1}b \in Z(R)$.

Proof Consider the generalized polynomial (3.1)

$$\Psi(x_1, x_2, y_1, y_2) =$$
$$\Big[a[x_1, x_2] + q[x_1, x_2]q^{-1}b, q[y_1, y_2]q^{-1}\Big] + \Big[q[x_1, x_2]q^{-1}, a[y_1, y_2] + q[y_1, y_2]q^{-1}b\Big]$$
$$- a\Big[[x_1, x_2], [y_1, y_2]\Big] - q\Big[[x_1, x_2], [y_1, y_2]\Big]q^{-1}b.$$

By our hypothesis, R satisfies this generalized polynomial identity. Replacing $[x_1, x_2]$ by $q^{-1}[x_1, x_2]q$ and $[y_1, y_2]$ by $q^{-1}[y_1, y_2]q$, we have that R satisfies the generalized polynomial identity

$$\Big[aq^{-1}[x_1, x_2]q + [x_1, x_2]b, [y_1, y_2]\Big]$$
$$\Big[[x_1, x_2], aq^{-1}[y_1, y_2]q + [y_1, y_2]b\Big]$$
$$- a\Big[q^{-1}[x_1, x_2]q, q^{-1}[y_1, y_2]q\Big] - \Big[[x_1, x_2], [y_1, y_2]\Big]b. \quad (3.5)$$

If $\{aq^{-1}, 1\}$ are linearly independent over C then (3.5) is a nontrivial generalized polynomial identity for R. Therefore, we may assume in all follows that $\{aq^{-1}, 1\}$ are linearly dependent over C, that is $aq^{-1} \in C$.

By (3.5) we have that R satisfies

$$[x_1, x_2](ab)[y_1, y_2] - [y_1, y_2](a + b)[x_1, x_2]$$

and by Remark 2.6 it follows that $a + b = 0$. Hence the generalized polynomial (3.1) reduces to

$$\Psi(x_1, x_2, y_1, y_2) =$$
$$\Big[a[x_1, x_2] - q[x_1, x_2]q^{-1}a, q[y_1, y_2]q^{-1}\Big] + \Big[q[x_1, x_2]q^{-1}, a[y_1, y_2] - q[y_1, y_2]q^{-1}a\Big]$$
$$- a\Big[[x_1, x_2], [y_1, y_2]\Big] + q\Big[[x_1, x_2], [y_1, y_2]\Big]q^{-1}a. \tag{3.6}$$

Left multiplying (3.6) by q^{-1} and by easy computations, one has that R satisfies

$$q^{-1}a\Big([x_1, x_2]q[y_1, y_2]q^{-1} + [y_1, y_2]q[x_1, x_2]q^{-1} - [x_1, x_2][y_1, y_2] + [y_1, y_2][x_1, x_2]\Big)$$
$$+ [x_1, x_2]q^{-1}a[y_1, y_2] - [y_1, y_2]q^{-1}a[x_1, x_2] + [y_1, y_2]a[x_1, x_2]q^{-1} - [x_1, x_2]a[y_1, y_2]q^{-1}. \tag{3.7}$$

If $\{q^{-1}a, 1\}$ are linearly dependent over C, then $q^{-1}a \in C$, that is $q^{-1}b \in C$ and we are done. On the other hand, if $\{q^{-1}a, 1\}$ are linearly independent over C and since (3.7) is a trivial generalized polynomial identity for R, then R satisfies

$$[x_1, x_2]q^{-1}a[y_1, y_2] - [y_1, y_2]q^{-1}a[x_1, x_2] + [y_1, y_2]a[x_1, x_2]q^{-1} - [x_1, x_2]a[y_1, y_2]q^{-1}. \tag{3.8}$$

Moreover, if $q \in C$, then the conclusion follows from Remark 3.3, so that we may assume $q \notin C$. In this last case, by (3.8) it follows that R satisfies the nontrivial generalized polynomial identity

$$[x_1, x_2]q^{-1}a[y_1, y_2] - [y_1, y_2]q^{-1}a[x_1, x_2]$$

which is a contradiction.

Proof of Proposition 3.1. The generalized polynomial $\Psi(x_1, x_2, y_1, y_2)$ is a generalized polynomial identity for R. By Lemma 3.7, we may assume that $\Psi(x_1, x_2, y_1, y_2)$ is a nontrivial generalized polynomial identity for R and, by [6] it follows that $\Psi(x_1, x_2, y_1, y_2)$ is a nontrivial generalized polynomial identity for Q_r. By the well-known Martindale's theorem of [20], Q_r is a primitive ring having nonzero socle with the field C as its associated division ring. By [15] (p.75) Q_r is isomorphic to a dense subring of the ring of linear transformations of a vector space V over C, containing nonzero linear transformations of finite rank. Assume first that $\dim_C V = k \geq 2$ is a finite positive integer, then $Q_r \cong M_k(C)$ and the conclusion follows from Lemma 3.6.

Let now $\dim_C V = \infty$.

Let $x_0, y_0 \in R$. By Litoff's theorem (see Theorem 4.3.11 in [2]) there exists an idempotent element $e \in R$ such that $x_0, y_0, a, b, q, q^{-1}a, q^{-1}b \in eRe \cong M_k(C)$ for some integer k. Of course

$$\Big[a[r_1,r_2]+q[r_1,r_2]q^{-1}b, q[s_1,s_2]q^{-1}\Big]+\Big[q[r_1,r_2]q^{-1}, a[s_1,s_2]+q[s_1,s_2]q^{-1}b\Big]$$
$$-a\Big[[r_1,r_2],[s_1,s_2]\Big]-q\Big[[r_1,r_2],[s_1,s_2]\Big]q^{-1}b=0, \quad \forall r_1,r_2,s_1,s_2 \in eRe. \quad (3.9)$$

For sake of clearness, here we write $F(x)=ax+qxq^{-1}b$, for any $x \in eRe$. By Lemma 3.6 one of the following holds:

(a) eRe is commutative, in particular $q^{-1}b, q$ are central elements of eRe. In this case $F(x)=(a+b)x$, for any $x \in eRe$, that is F is a generalized derivation of eRe. Moreover, since eRe satisfies (3.9) and q is a central element of eRe, we have that F acts as a Lie derivation on the set $[eRe, eRe]$. Thus, by Remark 2.7 it follows that F is an ordinary derivation of eRe, in particular $F(x_0y_0)= F(x_0)y_0+x_0F(y_0)$.

(b) $a+b=0$ and q is a central element of eRe. In this case $F(x)=ax-xa$, for any $x \in eRe$, that is F is an inner ordinary derivation of eRe and once again $F(x_0y_0)=F(x_0)y_0+x_0F(y_0)$ holds.

(c) $a+b=0$ and $q^{-1}b, q^{-1}a$ are central elements of eRe. In this case $F(x)=0$, for any $x \in eRe$, in particular $F(x_0)=F(y_0)=0$.

Therefore, in any case $F(x_0y_0)=F(x_0)y_0+x_0F(y_0)$ holds. Repeating this process for any $x, y \in R$, it follows that F satisfies the rule $F(xy)=F(x)y+xF(y)$ for any $x, y \in R$, that is F acts as a derivation on R, as required.

Now we extend the previous result to the case the automorphism α is not necessarily inner

Proposition 3.8 *Let R be a non-commutative prime ring of characteristic different from 2, Q_r be its right Martindale quotient ring and C be its extended centroid. Suppose that F is an inner generalized skew derivation of R, with associated automorphism α, defined as follows: $F(x)=ax+\alpha(x)b$ for all $x \in R$ and suitable fixed $a, b \in Q_r$. If $F \neq 0$ and*

$$F\Big([x,y]\Big)=[F(x),\alpha(y)]+[\alpha(x),F(y)]$$

for all $x, y \in [R,R]$, then α is the identity map on R and $a+b=0$.

In order to prove Proposition 3.8 we need to fix the following useful results:

Remark 3.9 Let R be a prime ring of characteristic different from 2.
If $\Big[[x_1,x_2],[y_1,y_2]\Big] \in Z(R)$, for all $x_1,x_2,y_1,y_2 \in R$, then R is commutative.

Proof Since R is a prime ring satisfying the polynomial identity

$$\Big[\Big[[x_1,x_2],[y_1,y_2]\Big],x_3\Big]$$

then there exists a field K such that R and $M_t(K)$, the ring of all $t \times t$ matrices over K, satisfy the same polynomial identities (see [16]).
Suppose $t \geq 2$. Let $x_1 = e_{11}$, $x_2 = e_{22}$, $y_1 = e_{22}$ and $y_2 = e_{21}$. By calculation we obtain $\big[[x_1, x_2], [y_1, y_2]\big] = e_{11} - e_{22} \notin Z(R)$, a contradiction. So $t = 1$ and R is commutative. □

Remark 3.10 Let R be a prime ring of characteristic different from 2 and $a \in R$. If $a\big[[x_1, x_2], [y_1, y_2]\big] = 0$ (respectively $\big[[x_1, x_2], [y_1, y_2]\big]a = 0$), for all $x_1, x_2, y_1, y_2 \in R$, then either $a = 0$ or R is commutative.

Proof By Remark 3.9 we may assume that the polynomial $\big[[x_1, x_2], [y_1, y_2]\big]$ is not central in R. Therefore $a = 0$ follows from [10]. □

Remark 3.11 Let R be a prime ring of characteristic different from 2 and $a \in R$. If $\big[a[x_1, x_2], [x_1, x_2]\big] = 0$ (respectively $\big[[x_1, x_2]a, [x_1, x_2]\big] = 0$), for all $x_1, x_2 \in R$, then either $a \in Z(R) = 0$.

Proof It is easy consequence of [1]. □

Proof of Proposition 3.8 If there exists an invertible element $q \in Q_r$ such that $\alpha(x) = qxq^{-1}$, for all $x \in R$, then the conclusion follows from Proposition 3.1. Hence, in what follows we assume that α is not an inner automorphism of R and prove that a contradiction follows. Thus, since R satisfies

$$\begin{aligned}&\Big[a[x_1, x_2] + \alpha([x_1, x_2])b, \alpha([y_1, y_2])\Big]\\ &\quad + \Big[\alpha([x_1, x_2]), a[y_1, y_2] + \alpha([y_1, y_2])b\Big]\\ &\quad - a\Big[[x_1, x_2], [y_1, y_2]\Big] - \alpha\Big(\Big[[x_1, x_2], [y_1, y_2]\Big]\Big)b \end{aligned} \tag{3.10}$$

then

$$\begin{aligned}&\Big[a[x_1, x_2] + [t_1, t_2]b, [z_1, z_2]\Big]\\ &\quad + \Big[[t_1, t_2]), a[y_1, y_2] + [z_1, z_2]b\Big]\\ &\quad - a\Big[[x_1, x_2], [y_1, y_2]\Big] - \Big[[t_1, t_2], [z_1, z_2]\Big]b \end{aligned} \tag{3.11}$$

is a generalized identity for R. In particular R satisfies $a\Big[[x_1, x_2], [y_1, y_2]\Big]$, which implies that $a = 0$ (see Remark 3.10). Hence (3.11) reduces to

$$[t_1, t_2]b[z_1, z_2] - [z_1, z_2]b[t_1, t_2]$$

and, by Remark 2.6, we get $b = 0$, which implies the contradiction $F = 0$.

4 The Proof of Theorem 1

As mentioned in the Introduction, we can write $F(x) = ax + f(x)$, for all $x \in R$, where $a \in Q_r$ and f is a skew derivation of R. Let α be the automorphism associated with f. That is $f(xy) = f(x)y + \alpha(x)f(y)$, for all $x, y \in R$.

Remark 4.1 Let S be the additive subgroup generated by the set

$$p(R) = \{p(r_1, \ldots, r_n) : r_1, \ldots, r_n \in R\} \neq 0.$$

Of course $F([x, y]) = [F(x), \alpha(y)] + [\alpha(x), F(y)]$, for all $x, y \in S$. Since $p(x_1, \ldots, x_n)$ is not central in R, by [5] and $char(R) \neq 2$, it follows that there exists a noncentral Lie ideal L of R such that $L \subseteq S$. Moreover it is well known that there exists a nonzero ideal I of R such that $[I, R] \subseteq L$ (see [14, pp. 4–5], [13, Lemma 2, Proposition 1], [17, Theorem 4]).

Proof of Theorem 1. By Remark 4.1 we assume there exists a noncentral ideal I of R such that

$$\Big[a[x_1, x_2] + f([x_1, x_2]), \alpha([y_1, y_2])\Big] + \Big[\alpha([x_1, x_2]), a[y_1, y_2] + f([y_1, y_2]),\Big] \\ - a\Big[[x_1, x_2], [y_1, y_2]\Big] - f\Big(\Big[[x_1, x_2], [y_1, y_2]\Big]\Big) \tag{4.1}$$

is satisfied by I. Since I and R satisfy the same generalized identities with derivations and automorphisms, then (4.1) is a generalized differential identity for R, that is R satisfies

$$\Big[a[x_1, x_2] + f(x_1)x_2 + \alpha(x_1)f(x_2) - f(x_2)x_1 - \alpha(x_2)f(x_1), \alpha([y_1, y_2])\Big] \\ \Big[\alpha([x_1, x_2]), a[y_1, y_2] + f(y_1)y_2 + \alpha(y_1)f(y_2) - f(y_2)y_1 - \alpha(y_2)f(y_1)\Big] \\ - a\Big[[x_1, x_2], [y_1, y_2]\Big] - \Big(f(x_1)x_2 + \alpha(x_1)f(x_2) - f(x_2)x_1 - \alpha(x_2)f(x_1)\Big)[y_1, y_2] \\ - \alpha([x_1, x_2])\Big(f(y_1)y_2 + \alpha(y_1)f(y_2) - f(y_2)y_1 - \alpha(y_2)f(y_1)\Big) \\ + \Big(f(y_1)y_2 + \alpha(y_1)f(y_2) - f(y_2)y_1 - \alpha(y_2)f(y_1)\Big)[x_1, x_2] \\ + \alpha([y_1, y_2])\Big(f(x_1)x_2 + \alpha(x_1)f(x_2) - f(x_2)x_1 - \alpha(x_2)f(x_1)\Big) \tag{4.2}$$

In all that follows we assume R is not commutative and $f \neq 0$, otherwise $F(x) = ax$, for all $x \in R$ and, by Remark 2.8, we have the contradiction $F = 0$. Moreover, we also assume α is not the identity map on R, if not $F(x) = ax + xb$, for all $x \in R$ and, by Remark 2.7, it follows $a = -b$ and F is an ordinary inner derivation of R.

In case f is an inner skew derivation of R, then we get the required conclusions by Proposition 3.8. Hence we now assume that f is not inner and show that a number of contradictions follows.

Since $0 \neq f$ is not inner, then, by relation (4.2), it follows that R satisfies

$$\begin{aligned}
&\Big[a[x_1, x_2] + t_1x_2 + \alpha(x_1)t_2 - t_2x_1 - \alpha(x_2)t_1, \alpha([y_1, y_2])\Big] \\
&\Big[\alpha([x_1, x_2]), a[y_1, y_2] + z_1y_2 + \alpha(y_1)z_2 - z_2y_1 - \alpha(y_2)z_1\Big] \\
&\quad - a\Big[[x_1, x_2], [y_1, y_2]\Big] - \Big(t_1x_2 + \alpha(x_1)t_2 - t_2x_1 - \alpha(x_2)t_1\Big)[y_1, y_2] \\
&\quad - \alpha([x_1, x_2])\Big(z_1y_2 + \alpha(y_1)z_2 - z_2y_1 - \alpha(y_2)z_1\Big) \\
&\quad + \Big(z_1y_2 + \alpha(y_1)z_2 - z_2y_1 - \alpha(y_2)z_1\Big)[x_1, x_2] \\
&\quad + \alpha([y_1, y_2])\Big(t_1x_2 + \alpha(x_1)t_2 - t_2x_1 - \alpha(x_2)t_1\Big) \qquad (4.3)
\end{aligned}$$

and in particular, by computation we get that

$$\Big(t_1x_2 - \alpha(x_2)t_1\Big)\alpha([y_1, y_2]) - \Big(t_1x_2 - \alpha(x_2)t_1\Big)[y_1, y_2] \qquad (4.4)$$

is satisfied by R.

If there exists an invertible element $q \in Q_r$ such that $q \notin C$ and $\alpha(x) = qxq^{-1}$, for all $x \in R$, we replace any t_i with qt_i in (4.4) and left multiplying by q^{-1}, one has that R satisfies $[t_1, x_2]\big(q[y_1, y_2]q^{-1} - [y_1, y_2]\big)$. Since R is not commutative and in light of Remark 2.5, the last relation implies $q[y_1, y_2]q^{-1} = [y_1, y_2]$, for any $y_1, y_2 \in R$, that is $q[y_1, y_2] = [y_1, y_2]q$, for any $y_1, y_2 \in R$. In this case, it is well known that the contradiction $q \in C$ follows.

On the other hand, in case α is not inner, then by (4.4) it follows that R satisfies the generalized identity

$$\Big(t_1x_2 - t_3t_1\Big)[z_1, z_2] - \Big(t_1x_2 - t_3t_1\Big)[y_1, y_2] \qquad (4.5)$$

and, for $y_1 = y_2 = 0$ and $x_2 = t_3$, we have that R satisfies the polynomial identity $[t_1, x_2][z_1, z_2]$, which implies again the contradiction that R is commutative.

References

1. Argac, N., Carini, L., De Filippis, V.: An Engel condition with generalized derivations on Lie ideals. Taiwan. J. Math. **12**(2), 419–433 (2008)
2. Beidar, K.I., Martindale III, W.S., Mikhalev, A.V.: Rings with Generalized Identities. Pure and Applied Mathematics. Dekker, New York (1996)
3. Bresar, M.: Commuting traces of biadditive mappings, commutativity preserving mappings and Lie mappings. Trans. Amer. Math. Soc. **335**, 525–546 (1993)
4. Chang, J.-C.: On the identity $h(x) = af(x) + g(x)b$. Taiwan. J. Math. **7**, 103–113 (2003)
5. Chuang, C.-L.: The additive subgroup generated by a polynomial. Isr. J. Math. **59**(1), 98–106 (1987)
6. Chuang, C.-L.: GPIs having coefficients in Utumi quotient rings. Proc. Amer. Math. Soc. **103**, 723–728 (1988)
7. Chuang, C.-L.: Differential identities with automorphisms and antiautomorphisms I. J. Algebra **149**, 371–404 (1992)
8. Chuang, C.-L.: Differential identities with automorphisms and antiautomorphisms II. J. Algebra **160**, 130–171 (1993)
9. Chuang, C.-L.: Identities with skew derivations. J. Algebra **224**, 292–335 (2000)
10. Chuang, C.-L., Lee, T.-K.: Rings with annihilators conditions on multilinear polynomials. Chin. J. Math. **24**(2), 177–185 (1996)
11. Chuang, C.-L., Lee, T.-K.: Identities with a single skew derivation. J. Algebra **288**, 59–77 (2005)
12. De Filippis, V.: A product of two generalized derivations on polynomials in prime rings. Collect. Math. **61**, 303–322 (2010)
13. Di Vincenzo, O.M.: On the n-th centralizer of a Lie ideal. Boll. UMI **3–A**, 77–85 (1989)
14. Herstein, I.N.: Topics in Ring Theory. University of Chicago Press, Chicago (1969)
15. Jacobson, N.: Structure of Rings. American Mathematical Society, Providence (1964)
16. Jacobson, N.: P.I. Algebras, An Introduction. Lecture Notes in Mathematics, vol. 44. Springer, Berlin (1975)
17. Lanski, C., Montgomery, S.: Lie structure of prime rings of characteristic 2. Pac. J. Math. **42**(1), 117–135 (1972)
18. Lee, T.-K.: Generalized derivations of left faithful rings. Comm. Algebra **27**(8), 4057–4073 (1999)
19. Lee, T.K., Shiue, W.K.: Derivations cocentralizing polynomials. Taiwan. J. Math. **2**(4), 457–467 (1998)
20. Martindale III, W.S.: Prime rings satisfying a generalized polynomial identity. J. Algebra **12**, 576–584 (1969)
21. Mathieu, M., Villena, A.R.: The structure of Lie derivations on C^*-algebras. J. Funct. Anal. **202**, 504–525 (2003)
22. Vukman, J.: On α-derivations of prime and semiprime rings. Demonstratio Math. **XXXVII**(2), 283–290 (2005)

Additive Representations of Elements in Rings: A Survey

Ashish K. Srivastava

Abstract This article presents a brief survey of the work done on various additive representations of elements in rings. In particular, we study rings where each element is a sum of units; rings where each element is a sum of idempotents; rings where each element is a sum of idempotents and units; and rings where each element is a sum of additive commutators. We have also included a number of open problems in this survey to generate further interest among readers in this topic.

Keywords Units · Idempotents · k-good rings · von Neumann regular ring · Unit sum number · Clean rings

2000 Mathematics Subject Classification 16U60 · 16D50

1 Additive Unit Representation

The historical origin of study of the additive unit structure of rings may be traced back to the work of Dieudonné on noncommutative Galois theory [11]. In [26], Hochschild studied additive unit representations of elements in simple algebras and proved that each element of a simple algebra over any field is a sum of units. Later, Zelinsky [53] proved that the ring of linear transformations is generated additively by its unit elements. Zelinsky showed that every linear transformation of a vector space V over a division ring is the sum of two invertible linear transformations, except when V is one-dimensional over $\mathbb{F}_2$, the field of two elements. Zelinsky also noted in his paper that this result follows from a previous result of Wolfson [52]. See [14] for another proof of this result.

Apart from the ring of linear transformations, there are several other natural classes of rings that are generated by their unit elements. Let X be a completely regular Hausdorff space. Then every element in the ring $C(X)$ of real-valued continuous

A.K. Srivastava (✉)
Department of Mathematics and Computer Science,
St. Louis University, St. Louis, MO 63103, USA
e-mail: asrivas3@slu.edu

S.T. Rizvi et al. (eds.), *Algebra and its Applications*, Springer Proceedings in Mathematics & Statistics 174, DOI 10.1007/978-981-10-1651-6_4

functions on X is the sum of two units. For any $f(x) \in C(X)$, we have $f(x) = [(f(x) \vee 0) + 1] + [(f(x) \wedge 0) - 1]$. Every element in a real or complex Banach algebra B is the sum of two units. For any $a \in B$, there exists a scalar $\lambda\,(\neq 0)$ such that $a - \lambda$ is a unit and $a = (a - \lambda) + \lambda$. On the other hand, any ring having a homomorphic image isomorphic to $\mathbb{F}_2 \times \mathbb{F}_2$ cannot be additively generated by its units because in $\mathbb{F}_2 \times \mathbb{F}_2$, the element $(1, 0)$ cannot be expressed as a sum of any number of units.

In this area of research, a lot of focus has been on representing ring elements as the sum of a fixed number of unit elements.

Definition 1 An element x in a ring R is called a k-good element if x can be written as the sum of k units in R. We say that a ring R is a k-good ring if each element $x \in R$ is a k-good element. The unit sum number of a ring R is defined as

$$usn(R) = \begin{cases} k & \textit{if k is the smallest integer such that R is k-good} \\ \omega & \textit{if every element of R is a sum of units but R is not k-good for any k} \\ \infty & \textit{there exists an element a in R that cannot be written as sum of units} \end{cases}$$

A natural question that one may think at this point is the following: given any positive integer $n \geq 2$, can we construct a ring whose unit sum number is exactly n. The answer is yes and it follows from a construction of Herwig and Ziegler [25]. Although Herwig and Ziegler stated their result in a weaker form, but a careful examination of their proof reveals that they actually prove more that we they have stated.

Theorem 2 (Herwig and Ziegler, [25]) *For each $n \geq 2$, there exists a domain R with $usn(R) = n$.*

The key point in proving this theorem is the observation that if R is an integral domain, $n \geq 2$, an integer and x, a nonzero element of R, then R is contained in a domain S satisfying the following properties:

(1) x is the sum of n units in S, and
(2) if an element of R is the sum of $k < n$ units in S, then it is the sum of k units in R as well.

Now that we know there exists domain with any given arbitrary unit sum number, it makes sense to ask if we can construct specific class of rings with any given unit sum number.

First, we consider matrix rings. The following result is due to Henriksen [24].

Theorem 3 *Let R be any ring.*

(1) Any diagonal matrix over R is a 2-good element.
(2) The matrix ring $\mathbb{M}_n(R)$ is 3-good for all $n \geq 2$.

In view of the above theorem, it immediately follows that if R is a ring that can be realized as a matrix ring $\mathbb{M}_n(S)$, $n \geq 2$ over some ring S, then the only possible values of unit sum number for R are 2 and 3.

Henriksen [24] gave examples of rings R for which $\mathbb{M}_n(R)$ is not 2-good. The example given by Henriksen was generalized by Vámos [47] in the next proposition.

Proposition 4 *Let R be a ring, $n \geq 2$ an integer and let $L = Ra_1 + \cdots + Ra_n$ be a left ideal of R generated by the elements $a_1, \ldots, a_n \in R$. Let A be the $n \times n$ matrix whose entries are all zero except for the first column which is $(a_1, \ldots, a_n)^T$. Suppose that*

(1) L cannot be generated by fewer than n elements, and
(2) zero is the only 2-good element in L.

Then A is not 2-good.

Example 1 If R is a commutative noetherian domain of Krull dimension greater than 1, then for any $n \geq 1$ there is an n-generated ideal of R which cannot be generated by fewer than n elements. It follows that if R is any of the rings $\mathbb{Z}[x]$ or $F[x, y]$, where F is a field, then condition (2) is also satisfied by any proper ideal of R. So for these rings, $\mathbb{M}_n(R)$ is not 2-good for all $n \geq 2$. Thus, $usn(\mathbb{M}_n(R)) = 3$ if $R = \mathbb{Z}[x]$ or $F[x, y]$.

Since for matrix rings $\mathbb{M}_n(R)$ the only possible values for unit sum number are 2 and 3, all the focus has been on finding class of rings R for which $\mathbb{M}_n(R)$ will have unit sum number 2. We give below a list of results in this direction.

We say that an $n \times n$ matrix A over a ring R *admits a diagonal reduction* if there exist invertible matrices $P, Q \in \mathbb{M}_n(R)$ such that PAQ is a diagonal matrix. Following Ara et al. [1], a ring R is called an *elementary divisor ring* if every square matrix over R admits a diagonal reduction. This definition is less stringent than the one proposed by Kaplansky in [29]. The class of elementary divisor rings includes right self-injective von Neumann regular rings, unit regular rings.

As we have already seen that if R is any ring, then any $n \times n$ (where $n \geq 2$) diagonal matrix over R is the sum of two invertible matrices. Thus it follows that

Lemma 5 *If R is an elementary divisor ring, then $\mathbb{M}_n(R)$ has unit sum number 2 for $n \geq 2$.*

A *permutation matrix* is a square matrix that has exactly one 1 in each row and column, and all other entries 0. An $n \times n$ matrix $A = [a_{ij}]$ is said to avoid a permutation matrix $P = [p_{ij}]$ if, for all i, j, such that $p_{ij} = 1$, $a_{ij} = 0$.

Theorem 6 (Vámos and Wiegand [48]) *Let R be any ring and $n \geq 2$. If $A \in \mathbb{M}_{n(R)}$ avoids a permutation matrix P, then A is 2-good.*

Call a matrix $A = [a_{ij}]$ to be b-banded if for each i, j with $|i - j| \geq b$, $a_{ij} = 0$. For example, a diagonal matrix is a 1-banded matrix. As a consequence of the above theorem, it follows that

Corollary 7 (Vámos and Wiegand [48]) *If $A \in \mathbb{M}_{n(R)}$ is a b-banded matrix with $n \geq 2b$, then A is 2-good.*

Define the *meeting number* $m(P, B)$ of a permutation matrix P with a matrix B as

$$m(P, B) := |\{\text{positions } i, j \text{ where both } B, P \text{ are nonzero}\}|.$$

Proposition 8 *Let $B = diag(B_1, B_2, \ldots, B_t)$ be an $n \times n$ matrix and each B_i is a matrix of size at most $\frac{n}{2}$. Then there exists an $n \times n$ permutation matrix Q that avoids the blocks of B.*

The above proposition yields that

Corollary 9 *If B is an $n \times n$ matrix in block diagonal form where the blocks have size $\leq n/2$, then B is 2-good.*

We have seen that if R is k-good, then $\mathbb{M}_n(R)$ is k-good. Clearly, the converse is not true. For example, $\mathbb{M}_2(\mathbb{F}_2)$ is 2-good, but $\mathbb{F}_2$ is not 2-good. This shows corner ring of a k-good ring need not be k-good. Also, R being k-good does not imply $R[x]$ is k-good. In fact, $usn(R[x]) = \infty$ as the only units in the polynomial ring are the units in R.

In case of infinite matrices too the situation is not bad. Wang and Chen [49] proved the following.

Theorem 10 *Let R be a 2-good ring. Then the ring $B(R)$ of all $\omega \times \omega$ row-and-column-finite matrices over R has unit sum number 2. However, if R is any arbitrary ring, then the ring $B(R)$ has unit sum number 2 or 3.*

In 1958 Skornyakov asked: *Is every von Neumann regular ring, which does not have a homomorphic image isomorphic to $\mathbb{F}_2 \times \mathbb{F}_2$, additively generated by its units?*

This question of Skornyakov was answered in the negative by George Bergman who constructed an example of a von Neumann regular ring in which not all elements are sums of units. Bergman's example given below was first reported in a paper by Handelman [21].

Example 2 Let k be any field, and $A = k[[x]]$ be the power series ring in one variable over k. Let K be the field of fractions of A. Let $R = \{r \in \text{End}(A_k) : \text{there exists } q \in K,$ a positive integer n, with $r(a) = qa$ for all $a \in (x^n)\}$. Then R is a von Neumann regular ring which is not generated by its units.

So the above example is an example of a von Neumann regular ring with unit sum number ∞. But, in general, we do not yet know what are all possible values for unit sum number of a von Neumann regular ring. For (von Neumann regular) right self-injective rings, the complete characterization of unit sum numbers was given by Khurana and Srivastava in [30] and [31]. If R is a right self-injective ring then $R/J(R)$ is a von Neumann regular right self-injective ring. Since unit sum number of R is same as the unit sum number of $R/J(R)$, in order to classify unit sum numbers of right self-injective rings, it suffices to do that for any von Neumann regular right self-injective ring. Kaplansky developed type theory as a classification tool for certain class of rings of operators and that theory applies to von Neumann regular right self-injective rings. A careful examination of the type theory leads us to the fact that a von Neumann regular right self-injective ring R is a direct product of an abelian regular ring and proper matrix rings over elementary divisor rings. It was this crucial observation that helped Khurana and Srivastava to extend the result of Zelinsky and give complete description of unit sum number of right self-injective rings.

Theorem 11 (Khurana and Srivastava, [30] and [31]) *The unit sum number of a nonzero right self-injective ring R is 2, ω or ∞. Moreover,*

(1) usn(R) $= 2$ if and only if R has no homomorphic image isomorphic to $\mathbb{F}_2$.
(2) usn(R) $= \omega$ if and only if R has a homomorphic image isomorphic to $\mathbb{F}_2$, but has no homomorphic image isomorphic to $\mathbb{F}_2 \times \mathbb{F}_2$. In this case every non-invertible element of R is a sum of either two or three units.
(3) usn(R) $= \infty$ if and only if R has a homomorphic image isomorphic to $\mathbb{F}_2 \times \mathbb{F}_2$.

As the above theorem shows that for right self-injective von Neumann regular rings, the only possibility of unit sum numbers is 2, ω and ∞, we propose the following problem.

Problem 12 Does there exist a von Neumann regular ring with unit sum number other than 2, ω and ∞?

The following conjecture proposed in [42] still remains open.

Conjecture 13 A unit regular ring R has unit sum number 2 if and only if R has no homomorphic image isomorphic to $\mathbb{F}_2$.

In [16] it is shown that if the identity in a ring R with stable range one is a sum of two units, then every von Neumann regular element in R is a sum of two units. Consequently, it follows that every element in a unit regular ring R is a sum of two units if the identity in R is a sum of two units.

The following problem was posed by Henriksen and it is still open.

Problem 14 (*Henriksen,* [24]) If R is a simple algebra over a field with more than 2 elements, and R has an idempotent $e \neq 0$ and 1, then R is generated additively by its units. Does R have a finite unit sum number?

Definition 15 The unit sum number of a module M, denoted by usn(M), is the unit sum number of its endomorphism ring.

In the sense of above definition, Zelinsky's result states that usn$(V_D) = 2$ for a vector space over a division ring D except when V is one-dimensional over D. Laszlo Fuchs raised the question of determining when an endomorphism ring is generated additively by automorphisms. For abelian groups, this question has been studied by many authors. In [20] Hill showed that if G is a totally projective p-group with $p \neq 2$, then any endomorphism of G is the sum of two automorphisms. As a consequence, Hill also obtained that if the primary group G is a direct sum of countable groups and has an odd prime associated with it, then any endomorphism of G is the sum of two automorphisms. For a primary group G having no elements of infinite height, Stringall [45] gave necessary and sufficient conditions for endomorphism ring of G to be additively generated by its automorphisms. Khurana and Srivastava studied this question for several classes of modules in [31]. They proved the following.

Theorem 16 *Let M be a quasi-continuous module with finite exchange property. Then the unit sum number of M is 2 if and only if* $\mathrm{End}(M)$ *has no homomorphic image isomorphic to* $\mathbb{F}_2$.

As a continuous module is quasi-continuous and also satisfies the exchange property, it follows that the unit sum number of a continuous (and hence also of injective and quasi-injective) module M is 2 if and only if $\mathrm{End}(M)$ has no homomorphic image isomorphic to $\mathbb{F}_2$. In [31] it is also shown that the unit sum number of a flat cotorsion (in particular, pure injective) module M is 2 if and only if $\mathrm{End}(M)$ has no homomorphic image isomorphic to $\mathbb{F}_2$.

Let M be a module and $\mathcal{X}$, a class of R-modules closed under isomorphisms. In [18], a module M is called *$\mathcal{X}$-automorphism-invariant* if there exists an $\mathcal{X}$-envelope $u : M \to X$ satisfying that for any automorphism $g : X \to X$ there exists an endomorphism $f : M \to M$ such that $u \circ f = g \circ u$.

Recently, Guil Asensio, Keskin Tütüncü and Srivastava [17] have shown that

Theorem 17 *If $\mathcal{X}$ is a class of modules closed under isomorphisms and M, an $\mathcal{X}$-automorphism-invariant module with $u : M \to X$, a monomorphic $\mathcal{X}$-envelope such that* $\mathrm{End}(X)/J(\mathrm{End}(X))$ *is a von Neumann regular right self-injective ring and idempotents lift modulo* $J(\mathrm{End}(X))$. *Then the unit sum number of M is 2 if and only if* $\mathrm{End}(M)$ *has no homomorphic image isomorphic to* $\mathbb{F}_2$.

In particular, as a consequence of the above theorem, we have the following.

Theorem 18 *Let M be a module that is invariant under automorphisms of its injective envelope or pure-injective envelope then the unit sum number of M is 2 if and only if* $\mathrm{End}(M)$ *has no homomorphic image isomorphic to* $\mathbb{F}_2$. *Also, if M is a flat module that is invariant under automorphisms of its cotorsion envelope then the unit sum number of M is 2 if and only if* $\mathrm{End}(M)$ *has no homomorphic image isomorphic to* $\mathbb{F}_2$.

1.1 Additive Unit Representation of Rings of Integers of Number Fields

Let $K = \mathbb{Q}(\xi)$ be a number field (that is, a finite extension of $\mathbb{Q}$) and let $\mathcal{O}_K$ be the ring of integers of K. The following theorem due to Frei [13] shows abundance of ring of integers of number field additively generated by units.

Theorem 19 *For any number field K, there exists a number field L containing K, such that the ring of integers of L is generated additively by its units, that is, $usn(\mathcal{O}_L) \leq \omega$.*

The following result of Jarden and Narkiewicz [28] is quite interesting as it shows that although there are so many ring of integers of number field additively generated by their units, but the ring of integers of any number field of finite degree cannot have a finite unit sum number.

Theorem 20 *If R is a finitely generated integral domain of characteristic zero and $n \geq 1$ is an integer, then there exists a constant $A_n(R)$ such that every arithmetic progression in R having more than $A_n(R)$ elements contains an element which is not a sum of n units.*

As a consequence, it follows that a finitely generated integral domain of zero characteristic cannot be n-good for any n. Thus, in particular

Theorem 21 *The ring of integers of any number field of finite degree cannot have finite unit sum number.*

In [3] and [4], the authors give conditions under which rings of integers of various number fields are generated additively by their units.

Theorem 22 *Let $K = \mathbb{Q}(\sqrt{d})$, where $d \in \mathbb{Z}$ is square-free. Then $usn(\mathcal{O}_K) = \omega$ if and only if*

(1) $d \in \{-1, -3\}$ or
(2) $d > 0, d \not\equiv 1 \bmod 4$, and $d+1$ or $d-1$ is a perfect square, or
(3) $d > 0, d \equiv 1 \bmod 4$, and $d+4$ or $d-1$ is a perfect square.

Theorem 23 *Let d be a cube-free integer and $K = \mathbb{Q}(\sqrt[3]{d})$. Then $usn(\mathcal{O}_K) = \omega$ if and only if*

(1) d is square-free, $d \not\equiv \pm 1 \bmod 9$, and $d+1$ or $d-1$ is a perfect cube, or
(2) $d = 28$.

1.2 Further Generalizations of Zelinsky's Result

Chen [10] has recently shown that if V is a countably generated right vector space over a division ring D where $|D| > 3$, then for each linear transformation T on V_D, there exist invertible linear transformations P and Q on V_D such that $T - P$, $T - P^{-1}$ and $T^2 - Q^2$ are invertible. Wang and Zhou continued this line of investigation in [50]. They considered the following properties

- A ring R is said to satisfy the property (P), if for all $a \in R$, there exists a unit $u \in R$ such that $a + u$, $a - u^{-1}$ are units.
- A ring R is said to satisfy the property (Q), if for all $a \in R$, there exists a unit $u \in R$ such that $a - u$, $a - u^{-1}$ are units.

Wang and Zhou showed the following.

Theorem 24 *Let* $\text{End}_D(V)$ *be the ring of linear transformations of a right vector space V over a division ring D.*

(1) If $|D| > 3$, then $End_D(V)$ satisfies (P).
(2) If $|D| > 2$, then $End_D(V)$ satisfies (Q).

In [40] a ring R is said to be a *twin-good ring* if for each $x \in R$ there exists a unit $u \in R$ such that both $x + u$ and $x - u$ are units in R. Clearly every twin-good ring is 2-good. However, there are numerous examples of 2-good rings which are not twin-good. For example, $\mathbb{F}_3$ is 2-good but not twin-good. Clearly, if D is a division ring such that $|D| \geq 4$, then D is twin-good.

Siddique and Srivastava [40] proved the following.

Lemma 25 *Let R be any ring.*

(1) Let R be an elementary divisor ring. Then the matrix ring $\mathbb{M}_n(R)$ is twin-good for each $n \geq 3$.
(2) If R is an abelian regular ring, then the matrix ring $\mathbb{M}_n(R)$ is twin-good for each $n \geq 2$. In particular, if D is a division ring, then the matrix ring $\mathbb{M}_n(D)$ is twin-good for each $n \geq 2$.

Using structure theory of von Neumann regular right self-injective rings and the above lemma, Siddique and Srivastava [40] obtained the following.

Theorem 26 *A right self-injective ring R is twin-good if and only if R has no homomorphic image isomorphic to $\mathbb{F}_2$ or $\mathbb{F}_3$.*

As a consequence, it follows that

Corollary 27 *For any linear transformation T on a right vector space V over a division ring D, there exists an invertible linear transformation S on V such that both $T - S$ and $T + S$ are invertible, except when V is one-dimensional over $\mathbb{F}_2$ or $\mathbb{F}_3$.*

Corollary 28 *Let M be a quasi-continuous module with finite exchange property and $R = \mathrm{End}(M)$. Then R is twin-good if and only if R has no homomorphic image isomorphic to $\mathbb{F}_2$ or $\mathbb{F}_3$.*

In particular, the endomorphism ring of a continuous module or a flat cotorsion (in particular, pure injective) or a Harada module is twin-good if and only if it has no homomorphic image isomorphic to $\mathbb{F}_2$ or $\mathbb{F}_3$.

We would like to raise the following problem.

Problem 29 Let R be a Dedekind domain with finite class number c. Let $n \geq 2c$. Is $\mathbb{M}_n(R)$ a twin-good ring?

The result of Siddique and Srivastava has recently been generalized in [32] where it is shown that

Theorem 30 *If no field of order less than $n + 2$ is a homomorphic image of a right self-injective ring R, then for any element $a \in R$ and central units $u_1, \ldots, u_n$ in R, there exists a unit $u \in R$, such that $a + u_i u$ is a unit in R for each i.*

In fact, in the above theorem, instead of assuming that units $u_1, \ldots, u_n$ are in center of ring R, it suffices to assume that the group of units $U(R)$ is abelian.

In [46] Tang and Zhou have shown that each linear transformation of a vector space V over a division ring D is a sum of two commuting invertible linear transformations if and only if V is finite-dimensional and $|D| \geq 3$. Recently, it has been generalized in [39] where it is shown that if E is a Σ-injective module such that each endomorphism of E is a sum of two commuting automorphisms then E is directly finite. We propose the following conjecture.

Conjecture 31 Let E be an injective module. Then each endomorphism of E can be expressed as a sum of two commuting automorphisms if and only if E is directly finite and $\text{End}(E)$ has no homomorphic image isomorphic to $\mathbb{M}_n(\mathbb{F}_2)$ for any n.

2 Additive Idempotent Representation

Hirano and Tominaga [27] studied rings in which each element is the sum of two idempotents. These rings may be seen as a generalization of Boolean rings. Let S and T be Boolean rings and M be an T-S-bimodule. Then the ring $R = \begin{bmatrix} S & 0 \\ M & T \end{bmatrix}$ satisfies the property that each element of R is the sum of two idempotents. However, this ring R is not Boolean. For $n \geq 2$, the matrix ring $\mathbb{M}_n(R)$ over any ring R contains an element which is not the sum of two idempotents.

Hirano and Tominaga proved the following.

Theorem 32 *The following conditions are equivalent for a ring R;*

(1) R is a commutative ring in which each element is the sum of two idempotents.
(2) R is a ring in which each element is the sum of two commuting idempotents.
(3) $x^3 = x$ for each element $x \in R$.

As a consequence of this theorem, they further deduced that

Theorem 33 *If R is a PI ring in which each element is the sum of two idempotents, then $R/N(R)$ satisfies the identity $x^3 = x$ where $N(R)$ denotes the prime radical of R.*

2.1 Clean Rings

An element x in a ring R is called a clean element if there exists a unit $u \in R$ and an idempotent $e \in R$ such that $x = e + u$. A ring in which every element is a clean element is called a *clean ring*. Clean rings were introduced by Nicholson as examples of exchange rings [35].

Šter gives another characterization for clean elements in [45].

Lemma 34 *Let a be any element in a ring R. Then a is a clean element if and only if there exists an idempotent $e \in R$ and a unit $u \in R$ such that $ua = eu + 1$.*

Examples of Clean Rings.

(1) The ring of linear transformations is a clean ring [38].
(2) Every unit regular ring is clean [6].
(3) If M is a module that is invariant under automorphisms of its injective envelope or pure-injective envelope then the endomorphism ring of M is a clean ring [17].
(4) If M is a flat module that is invariant under automorphisms of its cotorsion envelope then the endomorphism ring of M is a clean ring [17].
(5) The endomorphism ring of any continuous module is clean [7].

We give below some basic facts about clean rings.

Proposition 35 *Let R be any ring.*

(1) R is clean if and only if $R/J(R)$ is clean and idempotents lift modulo the Jacobson radical $J(R)$ [35].
(2) If R is clean, then R is an exchange ring [35]. Bergman's example mentioned in the first section is example of an exchange ring which is not clean.
(3) If R is an exchange ring with central idempotents, then R is clean [35].
(4) R is semiperfect if and only if R is clean and R does not contain an infinite set of orthogonal idempotents [8].

Burgess and Rapahel [5] have shown that every ring can be embedded in a clean ring as an essential ring extension. Recall that a ring extension $R \subseteq S$ is called an essential ring extension if for each nonzero ideal I of S, $I \cap R \neq 0$. This shows the abundance of clean rings.

Properties of Clean Rings.

(1) The center of a clean ring need not be clean [5].
(2) If R is a clean ring, then the matrix ring $\mathbb{M}_n(R)$ is also a clean ring [19].
(3) The corner ring of a clean ring need not be a clean ring [43].

The most important recent development in the theory of clean rings has been the work of Šter who constructed for $n \geq 2$, a ring R such that $\mathbb{M}_n(R)$ is clean but $\mathbb{M}_k(R)$ is not clean for $k < n$. This shows, in particular, that

Theorem 36 (Šter, [44]) *The property of being a clean ring is not a Morita-invariant property.*

The following question has been raised by Šter [44].

Problem 37 Does there exist a ring R such that $\mathbb{M}_n(R)$ is clean for every $n \geq 2$ but R is not clean?

There are numerous generalizations of clean rings available in the literature, but there are still some basic questions unanswered about clean rings. We list below some of them.

Problem 38 Characterize clean von Neumann regular rings.

The following conjecture proposed in [42] is still open.

Conjecture 39 Let R be a clean ring. Then the unit sum number of R is 2 if and only if R has no homomorphic image isomorphic to $\mathbb{F}_2$.

Very little is known about clean group rings. The following question is worth looking at.

Problem 40 Characterize rings R and groups G such that the group ring $R[G]$ is clean.

Let (X, d) be a locally finite metric space in the sense that all balls of finite radius are finite. Following Gromov [15], Ara et al. [2] defined the *translation ring* $T(X, R)$ of X over R to be the ring of all square matrices $[a(x, y)]$, indexed by $X \times X$ and with entries from R, such that $a(x, y) = 0$ whenever $d(x, y) > l$ for some constant l depending on the matrix. The least such l is called the *bandwidth* of the matrix. Thus the Gromov translation ring $T(X, R)$ is the ring of infinite matrices over R, indexed by $X \times X$, with constant bandwidth. The translation ring also makes sense when d is just a locally finite pseudo metric. Ara et al. [2] showed that if X is a discrete tree and R is any von Neumann regular ring then the translation ring $T(X, R)$ is an exchange ring.

We would like to propose following questions.

Problem 41 Is $T(X, F)$ a clean ring where F is a field?

Problem 42 Let R be a ring with unit sum number 2. Is the translation ring $T(X, R)$ also a ring with unit sum number 2?

If R is a clean ring then a set of idempotents $E \subset R$ such that each element in R can be expressed as $e + u$, where $e \in E$ and u is a unit in R, will be called a *clean-generator set* of R. If R is a clean ring, then the *clean-dimension* of R, denoted by clean-dim(R), is defined as clean-dim$(R) = \min\{|E| : E$ is a clean-generator set of $R\}$. It is easy to see that if D is a division ring, then clean-dim$(D) = 2$.

Problem 43 What is clean-dim$(\mathbb{M}_n(D))$, where D is a division ring?

3 Some Other Additive Representations

3.1 Additive Regular Representation

Chatters, Ginn and Robson studied rings that are additively generated by their regular elements (see [9], and [37]). Recall that an element x in a ring R is called a regular element if x is not a left or right zero-divisor. Here are the main results concerning additive regular representation of elements in a ring.

Theorem 44 *Let R be any ring.*

(1) If R is a prime right Goldie ring, then each element of R is the sum of at most two regular elements.
(2) If R is a semiprime right Goldie ring, then R is generated additively by its regular elements if and only if R does not have a direct summand isomorphic to $\mathbb{F}_2 \oplus \mathbb{F}_2$. Furthermore, each element of R is the sum of two regular elements if R does not have a direct summand isomorphic to $\mathbb{F}_2$.
(3) If R is a left and right noetherian ring in which 2 is a regular element, then R is generated additively by its regular elements.

3.2 Additive Commutator Representation

In [33] a ring R is called a *commutator ring* if each element of R is a sum of additive commutators. One of the twelve open problems asked by Kaplansky in 1956 was whether there exists a division ring which is a commutator ring. Harris [22] answered the question of Kaplansky in the affirmative by constructing a division ring in which element is a sum of additive commutators. All the results in this subsection are due to Mesyan ([33] and [34]).

Properties of commutator rings.

(1) Any homomorphic image of a commutator ring is again a commutator ring.
(2) Finite direct products of commutator rings are also commutator rings.
(3) If $R \subseteq S$ are rings such that R is a commutator ring and S is generated over R by elements centralizing R, then S is also a commutator ring. In particular, this means that matrix rings, group rings, and polynomial rings over commutator rings are also commutator rings.
(4) Over any ring, the ring of infinite matrices that are both row-finite and column-finite are commutator rings.
(5) A finite-dimensional algebra over any field can never be a commutator ring. This implies, in particular, that no PI ring can be a commutator ring as every PI ring has a homomorphic image that is finite-dimensional over a field.

Mesyan also proved that

Theorem 45 *Any ring can be embedded in a commutator ring.*

The idea behind the proof of the above theorem is to construct for any ring R and any set $\mathcal{I}$, a ring $A_{\mathcal{I}}(R) = R < \{x_i\}_{i \in \mathcal{I}}, \{y_i\}_{i \in \mathcal{I}} : [x_i, x_j] = [y_i, y_j] = [x_i, y_j] = 0$ for $i \neq j$, and $[x_i, y_i] = 1 >$ which is a commutator ring.

Let K be a field and E be an arbitrary directed graph. Let E^0 be the set of vertices, and E^1 be the set of edges of directed graph E. Consider two maps $r : E^1 \to E^0$ and $s : E^1 \to E^0$. For any edge e in E^1, $s(e)$ is called the source of e and $r(e)$ is called the range of e. If e is an edge starting from vertex v and pointing toward vertex w, then we imagine an edge starting from vertex w and pointing toward vertex v and call it the ghost edge of e and denote it by e^*. We denote by $(E^1)^*$, the set of all ghost edges of directed graph E. If $v \in E^0$ does not emit any edges, i.e. $s^{-1}(v) = \emptyset$, then v is called a sink and if v emits an infinite number of edges, i.e. $|s^{-1}(v)| = \infty$, then v is called an infinite emitter. If a vertex v is neither a sink nor an infinite emitter, then v is called a regular vertex.

The Leavitt path algebra of E with coefficients in K, denoted by $L_K(E)$, is the K-algebra generated by the sets E^0, E^1, and $(E^1)^*$, subject to the following conditions:

(A1) $v_i v_j = \delta_{ij} v_i$ for all $v_i, v_j \in E^0$.
(A2) $s(e)e = e = er(e)$ and $r(e)e^* = e^* = e^* s(e)$ for all e in E^1.
(CK1) $e_i^* e_j = \delta_{ij} r(e_i)$ for all $e_i, e_j \in E^1$.
(CK2) If $v \in E^0$ is any regular vertex, then $v = \sum_{\{e \in E^1 : s(e) = v\}} ee^*$.

Conditions (CK1) and (CK2) are known as the Cuntz-Krieger relations. If E^0 is finite, then $\sum\limits_{v_i \in E^0} v_i$ is an identity for $L_K(E)$ and if E^0 is infinite, then E^0 generates a set of local units for $L_K(E)$.

Let us denote the set of all paths in E by $P(E)$. Given two vertices $u, v \in E^0$ such that there is a path $p \in P(E)$ with $s(p) = u$ and $r(p) = v$, let $d(u, v)$ denote the length of the shortest such path. For each $u \in E^0$ and $m \in \mathbb{N}$, denote $D(u, m) = \{v \in E^0 : d(u, v) \leq m\}$.

Mesyan studied commutator Leavitt path algebras in [34] and proved the following.

Theorem 46 *The Leavitt path algebra $L_K(E)$ of a graph E over a field K is a commutator ring if and only if the following conditions hold;*

(1) E is acyclic.
(2) E^0 contains only regular vertices.
(3) The characteristic p of K is not zero.
(4) For each $u \in E^0$, there is an $m \in \mathbb{N}$ such that for all $w \in E^0$ satisfying $d(u, w) = m + 1$, the number of paths $q = e_1 e_2 \ldots e_k \in P(E)$ such that $s(q) = u$, $r(q) = w$ and $s(e_2), \ldots, s(e_k) \in D(u, m)$ is a multiple of p.

References

1. Ara, P., Goodearl, K.R., O'Meara, K.C., Pardo, E.: Diagonalization of matrices over regular rings. Linear Algbr. Appl. **265**, 147–163 (1997)
2. Ara, P., Meara, K.C.O., Perera, F.: Gromov Translation algebras over discrete trees are exchange rings. Trans. Amer. Math. Soc. **356**(5), 2067–2079 (2003)
3. Ashrafi, N., Vámos, P.: On the unit sum number of some rings. Quart. J. Math. **56**, 1–12 (2005)
4. Belcher, P.: A test for integers being sums of direct units applied to cubic fields. Bull. Lond. Math. Soc. **6**, 66–68 (1974)
5. Burgess, W., Raphael, R.: On embedding rings in clean rings. Comm. Algbr. **41**(2), 552–564 (2013)

6. Camillo, V.P., Khurana, D.: A characterization of unit regular rings. Comm. Algbr. **29**(5), 2293–2295 (2001)
7. Camillo, V.P., Khurana, D., Lam, T.Y., Nicholson, W.K., Zhou, Y.: Continuous modules are clean. J. Algebr. **304**, 94–111 (2006)
8. Camillo, V.P., Yu, H.P.: Exchange rings, units and idempotents. Comm. Algbr. **22**(12), 4737–4749 (1994)
9. Chatters, A.W., Ginn, S.M.: Rings generated by their regular elements. Glasgow Math. J. **25**, 1–5 (1984)
10. Chen, H.: Decompositions of Countable Linear Transformations. Glasgow Math. J. **52**, 427–433 (2010)
11. Dieudonné, J.: La théorie de Galois des anneux simples et semi-simples. Comment. Math. Helv. **21**, 154–184 (1948)
12. Fisher, J.W., Snider, R.L.: Rings generated by their units. J. Algebr. **42**, 363–368 (1976)
13. Frei, C.: On rings of integers generated by their units. Bull. Lond. Math. Soc. **44**(1), 167–182 (2012)
14. Goldsmith, B., Pabst, S., Scott, A.: Unit sum numbers of rings and modules, Quart. J. Math. Oxford, **49**(2), 331–344(1998)
15. Gromov, M.: Asymptotic invariants of infinite groups. In: Geometric Group Theory, vol. 2, Math. Soc. Lecture Notes Ser. 182, pp. 1–295, Cambridge University Press, London (1993),
16. Grover, H.K., Wang, Z., Khurana, D., Chen, J., Lam, T.Y.: Sums of units in rings. J. Algebr. Appl. **13**, 10 (2014)
17. Guil Asensio, P.A., Keskin Tütüncü, D., Srivastava, A.K.: Modules invariant under automorphisms of their covers and envelopes. Israel J. Math. **206**(1), 457–482 (2015)
18. Guil Asensio, P.A., Srivastava, A.K.: Automorphism-invariant modules satisfy the exchange property. J. Algebr. **388**, 101–106 (2013)
19. Han, J., Nicholson, W.K.: Extensions of clean rings. Comm. Algebr. **29**(6), 2589–2595 (2001)
20. Hochschild, G.: Automorphisms of simple algebras. Trans. Amer. Math. Soc. **69**, 292–301 (1950)
21. Handelman, D.: Perspectivity and cancellation in regular rings. J. Algebr. **48**, 1–16 (1977)
22. Harris, B.: Commutators in division rings. Proc. Amer. Math. Soc. **9**, 628–630 (1958)
23. Henriksen, M.: On a class of regular rings that are elementary divisor rings. Arch. Math. (Basel) **24**, 133–141 (1973)
24. Henriksen, M.: Two classes of rings generated by their units. J. Algebra **31**, 182–193 (1974)
25. Herwig, B., Ziegler, M.: A remark on sums of units. Arch. Math. **79**(6), 430–431 (2002)
26. Hill, P.: Endomorphism ring generated by units. Trans. Amer. Math. Soc. **141**, 99–105 (1969)
27. Hirano, Y., Tominaga, H.: Rings in which every element is the sum of two idempotents. Bull. Austral. Math. Soc. **37**, 161–164 (1988)
28. Jarden, M., Narkiewicz, W.: On sums of units. Monatsh. Math. **150**, 327–332 (2007)
29. Kaplansky, I.: Elementary divisors and modules. Trans. Amer. Math. Soc. **66**, 464–491 (1949)
30. Khurana, D., Srivastava, A.K.: Right self-injective rings in which each element is sum of two units. J. Algebr. Appl. **6**(2), 281–286 (2007)
31. Khurana, D., Srivastava, A.K.: Unit sum numbers of right self-injective rings. Bull. Austral. Math. Soc. **75**(3), 355–360 (2007)
32. Khurana, S., Khurana, D., Nielsen, P. P.: Sums of units in self-injective rings. J. Algebr. Appl. **13** 7 (2014)
33. Mesyan, Z.: Commutator rings. Bull. Austral. Math. Soc. **74**, 279–288 (2006)
34. Mesyan, Z.: Commutator Leavitt path algebras. Algebr. Represent. Theory **16**, 1207–1232 (2013)
35. Nicholson, W.K.: Lifting idempotents and exchange rings. Trans. Amer. Math. Soc. **229**, 269–278 (1977)
36. Raphael, R.: Rings which are generated by their units. J. Algebr. **28**, 199–205 (1974)
37. Robson, J.C.: Sums of two regular elements. Glasgow Math. J. **25**, 7–11 (1984)
38. Searcóid, M.: Perturbation of linear operators by idempotents. Irish Math. Soc. Bull., **39**, 10–13 (1997)

39. Siddique, F.: PhD. Dissertation, Saint Louis University, Saint Louis (2015)
40. Siddique, F., Srivastava, A. K.: Decomposing elements of a right self-injective ring. J. Algebr. Appl. **12**(6) 10 (2013)
41. Skornyakov, L.A.: Complemented Modular Lattices and Regular Rings. Oliver & Boyd, Edinburgh (1964)
42. Srivastava, A.K.: A survey of rings generated by units. Annales de la Faculté des Sciences de Toulouse Mathématiques **19**(S1), 203–213 (2010)
43. Šter, J.: Corner ring of a clean ring need not be clean. Comm. Algebr. **40**(5), 1595–1604 (2012)
44. Šter, J.: The clean property is not a Morita invariant. J. Algebr. **420**, 15–38 (2014)
45. Stringall, R.W.: Endomorphism rings of abelian groups generated by automorphism groups. Acta. Math. **18**, 401–404 (1967)
46. Tang, G., Zhou, Y.: When is every linear transformation a sum of two commuting invertible ones? Linear Algebr. Appl. **439**(11), 3615–3619 (2013)
47. Vámos, P.: 2-Good rings. Quart. J. Math. **56**, 417–430 (2005)
48. Vámos, P., Wiegand, S.: Block diagonalization and 2-unit sums of matrices over prüfer domains. Trans. Amer. Math. Soc. **363**, 4997–5020 (2011)
49. Wang, Z., Chen, J.L.: 2-Clean rings. Canad. Math. Bull. **52**, 145–153 (2009)
50. Wang, L., Zhou, Y.: Decomposing linear transformations. Bull. Aust. Math. Soc. **83**(2), 256–261 (2011)
51. Wang, L., Zhou, Y.: Corrections: Decomposing linear transformations. Bull. Aust. Math. Soc. **85**, 172–173 (2012)
52. Wolfson, K.G.: An ideal theoretic characterization of the ring of all linear transformations. Amer. J. Math. **75**, 358–386 (1953)
53. Zelinsky, D.: Every linear transformation is sum of nonsingular ones. Proc. Amer. Math. Soc. **5**, 627–630 (1954)

Notes on Commutativity of Prime Rings

Shuliang Huang

Abstract Let R be a prime ring with center $Z(R)$, J a nonzero left ideal, α an automorphism of R and R admits a generalized (α, α)-derivation F associated with a nonzero (α, α)-derivation d such that $d(Z(R)) \neq (0)$. In the present paper, we prove that if any one of the following holds: *(i)* $F([x, y]) - \alpha([x, y]) \in Z(R)$ *(ii)* $F([x, y]) + \alpha([x, y]) \in Z(R)$ *(iii)* $F(x \circ y) - \alpha(x \circ y) \in Z(R)$ *(iv)* $F(x \circ y) - \alpha(x \circ y) \in Z(R)$ for all $x, y \in J$, then R is commutative. Also some related results have been obtained.

Keywords Commutativity · Prime and semiprime rings · Generalized (α, α)-derivations

2000 Mathematics Subject Classification: 16N60 · 16W25 · 16U80 · 16D90

1 Introduction

In all that follows, unless stated otherwise, R will be an associative ring with the center $Z(R)$. For any $x, y \in R$, the symbol $[x, y]$ and $x \circ y$ stand for the Lie commutator $xy - yx$ and Jordan commutator $xy + yx$, respectively. A ring R is called 2-torsion free, if whenever $2x = 0$, with $x \in R$, then $x = 0$. If $S \subseteq R$, then we can define the left (resp. right) annihilator of S as $l(S) = \{x \in R \mid xs = 0 \ \textit{for all} \ \ s \in S\}$ (resp. $r(S) = \{x \in R \mid sx = 0 \ \textit{for all} \ \ s \in S\}$).

Recall that a ring R is prime if for any $a, b \in R$, $aRb = (0)$ implies $a = 0$ or $b = 0$, and is semiprime if for any $a \in R$, $aRa = (0)$ implies $a = 0$. An additive subgroup U of R is said to be a Lie ideal of R if $[u, r] \in U$ for all $u \in U$ and $r \in R$,

The paper is supported by the Anhui Provincial Natural Science Foundation (1408085QA08) and the Key University Science Research Project of Anhui Province (KJ2014A183) and also the Training Program of Chuzhou University (2014PY06) of China.

S. Huang (✉)
School of Mathematics and Finance, Chuzhou University,
Chuzhou 239000, People's Republic of China
e-mail: shulianghuang@163.com

S.T. Rizvi et al. (eds.), *Algebra and its Applications*, Springer Proceedings in Mathematics & Statistics 174, DOI 10.1007/978-981-10-1651-6_5

and a Lie ideal U is called square-closed if $u^2 \in U$ for all $u \in U$. By a derivation, we mean an additive mapping $d : R \longrightarrow R$ such that $d(xy) = d(x)y + xd(y)$ for all $x, y \in R$. Let α and β be endomorphisms of R, an additive mapping $d : R \longrightarrow R$ is said to be an (α, β)-derivation if $d(xy) = d(x)\alpha(y) + \beta(x)d(y)$ holds for all $x, y \in R$. Following [1], an additive mapping $F : R \longrightarrow R$ is called a generalized (α, β)-derivation on R if there exists an (α, β)-derivation $d : R \longrightarrow R$ such that $F(xy) = F(x)\alpha(y) + \beta(x)d(y)$ holds for all $x, y \in R$. Note that for I_R the identity map on R, this notion includes those of (α, β)-derivation when $F = d$, of derivation when $F = d$ and $\alpha = \beta = I_R$, and of generalized derivation, which is the case when $\alpha = \beta = I_R$.

Many results indicate that the global structure of a ring R is often tightly connected to the behavior of additive mappings defined on R. A well known result of Posner [10] states that if R is a prime ring and d a nonzero derivation of R such that $[d(x), x] \in Z(R)$ for all $x \in R$, then R must be commutative. Over the last few decades, several authors have investigated the relationship between the commutativity of the ring R and certain specific types of derivations of R (see [3–5, 7] where further references can be found).

Daif and Bell [6] showed that if in a semiprime ring R there exists a nonzero ideal I of R and a derivation d such that $d([x, y]) - [x, y] = 0$ or $d([x, y]) + [x, y] = 0$ for all $x, y \in I$, then $I \subseteq Z(R)$. In particular, if $I = R$ then R is commutative. At this point, the natural question is what happens in case the derivation is replaced by a generalized derivation. In [11], Quadri et al., proved that if R is a prime ring, I a nonzero ideal of R and F a generalized derivation associated with a nonzero derivation d such that any one of the following holds: (i) $F([x, y]) - [x, y] = 0$ (ii) $F([x, y]) + [x, y] = 0$ (iii) $F(x \circ y) - x \circ y = 0$ (iv) $F(x \circ y) + x \circ y = 0$ for all $x, y \in I$, then R is commutative. Following this line of investigation, Ali, Kumar and Miyan [2], explored the commutativity of a prime ring R admitting a generalized derivation F satisfying any one of the following conditions: (i) $F([x, y]) - [x, y] \in Z(R)$ (ii) $F([x, y]) + [x, y] \in Z(R)$ (iii) $F(x \circ y) - x \circ y \in Z(R)$ (iv) $F(x \circ y) + x \circ y \in Z(R)$ for all $x, y \in I$, a nonzero right ideal of R. On the other hand, Marubayashi et al. [8], established that if a 2-torsion free prime ring R admits a nonzero generalized (α, β)-derivation F associated with an (α, β)-derivation d such that either $F([u, v]) = 0$ or $F(u \circ v) = 0$ for all $u, v \in U$, where U is a nonzero square-closed Lie ideal of R, then $U \subseteq Z(R)$. In the present paper, our purpose is to prove the cited results for the case when the generalized (α, α)-derivation F acts on one sided ideal of R.

2 Main Results

In the remaining part of this paper, α and β will denote automorphisms of R. And we shall do a great deal of calculation with commutators and anti-commutators, routinely using the following basic identities: For all $x, y, z \in R$;

$$[xy, z] = x[y, z] + [x, z]y \text{ and } [x, yz] = y[x, z] + [x, y]z$$

$$xo(yz) = (xoy)z - y[x, z] = y(xoz) + [x, y]z$$

$$(xy)oz = x(yoz) - [x, z]y = (xoz)y + x[y, z].$$

Theorem 2.1 *Let R be a prime ring with center $Z(R)$ and J a nonzero left ideal of R. Suppose that R admits a generalized (α, α)-derivation F associated with a nonzero (α, α)-derivation d such that $d(Z(R)) \neq (0)$. If $F([x, y]) - \alpha([x, y]) \in Z(R)$ for all $x, y \in J$, then R is commutative.*

Proof It is easy to check that $d(Z(R)) \subseteq Z(R)$. Since $d(Z(R)) \neq (0)$, there exists $0 \neq c \in Z(R)$ such that $0 \neq d(c) \in Z(R)$. By assumption, we have

$$F([x, y]) - \alpha([x, y]) \in Z(R) \text{ for all } x, y \in J. \quad (1)$$

Replacing y by cy in (1), we get

$$(F([x, y]) - \alpha([x, y]))\alpha(c) + \alpha([x, y])d(c) \in Z(R) \text{ for all } x, y \in J. \quad (2)$$

Combining (1) and (2) and noting that the fact $\alpha(c) \in Z(R)$, we find that $\alpha([x, y])d(c) \in Z(R)$, which implies that $[\alpha([x, y])d(c), r] = 0 = [\alpha([x, y]), r]d(c)$ for all $x, y \in J$ and $r \in R$. Since R is prime and $0 \neq d(c) \in Z(R)$, we have

$$[\alpha([x, y]), r] = 0 \text{ for all } x, y \in J; r \in R. \quad (3)$$

Replacing y by yx in (3) and using (3), we get

$$\alpha([x, y])[\alpha(x), r] = 0 \text{ for all } x, y \in J; r \in R. \quad (4)$$

Replacing r by $r\alpha(s)$ in (4) and using (4), we arrive at $\alpha([x, y])r[\alpha(x), \alpha(s)] = 0$ for all $x, y \in J$ and $r, s \in R$. The primeness of R yields that for each $x \in J$, either $\alpha([x, y]) = 0$ or $[\alpha(x), \alpha(s)] = 0$. Equivalently, either $[x, J] = 0$ or $[x, R] = 0$. Set $J_1 = \{x \in J \mid [x, J] = 0\}$ and $J_2 = \{x \in J \mid [x, R] = 0\}$. Then, J_1 and J_2 are both additive subgroups of I such that $J = J_1 \cup J_2$. Thus, by Brauer's trick, we have either $J = J_1$ or $J = J_2$. If $J = J_1$, then $[J, J] = 0$, and if $J = J_2$, then $[J, R] = 0$. In both cases, we conclude that J is commutative and so, by a result of [9], R is commutative.

Corollary 2.2 *Let R be a prime ring with center $Z(R)$ and J a nonzero left ideal of R. Suppose that R admits a generalized (α, α)-derivation F associated with a nonzero (α, α)-derivation d such that $d(Z(R)) \neq (0)$. If $F(xy) - \alpha(xy) \in Z(R)$ for all $x, y \in J$, then R is commutative.*

Proof For any $x, y \in J$, we have $F([x, y]) - \alpha([x, y]) = (F(xy) - \alpha(xy)) - (F(yx) - \alpha(yx)) \in Z(R)$, and hence the result follows.

Theorem 2.3 *Let R be a prime ring with center $Z(R)$ and J a nonzero left ideal of R. Suppose that R admits a generalized (α, α)-derivation F associated with a nonzero (α, α)-derivation d such that $d(Z(R)) \neq (0)$. If $F([x, y]) + \alpha([x, y]) \in Z(R)$ for all $x, y \in J$, then R is commutative.*

Proof If $F([x, y]) + \alpha([x, y]) \in Z(R)$ for all $x, y \in J$, then the generalized (α, α)-derivation $-F$ satisfies the condition $(-F)([x, y]) - \alpha([x, y]) \in Z(R)$ for all $x, y \in J$. It follows from Theorem 2.1 that R is commutative.

Theorem 2.4 *Let R be a prime ring with center $Z(R)$ and J a nonzero left ideal of R. Suppose that R admits a generalized (α, α)-derivation F associated with a nonzero (α, α)-derivation d such that $d(Z(R)) \neq (0)$. If $F(x \circ y) - \alpha(x \circ y) \in Z(R)$ for all $x, y \in J$, then R is commutative.*

Proof We are given that

$$F(x \circ y) - \alpha(x \circ y) \in Z(R) \text{ for all } x, y \in J. \tag{5}$$

Since $d(Z(R)) \neq (0)$, there exists $0 \neq c \in Z(R)$ such that $0 \neq d(c) \in Z(R)$. Replacing y by cy in (5), we get

$$(F(x \circ y) - \alpha(x \circ y))\alpha(c) + \alpha(x \circ y)d(c) \in Z(R) \text{ for all } x, y \in J. \tag{6}$$

Combining (5) and (6), we find that $\alpha(x \circ y)d(c) \in Z(R)$ and hence $\alpha(x \circ y) \in Z(R)$. This implies that

$$[\alpha(x \circ y), r] = 0 \text{ for all } x, y \in J; r \in R. \tag{7}$$

Replacing yx for y in (7) and using (7), we have

$$\alpha(x \circ y)[\alpha(x), r] = 0 \text{ for all } x, y \in J; r \in R. \tag{8}$$

Replacing r by $r\alpha(s)$ in (8) and using (8), we have $\alpha(x \circ y)r[\alpha(x), \alpha(s)] = 0$ for all $x, y \in J$ and $r, s \in R$. The primeness of R yields that for each $x \in J$, either $\alpha(x \circ y) = 0$ or $[\alpha(x), \alpha(s)] = 0$. Now applying similar arguments as used in the proof of Theorem 2.1, we have either $x \circ y = 0$ for all $x, y \in J$; or $[J, R] = 0$. In the former case, replacing x by xz and using the fact $x \circ y = 0$ we find $[x, y]z = 0$ for all $x, y, z \in J$. This implies that $[x, y]J = 0$ and hence $[x, y]RJ = 0$. Since J is nonzero and R is prime, we get $[J, J] = 0$. Thus, J is commutative and so R. In the latter case, we have $[J, R] = 0$, in particular $[J, J] = 0$ and hence we get the required result.

Theorem 2.5 *Let R be a prime ring with center $Z(R)$ and J a nonzero left ideal of R. Suppose that R admits a generalized (α, α)-derivation F associated with a nonzero*

(α, α)-derivation d such that $d(Z(R)) \neq (0)$. If $F(x \circ y) + \alpha(x \circ y) \in Z(R)$ for all $x, y \in J$, then R is commutative.

Proof If $F(x \circ y) + \alpha(x \circ y) \in Z(R)$ for all $x, y \in J$, then the generalized (α, α)-derivation $-F$ satisfies the condition $(-F)(x \circ y) - \alpha(x \circ y) \in Z(R)$ for all $x, y \in J$. It follows from Theorem 2.4 that R is commutative.

Corollary 2.6 *Let R be a prime ring with center $Z(R)$ and J a nonzero left ideal of R. Suppose that R admits a generalized (α, α)-derivation F associated with a nonzero (α, α)-derivation d such that $d(Z(R)) \neq (0)$. If $F(xy) + \alpha(xy) \in Z(R)$ for all $x, y \in J$, then R is commutative.*

Proof For any $x, y \in I$, we have $F(x \circ y) + \alpha(x \circ y) = (F(xy) + \alpha(xy)) + (F(yx) + \alpha(yx)) \in Z(R)$, and hence our result follows.

Theorem 2.7 *Let R be a prime ring and J a nonzero left ideal of R such that $r(J) = 0$. If R admits a generalized (α, β)-derivation F associated with a nonzero (α, β)-derivation d such that $F([x, y]) = 0$ for all $x, y \in J$, then R is commutative.*

Proof By assumption, we have

$$F([x, y]) = 0 \text{ for all } x, y \in J. \tag{9}$$

Replacing y by yx in (9) and using (9), we get $\beta([x, y])d(x) = 0$, which implies

$$[x, y]\beta^{-1}(d(x)) = 0 \text{ for all } x, y \in J. \tag{10}$$

Now substituting ry for y in (10) and using (10), we obtain $[x, r]y\beta^{-1}(d(x)) = 0$ for all $x, y \in J$ and $r \in R$. In particular, $[x, R]RJ\beta^{-1}(d(x)) = 0$ for all $x \in J$. The primeness of R yields that for each $x \in J$, either $[x, R] = 0$ or $J\beta^{-1}(d(x)) = 0$, in this case $d(x) = 0$. In view of similar arguments as used in the proof of Theorem 2.1, we have either $[J, R] = 0$ or $d(J) = 0$. If $[J, R] = 0$, then J is commutative and we are done. If $d(J) = 0$, then $0 = d(RJ) = d(R)\alpha(J) + \beta(R)d(J)$, which reduces to $d(R)\alpha(J) = 0$. And hence $d(R)\alpha(RJ) = 0 = d(R)\alpha(R)\alpha(J) = d(R)R\alpha(I)$. Since J is nonzero and the last relation forces that $d = 0$, contradiction.

Using the same techniques with necessary variations, we can prove the following:

Theorem 2.8 *Let R be a prime ring and J a nonzero left ideal of R such that $r(J) = 0$. If R admits a generalized (α, β)-derivation F associated with a nonzero (α, β)-derivation d such that $F(x \circ y) = 0$ for all $x, y \in J$, then R is commutative.*

The following example demonstrates that R to be prime is essential in the hypothesis of Theorems 2.1, 2.3, 2.4 and 2.5.

Example 2.9 Let S be any ring. Next, let $R=\left\{\begin{pmatrix} 0 & a & b \\ 0 & 0 & c \\ 0 & 0 & 0 \end{pmatrix} |a, b, c \in S\right\}$ and $J=\left\{\begin{pmatrix} 0 & a & b \\ 0 & 0 & 0 \\ 0 & 0 & 0 \end{pmatrix} |a, b \in S\right\}$, a nonzero left ideal of R. Define maps $F,\ d\ ,\alpha : R \longrightarrow R$ as follows: $F\begin{pmatrix} 0 & a & b \\ 0 & 0 & c \\ 0 & 0 & 0 \end{pmatrix} = \begin{pmatrix} 0 & 0 & -c \\ 0 & 0 & 0 \\ 0 & 0 & 0 \end{pmatrix}$, $d\begin{pmatrix} 0 & a & b \\ 0 & 0 & c \\ 0 & 0 & 0 \end{pmatrix} = \begin{pmatrix} 0 & 0 & c \\ 0 & 0 & 0 \\ 0 & 0 & 0 \end{pmatrix}$, $\alpha\begin{pmatrix} 0 & a & b \\ 0 & 0 & c \\ 0 & 0 & 0 \end{pmatrix} = \begin{pmatrix} 0 & -a & b \\ 0 & 0 & -c \\ 0 & 0 & 0 \end{pmatrix}$, Then, it is straightforward to check that F is a generalized (α, α)-derivation associated with a nonzero (α, α)-derivation d such that $d(Z(R)) \neq (0)$. It is easy to see that (i) $F([x, y]) - \alpha([x, y]) \in Z(R)$ (ii) $F([x, y]) + \alpha([x, y]) \in Z(R)$ (iii) $F(x \circ y) - \alpha(x \circ y) \in Z(R)$ (iv) $F(x \circ y) - \alpha(x \circ y) \in Z(R)$ for all $x, y \in J$, however R is not commutative.

Acknowledgments The author would like to thank the referee for giving helpful comments and suggestions.

References

1. Ali, A., Kumar, D., Miyan, P.: On generalized derivations and commutativity of prime ang semiprime rings. Hacet. J. Math. Stat. **40**(3), 367–374 (2011)
2. Ashraf, M., Ali, A., Ali, S.: On Lie ideals and generalized (θ, ϕ)-derivations in prime rings. Commun. Algebr. **32**(8), 2977–2985 (2004)
3. Ashraf, M., Rehman, N.: On commutativity of rings with derivations. Results Math. **42**, 3–8 (2002)
4. Bell, H.E., Daif, M.N.: On commutativity and strong commutativity-preserving maps. Can. Math. Bull. **37**, 443–447 (1994)
5. Bell, H.E., Daif, M.N.: On derivations and commutativity in prime rings. Acta Math. Hung. **66**(4), 337–343 (1995)
6. Daif, M.N., Bell, H.E.: Remarks on derivations on semiprime rings. Int. J. Math. Math. Sci. **15**, 205–206 (1992)
7. Golbasi, O.: Commutativity of semiprime rings with genearlized derivations. Indian J. Pure Appl. Math. **40**, 191–199 (2009)
8. Marubayashi, H., Ashraf, M., Rehman, N., Ali, S.: On generalized (α, β)-derivations in prime rings. Algebr. Colloq. **17**(1), 865–874 (2010)
9. Mayne, J.H.: Centralizing mappings of prime rings. Can. Math. Bull. **27**, 122–126 (1984)
10. Posner, E.C.: Derivations in prime rings. Proc. Am. Math. Soc. **8**, 1093–1100 (1957)
11. Quadri, M.A., Khan, M.S., Rehman, N.: Generalized derivations and commutativity of prime rings. Indian J. Pure Appl. Math. **34**, 1393–1396 (2003)

Generalized Derivations on Rings and Banach Algebras

Shervin Sahebi and Venus Rahmani

Abstract Let R be a prime ring with Utumi quotient ring U. If R admits a generalized derivation F associated with a derivation d such that $F([x^m y, x]_k)^n - [x^m y, x]_k = 0$ for all $x, y \in R$ where $m \geq 0$ and $n, k \geq 1$ fixed integers, then R is commutative or $n = 1, d = 0$ and F is an identity map. Moreover, we also examine the case R is a semiprime ring. Finally, we apply the above result to noncommutative Banach algebras.

Keywords Prime ring · Semiprime ring · Generalized derivation · Utumi quotient ring · Banach algebra

1991 Mathematics Subject Classification. 16N60 · 16R50 · 16D60

1 Introduction

Let R be an associative ring with center $Z(R)$ and Utumi quotient ring U. The center of U, denoted by C, is called the extended centroid of R (we refer the reader to [1] for these objects). By a Banach algebra we shall mean complex normed algebra A whose underlying vector space is a Banach space. The Jacobson radical rad(A) of A is the intersection of all primitive ideals.

For any $x, y \in R$, we set $[x, y]_1 = [x, y] = xy - yx$, and $[x, y]_k = [[x, y]_{k-1}, y]$, where $k > 1$ is an integer. A linear mapping $d : R \to R$ is called a derivation if satisfies the Leibniz rule $d(xy) = d(x)y + xd(y)$ for all $x, y \in R$. In particular, d is an inner derivation induced by an element $a \in R$, if $d(x) = [a, x]$ for all $x \in R$.

S. Sahebi (✉) · V. Rahmani
Department of Mathematics, Islamic Azad University,
Central Tehran Branch, 13185/768 Tehran, Iran
e-mail: sahebi@iauctb.ac.ir

V. Rahmani
e-mail: venosrahmani@yahoo.com

S.T. Rizvi et al. (eds.), *Algebra and its Applications*, Springer Proceedings in Mathematics & Statistics 174, DOI 10.1007/978-981-10-1651-6_6

A first glance at the above definitions shows that

$$d(x^m) = \sum_{i=0}^{m-1} x^i d(x) x^{m-i-1}; \tag{1}$$

Moreover we have:

$$d[y, x]_k = [d(y), x]_k + \sum_{i=1}^{k} [[[y, x]_{i-1}, d(x)], x]_{k-i}, \tag{2}$$

for all $x, y \in R$ and k, m are fixed positive integers.

In [3], Bresar introduced the definition of generalized derivations. An additive mapping $F : R \to R$ is called a generalized derivation if there exists a derivation $d : R \to R$ such that $F(xy) = F(x)y + xd(y)$ holds for all $x, y \in R$, and d is called the associated derivation of F. Hence, the concept of generalized derivation covers the concept of derivation. In [12], Lee extended the definition of generalized derivation as follows: by a generalized derivation we mean an additive mapping $F : I \to U$ such that $F(xy) = F(x)y + xd(y)$ holds for all $x, y \in I$, where I is a dense left ideal of R and d is a derivation from I into U. Moreover, Lee also proved that every generalized derivation can be uniquely extended to a generalized derivation of U, and thus all generalized derivations of R will be implicitly assumed to be defined on the whole of U. Lee obtained the following: every generalized derivation F on a dense left ideal of R can be uniquely extended to U and assumes the form $F(x) = ax + d(x)$ for some $a \in U$ and a derivation d on U.

Let us introduce the background of our investigation. In [17], Singer and Werner obtained a fundamental result which started investigation into the ranges of derivations on Banach algebras. They proved that any continuous derivation on a commutative Banach algebra has its range in the Jacobson radical of the algebra. A very interesting question is how to obtain noncommutative version of Singer–Werner theorem. In [16] Sinclair obtained a fundamental result which started investigation into the ranges of derivations on a noncommutative Banach algebra. He proved that every continuous derivation of a Banach algebra leaves primitive ideals of the algebra invariant. Meanwhile many authors obtained more information about derivations of Banach algebras satisfying certain suitable conditions. For example, in [14] Park proved that if d is a linear continuous derivation of a noncommutative Banach algebra A such that $[[d(x), x], d(x)] \in \mathrm{rad}(A)$ for all $x \in A$ then $d(A) \subseteq \mathrm{rad}(A)$. In [7], Filippis extended the Park's result to generalized derivations.

Many results in the literature indicate that global structure of a prime ring R is often strongly connected to the behavior of additive mappings defined on R. In [5], Daif and Bell showed that if in a semiprime ring R there exist a nonzero ideal I of R and a derivation d such that $d([x, y]) = [x, y]$ for all $x, y \in I$, then $I \subseteq Z(R)$. At this point, a natural question is what happens in case the derivation is replaced by generalized derivation. In [15], Quadri, Khan and Rehman proved that if R is a prime ring, I a nonzero ideal of R and F a generalized derivation associated with a nonzero derivation d such that $F([x, y]) = [x, y]$ for all $x, y \in I$, then R is commutative.

More recently in [8], Filippis and Huang generalized the above as follows: Let R be a prime ring, I a nonzero ideal of R and n a fixed positive integer; If R admits a generalized derivation F with the property $(F([x, y]))^n = [x, y]$, for all $x, y \in I$, then either R is commutative or $n = 1$, $d = 0$ and F is the identity map on R. Moreover, they study the semiprime case.

The present article is our motivation by the previous results.

The main results of this paper are as follows:

Theorem 1.1 *Let R be a prime ring with extended centroid C, $m \geq 0$ and $n, k \geq 1$ fixed integers. If R admits a generalized derivation F associated with a derivation d such that $F([x^m y, x]_k)^n = [x^m y, x]_k$, then R is commutative or $n = 1$, $d = 0$ and F is identity map on R.*

In particular, we have the following theorem in semiprime case:

Theorem 1.2 *Let R be a semiprime ring with extended centroid C, $m \geq 0$ and $n, k \geq 1$ fixed integers. If R admits a generalized derivation F associated with a derivation d such that $F([x^m y, x]_k)^n = [x^m y, x]_k$, then there exists a central idempotent element e in U such that d vanishes identically on eU and the ring $(1 - e)U$ is commutative.*

Finally, we prove the following result regarding the noncommutative Banach algebra.

Theorem 1.3 *Let A be a noncommutative Banach algebra, $\zeta = L_a + d$ a continuous generalized derivations of A, $m \geq 0$ and $n, k \geq 1$ fixed integers. If $F([x^m y, x]_k)^n - [x^m y, x]_k \in rad(A)$, for all $x, y \in A$, then $d(A) \subseteq rad(A)$.*

2 Proof of the Main Results

The following lemmas are useful tools for the proof of Theorem 1.1.

Lemma 2.1 *Let $R = M_t(F)$, be the ring of all $t \times t$ matrices over a field F with $t \geq 2$, $a, b \in R$, $m \geq 0$ and $n, k \geq 1$ fixed integers. If $(a[x^m y, x]_k + [b, [x^m y, x]_k])^n - [x^m y, x]_k = 0$ for all $x, y \in R$, then $a, b \in F \cdot I_t$.*

Proof Let $a = (a_{ij})_{t\times t}$ and $b = (b_{ij})_{t\times t}$ where $a_{ij}, b_{ij} \in F$. Denote e_{ij} the usual matrix unit with 1 in (i, j)-entry and zero elsewhere. By choosing $x = e_{ii}$, $y = e_{ij}$ for any $i \neq j$, we have $[x^m y, x]_k = [e_{ij}, e_{ii}]_k = (-1)^k e_{ij}$ and hence

$$\begin{aligned} 0 &= (a[x^m y, x]_k + [b, [x^m y, x]_k])^n - [x^m y, x]_k \\ &= (-1)^{kn}((a + b)e_{ij} - e_{ij}b)^n - (-1)^k e_{ij} \end{aligned} \quad (3)$$

Right multiplying by e_{ij}, yields $b_{ji}^n = 0$ for any $i \neq j$. This implies b is diagonal matrix. Let $b = \sum_{i=1}^{t} b_{ii}e_{ii}$. For any F-automorphism θ of R, we have

$$(a^\theta[x^m y, x]_k + [b^\theta, [x^m y, x]_k])^n - [x^m y, x]_k = 0$$

for every $x, y \in R$. Hence b^θ must also be diagonal. We have $(1 + e_{ij})b(1 - e_{ij}) = \sum_{i=1}^{t} b_{ii}e_{ii} + (b_{jj} - b_{ii})e_{ij}$ diagonal. Therefore, $b_{jj} = b_{ii}$ and so $b \in F \cdot I_t$. Similarly, left multiplying (3) by e_{ij}, we can prove that $a + b \in F \cdot I_t$ and hence $a \in F \cdot I_t$ □

Lemma 2.2 *Let R be a prime ring with extended centroid C, and $a, b \in R$. Suppose that $(a[x^m y, x]_k + [b, [x^m y, x]_k])^n - [x^m y, x]_k = 0$ for any $x, y \in R$, where $m \geq 0$ and $n, k \geq 1$ are fixed integers. Then $a, b \in C$.*

Proof By assumption, R satisfies the generalized polynomial identity

$$f(x, y) = (a[x^m y, x]_k + [b, [x^m y, x]_k])^n - [x^m y, x]_k$$

We assume either $a \notin C$ or $b \notin C$. Then $f(x, y) = 0$ is a nontrivial (GPI) for R. By Martindale's Theorem [13], R is then primitive ring having nonzero $\mathrm{soc}(R)$ with C as the associated division ring. Hence by Jacobson's Theorem [9], R is isomorphic to a dense ring of linear transformations of vector space V over C. If $\dim_C V = t$, then $R \cong M_t(C)$. For $t = 1$, R is a commutative. If $t \geq 2$, then by Lemma 2.1, we have a contradiction. Assume next that $\dim_C V = \infty$. Let e and f be two orthogonal idempotent elements of $\mathrm{soc}(R)$. Then $fe = 0$. For $r \in \mathrm{soc}(R)$, we have

$$\begin{aligned} 0 &= (a[e^m(rf), e]_k + [b, [e^m(rf), e]_k)^n - [e^m(rf), e]_k \\ &= (-1)^{kn}((a + b)erf - erfb)^n - (-1)^k erf. \end{aligned}$$

Right multiplying by e yields $(erfb)^n e = 0$, i.e., $(fber)^{n+1} = 0$ for all $r \in \mathrm{soc}(R)$. By [6], $fber = 0$ implying $fbe = 0$. In particular, for any idempotent $e \in \mathrm{soc}(R)$, we have $(1 - e)be = 0 = eb(1 - e)$ that is $[b, e] = 0$. Therefore, $[b, E] = 0$, where E is the additive subgroup generated by all idempotents of $\mathrm{soc}(R)$. Since E is a noncentral Lie ideal of $\mathrm{soc}(R)$, this implies $b \in C$. Similarly, left multiplying by f, we can prove that $a + b \in C$ and hence $a \in C$, a contradiction. □

Now we are ready to prove Theorem 1.1.

Proof of Theorem 1.1. By assumption we get

$$(F([x^m y, x]_k))^n = (F(x^m)[y, x]_k + x^m d([y, x]_k))^n = x^m[y, x]_k. \tag{4}$$

Now since R is a prime ring and F is a generalized derivation of R, by Lee [12, Theorem 3], $F(x) = ax + d(x)$ for some $a \in U$ and a derivation d on U. Let d be the inner derivation induced by an element $b \in U$; that is, $d(x) = [b, x]$ and $d(y) = [b, y]$ for all $x, y \in U$. Thus by hypothesis, we have

$$(a[x^m y, x]_k + [b, [x^m y, x]_k])^n = [x^m y, x]_k.$$

In this case by Lemma 2.2, we have $a, b \in C$. Then $(a[x^m y, x]_k + [b, [x^m y, x]_k])^n - [x^m y, x]_k = 0$ becomes $a^n([x^m y, x]_k)^n - [x^m y, x]_k = 0$ which is polynomial identity for R. Then $R \subseteq M_t(K)$ for some field K and $M_t(K)$ satisfies $a^n([x^m y, x]_k)^n - [x^m y, x]_k = 0$. But by choosing $x = e_{ii}$, $y = e_{ij}$ we get

$$0 = a^n([x^m y, x]_k)^n - [x^m y, x]_k = a^n((-1)^k e_{ij})^n - (-1)^k e_{ij}.$$

If $n \geq 2$, it yields $0 = (-1)^k e_{ij}$, a contradiction. Thus $n = 1$ and then $0 = a^n([x^m y, x]_k)^n - [x^m y, x]_k = (a-1)(-1)^k e_i j$. This implies $a = 1$. Thus we have proved that R is commutative or $n = 1$, $d = 0$ and F is identity map on R. On the other hand, if d is outer derivation, then hypothesis, (1) and (2) gives

$$\begin{aligned}\{(a[x^m y, x]_k + [\textstyle\sum_{i=0}^{m-1} x^i d(x) x^{m-i-1} y + x^m d(y), x]_k \\ + \textstyle\sum_{i=1}^{k-1}[[[x^m y, x]_i, d(x)], x]_{k-i-1})^n = [x^m y, x]_k\}\end{aligned} \tag{5}$$

for all $x, y \in U$. Now by Kharchenko's Theorem [11],

$$\begin{aligned}\{(a[x^m y, x]_k + [\textstyle\sum_{i=0}^{m-1} x^i z x^{m-i-1} y + x^m u, x]_k \\ + \textstyle\sum_{i=1}^{k-1}[[[x^m y, x]_i, z], x]_{k-i-1})^n = [x^m y, x]_k\}\end{aligned}$$

for all $x, y, z, u \in U$. If U is noncommutative, then there exists $b' \in U$ such that $b' \notin C$. Now replacing z with $[b', x]$ and u with $[b', y]$, above identity becomes

$$(a[x^m y, x]_k + [b', [x^m y, x]_k])^n = [x^m y, x]_k$$

Then by above inner derivation case, it yields that $b' \in C$, a contradiction.

In particular of Theorem 1.1, for $m = 0$ and $k = 1$ we get the result of Fillipis [8].

Corollary 2.3 *Let R be a prime ring with extended centroid C and $n \geq 1$ fixed positive integer. If R admits a generalized derivation F associated with a derivation d such that $F([y, x])^n = [y, x]$, then R is commutative or $n = 1$, $d = 0$ and F is identity map on R.*

Now we are ready to prove the semiprime case.

The following result from [2], is a useful tool needed in the proof of the Theorem 1.2.

Lemma 2.4 *Let R be a semiprime ring and M a maximal ideal of C. Then MU is a prime ideal of U invariant under all derivations of U. Moreover, we have $\cap\{M | MU$ is maximal ideal of $C\} = 0$*

Proof of Theorem 1.2. Since R is semiprime and F is a generalized derivation of R we have $F(x) = ax + d(x)$ for some $a \in U$ and a derivation d on U [12, Theorem 3]. Now by the same argument in the proof of Theorem 1.1, for all $x, y \in U$, we have

$$\{(a[x^m y, x]_k + [\sum_{i=0}^{m-1} x^i d(x) x^{m-i-1} y + x^m d(y), x]_k$$

$$+ \sum_{i=1}^{k-1} [[[x^m y, x]_i, d(x)], x]_{k-i-1})^n = [x^m y, x]_k\}$$

Let M be any maximal ideal of C. Since U is a B-algebra orthogonal complete [4] and by Lemma 2.4, MU is a prime ideal of U invariant under d. Denote $\overline{U} = U/MU$ and $\bar{d}$ the derivation induced by d on $\overline{U}$, i.e., $\bar{d}(\bar{x}) = \overline{d(x)}$ for all $x \in U$. Therefore we get

$$\{(a[\overline{x}^m \overline{y}, \overline{x}]_k + [\sum_{i=0}^{m-1} \overline{x}^i d(\overline{x}) \overline{x}^{m-i-1} \overline{y} + \overline{x}^m d(\overline{y}), \overline{x}]_k$$

$$+ \sum_{i=1}^{k-1} [[[\overline{x}^m \overline{y}, \overline{x}]_i, d(\overline{x})], \overline{x}]_{k-i-1})^n = [\overline{x}^m \overline{y}, \overline{x}]_k\}$$

for all $\overline{x}, \overline{y} \in \overline{U}$. It is clear that $\overline{U}$ is prime. Therefore by Corollary 2.3, either $\overline{U}$ is commutative or $\bar{d} = 0$, that is either $d(U) \subseteq MU$ or $[U, U] \subseteq MU$. In light of previous argument, we have that both $d(U)[U, U] \subseteq MU$, where MU runs over all prime ideals of U. By Lemma 2.4, $\cap MU = 0$. Thus we get $d(U)[U, U] = 0$. Using the theory of orthogonal completion for semiprime rings (see [1, Chap. 3]), it follows that there exists a central idempotent element e in U such that on the direct sum decomposition $R = eU \oplus (1 - e)U$, d vanishes identically on eU and the ring $(1 - e)U$ is commutative.

Here A will denote a complex noncommutative Banach algebras. Our final result in this paper is about continuous generalized derivations on noncommutative Banach algebras.

The following results are useful tools needed in the proof of Theorem 1.3.

Remark 2.5 (see [16]*).* Any continuous derivation of Banach algebra leaves the primitive ideals invariant.

Remark 2.6 (see [17]*).* Any continuous linear derivation on a commutative Banach algebra maps the algebra into its radical.

Remark 2.7 (see [10]*).* Any linear derivation on semisimple Banach algebra is continuous.

Proof of Theorem 1.3. By the hypothesis ζ is continuous and since L_a, the left multiplication by some element $a \in A$, also continuous thus we have that the derivation d is continuous. By Remark 2.5, for any primitive ideal P of A we have $\zeta(P) \subseteq aP + d(P) \subseteq P$. It means that the continuous generalized derivation ζ leaves the primitive ideal invariant. Denote $\bar{A} = A/P$ for any primitive ideals P. Thus we can define the generalized derivations $\zeta_P : \bar{A} \to \bar{A}$ by $\zeta_P(\bar{x}) = \zeta_P(x + P) =$

$\zeta(x) + P$ for all $\bar{x} \in \bar{A}$, where $A/P = \bar{A}$. Since P is a primitive ideal, $\bar{A}$ is primitive and so it is prime. The hypothesis $F([x^m y, x]_k)^n - [x^m y, x]_k \in \text{rad}(A)$ yields that $F([\bar{x}^m \bar{y}, \bar{x}]_k)^n - [\bar{x}^m \bar{y}, \bar{x}]_k = \bar{0}$ for all $\bar{x}, \bar{y} \in \bar{A}$. Now by Corollary 2.3, it is immediate that either $\bar{A}$ is commutative or $d = \bar{0}$. That is $[A, A] \subseteq P$ or $d(A) \subseteq P$. Now we assume that P is a primitive ideal such that $\bar{A}$ is commutative. By Remarks 2.6 and 2.7, we know that there are no nonzero linear continuous derivations on commutative semisimple Banach algebras. Therefore, $d = \bar{0}$ in $\bar{A}$. Hence in any case we get $d(A) \subseteq P$ for all primitive ideal P of A. Thus we get $d(A) \subseteq \text{rad}(A)$, and we arrive the required conclusion.

Acknowledgments This paper is supported by Islamic Azad University Central Tehran Branch (IAUCTB). The authors want to thank the authority of IAUCTB for their support to complete this research. Also, the authors would like to thank the referee whose fruitful comments have improved this article.

References

1. Beidar, K.I., Martindale III, W.S., Mikhalev, A.V.: Rings with Generalized Identities, Monographs and Textbooks in Pure and Applied Math, vol. 196. Marcel Dekker Inc., New York (1996)
2. Beidar, K.I.: Rings of quotients of semiprime rings. Vestnik Moskovskogo Universiteta. **33**(5), 36–43 (1978)
3. Bresar, M.: On the distance of the composition of two derivations to be the generalized derivations. Glasg. Math. J. **33**, 89–93 (1991)
4. Chung, C.L.: GPIs having coefficients in Utumi quotient rings, proc. Am. Math. soc. **103**, 723–728 (1988)
5. Daif, M.N., Bell, H.E.: Remarks on derivations on semiprime rings. Int. J. Math. Math. Sci. **15**(1), 205–206 (1992)
6. Felzenszwalb, B.: On a result of Levitzki canad Math. Bull **21**, 241–242 (1978)
7. Filippis, V.: De., Generalized derivations on prime rings and noncommutative Banach algebras. Bull. Korean Math. Soc. **45**, 621–629 (2008)
8. Filippis, V., De Huang, S.: Generalized derivations on semiprime rings. Bull. Korean Math. Soc. **48**(6), 1253–1259 (2011)
9. Jacobson, N.: Structure of rings. Am. Math. Soc. Colloq. Pub. 37 (1964)
10. Jacobson, B.E., Sinclair, A.M.: Continuity of derivations and problem of kaplansky. Am. J. Math. **90**, 1067–1073 (1968)
11. Kharchenko, V.K.: Differential identity of prime rings. Algebr. Log. **17**, 155–168 (1978)
12. Lee, T.K.: Generalized derivations of left faithful rings. Commun. Algebr. **27**(8), 4057–4073 (1998)
13. Martindale III, W.S.: Prime rings satistying a generalized polynomial identity. J. Algebr. **12**, 576–584 (1969)
14. Park, K.H.: On derivations in non commutative semiprime rings and Banach algebras. Bull. Korean Math. Soc. **42**, 671–678 (2005)
15. Quadri, M.A., Khan, M.S., Rehman, N.: Generalized derivations and commutativity of prime rings. Indian J. Pure Appl. Math. **34**(98), 1393–1396 (2003)
16. Sinclair, A.M.: Continuous derivations on Banach algebras. Proc. Am. Math. Soc. **20**, 166–170 (1969)
17. Singer, I.M., Werner, J.: Derivations on commutative normed algebras. Math. Ann. **129**, 260–264 (1955)

A Study of Suslin Matrices: Their Properties and Uses

Ravi A. Rao and Selby Jose

Abstract We describe recent developments in the study of unimodular rows over a commutative ring by studying the associated group $SUm_r(R)$, generated by Suslin matrices associated to a pair of rows v, w with $\langle v, w\rangle = 1$. We also sketch some futuristic developments which we expect on how this association will help to solve a long standing conjecture of Bass–Suslin (initially in the metastable range, and later the entire expectation) regarding the completion of unimodular polynomial rows over a local ring, as well as how this study will lead to understanding the geometry and physics of the orbit space of unimodular rows under the action of the elementary subgroup.

Keywords Unimodular rows · Orthogonal transformations · Reflections

2000 Mathematics Subject Classification 13C10 · 15A63 · 19A13 · 19B14

1 Introduction

We begin by recapitulating the birth and early use of the Suslin matrices. The genesis is in the beautiful Sect. 5 of Suslin's paper [56]. He has said so much, with such fluency and consummate ease; it begets an area of mathematics rich in its connections with the rest of mathematics. The title of Sect. 5 'A procedure for constructing invertible matrices' is most intriguing. This section is also astounding in another sense; it is the first instance we know where Suslin has penned a flow of thoughts without much

---

R.A. Rao (✉) · S. Jose
School of Mathematics, Tata Institute of Fundamental Research,
1, Dr. Homi Bhabha Road, Mumbai 400005, India
e-mail: ravi@math.tifr.res.in

S. Jose
e-mail: selbyjose@gmail.com

R.A. Rao · S. Jose
Department of Mathematics, Institute of Science,
Madame Cama Road, Mumbai 400032, India

S.T. Rizvi et al. (eds.), *Algebra and its Applications*, Springer Proceedings
in Mathematics & Statistics 174, DOI 10.1007/978-981-10-1651-6_7

elaboration; as was his normal style. Naturally, it behoves his admirers to unearth the encrypted wisdom stored in it.

We intersperse this history with our own rambling thoughts of some of our immediate expectations. (A computer-algebra aided study, (especially wise with (perhaps) use of sparse matrices), will be helpful to ease some of our mendications.) We are prejudiced in choosing outlets which we feel will lead to a solution of two of the central problems in classical algebraic K-theory; both are questions regarding finding a procedure to complete a unimodular row to an invertible matrix, one of length d over a d-dimensional affine algebra over an algebraically closed field (posed by Suslin), and the other of a unimodular polynomial row of any length over a local ring (posed by Bass–Suslin). We have made some progress in these directions, using the compressed Suslin matrices, and we refer the reader to [19] for the first problem, and [42, 43] for the second one. But the reader will feel the stirrings that the subject of the study of unimodular rows will soon evolve far beyond the range of these important classical problems.

We proceed to detail the association of a composition of two reflections $\tau_{(v,w)} \circ \tau_{(e_1,e_1)}$ with a pair of rows v, w with $\langle v, w\rangle = 1$. This association enables one to study the orbit space of unimodular rows under elementary action. Moreover since $\tau_{(v,w)} \circ \tau_{(e_1,e_1)}$ is an orthogonal transformation, one gets a homomorphism from $SUm_r(R)$, the subgroup of the linear group generated by the Suslin matrices, to the special orthogonal group $SO_{2(r+1)}(R)$; which is a well studied object. This allows us to pull back useful information in the study of unimodular rows.

The group $SUm_r(R)$ has properties resembling those of classical spinor groups; and we feel that the further study of this group will lead to a better understanding of the geometry and physics of the orbit space of unimodular rows under the action of the elementary subgroup.

2 The Suslin Matrices

Given two rows $v, w \in M_{1,r+1}(R), r \geq 1$, in [56, Sect. 5] Suslin associates with them a matrix $S_r(v, w) \in M_{2^r}(R)$ of determinant $\langle v, w\rangle^{2^{r-1}} = (v \cdot w^t),^{2^{r-1}}$ whose entries are from the coordinates of v, w upto a sign. We call these the Suslin matrix w.r.t. v, w. They are particularly interesting to us when they are in $\mathrm{SL}_{2^r}(R)$, i.e. when $\langle v, w\rangle = v \cdot w^t$ is 1. The explicit construction of the Suslin matrix is defered for the moment.

Trimurthi of Suslin Matrices

So far the Suslin matrix has manifested in at least three different contexts.

- Establishing that the unimodular row $(a_0, a_1, a_2^2, \ldots, a_r^r)$ can be completed to an invertible matrix. See the seminal paper of Suslin [56]; especially Theorem 2, Proposition 1.6 and the beautiful Sect. 5.

- From studying the Koszul complex associated to a unimodular row. See [61, Sect. 2], especially Proposition 2.2, Corollary 2.5.
- As orthogonal transformations on a certain space. See [28, Corollary 4.2].

More Recent Developments

Two recent developments are briefly mentioned here. The reader should refer to the cited texts for notations which have not been explained here.

The Fundamental property of Suslin matrices in [27] led the referee to suspect a link between Suslin matrices and Spin groups. This connection was established in the thesis of Vineeth Chintala and appears in [15]. We sketch some of his ideas next.

For a commutative ring R, the hyperbolic space $H(R^n)$ is the module $R^n \times R^n$ endowed with a quadratic form q such that $q(v, w) = \langle v, w\rangle = v \cdot w^t$. To this structure one can associate the Clifford algebra $Cl_n(R)$ of the quadratic form, which is isomorphic to the matrix ring $M_{2^n}(R)$. Vineeth Chintala proved in [15] that the map $\varphi : H(R^n) \mapsto M_{2^n}(R)$ given by

$$\varphi(v, w) = \begin{pmatrix} 0 & S_{n-1}(v, w) \\ S_{n-1}(w, v)^t & 0 \end{pmatrix}$$

induces a R-algebra isomorphism. One can then derive the Jose–Rao fundamental property of Suslin matrices from this.

The map $S_{n-1}(v, w) \mapsto S_{n-1}(w, v)^t$ can be used to construct an involution $x \mapsto x^*$ on $Cl_n(R) = Cl_0(R) \oplus Cl_2(R)$. One defines the Spin groups

$$Spin_{2n}(R) = \{x \in Cl_0(R) \mid xx^* = 1 \text{ and } xH(R^n)x^{-1} = H(R^n)\}.$$

This involution on $Cl_n(R)$ corresponds to the standard involution on $M_{2^n}(R)$. One can define the groups

$$G_{n-1}(R) = \{g \in GL_{2^{n-1}} \mid gSg^* \text{is a Suslin matrix, for all Suslin matrices S}\}.$$

The subgroup of $G_{n-1}(R)$ consisting of those which preserve the quadratic form on $H(R^n)$ is denoted by $SG_{n-1}(R)$. Vineeth Chintala proves that there is an isomorphism $Spin_{2n}(R) \simeq SG_{n-1}(R)$.

The subgroup generated by the Suslin matrices is thus the rational points of a certain Spinor group.

The second new approach to Suslin matrices occurs in the work of Aravind Asok and Jean Fasel in [2]. Here there is an edge map interpretation for any regular algebra (with which 2 is invertible) in terms of Suslin matrices. We shall say a bit more about this later; but refer to [2] for more details of this approach.

Use of Suslin Matrices

The Suslin matrices have proved useful in several contexts. The main application of Suslin matrices, so far, have been in the following directions:

- A unimodular row of the form $(a_0, a_1, a_2^2, \ldots, a_r^r)$ can be completed to a matrix $\beta_r(v, w)$, with $v = (a_0, a_1, a_2, \ldots, a_r)$, and w any row with $\langle v, w\rangle = 1$, of determinant one. (We may also just write this as $\beta_r(v)$ for brevity.)

Suslin mentions in [56, Sect. 5] that a completion can be got by doing a series of row and column operations on the matrix $S_r(v, w)$ to reduce it to size $(r + 1)$. However, an explicit process (as suggested by Suslin, based on the sparseness of the Suslin matrix) is **far from clear**, even in small sizes. A different reasoning justifes this in [61, Sect. 2]. Undoubtedly, [56, Proposition 1.6] also gives a neat way of writing a completion, and also ties up with the Suslin matrix.

It would be both nice and useful if a good algorithm can be developed to get a $\beta_r(v, w)$ from a $S_r(v, w)$. We believe that an appropriate $\beta_r(v, w)$ will replicate the role played by $S_r(v, w)$. The actual use of a "nice" (and explicit) $\beta_2(v, w)$ can be seen in the works [44, Lemmas 2 and 3], [52, Sect. 5].

Note that it is unclear, and probably unjustified, to expect that any two $\beta_r(v, w)$ got from a $S_r(v, w)$ are equivalent in $E_{r+1}(R)$. Indeed, there seem to be completions β of (a^2, b, c) which may not arise from a $S_2(v, w)$: The first completion of a unimodular row of the form (a^2, b, c) comes from the theory of cancellation of projective modules in the paper [66] of Swan–Towber where an explicit completion is stated in [66, Theorem 2.1]. Here are the two completions: Let $aa' + bb' + cc' = 1$.

$$\begin{pmatrix} a^2 & b & c \\ b + ac' & -c'^2 + ba'c' & -a' + b'c' - a'bb' \\ c - ab' & a' + b'c' + a'cc' & -b'^2 - a'b'c \end{pmatrix}, \quad \begin{pmatrix} a^2 & b & c \\ -b - 2ac' & c'^2 & a' - b'c' \\ -c + 2ab' & -a' - b'c' & b'^2 \end{pmatrix}.$$

Can the Swan–Towber method of computation be extended to give completions of the universal factorial row, in view of Suslin's theorem in [56]? Is there some interpretation of those completions akin to the theory which Suslin has built? (Note that both approaches are derived from an explicit computation to show the transitivity of the group of automorphisms of a projective module $P \oplus R$ on its unimodular elements.)

Let us commence on a different tack. Bass observed that the projective module $P_v = ker(R^{2n} \xrightarrow{v} R)$ corresponding to a unimodular row $v = (v_1, v_2, \ldots, v_{2n})$ of even length always has a unimodular element, i.e. it splits of a free summand isomorphic to R: $w = (v_2, -v_1, v_4, -v_3, \ldots, -v_{2n}, v_{2n-1}) \in P_v$ and is a unimodular row.

Raja Sridharan and Ravi Rao observed that if $\chi_2(v) = (v_1^2, v_2, \ldots, v_{2n-1}) \in Um_{2n-1}(R)$ then the projective module $P_{\chi_2(v)}$ has a unimodular element. (See [36, p. 120, Theorem 5.6] for a more general statement).

S.M. Bhatwadekar commented on seeing this that a unimodular row of the form $(a_0^2, a_1, a_2, a_3^2, a_4, a_5)$ has two independent sections! T.Y. Lam (with inputs from R.G. Swan) also began the study of **Sectionable sequences** in [36, Sect. 5, p. 116] to make a preliminary study of this phenomenon.

Can one recover Sus lin's theorem on the completion of the 'universal factorial unimodular row' by using such an argument? In particular, to begin with, can one show that a unimodular row of the form $(a_0^6, a_1, \ldots, a_{2n-1})$ has two independent sections? etc.

- Suslin used it **in the computation of K-theory and $\mathcal{K}$-cohomology of group varieties** SL_n, GL_n, Sp_{2n}, etc. in [62]. We refer the reader to [62] where Suslin showed that
$$SK_1\left(\frac{\mathbb{Z}[x_1, \ldots, x_n, y_1, \ldots, y_n]}{(\sum_{i=1}^n x_i y_i - 1)}\right) \simeq \mathbb{Z},$$
with generator $[S_{n-1}((x_1, \ldots, x_n), (y_1, \ldots, y_n))]$. Is the group $SL_n(A)/E_n(A)$, for $A = \mathbb{Z}[x_1, \ldots, x_n, y_1, \ldots, y_n]/(\sum_{i=1}^n x_i y_i - 1)$, generated by $[\beta_{n-1}(v, w)]$, for $v, w \in Um_n(A)$, with $\langle v, w\rangle = 1$? (This may depend on n, but is it true atleast in the metastable range $n \leq 2d - 3$, where d is dimension of A?)

- **Patching information in set-theoretic complete intersection problems.**

  M. Boratyński showed in [11] that an ideal I in a polynomial ring R over a field can be generated upto radical by $m = \mu(I/I^2)$ elements, i.e. $\sqrt{I} = \sqrt{(f_1, \ldots, f_m)}$, for some $f_1, \ldots, f_m \in R$.

This is the first recorded use of the matrices $\beta_r(v, w)$ in the subject of Serre's program, followed by the Eisenbud–Evans program, which bridges properties of projective modules over a ring and the efficient generation of ideals in that ring. It replaces the homological methods used by Serre, and later by others like N. Mohan Kumar, M.P. Murthy in this context. The book [25] gives a nice introduction and survey of major previous literature on this topic.

Let us quickly recall M. Boratyński's idea: He says that if $\{x_1, \ldots, x_m\} \subset I$ with $\{\overline{x}_1, \ldots, \overline{x}_m\}$ generating the R/I-module I/I^2, and if J is the ideal generated by $(x_1, x_2, x_3^2, \ldots, x_m^{m-1})$, and $I^{(m-1)!}$, then $\sqrt{J} = \sqrt{I}$, and the projective R- module got by taking the fibre product
$$P = R_t^m \times_{\beta_{m-1}((x_1, \ldots, x_m))} R_{1-t}^m$$
maps onto J, for any $t \in R$ with $(1 - t)I \subset (x_1, \ldots, x_m)$. (Such a t is readily found, and the fact that J is locally generated by the obvious m elements on the open set $D(1 - t)$, and by one element on $D(t)$, is easily verified. This information is 'patched' via $\beta_{m-1}((x_1, \ldots, x_m))$.

By the Quillen–Suslin theorem [40, 53] P is free, and so J is generated by m elements.

Thus, M. Boratyński encoded Quillen's idea of local patching to ideals, and pushed forward Serre's program of projective generation of ideals; via a compressed version of a Suslin matrix.

- **Defining higher Mennicke symbols on orbits of unimodular rows.**

R. Fossum, H. Foxby, B. Iversen defined, for $n \geq 2$, a Mennicke n-symbol $Um_n(R) \overset{\text{wt}}{\to} SK_1 R$ using the theory of acyclic based complexes. (We refer the reader to [20]; a copy of which can be got by making a request.)

Let $v = (a_1, \ldots, a_n)$, $w = (b_1, \ldots, b_n) \in Um_n(R)$, with $\langle v, w\rangle = v \cdot w^t = 1$. The Koszul complex

$$X(v) = (\ldots \to \wedge^k(R^n) \overset{d_v}{\to} \wedge^{k-1}(R^n) \to \ldots)$$

is an acyclic based complex, with each $X_k(v) = \wedge^k(R^n)$ a free module with a canonical basis of exterior products $e_{i_1} \wedge \ldots \wedge e_{i_k}$, ordered lexicographically. External multiplication by w defines a contraction, say β for $X(v)$.

Since $(d+\beta)^2 = 1 + \beta^2$, and β is nilpotent, we get an isomorphism, independent of choice of the contraction,

$$X(v)_{odd} = \oplus X_{2i-1}(v) \longrightarrow \oplus X_{2i}(v) = X(v)_{even}.$$

$$\mathrm{wt}(v) = (-1)^{\binom{2n-1}{n}}[d+\beta] \in SK_1(R)$$

Suslin interprets this map in [61, Sect. 2] and showed that

$$\mathrm{wt}(v) = [S_{n-1}(v, w)] \in SK_1(R).$$

(The reader may consult [46] where details are worked out.)

- **Dual is not isomorphic**: Let $\sum_{i=1}^n x_i y_i = 1$. Let P be the projective module corresponding to the unimodular rows $(x_1, \ldots, x_n)$. Then the dual P^* of P, i.e. $\mathrm{Hom}_R(P, R)$, is isomorphic to the projective R-module corresponding to the unimodular row $w = (\overline{y}_1, \ldots, \overline{y}_n) = w$.

It can be seen easily that P and P^* are isomorphic when rank P is odd; in fact, the rows v, w are in the same elementary orbit by a lemma of M. Roitman in [51, Lemma 1].

However, if $n > 1$ is odd then there are several approaches due to M.V. Nori, R.G. Swan, who have independently shown (using topological arguments) that P, P^* are not isomorphic. For an exposition of this see the homepage of R.G. Swan at [64, 65].

Together with these approaches, we gave an approach via Suslin matrices following an argument of Suslin in [61]. We refer the reader to [39] where some of the approaches are collated. We mention the approach via Suslin matrices below: Let

$$R = \frac{\mathbb{Z}[x_1, \ldots, x_{2n-1}, y_1, \ldots, y_{2n-1}]}{\left(\sum_{i=1}^{2n-1} x_i y_i - 1\right)}.$$

Suppose that $v\sigma = w$, for some $\sigma \in GL_{2n-1}(R)$. Then

$$\mathrm{wt}\,(w) = \mathrm{wt}\,(v\sigma) = \mathrm{wt}\,(v) + \sum_{i=0}^{2n-1} (-1)^i[\wedge^i \sigma].$$

Since $SK_1(R) = \mathbb{Z}$, $[\sigma] = [S_{2n-2}(v, w)]^r$, for some r. Hence, $[\wedge^i \sigma] = r[\wedge^i S_{2n-2}(v, w)]$. Therefore,

$$\begin{aligned}\sum_{i=0}^{2n-1} (-1)^i [\wedge^i \sigma] &= r \sum_{i=0}^{2n-1} (-1)^i [\wedge^i S_{2n-2}(v, w)] \\ &= r\ \mathrm{wt}(\overline{x}_1, \overline{x}_2, \overline{x}_3^{-2}, \ldots, \overline{x}_{2n-1}^{2n-2}) \\ &= r(2n-2)!\ \mathrm{wt}(v).\end{aligned}$$

Thus,

$$\mathrm{wt}(w) = [S_{2n-2}(w, v)] = (1 + r(2n-2)!)\mathrm{wt}(v) = (1 + r(2n-2)!)[S_{2n-2}(v, w)].$$

But since v is of odd length, $[S_{2n-2}(w, v)] = [S_{2n-2}(w, v)^t] = [S_{2n-2}(w, v)]^t$, by the identities of Suslin (detailed a little later), and using the nomality of the elementary linear subgroup (see [55, Corollary 1.4]). But $S_{2n-2}(v, w) S_{2n-2}(w, v)^t = I$, and so $[S_{2n-2}(v, w)] = [S_{2n-2}(w, v)]^{-1}$.

Thus, one gets $(2 + r(2n-2)!)\mathrm{wt}\ (v) = 0$. A contradiction except when $n = 2, r = -1$. □

- The Suslin matrices can be used **to derive properties of the orbit space of unimodular rows.** Consider the following two principles:
  - ⋆ (Generalized Local Global Principle): Let $v(X), w(X) \in Um_r(R[X])$, $r \geq 3$. Suppose that $v(X)_\mathfrak{p} \in w(X)_\mathfrak{p} E_r(R_\mathfrak{p}[X])$, for all $\mathfrak{p} \in Spec(R)$, and $v(0) = w(0)$, then is $v(X) \in w(X) E_r(R[X])$?
  - ⋆ (Generalized Monic Inversion Principle): Let $v(X), w(X) \in Um_r(R[X])$, $r \geq 3$. Let $f(X) \in R[X]$ be a monic polynomial. Suppose that $v(X)_{f(X)} \in w(X)_{f(X)} E_r(R[X]_{f(X)})$, then is $v(X) \in w(X) E_r(R[X])$?

  Both the above questions were also raised by T.Y. Lam in [36, Chap. VIII, 5.6, 5.11]. We gave a partial answer in [46] where we showed that $\chi_2([v(X)]) = \chi_2([w(X)])$, if r is odd, and $\chi_4([v(X)]) = \chi_4([w(X)])$, if r is even. (Here if $v = (v_1, \ldots, v_r) \in Um_r(R)$ then $\chi_n([v])$ denotes the class of the row $(v_1^n, \ldots, v_r)$ (under elementary column operations. This is shown to be well defined in [74] by L.N. Vaserstein.)
- The Suslin matrices have thus been found useful for the study of unimodular rows; which are associated to 1-stably free projective modules. Can such a similar study also be done for any stably free projective module.

  It is natural to expect that an analogous Suslin theory will develop for a pair $(p, a) \in P \oplus R$, $(\psi, b) \in P^* \oplus R$, with $\psi(p) + ab = 1$.
- Suslin studied **the transitive action of the orthogonal group on rows of length one** in [56, Lemma 5.4]. The very existence of $S_3(v, w)$ implies that $O_8(R)$ acts transitively on the set of rows of length one, i.e. $\{(v, w), v, w \in Um_4(R), \langle v, w \rangle = 1\}$. In [29, Corollary 4.5] we showed that $SO_{2n}(R)$ acts transitively on pairs having the further property that

$$[v] = \begin{cases} \chi_2([v']) & \text{if } n \text{ is odd} \\ \chi_4([v']) & \text{if } n \text{ is even} \end{cases}$$

Consequently, in view of Lemma 25 which comes a little later, if R is an affine algebra of dimension d over a perfect C_1 field, or if $R = A[X]$, A a local ring in which 2 is invertible, in view of [42, Theorem 1], then $SO_{2(d+1)}(R)$ acts transitively on rows of length one.

- Bass–Milnor–Serre began the study of the stabilization for the linear group $GL_n(R)/E_n(R)$ for $n \geq 3$, where R is a commutative ring with identity. In [8], they showed that $K_1(R) = GL_{d+3}(R)/E_{d+3}(R)$, where d is the dimension of the maximum spectrum. In [70], L.N. Vaserstein proved their conjectured bound of $(d+2)$ for an associative ring with identity, where d is the stable dimension of the ring. After that, in [71], he introduced the orthogonal and the unitary K_1-functors, and obtained stabilization theorems for them. He showed that the natural map

$$\begin{cases} \varphi_{n,n+1} : \frac{S(n,R)}{E(n,R)} \longrightarrow \frac{S(n+1,R)}{E(n+1,R)} & \text{in the linear case} \\ \varphi_{n,n+2} : \frac{S(n,R)}{E(n,R)} \longrightarrow \frac{S(n+2,R)}{E(n+2,R)} & \text{otherwise} \end{cases}$$

(where $S(n, R)$ is the group of automorphisms of the projective, symplectic and orthogonal modules of rank n with determinant 1, and $E(n, R)$ is the elementary subgroup in the respective cases) is surjective for $n \geq d+1$ in the linear case, for $n \geq d$ in the symplectic case, and for $n \geq 2d+2$ in the orthogonal case, and is injective for $n \geq 2d+4$ in the symplectic and the orthogonal cases. Soon after, in [73], he studied stabilization for groups of automorphisms of modules over rings and modules with quadratic forms over rings with involution, and obtained similar stabilization results.

The Suslin matrices have been **found useful in the study of injective stabilization for the K_1-functor of the classical groups**:

Let A be a nonsingular affine algebra of dimension $d > 1$ over a perfect C_1-field. In [49] it is shown that the natural map $\frac{SL_n(A)}{E_n(A)} \longrightarrow \frac{SL_{n+1}(A)}{E_{n+1}(A)}$ is injective for $n \geq d+1$. In [9] it is shown that if $(d+1)!A = A$, then the natural map $\frac{Sp_n(A)}{E_n(A)} \longrightarrow \frac{Sp_{n+2}(A)}{ESp_{n+2}(A)}$ is injective for $n \geq d+1$. Similar results have also been obtained in the case of the classical modules in [9]. The completion of the universal factorial row, and H. Lindel–T. Vorst results in [37, 75] on the Bass–Quillen conjecture, played a crucial role in proving these results.

In the symplectic situation, in [12] these results have been simplified to some extent using a relative version of Quillen's Local Global Principle in [1], coupled with the Suslin completions of the factorial row. It is shown in [12] that $vE_{2n}(R, I) = vESp_{2n}(R, I)$, for any commutative ring R, and ideal I in R, and for any unimodular row $v \in Um_n(R, I)$, $n \geq 3$. Using this one can recapture the earlier results; and also show that if R be a finitely generated algebra of even dimension d over K, where $K = \mathbb{Z}$ or a finite field or its algebraic closure, and if $\sigma \in Sp_d(R)$ with $(I_2 \perp \sigma) \in ESp_{d+2}(R)$, then σ is (symplectic) homotopic to the identity. In fact, $\sigma = \rho(1)$ for some $\rho(X) \in Sp_d(R[X]) \cap ESp_{d+2}(R[X])$, with $\rho(0) = I_d$. Finally, all these results were improved in [22]; and optimal bounds were obtained there for smooth algebras over an algebraically closed field by using the Fasel–Rao–Swan theorem in [19]. Results of such type are also expected over a perfect field of cohomological dimension ≤ 1; but not over fields of cohomological dimension two, is demonstrated in [22], in view of N. Mohan Kumar's examples in [38] of non-free stably free modules of rank $d - 1$ over a field of cohomological dimension 1.

The relative strengthening of L.N. Vaserstein's famous lemma (in [54]) that $e_1 E_{2n}(R) = e_1 ESp_{2n}(R)$ done in [12] can also be deduced from it and the Excision theorem of W. van der Kallen in [67, Theorem 3.21], via the Key lemma for Suslin matrices. In fact, one can even get the stronger $vE_{2n}(R) = vESp_{\varphi_{2n}}(R)$, for any unimodular row $v \in Um_{2n}(R)$, and any invertible alternating matrix φ, for an appropriate definition of $ESp_{\varphi_{2n}}(R)$. It is an instructive exercise for the reader to figure this out using the material in this text.

The study of injective stabilization is useful to answer a question of Suslin in [59] regarding whether a stably free projective module of rank $(d - 1)$ over a (nonsingular) affine algebra of dimension d over an algebraically closed field, with some divisibility conditions, is free. This will be true for even dimensions if the injective stability estimate for K_1Sp falls to $d - 1$, over odd dimensional (nonsingular) affine algebras of dimension d over a perfect C_1-field. This will be true in any dimension if the injective stability for K_1 will fall to d over a d dimensional (nonsingular) affine algebras over a perfect C_1-field.

The latter was established in [19]; but as a *consequence* of establishing Suslin's question for nonsingular affine algebras over an algebraically closed field. (The contracted Suslin matrices played a vital role in its proof.)

- **The Suslin symbol**: In [56, Sect. 5] introduced the groups $G_r(A)$. $G_r(A)$ is the Witt group of nonsingular quadratic forms if $r \equiv 0 \bmod 4$; $G_r(A)$ is the symplectic K_1 functor of the ring A if $r \equiv 1 \bmod 4$; $G_r(A)$ is the Witt group of nonsingular skew-symmetric forms if $r \equiv 2 \bmod 4$; $G_r(A)$ is the orthogonal K_1 functor of the ring A if $r \equiv 3 \bmod 4$.

  One has the Suslin maps $S_r : Um_{r+1}(A) \longrightarrow G_r(A)$ defined as follows: Choose a w such that $\langle v, w\rangle = 1$, and set

$$S_r(v) = \begin{cases} [S_r(v, w)] \text{ if } r = 2k+1 \\ [S_r(v, w) \cdot I_r] \text{ if } r = 2k. \end{cases}$$

  For example, if $r = 1$ then the resulting map S_1 is precisely the well-known Mennicke symbol which had an important role in the solution of the congruence subgroup problem in [8]; for $r = 2$ S_2 is the Vaserstein symbol introduced in [54], and which was used to obtain some deep results on orbits of actions of $SL_3(A)$ on $Um_3(A)$. Suslin has asked for the meaning and properties of these maps. Our work in [29] was an initial attempt to understanding these maps and see if we could get some properties. We mention some progress on these questions below.

- **Hermitian K-theory**: One can reinterpret the groups $G_r(A)$ in the context of Hermitian K-theory as developed by M. Karoubi and, more recently, M. Schlichting. In [2], the authors show that these groups are avatars of higher Grothendieck–Witt groups. As said above, we have $G_1(A) = KSp_1(A)$ and $G_3(A) = KO_1(A)$. In Schlichting's notation, one writes $KSp_1(A) = GW_1^2(A)$ and $KO_1(A) = GW_1^0(A)$, where the letters GW stand for "Grothendieck–Witt" groups. These are bigraded abelian groups $GW_i^j(A)$ with $i \in \mathbb{Z}$ and $j \in \mathbb{Z}/4$. Suslin's symbol $Um_{r+1}(A) \to G_r(A)$ reads then as a collection of maps $Um_{r+1}(A) \to GW_1^{r+1}(A)$. In the same paper, A. Asok and J. Fasel show that Suslin's computation of the group SK_1 of the ring

$$A_n = \frac{\mathbb{Z}[x_1, \ldots, x_n, y_1, \ldots, y_n]}{(\sum_{i=1}^n x_i y_i - 1)}$$

refines in a computation of Grothendieck–Witt groups of A (with the price to consider $\mathbb{Z}[1/2]$-coefficients). Indeed, one finds

$$GW_1^{r+1}(A_{r+1}) = GW_0^0(\mathbb{Z}[1/2])$$

provided $r \geq 1$.
There is an analogue of Quillen's spectral sequence computing K-theory in terms of codimension of the support in the theory of Grothendieck–Witt groups (see e.g. [18]). Asok and Fasel show that in the case of the ring A_{r+1}, an edge map in the corresponding spectral sequence is indeed an isomorphism. This allows to compute this edge map for any regular algebra (with 2 invertible) of dimension $\leq r$ in terms of Suslin matrices.
- **Study of orbit spaces, and classifying spaces**: If $R = C(X)$ is the ring of continuous real valued functions on a topological space X then every unimodular row $v \in Um_n(C(X))$, $n \geq 2$, determines a map $arg(v) : X \longrightarrow \mathbb{R}^n \setminus \{0\} \longrightarrow S^{n-1}$. (The first is by evaluation, and the second is the standard homotopy equivalence.) We thus get an element $[arg(v)]$ of $[X, S^{n-1}]$. (As $n \geq 2$, we may ignore base points.) Clearly, rows in the same elementary orbit define homotopic maps. Thus, we have a natural map $Um_n(C(X))/E_n(C(X)) \longrightarrow [X, S^{n-1}] = \pi^{n-1}(X)$.

Note that J.F. Adams has shown that S^{n-1} is not a H-space, unless $n = 1, 2, 4$, or 8. It is classically known that this is equivalent to saying that there is no suitable way to multiply the two projection maps $S^{n-1} \times S^{n-1}$ in $[S^{n-1} \times S^{n-1}, S^{n-1}]$. However, under suitable restrictions on the 'dimension' of X we may expect to define a product.

Henceforth, let X be a finite CW-complex of dimension $d \geq 2$. L.N. Vaserstein has shown that the ring $C(X)$ has stable dimension d. Now let $n \geq 3$, so that S^{n-1} will be atleast 1-connected. By the Suspension Theorem, the suspension map $S : \ [X; S^{n-1}] \longrightarrow [SX; S^n]$ is surjective if $d \leq 2(n-2)+1$, and bijective if $d \leq 2(n-2)$. Moreover, we know that $[SX, S^n]$ is an abelian group. Hence, the orbit space has a structure of an abelian group. It is shown in [68, Theorem 7.7] that above map is a universal weak Mennicke symbol as defined by W. van der Kallen in [68].

We refer the recent preprints of Aravind Asok and Jean Fasel (in particular, see [3]) where there is a $\mathbb{A}^1$-homotopy interpretation of these results.

In the context of commutative rings, for $n = 3$ and d atmost 2, the orbit space of unimodular rows modulo elementary action was shown to be bijective to the elementary symplectic Witt group (denoted by $W_E(R)$) by L.N. Vaserstein in [54] and for $d \leq 2n - 4$, to the universal weak Mennicke symbol by W. van der Kallen in [68].

It would appear too strong to expect the bound to fall; and perhaps it is, but the article [47] encourages us, as it shows (using Suslin matrices) that there is a nice group structure on orbits of squares of unimodular rows when $\dim(R) \leq 2n - 3$.

We say that the orbit space $Um_r(R)/E_r(R)$ has a Mennicke-like (or nice) structure if

$$[(a, a_2, \ldots, a_r)] \star [(b, a_2, \ldots, a_r)] = [(ab, a_2, \ldots, a_r)].$$

In [21, Theorem 3.9] it is shown that if A is an affine algebra of dimension d over a perfect field k, of characteristic $\neq 2$, and with c.d.$_2(k) \leq 1$, then if $r = d + 1$, the van der Kallen group structure on it defined in [67] is Mennicke-like.

In [47] the Suslin matrix approach enables one to recapture this theorem when k is algebraically closed; and also to improve upon it for $r = d$, when k is a finite field. In fact, we realized later that the Suslin matrix approach in [47] would also enable us to recapture [21, Theorem 3.9]. We leave it to the reader to verify these details.

As pointed out in ([47], due to the strong results of J. Fasel in [17], for a smooth affine algebra over a field k, of characteristic $\neq 2$, and with c.d.$_2(k) \leq 2$, the group structure on the orbit space $Um_{d+1}(A)/E_{d+1}(A)$ is nice. Is this the optimal situation for smooth affine algebras over a field?

The recent progress we have made is to relate these two studies, via the Suslin symbol. We briefly sketch this next.

- **Defining group structures, Witt group structures on orbits of unimodular rows**

One can define a Witt group $W_{EUm}(R)$, and a map from the orbit space $Um_n(R)/E_n(R) \longrightarrow W_{EUm(R)}$ sending $[v]$ to $[S_{n-1}(v, w)]$, for any w, with $\langle v, w\rangle = 1$. This map is a homomorphism, and is a Steinberg symbol if $\dim(R) \leq 2n - 3$. It is also onto when $\dim(R) \leq 2n - 3$. One can commence here as the variant of the Mennicke–Newmann lemma as in [69, Lemma 3.2] is available. We expect it to also be injective under these conditions. This is mainly due to the inherent symmetry of the Suslin matrices.

Note that these would mean that the orbit space would then have a nice abelian Witt group structure under the condition $\dim(R) \leq 2n - 3$; which is an improvement on the condition $\dim(R) \leq 2n - 4$ in the theorem of van der Kallen in [68] stated above. More details will appear in [32].

- In [56, Sect. 3] Suslin points out that the fact that the universal factorial row can be completed can be used to find a completion of a linear unimodular row of length $(r+1)$, provided $r!$ is a unit. In fact he shows that there is a factorial row in the elementary orbit of any linear unimodular row. At the end of Sect. 5 he poses Problem 4 which reposes a question posed by Bass in [7], with an additional rider. We now know this as the Bass–Suslin conjecture; and it is one of the central open questions of classical algebraic K-theory. Let R be a local ring. Bass asked if $Um_r(R[X]) = e_1 SL_r(R[X])$. Suslin expects this if $1/(r-1)! \in R$. More generally, due to Suslin's example, one would expect to find a factorial row in the elementary orbit of any unimodular row over a polynomial ring over a local ring.

The results of M. Roitman in [51], and R.A. Rao in [41–43] bear testimony to this. In [41–43] unimodular polynomial rows are studied via the Vaserstein symbol. In [32] a similar study is undertaken via the Suslin symbol. This study promises to solve this question in the metastable range; however, one expects that if one couples this with the ideas developing in [48] then one could get a complete picture, based on the beautiful symmetry of the Suslin matrices. More precisely, the structure of the Suslin matrix forces a certain positioning; and the argument in [32] indicates that some positionings (enforced by the positioning of the coordinates of a Suslin matrix) are suitable to enable us to lift the yoke of restriction of injective stability estimates of K_1 so far.

Historical development often gives a clue to the route one should follow.

The study of completions of unimodular rows over a commutative noetherian ring R of dimension d gives a hint of things to come. It began with J-P. Serre, followed by H. Bass, *ideas of general position*; which were taken further by Eisenbud–Evans. L.N. Vaserstein started studying group structures on orbits of unimodular rows *using Witt groups*. But the paper [54] already contains enough of *non-stable* algebraic K-theory arguments on a unimodular row; which were expanded upon by W. van der Kallen in [67, 68]. Thus, the arguments of [42] give preliminary historical evidence of getting completion of unimodular polynomial rows in dimension three by a stable argument. Injective stabilization plays an important role here; but we suspect that this happens because we have not done the linearization in a proper way which preserves the anti-symmetry.

It is this combination of ideas that **we strongly advocate** in the polynomial case; doing stable linearization, preserving the inherent symmetry of the Suslin matrices, and taking nth roots, we believe should give a 'polynomial time' feedback completion algorithm at the non-stable level. We hope to be able to present these ideas in [48].

3 Study of the Suslin Matrix

We begin with the study of the alternating matrices; which gives a good role model to begin the topic.

The Alternating Matrix $V(v, w)$

Let $v = (a, b, c)$, $w = (a', b', c')$ with $\langle v, w \rangle = aa' + bb' + cc' = 1$.

We consider the 4×4 alternating matrix $V(v, w)$ of Pfaffian one

$$V(v,w) = \begin{pmatrix} 0 & a & b & c \\ -a & 0 & c' & -b' \\ -b & -c' & 0 & a' \\ -c & b' & -a' & 0 \end{pmatrix}$$

We wish to analyze the action of $\varepsilon \in E_4(R)$ on $V(v, w)$ by conjugation.

We first recall **the Cohn transformations** of a row below

Definition 1 Let $v = (a_0, a_1, \ldots, a_r)$, $w = (b_0, b_1, \ldots, b_r) \in R^{r+1}$ with $\langle v, w\rangle = 1$. We say that the row

$$v^* = vC_{ij}(\lambda) = (a_0, \ldots, a_i + \lambda b_j, \ldots, a_j - \lambda b_i, \ldots, a_r),$$

for $0 \le i \ne j \le r$, is a **Cohn transform** of v w.r.t. the row w.

P.M. Cohn in [16] had shown that the matrices $I_2 + \lambda \begin{pmatrix} a \\ b \end{pmatrix} (b\ -a)$ were not elementary matrices in general.

It was shown in [27, Lemma 2.1] that the Cohn orbit (got by a finite number of successive Cohn transforms) is the same as the elementary orbit when $r \ge 2$. Moreover, see [33, Theorem 3.6], if $\langle v, w\rangle = \langle v', w\rangle = 1$ then v' can be got from v by a finite number of Cohn transforms w.r.t. w.

Let us get back to analyzing the action of an elementary matrix on an alternating matrix.

One has the following identities:

$$\begin{aligned} E_{12}(\lambda)V(v,w)E_{12}(\lambda)^t &= V(vC_{12}(\lambda), w), \\ E_{13}(\lambda)V(v,w)E_{13}(\lambda)^t &= V(vC_{02}(-\lambda), w), \\ E_{14}(\lambda)V(v,w)E_{14}(\lambda)^t &= V(vC_{01}(\lambda), w), \\ E_{21}(\lambda)V(v,w)E_{21}(\lambda)^t &= V(v, wC_{21}(\lambda)), \\ E_{31}(\lambda)V(v,w)E_{31}(\lambda)^t &= V(v, wC_{20}(-\lambda)), \\ E_{41}(\lambda)V(v,w)E_{41}(\lambda)^t &= V(v, wC_{10}(\lambda)). \end{aligned} \tag{1}$$

Equation (1) describes completely the action of $E_4(R)$ on an alternating matrix $V(v, w)$.

We may consider the Vaserstein space V of dimension 6 consisting of all 4×4 alternating matrices over R. The above relations associates a linear transformation T_σ of V with any $\sigma \in SL_4(R)$ by $T_\sigma(V(v, w)) = \sigma V(v, w)\sigma^t$. The matrix of this linear transformation w.r.t the usual ordered basis $e_1, \ldots, e_6$ is not orthogonal. However, with respect to the following permutation of the standard basis $e_1, \ldots, e_6$ namely $e_1, e_2, e_3, e_6, -e_5, e_4$ we get

$$E_{12}(x) \to E_{62}(x)E_{53}(-x) \; E_{21}(x) \to E_{26}(x)E_{35}(-x)$$
$$E_{13}(x) \to E_{61}(-x)E_{43}(x) \; E_{31}(x) \to E_{16}(x)E_{34}(-x)$$
$$E_{14}(x) \to E_{51}(-x)E_{42}(x) \;\; E_{41}(x) \to E_{15}(x)E_{24}(x)$$

The images are all elementary orthogonal matrices. In particular, the matrix of T_σ will be an orthogonal matrix. One observes also that the map $E_4(R)(R)$ is onto $EO_4(R)$. This induces an injection of the quotient groups $SL_4(R)/E_4(R) \longrightarrow SO_4(R)/EO_4(R)$.

Let us compute T_σ. It is the matrix of $\wedge^2\sigma$. When $\sigma = V(v, w)$ something interesting is revealed: the matrix is $\left(I_4 - \begin{pmatrix} v^t \\ w^t \end{pmatrix} \begin{pmatrix} w & v \end{pmatrix}\right)\left(I_4 - \begin{pmatrix} e_1^t \\ e_1^t \end{pmatrix} \begin{pmatrix} e_1 & e_1 \end{pmatrix}\right)$. This is recognizable as the product of two reflections $\tau_{(v,w)} \circ \tau_{(e_1,e_1)}$. (See later for the definition.)

Is there a similar 'larger sizes' analogue? The observations above are replicated below with the Suslin matrix substituting for the alternating matrix $V(v, w)$.
Remark: When we did calculations with 6×6 alternating matrices of Pfaffian one we found that the corresponding linear transformations were not orthogonal, and so the theory is dissimilar. It seems worthwhile to investigate what is happening here.

The Suslin Matrix $S_r(v,w)$

We now describe the Suslin matrices in more detail.

The construction of the Suslin matrix $S_r(v, w)$ is possible once we have two rows v, w. These matrices will be invertible if their dot product $v \cdot w^t = 1$. (The rows are then automatically unimodular rows.) Suslin's inductive definition: Let

$$v = (a_0, a_1, \ldots, a_r) = (a_0, v_1),$$

with $v_1 = (a_1, \ldots, a_r)$,

$$w = (b_0, b_1, \ldots, b_r) = (b_0, w_1),$$

with $w_1 = (b_1, \ldots, b_r)$. Set $S_0(v, w) = a_0$, and set

$$S_r(v, w) = \begin{pmatrix} a_0 I_{2^{r-1}} & S_{r-1}(v_1, w_1) \\ -S_{r-1}(w_1, v_1)^t & b_0 I_{2^{r-1}} \end{pmatrix}.$$

Suslin noted that $S_r(v, w)S_r(w, v)^t = (v \cdot w^t)I_{2^r} = S_r(w, v)^t S_r(v, w)$, and $\det S_r(v, w) = (v \cdot w^t),^{2^{r-1}}$ for $r \geq 1$.

The positions of a_i and b_i in $S_r(v, w)$ are as follows: For $1 \leq i \leq r - 1$,

1. The positions of a_0 in $S_r(v, w)$ is given by $(k, k),\ 1 \leq k \leq 2,^{r-1}$ and the positions of b_0 in $S_r(v, w)$ is given by $(k, k),\ 2^{r-1} + 1 \leq k \leq 2^r$.
2. The positions of a_r in $S_r(v, w)$ is given by $(2k - 1, 2^r - 2k + 2), 1 \leq k \leq 2,^{r-1}$ and the positions of b_r in $S_r(v, w)$ is given by $(2k, 2^r - 2k + 1), 1 \leq k \leq 2^{r-1}$

3. The positions of $+a_i$ in $S_r(v, w)$ is given by

$$(2^2k2^{r-1-i} + j, (2 + (2^{i-1} - k - 1)2^2)2^{r-1-i} + j),$$

where $0 \le k \le 2^{i-1} - 1, 1 \le j \le 2^{r-1-i}_{.}$
4. The positions of $-a_i$ in $S_r(v, w)$ is given by

$$((3 + 2^2k)2^{r-1-i} + j, (1 + (2^{i-1} - k - 1)2^2)2^{r-1-i} + j),$$

where $0 \le k \le 2^{i-1} - 1, 1 \le j \le 2^{r-1-i}_{.}$
5. The positions of $+b_i$ in $S_r(v, w)$ is given by

$$((1 + (2^{i-1} - k - 1)2^2)2^{r-1-i} + j, (3 + 2^2k)2^{r-1-i} + j),$$

where $0 \le k \le 2^{i-1} - 1, 1 \le j \le 2^{r-1-i}_{.}$
6. The positions of $-b_i$ in $S_r(v, w)$ is given by

$$(2 + (2^{i-1} - k - 1)2^2)2^{r-1-i} + j, 2^2k2^{r-1-i} + j),$$

where $0 \le k \le 2^{i-1} - 1, 1 \le j \le 2^{r-1-i}_{.}$

The Suslin Forms J_r

To understand the nature of the shape of the Suslin matrices we recall Suslin's sequence of forms $J_r \in \mathrm{M}_{2^r}(R)$ given by the recurrence formulae

$$J_r = \begin{cases} 1 & \text{for r = 0} \\ J_{r-1} \perp -J_{r-1}, & \text{for r even ,} \\ J_{r-1} \top - J_{r-1}, & \text{for r odd.} \end{cases}$$

(The English translation wrongly says $J_r = J_{r-1} \perp J_{r-1}$ when r is even.)
(Here $\alpha \perp \beta = \begin{pmatrix} \alpha & 0 \\ 0 & \beta \end{pmatrix}$, while $\alpha \top \beta = \begin{pmatrix} 0 & \alpha \\ \beta & 0 \end{pmatrix}$.)

How did Suslin think of these forms? What will the form be if the 'Suslin matrix' is constructed by a slightly different basis; say by the usual lexicographic ordering of the basis to describe the map $\oplus_{i \text{odd}} \wedge^i R^r \to \oplus_{i \text{even}} \wedge^i R^r$ in the earlier construction. We give a possible approach: Observe that $J_r = \prod_{i=1}^{r+1} S_r(e_i, e_i)$. The reader can verify this by an easy induction on r. (Or we refer to [46] where it is proved.)

It is easy to see that det $J_r = 1$, for all r, and that $J_r^t = J_r^{-1} = (-1)^{\frac{r(r+1)}{2}} J_r$. Moreover, J_r is antisymmetric if $r = 4k + 1$ and $r = 4k + 2$, whereas J_r is symmetric for $r = 4k$ and $r = 4k + 3$.

We know from Suslin that he was unaware of M. Krusemeyer's explanations in [34, 35] for the Swan–Towber completion of (a^2, b, c). The explanations of M. Krusemeyer seem to be adequate only in the case of alternating forms. (Are we wrong in saying this?)

Suslin recognized the need to analyze the shapes of the Suslin matrices $S_r(v, w)$. He realized that the shapes satisfied similar properties according to the length $(r + 1)$ of the row.

In [56, Lemma 5.3], it is noted that the following formulae are valid

$$\begin{aligned} \text{for } r = 4k : (S_r(v, w)J_r)^t &= S_r(v, w)J_r; \\ \text{for } r = 4k+1 : S_r(v, w)J_r S_r(v, w)^t &= (v \cdot w^t)J_r; \\ \text{for } r = 4k+2 : (S_r(v, w)J_r)^t &= -S_r(v, w)J_r; \\ \text{for } r = 4k+3 : S_r(v, w)J_r S_r(v, w)^t &= (v \cdot w^t)J_r. \end{aligned}$$

We call these **the Suslin identities**. These identities are the core of the underlying four physical configuration spaces in which unimodular rows live.

These identities may be easily verified by induction on r. Alternatively, one can also observe it after noting that for $r \geq 1$, and $2 \leq i \leq r+1$, $S_r(e_i, e_i)^{-1} = S_r(e_i, e_i)^t = -S_r(e_i, e_i)$, $S_r(e_i, e_i)^2 = -I_{2^r}$, and $\det S_r(e_i, e_i) = 1$, and the following lemma:

Lemma 2 *Let* $v = (a_0, a_1, \ldots, a_r)$, $w = (b_0, b_1, \ldots, b_r) \in M_{1r+1}(R)$, $r \geq 1$. *Then for* $2 \leq i \leq r+1$,

$$S_r(e_i, e_i)S_r(v, w)S_r(e_i, e_i)^{-1} = S_r(v', w'),$$

where

$$\begin{aligned} v' &= (b_0, -a_1, \ldots, -a_{i-2}, b_{i-1}, -a_i, \ldots, -a_r), \text{ and} \\ w' &= (a_0, -b_1, \ldots, -b_{i-2}, a_{i-1}, -b_i, \ldots, -b_r). \end{aligned}$$

Thus, one has

$$J_r S_r(v, w) J_r^{-1} = \begin{cases} S_r(v, w)^t & \text{if } r \text{ even} \\ S_r(w, v) & \text{if } r \text{ odd.} \end{cases}$$

The Suslin identities show that unimodular rows of length $r + 1$ will have properties depending on $[r]$ modulo 4. We have already seen an instance of a property which depends on the parity of r when discussing the isomorphism of a projective module corresponding to a row and its dual projective module. Is there such an example of a property for unimodular rows which depends on the $[r]$ modulo 4?

When searching for an algorithm to create a $\beta_r(v, w)$ from $S_r(v, w)$ one should also keep the following question in mind. One knows that there is a $\beta_r(v, w) \in S_r(v, w)E_{2^r}(R)$. When r is odd, is there a $\beta_r(v, w) \in S_r(v, w)GE_{2^r}(R)$, where $GE = ESp$ when $r = 4k + 1$, and $GE = EO$ when $r = 4k + 3$? (The "right" ESp, EO is part of the query.)

The Fundamental Property and the Key Lemma

We give a simple proof of the Fundamental property of Suslin matrices, which first appeared in [27].

Lemma 3 *Let R be a ring with* 1. *Let S be a subset of R satisfying*

1. $a \in S$ *implies* $-a \in S$.
2. $a, b \in S$ *implies* $a + b \in S$.
3. $a \in S$ *implies* $a^2 \in S$.

Then $a, b \in S$ *implies* $ab + ba \in S$, $2abc \in S$.

Proof $ab + ba = \{(a + b)^2 - a^2\} - b^2 \in S$. Hence,

$$\{a(ab + ba) + (ab + ba)a\} - (a^2b + ba^2) = 2aba \in S.$$

We now state and prove the important Fundamental property satisfied by the Suslin matrices.

Corollary 4 (Fundamental property) *Let* $S_r(s, t)$, $S_r(v, w)$ *be Suslin matrices. Then*

$$S_r(s, t)S_r(v, w)S_r(s, t) = S_r(v', w')$$
$$S_r(t, s)S_r(w, v)S_r(t, s) = S_r(w', v'),$$

for some $v', w' \in M_{1,r+1}(R)$, *which depend linearly on* v, w *and quadratically on* s, t. *Consequently,* $v' \cdot w'^{t} = (s \cdot t^t)^2(v \cdot w^t)$.

Proof Take $R = \mathrm{M}_{2^r}(R)$, and let S be the subset of all Suslin matrices above. Take $a = S_r(s, t)$, $b = S_r(v, w)$. Then $2aba \in S$. A generic argument will enable us to assume that 2 is a non-zero-divisor, and allow us to conclude that $aba \in S$.

The last two assertions will need the more specific argument given in [29, Lemma 2.5].

Remark 5 L. Avramov has independently observed a similar argument to prove the Fundamental property of Suslin matrices.

The Key Lemma

Recall that we were led to the above Fundamental property in [27, Corollary 3.3] by the Key Lemma via the methods of commutative algebra. We next recall the Key Lemma which is actually equivalent to the Fundamental Property. (We refer the reader to the thesis of Selby Jose [26, Chap. 4, Lemma 4.3.16] where this equivalence has been detailed).

The Cohn transforms were first sighted in the work of L.N. Vaserstein in [54] when he considered the action of an elementary matrix on a 4×4 invertible alternating matrix as described earlier. His analysis led us to the key lemma below

Notation. For a matrix $\alpha \in M_k(R)$, we define α^{top} as the matrix whose entries are the same as that of α above the diagonal, and on the diagonal, and is zero below the diagonal. Similarly, we define α^{bot}.

For simplicity we may write α^t for $\alpha,^{top}$ α^b for $\alpha,^{bot}$ and α^t for α transpose. Moreover, we use α^{tb} for α^{top} or $\alpha^{bot}_.$

Lemma 6 (Key Lemma) *Let $v,\ w \in M_{1,r+1}(R)$. Then, for $r \geq 2$, $2 \leq i \leq r+1$, $\lambda \in R$,*

$$\begin{aligned}
S_r(vE_{i1}(-\lambda), wE_{1i}(\lambda)) &= S_r(e_1, e_1E_{1i}(\lambda))^{top} S_r(v, w) S_r(e_1, e_1E_{1i}(\lambda)),^{bot} \\
S_r(vE_{1i}(\lambda), wE_{i1}(-\lambda)) &= S_r(e_1E_{1i}(\lambda), e_1)^{bot} S_r(v, w) S_r(e_1E_{1i}(\lambda), e_1),^{top} \\
S_r(vC_{0i-1}(-\lambda), w) &= S_r(e_1E_{1i}(\lambda), e_1)^{top} S_r(v, w) S_r(e_1E_{1i}(\lambda), e_1),^{bot} \\
S_r(v, wC_{0i-1}(-\lambda)) &= S_r(e_1, e_1E_{1i}(\lambda))^{bot} S_r(v, w) S_r(e_1, e_1E_{1i}(\lambda))^{top}_.
\end{aligned}$$

In view of its own importance we wish to record the useful observation which led us to the proof of the Key Lemma given in [27, Lemma 3.1]:

Lemma 7 *Let $v, w, s, t \in R^{r+1}$ and let $v = (a_0, a_1, \ldots, a_r)$, $w = (b_0, b_1, \ldots, b_r)$. Then*

$$\begin{aligned}
S_r(v, w) + S_r(w, v)^t &= \{a_0 + b_0\} I_{2^r}. \\
S_r(s, t) S_r(w, v)^t + S_r(v, w) S_r(t, s)^t &= \{\langle s, w\rangle + \langle v, t\rangle\} I_{2^r}. \\
S_r(w, v)^t S_r(s, t) + S_r(t, s)^t S_r(v, w) &= \{\langle s, w\rangle + \langle v, t\rangle\} I_{2^r}.
\end{aligned}$$

Proof This is the usual bilinear consequence of the quadratic relation

$$\begin{aligned}
S_r(v+s, w+t) S_r(w+t, v+s)^t &= \langle v+s, w+t\rangle I_{2^r} \\
&= \{\langle v, w\rangle + \langle v, t\rangle + \langle s, w\rangle + \langle s, t\rangle\} I_{2^r}.
\end{aligned}$$

The relations above reminds one of the relations in a Clifford algebra.

Commutator Calculus

Finally, we record a few interesting relations we got in [27, Lemma 3.6] by use of the Key Lemma: This is the Yoga of commutators in the elementary unimodular vector group. As is known, a proper handle of this, can lead one to understand the quotient group of the Suslin unimodular vector group by its elementary unimodular vector group, better. In fact, it is eventually shown that this is a solvable group, using the methods of A. Bak in [6]. (See below for the definitions, and indications of a proof.)

Lemma 8 *Let $2 \leq i \neq j \leq r+1$, and let $\lambda = -2xy$. If*

$$\begin{aligned}
\alpha &= [S_r(e_1E_{1i}(x), e_1), S_r(e_1E_{1j}(y), e_1)], \\
\alpha^* &= [S_r(e_1E_{1j}(-y), e_1), S_r(e_1E_{1i}(-x), e_1)]
\end{aligned}$$

then $\alpha^* = \alpha^{-1}$, *and* $S_r(vC_{i-1j-1}(\lambda), w) = \alpha S_r(v, w)\alpha^{-1}$;

$$\beta = [S_r(e_1, e_1E_{1i}(x)), S_r(e_1, e_1E_{1j}(y))],$$
$$\beta^* = [S_r(e_1E_{1j}(-y), e_1), S_r(e_1, e_1E_{1i}(-x))],$$

then $\beta^* = \beta^{-1}$, *and* $S_r(v, wC_{i-1j-1}(\lambda)) = \beta S_r(v, w)\beta^{-1}$;

$$\gamma = [S_r(e_1E_{1j}(x), e_1), S_r(e_1, e_1E_{1i}(y))],$$
$$\gamma^* = [S_r(e_1, e_1E_{1i}(-y)), S_r(e_1E_{1j}(-x), e_1)],$$

then $\gamma^* = \gamma^{-1}$, *and* $S_r(vE_{ij}(\lambda), wE_{ji}(-\lambda)) = \gamma S_r(v, w)\gamma.^{-1}$

The Suslin Vector Space

It is easy to see that the set

$$S = \{S_r(v, w) | v, w \in M_{1r+1}(R)\}$$

is a free R-module or rank $2(r+1)$. For a basis one can take $se_0, \ldots, se_{r+1}, se_0^*, \ldots, se_{r+1}^*$, where $se_i = S_r(e_i, 0)$, $se_i^* = S_r(0, e_i)$, for $0 \le i \le r$. We shall call this the **Suslin space**.

The Suslin Matrix Groups

Definition: The *Special Unimodular Vector group* $SUm_r(R)$ is the subgroup of $SL_{2^r}(R)$ generated by the Suslin matrices $S_r(v, w)$ w.r.t. the pair (v, w), with $v \in Um_{r+1}(R)$, for some w with $\langle v, w\rangle = v \cdot w^t = 1$.

Remark 9 One can analogous to the linear case, define the Elementary Unimodular vector subgroup $EUm_r(R)$ of $SUm_r(R)$ generated by the Suslin matrices $S_r(v, w)$, with $v = e_1\varepsilon$, for some $\varepsilon \in E_{r+1}(R)$, and with $v \cdot w^t = 1$.

Proposition 10 (Center of $SUm_r(R)$ [29, Corollary 3.5]) *Let R be a commutative ring. The center $Z(SUm_r(R))$ of the Special Unimodular vector group $SUm_r(R)$ consists of scalar matrices uI_{2^r}. Moreover,*

$$Z(SUm_r(R)) = \begin{cases} \{uI_{2^r} : u \in R, u^2 = 1\}, & \text{if } r \text{ odd} \\ \{uI_{2^r} : u \in R, u^4 = 1\}, & \text{if } r \text{ even.} \end{cases}$$

Hence $Z(SUm_r(R)) \subseteq EUm_r(R)$.

Commutator Calculus (contd.)

There is yet another set of generators for $\mathrm{EUm}_r(R)$, *viz.* $S_r(e_1E_{1i}(x), e_1)$, $S_r(e_1, e_1E_{1i}(y))$, and $S_r(e_i, e_iE_{i1}(a))$, $S_r(e_iE_{i1}(b), e_i)$, for $2 \le i \le r+1$, $x, y, a, b \in R$. This was shown in [27], via the Key Lemmas 6 and 13.

We record the commutator formulae in $\mathrm{EUm}_r(R)^{tb}$ next: We use the convenient notation that for $r \geq 1$, $1 \leq i \leq r+1$, $\lambda \in R$,

$$E(e_1)(\lambda) = I_{2^r} = E(e_1^*)(\lambda)$$
$$E(e_i)(\lambda) = S_r(e_1 E_{1i}(\lambda), e_1);\ i > 1$$
$$E(e_i^*)(\lambda) = S_r(e_1, e_1 E_{1i}(\lambda));\ i > 1$$

(If we wish to stress the size we will write $E_r(e_i)(\lambda)$, $E_r(e_i^*)(\lambda)$).[1]

Proposition 11 *For $r \geq 2$, $\lambda, \mu \in R$, $c_i = e_i$ or e_i^*, $d_j = e_j$ or e_j^*, we have, for $2 \leq i < j \leq r+1$,*

$$\begin{aligned}&[E_r(c_i)(\lambda)^t, E_r(d_j)(\mu)^b]\\ &= [E_{r-1}(c_{i-1})(\lambda)^t, E_{r-1}(d_{j-1})(\mu)^b] \perp [E_{r-1}(c_{i-1})(\lambda)^t, E_{r-1}(d_{j-1})(\mu)^b]\\ &= \underbrace{\alpha \perp \cdots \perp \alpha}_{2^{i-2}\text{ times}},\end{aligned}$$

where

$$\alpha = \begin{cases} \{E_{r-i+1}(d_{j-i+1})(\lambda\mu)^{top} \perp E_{r-i+1}(d_{j-i+1})(-\lambda\mu)^{bot}\} \text{ if } c_i = e_i, \\ \{E_{r-i+1}(d_{j-i+1})(\lambda\mu)^{bot} \perp E_{r-i+1}(d_{j-i+1})(-\lambda\mu)^{top}\} \text{ if } c_i = e_i^*. \end{cases}$$

We next calculate the triple commutators

Lemma 12 *For $r \geq 2$, $2 \leq i \neq j \leq r+1$, $\lambda, \mu, \nu \in R$,*

(i) $\big[[E(e_i)(\lambda)^{top}, E(e_j)(\mu)^{bot}], E(e_i^*)(\nu)^{tb}\big] = E(e_j)(\lambda\mu\nu)^{tb}_{,}$

(ii) $\big[[E(e_i^*)(\lambda)^{top}, E(e_j)(\mu)^{bot}], E(e_i)(\nu)^{tb}\big] = E(e_j)(\lambda\mu\nu)^{tb}_{,}$

(iii) $\Big[[E(e_i)(\lambda)^{top}, E(e_j^*)(\mu)^{bot}], E(e_i^*)(\nu)^{tb}\Big] = E(e_j^*)(\lambda\mu\nu)^{tb}_{,}$

(iv) $\Big[[E(e_i^*)(\lambda)^{top}, E(e_j^*)(\mu)^{bot}], E(e_i)(\nu)^{tb}\Big] = E(e_j^*)(\lambda\mu\nu)^{tb}_{.}$

The Key Lemma makes us consider the subgroup $\mathrm{EUm}_r(R)^{tb}$ of $\mathrm{E}_{2^r}(R)$ generated by elements of the type $S_r(e_1 E_{1i}(x), e_1)^{tb}$, $S_r(e_1, e_1 E_{1i}(x))^{tb}$. In view of the Key Lemma it is clear that $\mathrm{EUm}_r(R) \subset EUm_r(R)^{tb}$.

Via the triple commutator laws, one gets the following relations, which prove the fact that $EUm_r(R) = EUm_r(R)^{tb}$.

[1]The definition of $E(e_1)(\lambda)$ was erroneously defined as $\lambda I_{2^{r-1}} \perp \lambda^{-1} I_{2^{r-1}}$ in [27].

Lemma 13 ([29, Lemma 4.9]
For $r \geq 2$, $2 \leq i \neq j \leq r+1$, and $\lambda \in R$.

$$\begin{aligned}
S_r(e_1E_{1i}(\lambda), e_1)^{top} &= S_r(e_1E_{1j}(\lambda), e_1)S_r(e_1 - \lambda e_j + e_i, e_1)S_r(e_1E_{1i}(-1), e_1)\\
&\quad S_r((1-\lambda)e_1 + \lambda e_j + e_i, e_1 + e_j)S_r(e_1 - e_i, e_1 - e_j)\\
&\quad S_r((1+\lambda)e_1 - \lambda e_j, e_1 + e_j)S_r(e_1, e_1E_{1j}(-1)),\\
S_r(e_1E_{1i}(\lambda), e_1)^{bot} &= S_r(e_1, e_1E_{1j}(-1))S_r((1+\lambda)e_1 - \lambda e_j, e_1 + e_j)\\
&\quad S_r(e_1 - e_i, e_1 - e_j)S_r((1-\lambda)e_1 + \lambda e_j + e_i, e_1 + e_j)\\
&\quad S_r(e_1E_{1i}(-1), e_1)S_r(e_1 - \lambda e_j + e_i, e_1)S_r(e_1E_{1j}(\lambda), e_1)\\
S_r(e_1, e_1E_{1i}(\lambda))^{top} &= S_r(e_1E_{1j}(-1), e_1)S_r(e_1 + e_j, (1+\lambda)e_1 - \lambda e_j)\\
&\quad S_r(e_1 - e_j, e_1 - e_i)S_r(e_1 + e_j, (1-\lambda)e_1 + \lambda e_j + e_i)\\
&\quad S_r(e_1, e_1E_{1i}(-1))S_r(e_1, e_1 - \lambda e_j + e_i)S_r(e_1, e_1E_{1j}(\lambda)),\\
S_r(e_1, e_1E_{1i}(\lambda))^{bot} &= S_r(e_1, e_1E_{1j}(\lambda))S_r(e_1, e_1 - \lambda e_j + e_i)S_r(e_1, e_1E_{1i}(-1))\\
&\quad S_r(e_1 + e_j, (1-\lambda)e_1 + \lambda e_j + e_i)S_r(e_1 - e_j, e_1 - e_i)\\
&\quad S_r(e_1 + e_j, (1+\lambda)e_1 - \lambda e_j)S_r(e_1E_{1j}(-1), e_1)
\end{aligned}$$

(Note that the alternate relations are got by reversing the order).

The first step in computing the center $Z(SUm_r(R))$ is to show that it consists of scalars. We give a different proof than in [27] of the fact that $Z(SUm_r(R))$ consists of scalar matrices. We use the fact here that $EUm_r(R)^{tb} = EUm_r(R)$, for $r > 1$.

Lemma 14 *Let $A \in M_{2^s}(M_{2^t}(R))$, $t \geq 1$, $s + t = r$ be a diagonal block matrix, where the alternating diagonal blocks are the same. If A commutes with $E_r(e_{s+1})(1)^{top}$ and $E_r(e^*_{s+1})(1)$,^{top} then $A \in M_{2^{s+1}}(M_{2^{t-1}}(R))$ is a diagonal block matrix whose alternating diagonal block entries are same.*

Proof Let $\begin{pmatrix} a_{11} & a_{12} \\ a_{21} & a_{22} \end{pmatrix}$, $\begin{pmatrix} a_{33} & a_{34} \\ a_{43} & a_{44} \end{pmatrix} \in M_{2^{t-1}}(M_2(R))$ be the two, perhaps different, diagonal blocks of A. Compare the $(1, 2^s)$th, $(1, 2^s - 1)$th, and $(2, 2^s - 1)$th block entries of $AE_r(e_{s+1})(1)^{top}$ and $E_r(e_{s+1})(1)^{top}A$ we get $a_{21} = 0$, $a_{34} = 0$, and $a_{11} = a_{33}$ respectively. Compare the $(1, 2^s)$th, $(2, 2^s)$th, and $(2, 2^s - 1)$th block entries of $AE_r(e^*_{s+1})(1)^{top}$ and $E_r(e^*_{s+1})(1)^{top}A$ we get $a_{12} = 0$, $a_{22} = a_{44}$, and $a_{43} = 0$ respectively. Hence $A \in M_{2^{s+1}}(M_{2^{t-1}}(R))$ and is a diagonal matrix with alternating entries equal.

Lemma 15 *Let $A \in M_{2^r}(R)$ be a diagonal matrix with equal alternating diagonal entries. If A commutes with $E_r(e_{r+1})(1)$,^{top} then A is a scalar matrix.*

Proof Let a_{11} and a_{22} be the two different diagonal entries of the matrix A. Compare the $(1, 2^r)$th entry of $AE_r(e_{r+1})(1)^{top}$ and $E_r(e_{r+1})(1)^{top}A$, we get $a_{11} = a_{22}$. Hence A is a scalar matrix.

Proposition 16 (Center of $SUm_r(R)$) *Let $A \in M_{2^r}(R)$. If A commutes with every element of $SUm_r(R)$, then A is a scalar matrix.*

Proof Since $SUm_2(R) = SL_2(R)$, the result is clear for $r = 1$. So let $r \geq 2$. Let us write $A = (a_{ij})_{1 \leq i,j \leq 4}$ in block form. By comparing entries we observe that

(1) $E_r(e_2^*)(1)^{top} A = A E_r(e_2^*)(1)^{top}$ implies $a_{12} = a_{32} = a_{41} = a_{42} = a_{43} = 0$, $a_{22} = a_{44}$,
(2) $E_r(e_2)(1)^{top} A = A E_r(e_2)(1)^{top}$ implies $a_{21} = a_{31} = a_{34} = 0$, $a_{11} = a_{33}$,
(3) $E_r(e_2^*)(1)^{bot} A = A E_r(e_2^*)(1)^{bot}$ implies $a_{13} = a_{14} = a_{23} = 0$, and
(4) $E_r(e_2)(1)^{bot} A = A E_r(e_2)(1)^{bot}$ implies $a_{24} = 0$.

Hence $A \in M_{2^2}(M_{2^{r-2}}(R))$ is a diagonal block matrix with alternating diagonal blocks same. Apply Lemma 14 $r - 2$ times and conclude that A is a diagonal matrix with alternating entries same. Now apply Lemma 15 to get the desired result.

Corollary 17 *An element in $M_{2^r}(R)$ which commutes with $E(c)(1)_r^{top}$ for $c = e_i$ or e_i^*, $3 \leq i \leq r + 1$, and $E(d)(1)_r^{tb}$ $d = e_2$ or e_2^*, is a scalar matrix.*

Proof Obvious from the proof of Proposition 16.

An Involution on $SUm_{r(R)}$, r Even

The case when r is even; where the involution can be defined.

Let $\alpha = \prod_{i=1}^{n} S_i$ be a product of Suslin matries $S_i = S_r(v_i, w_i)$, and let α^* denote $\prod_{i=n}^{1} S_i$. If r is even, then $\alpha \mapsto \alpha^*$ is a well defined anti-involution of $SUm_r(R)$: By Suslin's identities,

$$S_r(v, w) = J_r S_r(v, w)^t J_r^{-1}.$$

Hence, $\alpha^* = J_r \alpha^t J_r^{-1}$, and we are done. From this, it follows that $Z(SUm_r(R)) = \{uI_{2^r} | u^2 = 1\}$, when r is even.

We now discuss the case when r is odd; where we showed that there is an ambiguity to define the involution.

In [29, Corollary 3.2] we show that if $I_{2^r} = S_r(v_1, w_1) \ldots S_r(v_k, w_k)$, for some $\langle v_i, w_i \rangle = 1$, for $1 \leq i \leq k$, then $S_r(v_k, w_k) \ldots S_r(v_1, w_1) = uI_{2^r}$, for some unit u with $u^2 = 1$.

Moreover, in [29, Sect. 5] we show that given a unit u with $u^2 = 1$, we can find $S_r(v_i, w_i)$, with $\langle v_i, w_i \rangle = 1$, $1 \leq i \leq k$, for some k, such that

$$\begin{aligned} I_{2^r} &= S_r(v_1, w_1) \ldots S_r(v_k, w_k) \\ uI_{2^r} &= S_r(v_k, w_k) \ldots S_r(v_1, w_1). \end{aligned}$$

Thus, α^* is defined upto a unit factor when r is odd. This fact is useful to compute $Z(SUm_r(R))$ when r is odd.

Suslin Matrices, Orthogonal Transformations

The Fundamental property of Suslin matrices enables one to define an action of the group $SUm_r(R)$ on the Suslin space. One associates a linear transformation T_g of the Suslin space with a Suslin matrix **g**, via

$$T_g(x, y) = (x', y'),$$

where $\mathbf{g}S_r(x, y)\mathbf{g}^* = S_r(x', y')$. Moreover, if $\mathbf{g}$ is a product of Suslin matrices $S_r(v_i, w_i)$, with $v_i \cdot w_i^t = 1$, for all i, then $T_g \in O_{2(r+1)}(R)$, i.e.

$$\langle T_g(v, w), T_g(s, t)\rangle = \langle (v, w), (s, t)\rangle = v \cdot w^t + s \cdot t^t.$$

Translating the Fundamental Identities

Theorem 18 [29, Corollary 4.2] *The above action induces a canonical homomorphism $\varphi : SUm_r(R) \to SO_{2(r+1)}(R)$, with*

$$\varphi(S_r(v, w)) = T_{S_r(v,w)} = \tau_{(v,w)} \circ \tau_{(e_1 e_1)},$$

where $\tau_{(v,w)}$ is the standard reflection with respect to the vector $(v, w) \in R^{2(r+1)}$ given by the formula

$$\tau_{(v,w)}(s, t) = \langle v, w\rangle(s, t) - (\langle v, t\rangle + \langle s, w\rangle)(v, w).$$

The matrix of the linear transformation was also calculated in [26, Chap. 5, Lemma 5.2.1].

Lemma 19 *Let R be a commutative ring with identity. Let $v, w \in Um_{r+1}(R)$, then the matrix of the linear transformation $T_{S_r(v,w)}$ with respect to the (ordered) basis $\{S_r(e_1, 0), S_r(e_2, 0), \dots, S_r(e_{r+1}, 0), S_r(0, e_1), S_r(0, e_2), \dots, S_r(0, e_{r+1})\}$ is*

$$\left(I - (v, w)^t(w, v)\right)\left(I - (e_1, e_1)^t(e_1, e_1)\right).$$

In particular, for $v = e_1\varepsilon$, $w = e_1{\varepsilon^t}^{-1}$ for some $\varepsilon \in SL_{r+1}(R)$, the matrix of $T_{S_r(v,w)}$ is the commutator $\left[\varepsilon^t \perp \varepsilon^{-1}, \left(I - (e_1, e_1)^t(e_1, e_1)\right)\right]$.

Elementary Orthogonal Matrices and Reflections

Let π denote the permutation $(1\ r+1)\dots(r\ 2r)$ corresponding to the form $I_r \top I_r$. The **elementary orthogonal matrices** over R is defined as

$$oe_{ij}(z) = I_{2r} + ze_{ij} - ze_{\pi(j)\pi(i)}, \text{ if } i \neq \pi(j) \text{ and } i < j,$$

where $1 \le i \ne j \le 2r$, and $z \in R$.

The **elementary orthogonal group** $EO_{2r}(R)$ is a subgroup of $SO_{2r}(R)$ generated by the matrices $oe_{ij}(z)$, where $1 \le i \ne \pi(i) \ne j \le 2r$, and $z \in R$.

We showed in [27] that every elementary orthogonal transformation can be written as a product of reflections. In fact, the standard generators of $EUm_r(R)^{tb}$ map onto the standard generators of $EO_{2(r+1)}(R)$, when r is even. Now apply:

Proposition 20 *Let $\lambda \in R$. For $r \ge 2$, $2 \le i \ne j \le r+1$, and $j \ne \pi(i)$, one has, w.r.t. the splitting given in Lemma 13,*

$$\begin{aligned}
oe_{1i}(\lambda) &= \mathcal{T}_{(e_1-e_j,e_1)} \circ \mathcal{T}_{(-(1-\lambda)e_1+e_j,-e_1+\lambda e_j)} \circ \mathcal{T}_{(e_1-e_j,e_1-e_i)} \\
&\circ \mathcal{T}_{(-(1+\lambda)e_1+e_j,-e_1-\lambda e_j+e_i)} \circ \mathcal{T}_{(e_1,e_1-e_i)} \circ \mathcal{T}_{(-e_1,-e_1+\lambda e_j+e_i)} \\
&\circ \mathcal{T}_{(e_1,e_1-\lambda e_j)} \circ \mathcal{T}_{(e_1,e_1)} = T_{S_r(e_1,e_1E_{1i}(-\lambda)),top} \\
oe_{i1}(\lambda) &= \mathcal{T}_{(e_1,e_1-e_j)} \circ \mathcal{T}_{(-e_1-\lambda e_j,-(1+\lambda)e_1+e_j)} \circ \mathcal{T}_{(e_1-e_i,e_1-e_j)} \\
&\circ \mathcal{T}_{(-e_1+\lambda e_j+e_i,(\lambda-1)e_1+e_j)} \circ \mathcal{T}_{(e_1-e_i,e_1)} \circ \mathcal{T}_{(-e_1-\lambda e_j+e_i,-e_1)} \\
&\circ \mathcal{T}_{(e_1+\lambda e_j,e_1)} \circ \mathcal{T}_{(e_1,e_1)} = T_{S_r(e_1E_{1i}(\lambda),e_1),bot} \\
oe_{\pi(1)i}(\lambda) &= \mathcal{T}_{(e_1,e_1-\lambda e_j)} \circ \mathcal{T}_{(-e_1,-e_1+\lambda e_j+e_i)} \circ \mathcal{T}_{(e_1,e_1-e_i)} \\
&\circ \mathcal{T}_{(-(1+\lambda)e_1+e_j,-e_1-\lambda e_j+e_i)} \circ \mathcal{T}_{(e_1-e_j,e_1-e_i)} \\
&\circ \mathcal{T}_{((\lambda-1)e_1+e_j,-e_1+\lambda e_j)} \circ \mathcal{T}_{(e_1-e_j,e_1)} \circ \mathcal{T}_{(e_1,e_1)} = T_{S_r(e_1,e_1E_{1i}(-\lambda)),bot} \\
oe_{i\pi(1)}(\lambda) &= \mathcal{T}_{(e_1+\lambda e_j,e_1)} \circ \mathcal{T}_{(-e_1-\lambda e_j+e_i,-e_1)} \circ \mathcal{T}_{(e_1-e_i,e_1)} \\
&\circ \mathcal{T}_{(-e_1+\lambda e_j+e_i,(\lambda-1)e_1+e_j)} \circ \mathcal{T}_{(e_1-e_i,e_1-e_j)} \circ \mathcal{T}_{(-e_1-\lambda e_j,-(1+\lambda)e_1+e_j)} \\
&\circ \mathcal{T}_{(e_1,e_1-ej)} \circ \mathcal{T}_{(e_1,e_1)} = T_{S_r(e_1E_{1i}(\lambda),e_1).top}
\end{aligned}$$

We refer the reader to the Appendix where we show how the mathematical software MuPAD helps in the computation of composition of reflections.

Kernel of $\varphi : SUm_{r(R)} \longrightarrow SO_{2(r+1)}(R)$

We compute the kernel of the map φ, and show that it consists of scalars uI_{2^r}, with $u^2 = 1$. This follows from:

Lemma 21 *Let R be a commutative ring in which 2 is invertible. Let $\alpha \in SUm_r(R)$. Suppose that $\alpha S_r(v, w)\alpha^* = S_r(v, w)$, for all $S_r(v, w) \in EUm_r(R)$. Then α^* centralizes $EUm_r(R)$. Consequently, α is a scalar uI_{2^r}, for some unit $u \in R$, and $\alpha \in Z(SUm_r(R))$.*

Note that in the above statement we have replaced $SUm_r(R)$ by $EUm_r(R)$ in [29, Lemma 4.7]. This is possible due to the Corollary 17.

The above lemma is really the key to verifying formulas relating to the action of an element of $SUm_r(R)$ on a Suslin matrix $S_r(v, w)$.

Computational Techniques in $SUm_r(R)$

We illustrate different computational techniques which help to prove the relations in the group $EUm_r(R)$, etc. Each method has its own merit. Here we collate **five** such methods.

(1) **Direct Computational Method**: In this method we directly evaluate both sides of the relation using the properties of Suslin matrices as in Lemma 6, and show that they are equal.
(2) **Circle Type Method**: In this method we arrange the matrix block entries in a particular way. The arrangement helps us to do the matrix multiplication easily; as well as gives an inductive framework. We now define this particular type of arrangement in the following definition.

Definition 22 Let R be a commutative ring with 1. For $\alpha, \beta \in M_2(M_n(R))$, $r \geq 1$, say $\alpha = \begin{pmatrix} \alpha_{11} & \alpha_{12} \\ \alpha_{21} & \alpha_{22} \end{pmatrix}$, $\beta = \begin{pmatrix} \beta_{11} & \beta_{12} \\ \beta_{21} & \beta_{22} \end{pmatrix}$, where each $\alpha_{ij}, \beta_{ij} \in M_n(R)$, $1 \leq i, j \leq 2$. We denote by $\alpha \odot \beta$ (read as 'α circles β') the matrix

$$\begin{pmatrix} \alpha_{11} & 0 & 0 & \alpha_{12} \\ 0 & \beta_{11} & \beta_{12} & 0 \\ 0 & \beta_{21} & \beta_{22} & 0 \\ \alpha_{21} & 0 & 0 & \alpha_{22} \end{pmatrix} \in M_4(M_n(R)).$$

Definition 23 A matrix $\alpha \in M_{2r}(R)$ is said to be **circled type** if there eixsts $\beta, \gamma \in M_r(R)$ such that $\alpha = \beta \odot \gamma$.

(Suslin Matrices of Circled Type)

Let R be a commutative ring with 1 and let $v = (a_0, a_1, \ldots, a_r)$, $w = (b_0, b_1, \ldots, b_r) \in M_{1r+1}(R)$. The Suslin matrix $S_r(v, w)$ is of circled type if and only if $a_1 = 0 = b_1$.

In this case, one observes that

$$S_r(v, w) = S_{r-1}(v_1, w_1) \odot S_{r-1}(v_1^{\odot}, w_1^{\odot}),$$

where $v_1 = (a_0, a_2, \ldots, a_r)$, $w_1 = (b_0, b_2, \ldots, b_r)$, $v_1^{\odot} = (a_0, -b_2, a_3, \ldots, a_r)$, and $w_1^{\odot} = (b_0, -a_2, b_3, \ldots, b_r)$.

(3) **Action Method**: Suppose we expect some relation LHS = RHS. In this method we first show that the action of LHS on $S_r(v, w)$ and the action of RHS on $S_r(v, w)$ are same. Using this equality, one can show via Lemma 20 that LHS·RHS$^{-1} \in Z(SUm_r(R))$ and then by some argument (generic argument), one can show that both LHS and RHS are equal.
(4) **Method of Reflection**: This is a refinement of the previous method. Again see the action on $S_r(v, w)$ as above. However, do show the equality compute via the homomorphism in Theorem 16, $\varphi : SUm_r(R) \to SO_{2(r+1)}(R)$ whose image is a composition of two reflections. So instead of multiplying matrices, one plays with pairs of rows (v, w) of unit length. Finally to check equality, use the fact that $\varphi : EUm_r(R) \to EO_{2(r+1)}(R)$ is surjective and $\ker \varphi \subset Z(SUm_r(R))$.
(5) **Orthogonal Matrix Method**: In this method, we evaluate the image of both LHS and RHS under φ in the matrix form and do the computation in $EO_{2(r+1)}(R)$ and show that both the images are the same. Using the surjectivity of φ, one can come back to $EUm_r(R)$ and using some argument as in the previous method one can say that both sides are equal.

Quillen–Suslin Theory for $EUm_r(R[X])$

The image of φ contains all even products of reflections, and hence, in particular, all elementary orthogonal matrices.

Thus, all questions concerning the group $EUm_r(R)$ can be reduced to the corresponding questions regarding elementary orthogonal matrices. For example, one has a Quillen–Suslin theory for the elementary orthogonal groups $EO_{2n}(R[X])$ due to results of Suslin–Kopeiko in [57]—both the Local Global Principle and the Monic Inversion Principle of Quillen–Suslin hold for the Elementary Unimodular vector group $EUm_r(R[X])$. From the Local Global Principle, or otherwise, one can conclude that $\mathrm{EUm}_r(R[X])$ is a normal subgroup of $SUm_r(R[X])$, for $r > 1$.

$SUm_r(R)/EUm_r(R) \hookrightarrow SO_{2(r+1)}(R)/EO_{2(r+1)}(R)$

In this subsection, we recall the main work of Jose–Rao in [28] where they show how the Fundamental property led to showing that the quotient of the Special Unimodular vector group by its Elementary unimodular vector group sits inside the orthogonal quotient; *viz.* it was shown in [29, Theorem 4.14] that the induced map φ on the quotients is an injection, whence $SUm_r(R)/EUm_r(R)$ is a subgroup of the orthogonal quotient $SO_{2(r+1)}(R)/EO_{2(r+1)}(R)$.

This is clear from Proposition 20 which shows that φ maps $EUm_r(R) \to EO_{2(r+1)}(R)$ given by $\varphi(S_r(v, w)) = T_{S_r(v,w)}$ is surjective. Moreover, one has the kernel of the map $\varphi : SUm_r(R) \to SO_{2(r+1)}(R)$ is contained in $Z(SUm_r(R))$.

R. Hazrat and N. Vavilov, using ideas of A. Bak in [6], have shown in [24] that the orthogonal quotient group is nilpotent. Hence, the unimodular vector group quotient $SUm_r(R)/EUm_r(R)$ is a nilpotent group, for $r > 1$.

Injective Stability for the K_1 Orthogonal Functor

We used results in [48, Sect. 4] in [50, Corollary 2.7] to show that the injective stability for the orthogonal K_1O functor cannot fall, in general for an affine algebra. We recapitulate that result here. Thus the Suslin matrices have been found useful in the context of injective stability bounds of the orthogonal K_1O functors.

Before that we recall yet another lemma from [29].

Lemma 24 *Let $S_r(v, w)$, $S_r(v', w')$, $r > 1$, $\langle v, w\rangle = \langle v', w'\rangle = 1$, be Suslin matrices. If $S_r(v, w) \in S_r(v', w')EUm_r(R)$, then*

(i) if r is even $\chi_2(v) \overset{E}{\sim} \chi_2(v')$,
(ii) if r is odd $\chi_4(v) \overset{E}{\sim} \chi_4(v')$.

Lemma 25 *Let A be a an affine algebra of dimension d over a perfect field k, of characteristic $\neq 2$, and with $c.d._2(k) \leq 1$. Assume that $mA = A$ for some $m > 0$. If $v \in Um_{d+1}(A)$ then there is a row of the form $(v_1^m, \ldots, v_{d+1})$ in the elementary orbit of v.*

Theorem 26 ([50, Theorem 2]) *Let A be a an affine algebra of dimension d over an algebraically clsoed field, or a nonsingular one over a perfect C_1-field. Assume $2A = A$. If the natural map*

$$\rho_O : \frac{SO_{2(d+1)}(A)}{EO_{2(d+1)}(A)} \leftrightarrow \frac{SO_{2(d+2)}(A)}{EO_{2(d+2)}(A)}$$

is an isomorphism, then every unimodular $(d+1)$*-row over* A *can be completed to an elementary matrix. However,* $Um_{d+1}(A) = e_1 E_{d+1}(A)$ *does not hold in general.*

Proof Let d be odd. Let $v \in Um_{d+1}(A)$. Choose any w with $v \cdot w^t = 1$. By Lemma 19 the matrix of the linear transformation $T_{S_d(v,w)}$ is a commutator, hence stably elementary orthogonal. The hypothesis enables us to conclude that it is elementary orthogonal. By Lemma 24, $S_d(v, w) \in EUm_d(A)$. Moreover, by Lemma 24, $\chi_4(v) = 1$.

By Lemma 25 as $2A = A$, every row $v \in Um_{d+1}(A)$ is a $\chi_4(v')$, for some $v' \in Um_{d+1}(A)$. The result follows.

A similar argument can be given when d is even. Using the corresponding results of [29].

Corollary 27 *There exist affine algebras* A *of dimension* $d \geq 2$ *over a perfect* C_1*-field* k *for which the injective stability estimate for* $K_1 O(A)$ *is not less than* $2(d+2)$.

Proof. We recall the argument in [50] for completeness. If not, then by previous theorem, $e_1 E_{d+1}(A) = Um_{d+1}(A)$, for all such A. This would imply by [58, Lemmas 8.5 and 8.9] that stable rank of $A \leq$ stable rank of $k[X_0, \ldots, X_d] \leq d+1$. By [58, Corollary 8.6], $MS_d(A, (a)) = 0$. But this contradicts the examples in [58, Sect. 6], where Suslin shows that there are affine k-algebras A, and principal ideals (a), with $MS_d(A, (a)) \neq 0$.

Theorem 28 *Let* A *be a local ring of dimension* d*, with* $2A = A$*. If the natural map* $SO_{2(d+1)}(A[X])/EO_{2(d+1)}(A[X]) \longrightarrow K_1 O(A)$ *is an isomorphism, then every unimodular* $(d+1)$*-row over* $A[X]$ *can be completed to an elementary matrix.*

Proof: Arguing as above one can deduce, via the divisibility results in [42], that $Um_{d+1}(R[X]) = e_1 E_{d+1}(A[X])$. This contradicts the result in [48].

Corollary 29 *There exists an affine algebras* A *of dimension* 3*, and a maximal ideal* $\mathfrak{m}$ *of* A*, for which the injective stability estimate for* $K_1 O(A_{\mathfrak{m}}[X])$ *is not* 8.

Proof We revisit the proof in [50]. In [48, Sect. 4], it is shown that if $A = k[X, Y, Z]/(Z^7 - X^2 - Y^3)$, where $k = \mathbb{C}$ or a sufficiently large field, then $Um_3(A[T, T^{-1}][X], (X)) \neq e_1 E_3(A[T, T^{-1}][X])$. Note that A is regular except at the maximal ideal $\mathfrak{m} = (X, Y, Z)$. Hence, by Suslin's version of the Local Global Principle in [55], and T. Vorst's theorem in [75], it follows that there is a maximal ideal $\mathfrak{M}$ containing $\mathfrak{m}[T, T^{-1}]$ such that $e_1 E_3(A[T, T^{-1}]_{\mathfrak{M}}[X] \neq Um_3(A[T, T^{-1}]_{\mathfrak{M}}[X])$. Now apply Theorem 28.

Acknowledgments The second named author would like to thank I.C.T.P., Trieste for its hospitality during which parts of this survey was studied. The authors thank Jean Fasel for his inputs on recent developments on Suslin matrices.

Appendix: Reflections via MuPAD

We define the reflection function $\tau_{(x,y)}(z, w)$ via MuPAD for $r = 4$, where x, y, z, w are vectors of length 5 as follows: In all the commands given below, we suppress the output by putting colon (:) at the end of each input statement.

To define the function $\tau_{(x,y)}(z, w)$, we need to define the vectors x, y, z, w. The vectors `x, y, z, w` are defined as

- `x := matrix([[x0,x1,x2,x3,x4]]):`
- `y := matrix([[y0,y1,y2,y3,y4]]):`
- `z := matrix([[z0,z1,z2,z3,z4]]):`
- `w := matrix([[b0,b1,b2,b3,b4]]):`
- `assume(Type::Real):`
- ```
  f:=(x,y,z,w) -> linalg::scalarProduct(x,y) * matrix
  ([z,w])
  -(linalg::scalarProduct(x,w) + linalg::scalarProduct
  (y,z))
  * matrix([x,y]):
  ```

The above statement defines the function

$$f(x, y, z, w) = \langle x, y\rangle(z, w) - (\langle x, w\rangle + \langle y, z\rangle)(x, y).$$

Thus $f(x, y, z, w)$ will give the value of $\tau_{(x,y)}(z, w)$.

As an illustration, we give the computation we did in the proof of Proposition 20 for $i = 5$, $j = 3$. The computation uses the following vectors:

- `v := matrix([[a0,a1,a2,a3,a4]]):`
- `w := matrix([[b0,b1,b2,b3,b4]]):`
- `e1 := matrix([[1,0,0,0,0]]):`
- `ei := matrix([[0,0,0,0,1]]):`
- `ej := matrix([[0,0,1,0,0]]):`

In the following input statements, we use `L` for λ. We first evaluate $\tau_{(e_1-e_j,e_1)} \circ \tau_{(-(1-\lambda)e_1+e_j,-e_1+\lambda e_j)} \circ \tau_{(e_1-e_j,e_1-e_i)} \circ \tau_{(-(1+\lambda)e_1+e_j,-e_1-\lambda e_j+e_i)} \circ \tau_{(e_1,e_1-e_i)} \circ \tau_{(-e_1,-e_1+\lambda e_j+e_i)} \circ \tau_{(e_1,e_1-\lambda e_j)} \circ \tau_{(e_1,e_1)}$ at (v, w).

- `AA := simplify(f(e1,e1,v,w))`

Output:

$$v_1 = (-b_0, a_1, a_2, a_3, a_4) \text{ and}$$
$$w_1 = (-a_0, b_1, b_2, b_3, b_4)$$

- `AB := simplify(f(e1,e1-L*ej,AA[1],AA[2]))`
  ```

Output:

$$v_2 = (a_0 + LPa_2, a_1, a_2, a_3, a_4) \text{ and}$$
$$w_2 = (b_0 + LPa_2, b_1, b_2 - L^2Pa_2 - LPa_0 - LPb_0, b_3, b_4)$$

- `AC := simplify(f(-e1,-e1+L*ej+ei,AB[1],AB[2]))`

Output:

$$v_3 = (a_4 - b_0, a_1, a_2, a_3, a_4) \text{ and}$$
$$w_3 = (-a_0 + a_4, b_1, b_2 - LPa_4, b_3, a_0 - a_4 + b_0 + b_4 + LPa_2)$$

- `AD := simplify(f(e1,e1-ei,AC[1],AC[2]))`

Output:

$$v_4 = (a_0, a_1, a_2, a_3, a_4) \text{ and}$$
$$w_4 = (b_0, b_1, b_2 - LPa_4, b_3, b_4 + LPa_2)$$

- `AE := simplify(f(-(1+L)*e1+ej,-e1-L*ej+ei,AD[1], AD[2]))`

Output:

$$v_5 = (a_4 - b_0 + b_2 - L^2Pa_2 - L^2Pa_4 - L^2Pb_0 - LPa_0 - LPa_2 - 2PLPb_0 + LPb_2,$$
$$a_1, a_0 + a_2 - a_4 + b_0 - b_2 + L.a_2 + L.a_4 + L.b_0, a_3, a_4) \text{ and}$$
$$w_5 = (-a_0 + a_4 + b_2 - LPa_2 - LPa_4 - LPb_0, b_1,$$
$$b_2 - L^2Pa_2 - L^2Pa_4 - L^2Pb_0 - LPa_0 - LPb_0 + LPb_2, b3,$$
$$a_0 - a_4 + b_0 - b_2 + b_4 + 2.L.a_2 + L.a_4 + L.b_0)$$

- `AF := simplify(f(e1-ej,e1-ei,AE[1],AE[2]))`

Output:

$$v_6 = (a_0 - L^2Pa_2 - L^2Pa_4 - L^2Pb_0 - LPa_0 + LPa_2 + LPa_4 + LPb_2, a_1,$$
$$a_2 - LPa_2 - LPb_0, a_3, a_4) \text{ and}$$
$$w_6 = (b_0 + LPa_2 + LPb_0, b_1,$$
$$b_2 - L^2Pa_2 - L^2Pa_4 - L^2Pb_0 - LPa_0 - LPb_0 + LPb_2, b_3, b_4 - LPb_0)$$

- `AG := simplify(f(-(1-L)*e1+ej,-e1+L*ej,AF[1],AF[2]))`

Output:

$$v_7 = (-b_0 + b_2, a_1, a_0 + a_2 + b_0 - b_2 + L.a_4, a_3, a_4) \text{ and}$$
$$w_7 = (-a_0 + b_2 - L.a_4, b_1, b_2, b_3, b_4 - L.b_0)$$

- `AH := simplify(f(e1-ej,e1,AG[1],AG[2]))`

Output:

$$v_8 = (a_0 + LPa_4, a_1, a_2, a_3, a_4) \text{ and}$$
$$w_8 = (b_0, b_1, b_2, b_3, b_4 - LPb_0)$$

Note that this value is same as $oe_{15}(\lambda)\begin{pmatrix} v^t \\ w^t \end{pmatrix}$, where

$$oe_{15}(\lambda) = \begin{pmatrix} 1 & 0 & 0 & 0 & \lambda & 0 & 0 & 0 & 0 & 0 \\ 0 & 1 & 0 & 0 & 0 & 0 & 0 & 0 & 0 & 0 \\ 0 & 0 & 1 & 0 & 0 & 0 & 0 & 0 & 0 & 0 \\ 0 & 0 & 0 & 1 & 0 & 0 & 0 & 0 & 0 & 0 \\ 0 & 0 & 0 & 0 & 1 & 0 & 0 & 0 & 0 & 0 \\ 0 & 0 & 0 & 0 & 0 & 1 & 0 & 0 & 0 & 0 \\ 0 & 0 & 0 & 0 & 0 & 0 & 1 & 0 & 0 & 0 \\ 0 & 0 & 0 & 0 & 0 & 0 & 0 & 1 & 0 & 0 \\ 0 & 0 & 0 & 0 & 0 & 0 & 0 & 0 & 1 & 0 \\ 0 & 0 & 0 & 0 & 0 & -\lambda & 0 & 0 & 0 & 1 \end{pmatrix}.$$

References

1. Apte, H., Chattopadhyay, P., Rao, R.A.: A local global theorem for extended ideals. J. Ramanujan Math. Soc. **27**(1), 1–20 (2012)
2. Asok, A., Fasel, J.: An explicit KO-degree map and applications. arXiv:1403.4588
3. Asok, A., Fasel, J.: Euler Class Groups and Motivic Stable Cohomotopy, Preprint. http://arxiv.org/pdf/1601.05723v1.pdf
4. Asok, A., Fasel, J.: Algebraic vector bundles on spheres. J. Topol. **7**(3), 894–926 (2014)
5. Asok, A., Fasel, J.: Splitting vector bundles outside the stable range and $\mathbb{A}^1$-homotopy sheaves of punctured affine spaces. J. Am. Math. Soc. **28**(4), 1031–1062 (2015)
6. Bak, A.: Nonabelian K-Theory: The Nilpotent Class of K_1 and General Stability. K-Theory **4**(4), 363–397 (1991)
7. Bass, H.: Some problems in "classical" algebraic K-theory, Algebraic K-theory, II: "Classical" algebraic K-theory and connections with arithmetic. Proceedings of Conference, Battelle Memorial Institute, Seattle, Washington, 1972. Lecture Notes in Mathematics, vol. 342, pp. 3–73. Springer, Berlin (1973)
8. Bass, H., Milnor, J., Serre, J.-P.: Solution of the congruence subgroup problem for SL_n ($n \geq 2$). Publ. Math. IHES **33**, 59–137 (1967)
9. Basu, R., Rao, R.A.: Injective stability for K_1 of classical modules. J. Algebra **323**(4), 867–877 (2010)
10. Basu, R., Chattopadhyay, P., Rao, R.A.: Some remarks on symplectic injective stability. Proc. Am. Math. Soc. **139**(7), 2317–2325 (2011)
11. Boratyński, M.: A note on set-theoretic complete intersection ideals. J. Algebra **54**(1), 1–5 (1978)
12. Chattopadhyay, P., Rao, R.A.: Elementary symplectic orbits and improved K_1-stability. J. K-Theory **7**, 389–403 (2011)
13. Chattopadhya, P., Rao, R.A.: Equality of linear and symplectic orbits (with an Appendix by Alpesh Dhorajia). J. Pure Appl. Algebra **219**, 5363–5386 (2015)

14. Chattopadhyay, P., Rao, R.A.: Equality of elementary linear and symplectic orbits with respect to an alternating form. J. Algebra **451**, 46–64 (2016)
15. Chintala, V.R.: On Suslin matrices and their connection to spin groups. Doc. Math. **20**, 531–550 (2015)
16. Cohn, P.M.: On the structure of the GL_2 of a ring. Publ. Math. IHES **30**, 5–53 (1966)
17. Fasel, J.: Mennicke symbols, K-cohomology and a Bass-Kubota theorem. Trans. Am. Math. Soc. **367**(1), 191–208 (2015)
18. Fasel, J., Srinivas, V.: Chow-Witt groups and Grothendieck-Witt groups of regular schemes. Adv. Math **221**(1), 302–329 (2009)
19. Fasel, J., Rao, R.A., Swan, R.G.: On stably free modules over affine algebras. Inst. Hautes Études Scientifiques **116**(1), 223–243 (2012)
20. Fossum, R., Foxby, H.B., Iversen, B: A Characteristic class in Algebraic K-theory, Aarhus University, 1978–1979, Preprint Number 29 (Unpublished)
21. Garge, A.S., Rao, R.A.: A nice group structure on the orbit space of unimodular rows. K-Theory **38**(2), 113–133 (2008)
22. Gupta, A.: Optimal injective stability for the symplectic K1Sp group. J. Pure Appl. Algebra **219**(5), 1336–1348 (2015)
23. Gupta, A., Garge, A., Rao, R.A.: A nice group structure on the orbit space of unimodular rowsII. J. Algebra **407**, 201–223 (2014)
24. Hazrat, R., Vavilov, N.: K_1 of Chevalley groups are nilpotent. J. Pure Appl. Algebra **179**(1–2), 99–116 (2003)
25. Ischebeck, I., Rao, R.A.: Ideals and reality, Projective modules and number of generators of ideals. Springer Monographs in Mathematics. Springer, Berlin (2005). ISBN: 3-540-23032-7
26. Jose, S.: Quillen–Suslin theory, Ph.D. thesis submitted to the University of Pune (2007)
27. Jose, S., Rao, R.A.: A structure theorem for the elementary unimodular vector group. Trans. Am. Math. Soc. **358**(7), 3097–3112 (2005)
28. Jose, S., Rao, R.A.: A local global principle for the elementary unimodular vector group. Commutative Algebra and Algebraic Geometry (Bangalore, India, 2003). Contemporary Mathematics, pp. 119–125. American Mathematical Society, Providence (2005)
29. Jose, S., Rao, R.A.: A fundamental property of Suslin matrices. J. K-Theory **5**(3), 407–436 (2010)
30. Jose, S., Rao, R.A.: Suslin forms and the Hodge star operator. Linear Algebra Appl. **452**, 328–344 (2014)
31. Jose, S., Rao, R.A.: The orbit set and the orthogonal quotient. J. Algebra Appl. **15**(6), 1650109 (2016)
32. Jose, S., Rao, R.A.: A Witt group structure on orbit spaces of unimodular rows (in preparation)
33. Katre, S.A., Rao, R.A., Sheth, D.N.: Solving linear systems via Pfaffians. Linear Algebra Appl. **430**(4), 968–975 (2009)
34. Krusemeyer, M.: Completing $(\alpha^2, \beta, \gamma)$. Conference on Commutative Algebra-1975. Queen's Papers on Pure and Applied Mathematics, vol. 42, pp. 253–254. Queen's University, Kingston, Ontario (1975)
35. Krusemeyer, M.: Skewly completable rows and a theorem of Swan and Towber. Commun. Algebra **4**(7), 657–663 (1976)
36. Lam, T.Y.: Springer Monographs in Mathematics. Springer, Berlin (2006)
37. Lindel, H.: On the Bass–Quillen conjecture concerning projective modules over polynomial rings. Invent. Math. **65**(2), 319–323 (1981–1982)
38. Mohan Kumar, N.: Stably free modules. Am. J. Math. **107**(6), 1439–1444 (1985, 1986)
39. Nori, M.V., Rao, R.A., Swan, R.G.: Non-self-dual stably free modules. Quadratic forms, linear algebraic groups, and cohomology. Developmental Mathematics, vol. 18, pp. 315–324. Springer, New York (2010)
40. Quillen, D.: Projective modules over polynomial rings. Invent. Math. **36**, 167–171 (1976)
41. Rao, R.A.: Two examples of the Bass-Quillen-Suslin conjectures. Math. Ann. **279**(2), 227–238 (1987)

42. Rao, R.A.: The Bass-Quillen conjecture in dimension three but characteristic $\neq 2, 3$ via a question of A. Suslin. Invent. Math. **93**(3), 609–618 (1988)
43. Rao, R.A.: On completing unimodular polynomial vectors of length three. Trans. Am. Math. Soc. **325**(1), 231–239 (1991)
44. Rao, R.A.: An abelian group structure on orbits of "unimodular squares" in dimension 3. J. Algebra **210**(1), 216–224 (1998)
45. Rao, R.A.: A stably elementary homotopy. Proc. Am. Math. Soc. **137**(11), 3637–3645 (2009)
46. Rao, R.A., Jose, S.: Quillen-Suslin theory revisited. J. Pure Appl. Algebra **211**(2), 541–546 (2007)
47. Rao, R.A., Jose, S.: A group structure on squares. Proc. Am. Math. Soc. **136**(4), 1181–1191 (2008)
48. Rao, R.A., Swan, R.G.: A Regenerative property of a fibre of Invertible Alternating Polynomial Matrices, in progress. www.math.uchicago.edu/swan/regenerative.pdf
49. Rao, R.A., van der Kallen, W.: Improved stability for SK_1 and WMS_d of a non-singular affine algebra. K-theory (Strasbourg, 1992). Astérisque **226**, 411–420 (1994)
50. Rao, R.A., Basu, R., Jose, S.: Injective stability for K_1 of the orthogonal group. J. Algebra **323**(2), 393–396 (2010)
51. Roitman, M.: On stably extended projective modules over polynomial rings. Proc. Am. Math. Soc. **97**(4), 585–589 (1986)
52. Sridharan, R.: Homotopy and sections of algebraic vector bundles over real affine algebras. Comment. Math. Helv. **71**(3), 435–452 (1996)
53. Suslin, A.A.: Projective modules over polynomial rings are free. (Russian). Dokl. Akad. Nauk SSSR **229**(5), 1063–1066 (1976)
54. Suslin, A.A., Vaserstein, L.N.: Serre's problem on projective modules over polynomial rings and algebraic K-theory. Math. USSR Izv. **10**, 937–1001 (1976)
55. Suslin, A.A.: On the structure of the special linear group over polynomial rings. Math. USSR Izv. **11**, 221–238 (1977)
56. Suslin, A.A.: On stably free modules. Math. USSR Sbornik **31**, 479–491 (1977)
57. Suslin, A.A., Kopeiko, V.I.: Quadratic modules and the orthogonal group over polynomial rings. (Russian) Modules and representations. Zap. Naun. Sem. Leningrad. Otdel. Mat. Inst. Steklov. (LOMI) **71**, 216–250 (1977)
58. Suslin, A.A.: Reciprocity laws and the stable rank of polynomial rings. (Russian) Izv. Akad. Nauk SSSR Ser. Mat. **43**(6), 1394–1429 (1979)
59. Suslin, A.A.: The cancellation problem for projective modules, and related questions. (Russian) Proceedings of the International Congress of Mathematicians (Helsinki, 1978), pp. 323–330. Academic Science, Fennica, Helsinki (1980)
60. Suslin, A.A.: Cancellation for affine varieties. (Russian) Modules and algebraic groups. Zap. Nauchn. Sem. Leningrad. Otdel. Mat. Inst. Steklov. (LOMI) **114**, 187–195 (1982)
61. Suslin, A.A.: Mennicke symbols and their applications in the K-theory of fields. Algebraic K-theory, Part I (Oberwolfach, 1980). Lecture Notes in Mathematics, vol. 966, pp.334–356. Springer, Berlin (1982)
62. Suslin, A.A.: K-theory and K-cohomology of certain group varieties. Algebraic K-Theory. Advances in Soviet Mathematics, vol. 4, pp. 53–74. American Mathematical Society, Providence (1991)
63. Swan, R.G.: A cancellation theorem for projective modules in the metastable range. Invent. Math. **27**, 23–43 (1974)
64. Swan, R.G.: On a Theorem of Mohan Kumar and Nori. http://www.math.uchicago.edu/swan/MKN.pdf
65. Swan, R.G.: Some Stably Free Modules which are not Self Dual. http://www.math.uchicago.edu/swan/stablyfree.pdf
66. Swan, R.G., Towber, J.: A class of projective modules which are nearly free. J. Algebra **36**(3), 427–434 (1975)
67. van der Kallen, W.: A group structure on certain orbit sets of unimodular rows. J. Algebra **82**(2), 363–397 (1983)

68. van der Kallen, W.: A module structure on certain orbit sets of unimodular rows. J. Pure Appl. Algebra **57**(3), 281–316 (1989)
69. van der Kallen, W.: From Mennicke symbols to Euler class groups. Algebra, arithmetic and geometry, Part I, II (Mumbai, 2000), 341–354, Tata Inst. Fund. Res. Stud. Math., 16: Tata Inst. Fund. Res, Bombay (2002)
70. Vaserstein, L.N.: On the stabilization of the general linear group over a ring. Mat. Sbornik (N.S.) **79**(121), 405–424 (Russian). English translated in Math. USSR-Sbornik. **8**, 383–400 (1969)
71. Vaserstein, L.N.: Stabilization of unitary and orthogonal groups over a ring with involution. Mat. Sbornik, Tom **81**(123)(3), 307–326 (1970)
72. Vaserstein, L.N.: Stable range of rings and dimensionality of topological spaces. Fuct. Anal. Appl. **5**(2), 102–110 (1971)
73. Vaserstein, L.N.: Stabilization For Classical Groups over Rings. (Russian) Mat. Sb. (N.S.) **93**(135), 268–295 (1974)
74. Vaserstein, L.N.: Operations on orbits of unimodular vectors. J. Algebra **100**(2), 456–461 (1986)
75. Vorst, T.: The general linear group of polynomial rings over regular rings. Commun. Algebra **9**(5), 499–509 (1981)

Variations on the Grothendieck–Serre Formula for Hilbert Functions and Their Applications

Shreedevi K. Masuti, Parangama Sarkar and J.K. Verma

Abstract In this expository paper, we present proofs of Grothendieck–Serre formula for multi-graded algebras and Rees algebras for admissible multi-graded filtrations. As applications, we derive formulas of Sally for postulation number of admissible filtrations and Hilbert coefficients. We also discuss a partial solution of Itoh's conjecture by Kummini and Masuti. We present an alternate proof of Huneke–Ooishi Theorem and a generalisation for multi-graded filtrations.

Keywords Hilbert polynomial · Admissible filtration · Normal Hilbert polynomial · Joint reduction · Local cohomology · Rees algebra · Multi-graded filtration · Grothendieck–Serre formula

1 Introduction

The objective of this expository paper is to collect together several fundamental results about Hilbert coefficients of admissible filtrations of ideals which can be proved using various avatars of the Grothendieck–Serre formula for the difference of the Hilbert function and Hilbert polynomial of a finite graded module of a standard graded Noetherian ring. The proofs presented here provide a unified way of approach-

---

The first author is partially supported by a grant from Infosys Foundation; The second author is supported by CSIR Fellowship of Government of India.

S.K. Masuti (✉)
Chennai Mathematical Institute,
Siruseri, Kelambakkam, Chennai 603103, Tamil Nadu, India
e-mail: shreedevikm@cmi.ac.in

P. Sarkar · J.K. Verma
Department of Mathematics, Indian Institute of Technology Bombay,
Mumbai 400076, India
e-mail: parangama@math.iitb.ac.in

J.K. Verma
e-mail: jkv@math.iitb.ac.in

S.T. Rizvi et al. (eds.), *Algebra and its Applications*, Springer Proceedings in Mathematics & Statistics 174, DOI 10.1007/978-981-10-1651-6_8

ing these results. Some of these results are not known in the multi-graded case. We hope that the unified approach presented here could lead to suitable multi-graded analogues of these results.

We begin by recalling the Grothendieck–Serre formula. For the sake of simplicity, we assume that the graded rings considered in this paper are standard and Noetherian. Let $R = \bigoplus_{n=0}^{\infty} R_n$ be a standard graded Noetherian ring where R_0 is an Artinian local ring. Let $M = \bigoplus_{n\in\mathbb{Z}} M_n$ be a finite graded R-module of dimension d. The Hilbert function of M is the function $H(M,n) = \lambda_{R_0}(M_n)$ for all $n \in \mathbb{Z}$. Here λ denotes the length function. Serre showed that there exists an integer m so that $H(M,n)$ is given by a polynomial $P(M,x) \in \mathbb{Q}[x]$ of degree $d-1$ such that $H(M,n) = P(M,n)$ for all $n > m$. The smallest such m is called the postulation number of M. Let R_+ denote the homogeneous ideal of R generated by elements of positive degree and $[H^i_{R_+}(M)]_n$ denote the nth graded component of the ith local cohomology module $H^i_{R_+}(M)$ of M with respect to the ideal R_+. We put $\lambda_{R_0}([H^i_{R_+}(M)]_n) = h^i_{R_+}(M)_n$.

Theorem 1.1 (Grothendieck–Serre) *For all $n \in \mathbb{Z}$, we have*

$$H(M,n) - P(M,n) = \sum_{i=0}^{d}(-1)^i h^i_{R_+}(M)_n.$$

The GSF was proved in the fundamental paper [37] of J.-P. Serre. We quote from [6]: "In this paper, Serre introduced the theory of coherent sheaves over algebraic varieties over an algebraically closed field and a cohomology theory of such varieties with coefficients in coherent sheaves. He did speak of algebraic coherent sheaves, as at the first time he managed to introduce these theories with purely algebraic tools, using consequently the Zariski topology instead of the complex topology and homological methods instead of tools from multivariate complex analysis. Since then, the cohomology theory introduced in Serre's paper is often called Serre cohomology or sheaf cohomology.

One of the achievement of Serre's paper is the Grothendieck–Serre Formula, which is given there in terms of sheaf cohomology and showed in this way that sheaf cohomology gives a functorial understanding of the so called postulation problem of algebraic geometry, the problem which classically consisted in understanding the difference between the Hilbert function and the Hilbert polynomial of the coordinate ring of a projective variety."

The Grothendieck–Serre Formula (GSF) is valid for nonstandard graded rings also if the Hilbert polynomial $P(M,x)$ is replaced by the Hilbert quasi-polynomial [4, Theorem 4.4.3]. The GSF has been generalised in several directions. For some of the applications, we need it in the context of $\mathbb{Z}^s$-graded modules over standard $\mathbb{N}^s$-graded rings. In order to state the GSF for $\mathbb{Z}^s$-graded module, first we set up notation and recall some definitions. Let $(R,\mathfrak{m})$ be a Noetherian local ring and $I_1,\dots,I_s$ be $\mathfrak{m}$-primary ideals of R. We put $e = (1,\dots,1)$, $\underline{0} = (0,\dots,0) \in \mathbb{Z}^s$ and for all $i = 1,\dots,s$, $e_i = (0,\dots,1,\dots,0) \in \mathbb{Z}^s$ where 1 occurs at ith position. For $\underline{n} = (n_1, \ldots, n_s) \in \mathbb{Z}^s$, we write $\underline{I}^{\underline{n}} = I_1^{n_1} \cdots I_s^{n_s}$ and $\underline{n}^+ = (n_1^+, \ldots, n_s^+)$ where $n_i^+ = \max\{0, n_i\}$ for all $i = 1, \ldots, s$. For $\alpha = (\alpha_1, \ldots, \alpha_s) \in \mathbb{N}^s$, we put $|\alpha| = \alpha_1 + \cdots + \alpha_s$. We define $\underline{m} = (m_1, \ldots, m_s) \geq \underline{n} = (n_1, \ldots, n_s)$ if $m_i \geq n_i$ for all $i = 1, \ldots, s$. By the phrase "for all large $\underline{n}$" we mean $\underline{n} \in \mathbb{N}^s$ and $n_i \gg 0$ for all $i = 1, \ldots, s$. For an $\mathbb{N}^s$ (or a $\mathbb{Z}^s$)-graded ring T, the ideal generated by elements of degree e is denoted by T_{++}.

Definition 1.2 A set of ideals $\mathcal{F} = \{\mathcal{F}(\underline{n})\}_{\underline{n} \in \mathbb{Z}^s}$ is called a $\mathbb{Z}^s$-graded $\underline{I} = (I_1, \ldots, I_s)$-**filtration** if for all $\underline{m}, \underline{n} \in \mathbb{Z}^s$, (i) $\underline{I}^{\underline{n}} \subseteq \mathcal{F}(\underline{n})$, (ii) $\mathcal{F}(\underline{n})\mathcal{F}(\underline{m}) \subseteq \mathcal{F}(\underline{n} + \underline{m})$ and (iii) if $\underline{m} \geq \underline{n}$, $\mathcal{F}(\underline{m}) \subseteq \mathcal{F}(\underline{n})$.

Let $R = \bigoplus_{\underline{n} \in \mathbb{N}^s} R_{\underline{n}}$ be a standard Noetherian $\mathbb{N}^s$-graded ring defined over a local ring $(R_{\underline{0}}, \mathfrak{m})$ and $R_{++} = \bigoplus_{\underline{n} \geq e} R_{\underline{n}}$. Let $\operatorname{Proj}(R)$ denote the set of all homogeneous prime ideals P in R such that $R_{++} \not\subseteq P$. For a finitely generated module M, set $\operatorname{Supp}_{++}(M) = \{P \in \operatorname{Proj}(R) \mid M_P \neq 0\}$. Note that $\operatorname{Supp}_{++}(M) = V_{++}(\operatorname{Ann}(M))$ [7, Lemma 2.2.5], [15].

Definition 1.3 The **relevant dimension** of M is

$$\operatorname{rel.dim}(M) = \begin{cases} s - 1 & \text{if } \operatorname{Supp}_{++}(M) = \emptyset \\ \max\{\dim(R/P) \mid P \in \operatorname{Supp}_{++}(M)\} & \text{if } \operatorname{Supp}_{++}(M) \neq \emptyset. \end{cases}$$

By [15, Lemma 1.1], $\dim \operatorname{Supp}_{++}(M) = \operatorname{rel.dim}(M) - s$. M. Herrmann, E. Hyry, J. Ribbe and Z. Tang [15, Theorem 4.1] proved that if $R = \bigoplus_{\underline{n} \in \mathbb{N}^s} R_{\underline{n}}$ is a standard Noetherian $\mathbb{N}^s$-graded ring defined over an Artinian local ring $(R_{\underline{0}}, \mathfrak{m})$ and $M = \bigoplus_{\underline{n} \in \mathbb{Z}^s} M_{\underline{n}}$ is a finitely generated $\mathbb{Z}^s$-graded R-module then there exists a polynomial, called the Hilbert polynomial of M, $P_M(x_1, x_2, \ldots, x_s) \in \mathbb{Q}[x_1, \ldots, x_s]$ of total degree $\dim \operatorname{Supp}_{++}(M)$ satisfying $P_M(\underline{n}) = \lambda(M_{\underline{n}})$ for all large $\underline{n}$. Moreover all monomials of highest degree in this polynomial have nonnegative coefficients.

The next two results are due to G. Colomé-Nin [7, Propositions 2.4.2 and 2.4.3] for nonstandard multi-graded rings. In Sect. 2, we present her proofs to prove the same results for standard multigraded rings for the sake of simplicity. These results were proved in the bigraded case by A.V. Jayanthan and J.K. Verma [19].

Proposition 1.4 *Let $R = \bigoplus_{\underline{n} \in \mathbb{N}^s} R_{\underline{n}}$ be a standard Noetherian $\mathbb{N}^s$-graded ring defined over a local ring $(R_{\underline{0}}, \mathfrak{m})$ and $M = \bigoplus_{\underline{n} \in \mathbb{Z}^s} M_{\underline{n}}$ a finitely generated $\mathbb{Z}^s$-graded R-module. Then*

(1) For all $i \geq 0$ and $\underline{n} \in \mathbb{Z}^s$, $[H^i_{R_{++}}(M)]_{\underline{n}}$ is finitely generated $R_{\underline{0}}$-module.
(2) For all large $\underline{n}$ and $i \geq 0$, $[H^i_{R_{++}}(M)]_{\underline{n}} = 0$.

Theorem 1.5 (Grothendieck–Serre formula for $\mathbb{Z}^s$-graded modules) *Let* $R = \bigoplus_{\underline{n} \in \mathbb{N}^s} R_{\underline{n}}$ *be a standard Noetherian* $\mathbb{N}^s$*-graded ring defined over an Artinian local ring* $(R_{\underline{0}}, \mathfrak{m})$ *and* $M = \bigoplus_{\underline{n} \in \mathbb{Z}^s} M_{\underline{n}}$ *a finitely generated* $\mathbb{Z}^s$*-graded* R*-module. Let* $H_M(\underline{n}) = \lambda(M_{\underline{n}})$ *and* $P_M(x_1, \ldots, x_s)$ *be the Hilbert polynomial of* M*. Then for all* $\underline{n} \in \mathbb{Z}^s$,

$$H_M(\underline{n}) - P_M(\underline{n}) = \sum_{j \geq 0} (-1)^j h^j_{R_{++}}(M)_{\underline{n}}.$$

The above form of the GSF leads to another version of it which gives the difference between the Hilbert polynomial and the function of $\mathbb{Z}^s$-graded filtrations of ideals in terms of local cohomology modules of various forms of Rees rings and associated graded rings of ideals. To define these, let $t_1, t_2, \ldots, t_s$ be indeterminates and $\underline{t}^{\underline{n}} = t_1^{n_1} \cdots t_s^{n_s}$. We put

$\mathcal{R}(\mathcal{F}) = \bigoplus_{\underline{n} \in \mathbb{N}^s} \mathcal{F}(\underline{n}) \underline{t}^{\underline{n}}$ the Rees ring of $\mathcal{F}$,

$\mathcal{R}'(\mathcal{F}) = \bigoplus_{\underline{n} \in \mathbb{Z}^s} \mathcal{F}(\underline{n}) \underline{t}^{\underline{n}}$ the extended Rees ring of $\mathcal{F}$,

$G(\mathcal{F}) = \bigoplus_{\underline{n} \in \mathbb{N}^s} \dfrac{\mathcal{F}(\underline{n})}{\mathcal{F}(\underline{n} + e)}$ the associated multigraded ring of $\mathcal{F}$ with respect to $\mathcal{F}(e)$,

$G_i(\mathcal{F}) = \bigoplus_{\underline{n} \in \mathbb{N}^s} \dfrac{\mathcal{F}(\underline{n})}{\mathcal{F}(\underline{n} + e_i)}$ the associated graded ring of $\mathcal{F}$ with respect to $\mathcal{F}(e_i)$.

For $\mathcal{F} = \{\underline{I}^{\underline{n}}\}_{\underline{n} \in \mathbb{Z}^s}$, we set $\mathcal{R}(\mathcal{F}) = \mathcal{R}(\underline{I})$ and $\mathcal{R}'(\mathcal{F}) = \mathcal{R}'(\underline{I})$, $G(\mathcal{F}) = G(\underline{I})$ and $G_i(\mathcal{F}) = G_i(\underline{I})$ for all $i = 1, \ldots, s$.

Definition 1.6 A $\mathbb{Z}^s$-graded $\underline{I}$-filtration $\mathcal{F} = \{\mathcal{F}(\underline{n})\}_{\underline{n} \in \mathbb{Z}^s}$ of ideals in R is called an $\underline{I}$**-admissible filtration** if $\mathcal{F}(\underline{n}) = \mathcal{F}(\underline{n}^+)$ and $\mathcal{R}'(\mathcal{F})$ is a finite $\mathcal{R}'(\underline{I})$-module. For $s = 1$, if a filtration $\mathcal{F}$ is I-admissible for some $\mathfrak{m}$-primary ideal I then it is also I_1-admissible.

Primary examples of $\underline{I}$-admissible filtrations are $\{\underline{I}^{\underline{n}}\}_{\underline{n} \in \mathbb{Z}^s}$ in a Noetherian local ring and $\{\overline{\underline{I}^{\underline{n}}}\}_{\underline{n} \in \mathbb{Z}^s}$ in an analytically unramified local ring. Recall that for an ideal I in R, the integral closure of I is the ideal

$$\begin{aligned} \overline{I} := \{x \in R \mid\ & x^n + a_1 x^{n-1} + \cdots + a_{n-1} x + a_n = 0 \text{ for some } n \in \mathbb{N} \\ & \text{and } a_i \in I^i \text{ for } i = 1, 2 \ldots, n\}. \end{aligned}$$

We now set up the notation for a variety of Hilbert polynomials associated to filtrations of ideals. Let I be an $\mathfrak{m}$-primary ideal of a Noetherian local ring $(R, \mathfrak{m})$ of dimension d. For a $\mathbb{Z}$-graded I-admissible filtration $\mathcal{I} = \{I_n\}_{n \in \mathbb{Z}}$, Marley [23] proved existence of a polynomial $P_{\mathcal{I}}(x) \in \mathbb{Q}[x]$ of degree d, written in the form,

$$P_{\mathcal{I}}(n) = e_0(\mathcal{I})\binom{n+d-1}{d} - e_1(\mathcal{I})\binom{n+d-2}{d-1} + \cdots + (-1)^d e_d(\mathcal{I})$$

such that $P_{\mathcal{I}}(n) = H_{\mathcal{I}}(n)$ for all large n, where $H_{\mathcal{I}}(n) = \lambda(R/I_n)$ is the **Hilbert function** of the filtration $\mathcal{I}$. The coefficients $e_i(\mathcal{I})$ for $i = 0, 1, \ldots, d$ are integers, called the **Hilbert coefficients** of $\mathcal{I}$. The coefficient $e_0(\mathcal{I})$ is called the multiplicity of $\mathcal{I}$. P. Samuel [36] showed existence of this polynomial for the I-adic filtration $\{I^n\}_{n\in\mathbb{Z}}$. Many results about Hilbert polynomials for admissible filtrations were proved in [9, 33].

For $\mathfrak{m}$-primary ideals $I_1, \ldots, I_s$, B. Teissier [38] proved that for all $\underline{n}$ sufficiently large, the **Hilbert function** $H_{\underline{I}}(\underline{n}) = \lambda\left(R/\underline{I}^{\underline{n}}\right)$ coincides with a polynomial

$$P_{\underline{I}}(\underline{n}) = \sum_{\substack{\alpha=(\alpha_1,\ldots,\alpha_s)\in\mathbb{N}^s \\ |\alpha|\le d}} (-1)^{d-|\alpha|} e_\alpha(\underline{I}) \binom{n_1+\alpha_1-1}{\alpha_1} \cdots \binom{n_s+\alpha_s-1}{\alpha_s}$$

of degree d, called the **Hilbert polynomial** of $\underline{I}$. Here we assume that $s \geq 2$ in order to write $P_{\underline{I}}(\underline{n})$ in the above form. This was proved by P.B. Bhattacharya for $s = 2$ in [1]. Here $e_\alpha(\underline{I})$ are integers which are called the **Hilbert coefficients** of $\underline{I}$. D. Rees [31] showed that $e_\alpha(\underline{I}) > 0$ for $|\alpha| = d$. These are called the **mixed multiplicities** of $\underline{I}$.

For an $\underline{I}$-admissible filtration $\mathcal{F} = \{\mathcal{F}(\underline{n})\}_{\underline{n}\in\mathbb{Z}^s}$ in a Noetherian local ring $(R, \mathfrak{m})$ of dimension d, Rees [31] showed the existence of a polynomial

$$P_{\mathcal{F}}(\underline{n}) = \sum_{\substack{\alpha=(\alpha_1,\ldots,\alpha_s)\in\mathbb{N}^s \\ |\alpha|\le d}} (-1)^{d-|\alpha|} e_\alpha(\mathcal{F}) \binom{n_1+\alpha_1-1}{\alpha_1} \cdots \binom{n_s+\alpha_s-1}{\alpha_s}$$

of degree d which coincides with the **Hilbert function** $H_{\mathcal{F}}(\underline{n}) = \lambda\left(R/\mathcal{F}(\underline{n})\right)$ for all large $\underline{n}$ [31]. This polynomial is called the **Hilbert polynomial** of $\mathcal{F}$. Rees [31, Theorem 2.4] proved that $e_\alpha(\mathcal{F}) = e_\alpha(\underline{I})$ for all $\alpha \in \mathbb{N}^s$ such that $|\alpha| = d$.

In Sect. 2, we prove the following version of the GSF for the extended Rees algebras. It was proved for I-adic filtration and for nonnegative integers by Johnston–Verma [20] and for $\mathbb{Z}$-graded admissible filtration of ideals by C. Blancafort for all integers [2].

Theorem 1.7 ([25, Theorem 4.3]) *Let $(R, \mathfrak{m})$ be a Noetherian local ring of dimension d and $I_1, \ldots, I_s$ be $\mathfrak{m}$-primary ideals of R. Let $\mathcal{F} = \{\mathcal{F}(\underline{n})\}_{\underline{n}\in\mathbb{Z}^s}$ be an $\underline{I}$-admissible filtration of ideals in R. Then*

(1) $h^i_{\mathcal{R}_{++}}(\mathcal{R}'(\mathcal{F}))_{\underline{n}} < \infty$ for all $i \geq 0$ and $\underline{n} \in \mathbb{Z}^s$.
(2) $P_{\mathcal{F}}(\underline{n}) - H_{\mathcal{F}}(\underline{n}) = \sum_{i\geq 0}(-1)^i h^i_{\mathcal{R}_{++}}(\mathcal{R}'(\mathcal{F}))_{\underline{n}}$ for all $\underline{n} \in \mathbb{Z}^s$.

In Sect. 3, we derive explicit formulas in terms of the Ratliff–Rush closure filtration of a multi-graded filtration of ideals for the graded components of the local cohomology modules of certain Rees rings and associated graded rings. For an ideal I in a Noetherian ring R, L.J. Ratliff and D. Rush [30] introduced the ideal

$$\tilde{I} = \bigcup_{k \geq 1} (I^{k+1} : I^k),$$

called the **Ratliff–Rush closure** of I. If I has a regular element then the ideal $\tilde{I}$ has some nice properties such as for all large n, $(\tilde{I})^n = I^n$, $\widetilde{I^n} = I^n$ etc. If I is an $\mathfrak{m}$-primary regular ideal then $\tilde{I}$ is the largest ideal with respect to inclusion having the same Hilbert polynomial as that of I. Blancafort [2] introduced Ratliff–Rush closure filtration of an $\mathbb{N}$-graded good filtration. Let $(R, \mathfrak{m})$ be a Noetherian local ring and $I_1, \ldots, I_s$ be $\mathfrak{m}$-primary ideals of R. Let $\mathcal{F} = \{\mathcal{F}(\underline{n})\}_{\underline{n} \in \mathbb{Z}^s}$ be an $\underline{I}$-admissible filtration of ideals in R. We need the concept of the Ratliff–Rush closure of $\mathcal{F}$ in order to find formulas for certain local cohomology modules.

Definition 1.8 The **Ratliff–Rush closure filtration** of $\mathcal{F} = \{\mathcal{F}(\underline{n})\}_{\underline{n} \in \mathbb{Z}^s}$ is the filtration of ideals $\breve{\mathcal{F}} = \{\breve{\mathcal{F}}(\underline{n})\}_{\underline{n} \in \mathbb{Z}^s}$ given by

(1) $\breve{\mathcal{F}}(\underline{n}) = \bigcup_{k \geq 1} (\mathcal{F}(\underline{n} + ke) : \mathcal{F}(e)^k)$ for all $\underline{n} \in \mathbb{N}^s$,

(2) $\breve{\mathcal{F}}(\underline{n}) = \breve{\mathcal{F}}(\underline{n}^+)$ for all $\underline{n} \in \mathbb{Z}^s$.

The next three results to be proved in Sect. 3 are needed to prove several results about Hilbert coefficients in Sect. 5.

Proposition 1.9 ([25, Proposition 3.5]) *Let $(R, \mathfrak{m})$ be a Cohen–Macaulay local ring of dimension two with infinite residue field and $I_1, \ldots, I_s$ be $\mathfrak{m}$-primary ideals in R. Let $\mathcal{F} = \{\mathcal{F}(\underline{n})\}_{\underline{n} \in \mathbb{Z}^s}$ be an $\underline{I}$-admissible filtration of ideals in R. Then for all $\underline{n} \in \mathbb{N}^s$,*

$$[H^1_{\mathcal{R}(\mathcal{F})_{++}}(\mathcal{R}(\mathcal{F}))]_{\underline{n}} \cong \frac{\breve{\mathcal{F}}(\underline{n})}{\mathcal{F}(\underline{n})}.$$

Proposition 1.10 ([2, Theorem 3.5]) *Let $(R, \mathfrak{m})$ be a Cohen–Macaulay local ring of dimension two with infinite residue field, I an $\mathfrak{m}$-primary ideal of R and $\mathcal{F} = \{I_n\}_{n \in \mathbb{Z}}$ be an I-admissible filtration of ideals in R. Then*

$$[H^1_{\mathcal{R}(\mathcal{F})_+}(\mathcal{R}'(\mathcal{F}))]_n = \begin{cases} \breve{I}_n / I_n & \text{if } n \geq 0 \\ 0 & \text{if } n < 0. \end{cases}$$

Theorem 1.11 ([25, Theorem 3.3]) *Let $(R, \mathfrak{m})$ be a Noetherian local ring of dimension $d \geq 1$ with infinite residue field and $I_1, \ldots, I_s$ be $\mathfrak{m}$-primary ideals in R such that grade$(I_1 \cdots I_s) \geq 1$. Let $\mathcal{F} = \{\mathcal{F}(\underline{n})\}_{\underline{n} \in \mathbb{Z}^s}$ be an $\underline{I}$-admissible filtration of ideals in R. Then for all $\underline{n} \in \mathbb{N}^s$ and $i = 1, \ldots, s$,*

$$[H^0_{G_i(\mathcal{F})_{++}}(G_i(\mathcal{F}))]_{\underline{n}} = \frac{\breve{\mathcal{F}}(\underline{n} + e_i) \cap \mathcal{F}(\underline{n})}{\mathcal{F}(\underline{n} + e_i)}.$$

In Sect. 4, we present several applications of the GSF for Rees algebra and associated graded ring of an ideal. The first application due to J.D. Sally, who pioneered these techniques for the study of Hilbert–Samuel coefficients, shows the connection of the postulation number with reduction number. Let $(R, \mathfrak{m})$ be a Noetherian local ring, I be an $\mathfrak{m}$-primary ideal and $\mathcal{F} = \{I_n\}_{n\in\mathbb{Z}}$ be an admissible I-filtration of ideals in R.

Definition 1.12 A **reduction** of an I-admissible filtration $\mathcal{F} = \{I_n\}_{n\in\mathbb{Z}}$ is an ideal $J \subseteq I_1$ such that $JI_n = I_{n+1}$ for all large n. A **minimal reduction** of $\mathcal{F}$ is a reduction of $\mathcal{F}$ minimal with respect to inclusion. For a minimal reduction J of $\mathcal{F}$, we set

$$r_J(\mathcal{F}) = \min\{m : JI_n = I_{n+1} \text{ for } n \geq m\} \text{ and}$$
$$r(\mathcal{F}) = \min\{r_J(\mathcal{I}) : J \text{ is a minimal reduction of } \mathcal{F}\}.$$

For $\mathcal{F} = \{I^n\}_{n\in\mathbb{Z}}$, we set $r_J(\mathcal{F}) = r_J(I)$ and $r(\mathcal{F}) = r(I)$.

Definition 1.13 An integer $n \in \mathbb{Z}$ is called the **postulation number** of $\mathcal{F}$, denoted by $n(\mathcal{F})$, if $P_{\mathcal{F}}(m) = H_{\mathcal{F}}(m)$ for all $m > n$ and $P_{\mathcal{F}}(n) \neq H_{\mathcal{F}}(n)$. It is denoted by $n(\mathcal{F})$.

The next result was proved by J.D. Sally [35] for the $\mathfrak{m}$-adic filtration. Her proof remains valid for any admissible filtration.

Theorem 1.14 *Let $(R, \mathfrak{m})$ be a Cohen–Macaulay local ring of dimension $d \geq 1$ with infinite residue field, I an $\mathfrak{m}$-primary ideal and $\mathcal{F} = \{I_n\}$ be an I-admissible filtration of ideals in R. Let $H_R(n) = \lambda(I_n/I_{n+1})$ and $P_R(X) \in \mathbb{Q}[X]$ such that $P_R(n) = H_R(n)$ for all large n. Suppose* grade $G(\mathcal{F})_+ \geq d-1$. *Then for a minimal reduction $J = (x_1, \ldots, x_d)$ of $\mathcal{F}$, $H_R(r_J(\mathcal{F}) - d) \neq P_R(r_J(\mathcal{F}) - d)$ and $H_R(n) = P_R(n)$ for all $n \geq r_J(\mathcal{F}) - d + 1$.*

The following result is due to Marley [23, Corollary 3.8]. We give another proof which follows from the above theorem.

Theorem 1.15 ([23, Corollary 3.8]) *Let $(R, \mathfrak{m})$ be a Cohen–Macaulay local ring of dimension $d \geq 1$ with infinite residue field, I an $\mathfrak{m}$-primary ideal and $\mathcal{F} = \{I_n\}$ be an I-admissible filtration of ideals in R. Let* grade $G(\mathcal{F})_+ \geq d-1$. *Then $r(\mathcal{F}) = n(\mathcal{F}) + d$.*

In Sect. 5, we discuss several results about nonnegativity of Hilbert coefficients of multi-graded filtrations of ideals as easy consequences of the GSF for such filtrations. We prove the following result which implies earlier results of Northcott, Narita, and Marley.

Theorem 1.16 ([25, Theorem 5.6]) *Let $(R, \mathfrak{m})$ be a Cohen–Macaulay local ring of dimension $d \geq 1$ and $I_1, \ldots, I_s$ be $\mathfrak{m}$-primary ideals of R. Let $\mathcal{F} = \{\mathcal{F}(\underline{n})\}_{\underline{n}\in\mathbb{Z}^s}$ be an $\underline{I}$-admissible filtration of ideals in R. Then*

(1) $e_\alpha(\mathcal{F}) \geq 0$ *where* $\alpha = (\alpha_1, \ldots, \alpha_s) \in \mathbb{N}^s$, $|\alpha| \geq d-1$.
(2) $e_\alpha(\mathcal{F}) \geq 0$ *where* $\alpha = (\alpha_1, \ldots, \alpha_s) \in \mathbb{N}^s$, $|\alpha| = d-2$ *and* $d \geq 2$.

We also discuss the results of S. Itoh about nonnegativity and vanishing of the third coefficient of the normal Hilbert polynomial of the filtration $\{\overline{I^n}\}_{n\in\mathbb{Z}}$ in an analytically unramified Cohen–Macaulay local ring. We prove an analogue of a theorem due to Sally for admissible filtrations in two-dimensional Cohen–Macaulay local rings which gives explicit formulas for all the coefficients of their Hilbert polynomial. Here again we show that these formulas follow in a natural way from the variant of GSF for Rees algebra of the filtration.

Proposition 1.17 *Let* $(R, \mathfrak{m})$ *be a two-dimensional Cohen–Macaulay local ring,* I *be any* $\mathfrak{m}$*-primary ideal of* R *and* $\mathcal{F} = \{I_n\}_{n\in\mathbb{Z}}$ *an admissible* I*-filtration of ideals in* R*. Then*

(1) $\lambda\left(H^2_{\mathcal{R}(\mathcal{F})_+}(\mathcal{R}(\mathcal{F})_0\right) = e_2(\mathcal{F})$,
(2) $\lambda\left(H^2_{\mathcal{R}(\mathcal{F})_+}(\mathcal{R}(\mathcal{F}))_1\right) = e_0(\mathcal{F}) - e_1(\mathcal{F}) + e_2(\mathcal{F}) - \lambda\left(\dfrac{R}{\tilde{I_1}}\right)$,
(3) $\lambda\left(H^2_{\mathcal{R}(\mathcal{F})_+}(\mathcal{R}(\mathcal{F}))_{-1}\right) = e_1(\mathcal{F}) + e_2(\mathcal{F})$.

C. Huneke [14] and A. Ooishi [28] independently proved that if $(R, \mathfrak{m})$ is a Cohen–Macaulay local ring of dimension $d \geq 1$ and I is an $\mathfrak{m}$-primary ideal then $e_0(I) - e_1(I) = \lambda(R/I)$ if and only if $r(I) \leq 1$. Huckaba and Marley [13] proved this result for $\mathbb{Z}$-graded admissible filtrations. In Sect. 6, we present a proof, due to Blancafort, for $\mathbb{Z}$-graded admissible filtrations of Huneke–Ooishi Theorem. The original proofs due to Huneke and Ooishi did not employ local cohomology and relied on use of superficial sequences. Our purpose in presenting the alternative proof using the GSF for Rees algebras is to motivate the proof of an analogue of the Huneke–Ooishi Theorem for multi-graded filtrations of ideals.

Theorem 1.18 ([3, Theorem 4.3.6]) *Let* $(R, \mathfrak{m})$ *be a Cohen–Macaulay local ring with infinite residue field of dimension* $d \geq 1$, I_1 *an* $\mathfrak{m}$*-primary ideal and* $\mathcal{F} = \{I_n\}_{n\in\mathbb{Z}}$ *be an* I_1*-admissible filtration of ideals in* R*. Then the following are equivalent:*

(1) $e_0(\mathcal{F}) - e_1(\mathcal{F}) = \lambda(R/I_1)$,
(2) $r(\mathcal{F}) \leq 1$.

In this case, $e_2(\mathcal{F}) = \cdots = e_d(\mathcal{F}) = 0$, $G(\mathcal{F})$ *is Cohen–Macaulay,* $n(\mathcal{F}) \leq 0$, $r(\mathcal{F})$ *is independent of the reduction chosen and* $\mathcal{F} = \{I_1^n\}$.

Using the GSF for multi-graded Rees algebras we prove the following analogue of the Huneke–Ooishi Theorem for multi-graded admissible filtrations.

Theorem 1.19 ([25, Theorem 5.5]) *Let* $(R, \mathfrak{m})$ *be a Cohen–Macaulay local ring of dimension* $d \geq 1$ *and* $I_1, \ldots, I_s$ *be* $\mathfrak{m}$*-primary ideals of* R*. Let* $\mathcal{F} = \{\mathcal{F}(\underline{n})\}_{\underline{n}\in\mathbb{Z}^s}$ *be an* $\underline{I}$*-admissible filtration of ideals in* R*. Then for all* $i = 1, \ldots, s$,

(1) $e_{(d-1)e_i}(\mathcal{F}) \geq e_1(\mathcal{F}^{(i)})$,
(2) $e(I_i) - e_{(d-1)e_i}(\mathcal{F}) \leq \lambda(R/\mathcal{F}(e_i))$,
(3) $e(I_i) - e_{(d-1)e_i}(\mathcal{F}) = \lambda(R/\mathcal{F}(e_i))$ *if and only if* $r(\mathcal{F}^{(i)}) \leq 1$ *and* $e_{(d-1)e_i}(\mathcal{F}) = e_1(\mathcal{F}^{(i)})$, *where* $\mathcal{F}^{(i)} = \{\mathcal{F}(ne_i)\}_{n\in\mathbb{Z}}$ *is an* I_i*-admissible filtration.*

The vanishing of the constant term of the Hilbert polynomial of a filtration gives insight into the filtration as well as the local ring. For any $\mathfrak{m}$-primary ideal I in an analytically unramified local ring $(R, \mathfrak{m})$ of dimension d, the **normal Hilbert function** of I is defined to be the function $\overline{H}(I, n) = \lambda(R/\overline{I^n})$. Rees showed that for large n, it is given by the **normal Hilbert polynomial**

$$\overline{P}(I, x) = \overline{e}_0(I)\binom{x+d-1}{d} - \overline{e}_1(I)\binom{x+d-2}{d-1} + \cdots + (-1)^d\overline{e}_d(I).$$

The integers $\overline{e}_0(I), \overline{e}_1(I), \ldots, \overline{e}_d(I)$ are called the **normal Hilbert coefficients** of I. Rees defined a 2-dimensional normal analytically unramified local ring $(R, \mathfrak{m})$ to be **pseudo-rational** if $\overline{e}_2(I) = 0$ for all $\mathfrak{m}$-primary ideals. It can be shown that two-dimensional local rings having a rational singularity are pseudo-rational. It is natural to characterise $\overline{e}_2(I) = 0$ in terms of computable data. This was considered by Huneke [14] in which he proved.

Theorem 1.20 ([14, Theorem 4.5]) *Let* $(R, \mathfrak{m})$ *be a two-dimensional analytically unramified Cohen–Macaulay local ring. Let* I *be an* $\mathfrak{m}$*-primary ideal. Then* $\overline{e}_2(I) = 0$ *if and only if* $\overline{I^n} = (x, y)\overline{I^{n-1}}$ *for* $n \geq 2$ *and for any minimal reduction* (x, y) *of* I.

A similar result was proved by Itoh [18] about vanishing of $\overline{e}_2(I)$. Using the GSF for multi-graded filtrations, we prove the following theorem which characterises the vanishing of the constant term of the Hilbert polynomial of a multi-graded admissible filtration and derive results of Itoh and Huneke as consequences.

Theorem 1.21 ([25, Theorem 5.7]) *Let* $(R, \mathfrak{m})$ *be a Cohen–Macaulay local ring of dimension two and* $I_1, \ldots, I_s$ *be* $\mathfrak{m}$*-primary ideals of* R*. Let* $\mathcal{F} = \{\mathcal{F}(\underline{n})\}_{\underline{n}\in\mathbb{Z}^s}$ *be an* $\underline{I}$*-admissible filtration of ideals in* R*. Then* $e_{\underline{0}}(\mathcal{F}) = 0$ *implies* $e(I_i) - e_{e_i}(\mathcal{F}) = \lambda\left(\frac{R}{\mathcal{F}(e_i)}\right)$ *for all* $i = 1, \ldots, s$*. Suppose* $\check{\mathcal{F}}$ *is* $\underline{I}$*-admissible filtration, then the converse is also true.*

2 Variations on the Grothendieck–Serre Formula

The main aim of this section is to prove the Grothendieck–Serre formula (Theorem 2.3) and its variations. In [7, Propositions 2.4.2 and 2.4.3], Colomé-Nin proved the Grothendieck–Serre formula for nonstandard multi-graded rings. For the sake of simplicity, we present her proof for standard multi-graded rings. As a consequence we prove [25, Theorem 4.3] (Theorem 2.5) which relates the difference of Hilbert polynomial and Hilbert function of an $\underline{I}$-admissible filtration to the Euler characteristic of the extended multi-Rees algebra.

We recall the following Lemma from [7] which is needed to prove Theorem 2.3.

Lemma 2.1 ([7, Lemma 2.2.8]) *Let* $R = \bigoplus_{\underline{n}\in\mathbb{N}^s} R_{\underline{n}}$ *be a standard Noetherian* $\mathbb{N}^s$*-graded ring defined over a local ring* $(R_{\underline{0}}, \mathfrak{m})$ *and* $M = \bigoplus_{\underline{n}\in\mathbb{Z}^s} M_{\underline{n}}$ *a finitely generated* $\mathbb{Z}^s$*-graded* R*-module. Let* $x \in R_{\underline{n}}$ *where* $\underline{n} \geq e$ *and* $x \notin \bigcup_{P\in Ass(M)} P$*. Then* $\operatorname{rel.dim}(M/xM) = \operatorname{rel.dim}(M) - 1$.

Proposition 2.2 *Let* $R = \bigoplus_{\underline{n}\in\mathbb{N}^s} R_{\underline{n}}$ *be a standard Noetherian* $\mathbb{N}^s$*-graded ring defined over a local ring* $(R_{\underline{0}}, \mathfrak{m})$ *and* $M = \bigoplus_{\underline{n}\in\mathbb{Z}^s} M_{\underline{n}}$ *a finitely generated* $\mathbb{Z}^s$*-graded* R*-module. Then*

(1) For all $i \geq 0$ *and* $\underline{n} \in \mathbb{Z}^s$, $[H^i_{R_{++}}(M)]_{\underline{n}}$ *is finitely generated* $R_{\underline{0}}$*-module.*
(2) For all large $\underline{n}$ *and* $i \geq 0$, $[H^i_{R_{++}}(M)]_{\underline{n}} = 0$.

Proof Note that R_{++} is finitely generated. We prove both (1) and (2) together by induction on i. Suppose $i = 0$. Note that $H^0_{R_{++}}(M) \subseteq M$ and hence $H^0_{R_{++}}(M)$ is finitely generated R-module. Let $\{\gamma_1, \ldots, \gamma_q\}$ be a generating set of $H^0_{R_{++}}(M)$ as an R-module and $deg(\gamma_j) = p(j) = (p(j1), \ldots, p(js))$ for all $j = 1, \ldots, q$. Let $\alpha_i = \max\{|p(ji)| : j = 1, \ldots, q\}$ for all $i = 1, \ldots, s$ and $\alpha = (\alpha_1, \ldots, \alpha_s)$. Since $H^0_{R_{++}}(M)$ is R_{++}-torsion, there exists an integer $t \geq 1$ such that $R^t_{++}H^0_{R_{++}}(M) = 0$. Then for all $\underline{n} \geq \alpha + te$,

$$[H^0_{R_{++}}(M)]_{\underline{n}} = R_{\underline{n}-p(1)}\gamma_1 + \cdots + R_{\underline{n}-p(q)}\gamma_q \subseteq R^t_{++}H^0_{R_{++}}(M) = 0.$$

Fix $\underline{n} \in \mathbb{Z}^s$. Since R is a standard Noetherian $\mathbb{N}^s$-graded ring defined over $R_{\underline{0}}$, there exist elements $a_{i1}, \ldots, a_{ik_i} \in R_{e_i}$ for all $i = 1, \ldots, s$ such that each nonzero element of $[H^0_{R_{++}}(M)]_{\underline{n}}$ can be written as sum of monomials $\prod_{1\leq i\leq s} a_{i1}^{t_{i1}} \cdots a_{ik_i}^{t_{ik_i}}\gamma_j$ of degree $\underline{n}$ with coefficients from $R_{\underline{0}}$ where $j = 1, \ldots, q$, $t_{i1}, \ldots, t_{ik_i} \geq 0$. Since $0 \leq t_{i1}, \ldots, t_{ik_i} \leq n_i - p(ji)$, the number of monomial generators are finite. Hence $[H^0_{R_{++}}(M)]_{\underline{n}}$ is finitely generated $R_{\underline{0}}$-module.

Now assume $i > 0$. Let M' denote $M/H^0_{R_{++}}(M)$. Consider the short exact sequence of R-modules

$$0 \longrightarrow H^0_{R_{++}}(M) \longrightarrow M \longrightarrow M' \longrightarrow 0$$

which gives long exact sequence of local cohomology modules

$$\cdots \longrightarrow H^i_{R_{++}}(H^0_{R_{++}}(M)) \longrightarrow H^i_{R_{++}}(M) \longrightarrow H^i_{R_{++}}(M') \longrightarrow \cdots.$$

Since $H^0_{R_{++}}(M)$ is R_{++}-torsion, $H^i_{R_{++}}(H^0_{R_{++}}(M)) = 0$ for all $i \geq 1$. Thus

$$H^i_{R_{++}}(M) \simeq H^i_{R_{++}}(M') \text{ for all } i \geq 1. \tag{2.2.1}$$

By [7, Lemma 2.4.1], there exists an element $x \in R_{\underline{p}}$ for some $\underline{p} \geq \underline{e}$ such that $x \notin P$ for all $P \in Ass(M') = Ass(M)\backslash V(R_{++})$. Fix $i \geq 1$. Consider the short exact sequence of R-modules

$$0 \longrightarrow M'(-\underline{p}) \xrightarrow{.x} M' \longrightarrow M'/xM' \longrightarrow 0$$

which gives long exact sequence of local cohomology modules whose $\underline{r}$th component is

$$\cdots \longrightarrow \left[H^{i-1}_{R_{++}}(M'/xM')\right]_{\underline{r}} \longrightarrow [H^i_{R_{++}}(M')]_{\underline{r}-\underline{p}} \xrightarrow{.x} [H^i_{R_{++}}(M')]_{\underline{r}} \longrightarrow [H^i_{R_{++}}(M'/xM')]_{\underline{r}} \longrightarrow \cdots.$$

By inductive hypothesis $\left[H^{i-1}_{R_{++}}(M'/xM')\right]_{\underline{m}} = 0$ for all large $\underline{m}$, say, for all $\underline{m} \geq \underline{k}$ for some $\underline{k} \in \mathbb{N}^s$. Then for all $\underline{n} \geq \underline{k}$, we have the exact sequence

$$0 \longrightarrow [H^i_{R_{++}}(M')]_{\underline{n}-\underline{p}} \xrightarrow{.x} [H^i_{R_{++}}(M')]_{\underline{n}}.$$

Since $H^i_{R_{++}}(M')$ is R_{++}-torsion and $x \in R_{++}$, we have $\left[H^i_{R_{++}}(M')\right]_{\underline{m}} = 0$ for all $\underline{m} \geq \underline{k} - \underline{p}$. Hence we prove part (2).

Fix $i > 0$ and $\underline{n} \in \mathbb{Z}^s$. By [7, Lemma 2.4.1], there exists an element $y \in R_{++}$ such that $y \notin P$ for all $P \in Ass(M') = Ass(M)\backslash V(R_{++})$ and we can assume degree$(y) = \underline{m}$ such that $[H^i_{R_{++}}(M')]_{\underline{r}} = 0$ for all $\underline{r} \geq \underline{n} + \underline{m}$. Consider the short exact sequence of R-modules

$$0 \longrightarrow M'(-\underline{m}) \xrightarrow{.y} M' \longrightarrow M'/yM' \longrightarrow 0$$

which gives long exact sequence of cohomology modules whose $(\underline{m} + \underline{n})$th component is

$$\cdots \longrightarrow \left[H^{i-1}_{R_{++}}(M'/yM')\right]_{\underline{m}+\underline{n}} \longrightarrow [H^i_{R_{++}}(M')]_{\underline{n}} \xrightarrow{.y} [H^i_{R_{++}}(M')]_{\underline{m}+\underline{n}} \longrightarrow \cdots.$$

Since $[H^i_{R_{++}}(M')]_{\underline{m}+\underline{n}} = 0$ and by induction hypothesis $\left[H^{i-1}_{R_{++}}(M'/yM')\right]_{\underline{m}+\underline{n}}$ is finitely generated $R_{\underline{0}}$-module, from the above exact sequence, we get $[H^i_{R_{++}}(M')]_{\underline{n}}$ is finitely generated $R_{\underline{0}}$-module. Hence by Eq. (2.2.1), we get the required result. □

Theorem 2.3 (Grothendieck–Serre formula for multi-graded modules) *Let* $R = \bigoplus_{\underline{n} \in \mathbb{N}^s} R_{\underline{n}}$ *be a standard Noetherian* $\mathbb{N}^s$*-graded ring defined over an Artinian local ring* $(R_{\underline{0}}, \mathfrak{m})$ *and* $M = \bigoplus_{\underline{n} \in \mathbb{Z}^s} M_{\underline{n}}$ *a finitely generated* $\mathbb{Z}^s$*-graded* R*-module. Let* $H_M(\underline{n}) = \lambda(M_{\underline{n}})$ *and* $P_M(x_1, \ldots, x_s)$ *be the Hilbert polynomial of* M*. Then for all* $\underline{n} \in \mathbb{Z}^s$,

$$H_M(\underline{n}) - P_M(\underline{n}) = \sum_{j \geq 0} (-1)^j h^j_{R_{++}}(M)_{\underline{n}}.$$

Proof For all $\underline{n} \in \mathbb{Z}^s$, we define $\chi_M(\underline{n}) = \sum_{j \geq 0} (-1)^j h^j_{R_{++}}(M)_{\underline{n}}$ and $f_M(\underline{n}) = H_M(\underline{n}) - P_M(\underline{n})$. We use induction on $\operatorname{rel.dim}(M)$. Suppose $\operatorname{rel.dim}(M) = s - 1$. Then $\operatorname{Supp}_{++}(M) = V_{++}(\operatorname{Ann}(M)) = \emptyset$. Therefore there exists an integer $k \geq 1$ such that $R^k_{++} M = 0$. Hence $H^0_{R_{++}}(M) = M$ and $H^i_{R_{++}}(M) = 0$ for all $i \geq 1$. Since $P_M(X_1, \ldots, X_s)$ has degree -1, we have $P_M(\underline{n}) = 0$ for all $\underline{n} \in \mathbb{Z}^s$. Thus we get the required equality.

Assume that $\operatorname{rel.dim}(M) \geq s$. Let M' denote $M / H^0_{R_{++}}(M)$. Consider the short exact sequence of R-modules

$$0 \longrightarrow H^0_{R_{++}}(M) \longrightarrow M \longrightarrow M' \longrightarrow 0$$

which gives long exact sequence of local cohomology modules

$$\cdots \longrightarrow H^i_{R_{++}}(H^0_{R_{++}}(M)) \longrightarrow H^i_{R_{++}}(M) \longrightarrow H^i_{R_{++}}(M') \longrightarrow \cdots .$$

Note that $H^0_{R_{++}}(M)$ is R_{++}-torsion. Hence for all $i \geq 1$, $H^i_{R_{++}}(H^0_{R_{++}}(M)) = 0$ and

$$H^i_{R_{++}}(M) \simeq H^i_{R_{++}}(M'). \tag{2.3.1}$$

Since $H_M(\underline{n}) = H_{M'}(\underline{n}) + h^0_{R_{++}}(M)_{\underline{n}}$ and hence by Proposition 2.2 part (2), $P_M(\underline{n}) = P_{M'}(\underline{n})$. Thus

$$H_M(\underline{n}) - P_M(\underline{n}) = H_{M'}(\underline{n}) + h^0_{R_{++}}(M)_{\underline{n}} - P_{M'}(\underline{n}) = H_{M'}(\underline{n}) - P_{M'}(\underline{n}) + h^0_{R_{++}}(M)_{\underline{n}}.$$

Therefore by the Eq. (2.3.1), it is enough to prove the result for M'. By [7, Lemma 2.4.1], there exists an element $x \in R_{\underline{p}}$ for some $\underline{p} \geq e$ such that $x \notin P$ for all $P \in Ass(M') = Ass(M) \backslash V(R_{++})$. Consider the short exact sequence of R-modules

$$0 \longrightarrow M'(-\underline{p}) \xrightarrow{.x} M' \longrightarrow M'/xM' \longrightarrow 0$$

which gives long exact sequence of cohomology modules whose $\underline{r}$th component is

$$\cdots \longrightarrow \left[H^{i-1}_{R_{++}}(M'/xM')\right]_{\underline{r}} \longrightarrow \left[H^{i}_{R_{++}}(M')\right]_{\underline{r}-\underline{p}} \xrightarrow{.x} \left[H^{i}_{R_{++}}(M')\right]_{\underline{r}}$$
$$\longrightarrow \left[H^{i}_{R_{++}}(M'/xM')\right]_{\underline{r}} \longrightarrow \cdots.$$

Thus for all $\underline{n} \in \mathbb{Z}^s$, $H_{M'/xM'}(\underline{n}) = H_{M'}(\underline{n}) - H_{M'}(\underline{n}-\underline{p})$. Hence $P_{M'/xM'}(\underline{n}) = P_{M'}(\underline{n}) - P_{M'}(\underline{n}-\underline{p})$. By Lemma 2.1, $\operatorname{rel.dim}(M'/xM') < \operatorname{rel.dim}(M')$. Therefore for all $\underline{n} \in \mathbb{Z}^s$,

$$f_{M'}(\underline{n}) - f_{M'}(\underline{n}-\underline{p}) = f_{M'/xM'}(\underline{n}) = \chi_{M'/xM'}(\underline{n}) = \chi_{M'}(\underline{n}) - \chi_{M'}(\underline{n}-\underline{p}).$$

Hence $f_{M'}(\underline{n}) - \chi_{M'}(\underline{n}) = f_{M'}(\underline{n}-\underline{p}) - \chi_{M'}(\underline{n}-\underline{p})$. Since for all large $\underline{n}$, $f_{M'}(\underline{n}) - \chi_{M'}(\underline{n}) = 0$, we get the required result. □

Proposition 2.4 *Let S' be a $\mathbb{Z}^s$-graded ring and $S = \bigoplus_{\underline{n}\in\mathbb{N}^s} S'_{\underline{n}}$. Then $H^i_{S_{++}}(S') \cong H^i_{S_{++}}(S)$ for all $i > 1$ and we have the exact sequence*

$$0 \longrightarrow H^0_{S_{++}}(S) \longrightarrow H^0_{S_{++}}(S') \longrightarrow \frac{S'}{S} \longrightarrow H^1_{S_{++}}(S) \longrightarrow H^1_{S_{++}}(S') \longrightarrow 0.$$

Proof Consider the short exact sequence of S-modules

$$0 \longrightarrow S \longrightarrow S' \longrightarrow \frac{S'}{S} \longrightarrow 0.$$

This gives the long exact sequence of S-modules

$$\cdots \longrightarrow H^i_{S_{++}}(S) \longrightarrow H^i_{S_{++}}(S') \longrightarrow H^i_{S_{++}}\left(\frac{S'}{S}\right) \longrightarrow \cdots.$$

Since $\frac{S'}{S}$ is S_{++}-torsion, $H^0_{S_{++}}\left(\frac{S'}{S}\right) = \frac{S'}{S}$ and $H^i_{S_{++}}\left(\frac{S'}{S}\right) = 0$ for all $i > 0$. Hence the result follows. □

The GSF for multi-graded Rees algebras proved below generalises the theorems [19, Theorem 5.1], [24, Theorem 1] and [2, Theorem 4.1].

Theorem 2.5 ([25, Theorem 4.3]) *Let $(R, \mathfrak{m})$ be a Noetherian local ring of dimension d and $I_1, \ldots, I_s$ be $\mathfrak{m}$-primary ideals of R. Let $\mathcal{F} = \{\mathcal{F}(\underline{n})\}_{\underline{n}\in\mathbb{Z}^s}$ be an $\underline{I}$-admissible filtration of ideals in R. Then*

(1) $h^i_{\mathcal{R}_{++}}(\mathcal{R}'(\mathcal{F}))_{\underline{n}} < \infty$ *for all* $i \geq 0$ *and* $\underline{n} \in \mathbb{Z}^s$.
(2) $P_{\mathcal{F}}(\underline{n}) - H_{\mathcal{F}}(\underline{n}) = \sum_{i\geq 0}(-1)^i h^i_{\mathcal{R}_{++}}(\mathcal{R}'(\mathcal{F}))_{\underline{n}}$ *for all* $\underline{n} \in \mathbb{Z}^s$.

Proof (1) Denote $\dfrac{\mathcal{R}'(\mathcal{F})}{\mathcal{R}'(\mathcal{F})(e_i)}$ by $G_i'(\mathcal{F})$. By the change of ring principle, $H^j_{G_i(\underline{I})_{++}}(G_i'(\mathcal{F})) \cong H^j_{\mathcal{R}_{++}}(G_i'(\mathcal{F}))$ for all $i = 1, \ldots, s$ and $j \geq 0$. For a fixed i, consider the short exact sequence of $\mathcal{R}(\underline{I})$-modules

$$0 \longrightarrow \mathcal{R}'(\mathcal{F})(e_i) \longrightarrow \mathcal{R}'(\mathcal{F}) \longrightarrow G_i'(\mathcal{F}) \longrightarrow 0. \tag{2.5.1}$$

This induces the long exact sequence of R-modules

$$0 \longrightarrow [H^0_{\mathcal{R}_{++}}(\mathcal{R}'(\mathcal{F}))]_{\underline{n}+e_i} \longrightarrow [H^0_{\mathcal{R}_{++}}(\mathcal{R}'(\mathcal{F}))]_{\underline{n}} \longrightarrow [H^0_{\mathcal{R}_{++}}(G_i'(\mathcal{F}))]_{\underline{n}} \longrightarrow [H^1_{\mathcal{R}_{++}}(\mathcal{R}'(\mathcal{F}))]_{\underline{n}+e_i} \longrightarrow \cdots.$$

By Propositions 2.2 and 2.4, $[H^j_{\mathcal{R}_{++}}(\mathcal{R}'(\mathcal{F}))]_{\underline{n}} = 0$ for all large $\underline{n}$ and $j \geq 0$. Since $\left(\frac{G_i'(\mathcal{F})}{G_i(\mathcal{F})}\right)_{\underline{n}} = 0$ for all $\underline{n} \in \mathbb{N}^s$ or $n_i < 0$, by Propositions 2.2 and 2.4, $[H^j_{\mathcal{R}_{++}}(G_i'(\mathcal{F}))]_{\underline{n}}$ is finitely generated $(G_i(\underline{I}))_{\underline{0}}$-module for all $\underline{n} \in \mathbb{N}^s$ or $n_i < 0$ and $j \geq 0$. Since $(G_i(\underline{I}))_{\underline{0}}$ is Artinian, $[H^j_{\mathcal{R}_{++}}(G_i'(\mathcal{F}))]_{\underline{n}}$ has finite length for all $\underline{n} \in \mathbb{N}^s$ or $n_i < 0$ and $j \geq 0$. Hence using decreasing induction on $\underline{n}$, we get that $h^j_{\mathcal{R}_{++}}(\mathcal{R}'(\mathcal{F}))_{\underline{n}} < \infty$ for all $j \geq 0$ and $\underline{n} \in \mathbb{Z}^s$.

(2) Let $\chi_M(\underline{n}) = \sum\limits_{i \geq 0}(-1)^i h^i_{\mathcal{R}_{++}}(M)_{\underline{n}}$ where M is an $\mathcal{R}(\underline{I})$-module. Then from the short exact sequence (2.5.1), Theorem 2.3 and Proposition 2.4, for each $i = 1, \ldots, s$ and $\underline{n} \in \mathbb{N}^s$ or $n_i < 0$,

$$\begin{aligned}
\chi_{\mathcal{R}'(\mathcal{F})}(\underline{n} + e_i) - \chi_{\mathcal{R}'(\mathcal{F})}(\underline{n}) &= -\chi_{G_i'(\mathcal{F})}(\underline{n}) \\
&= -\chi_{G_i(\mathcal{F})}(\underline{n}) \\
&= P_{G_i(\mathcal{F})}(\underline{n}) - H_{G_i(\mathcal{F})}(\underline{n}) \\
&= (P_{\mathcal{F}}(\underline{n} + e_i) - P_{\mathcal{F}}(\underline{n})) - (H_{\mathcal{F}}(\underline{n} + e_i) - H_{\mathcal{F}}(\underline{n})).
\end{aligned}$$

Let $h(\underline{n}) = \chi_{\mathcal{R}'(\mathcal{F})}(\underline{n}) - (P_{\mathcal{F}}(\underline{n}) - H_{\mathcal{F}}(\underline{n}))$. Then $h(\underline{n} + e_i) = h(\underline{n})$ for all $\underline{n} \in \mathbb{N}^s$ or $n_i < 0$ and $i = 1, \ldots, s$. Since $h(\underline{n}) = 0$ for all large $\underline{n}$, $h(\underline{n}) = 0$ forall $\underline{n} \in \mathbb{Z}^s$.

□

3 Formulas for Local Cohomology Modules

In this section, we derive formulas for the graded components of the local cohomology modules of certain Rees rings and associated graded rings in terms of the Ratliff–Rush closure filtration of a multi-graded filtration of ideals. These generalise [2, Proposition 2.5 and Theorem 3.5]. We use these formulas to derive various properties of the Hilbert coefficients in further sections.

In the following proposition we derive a formula for $H^d_{G(\mathcal{F})_+}(G(\mathcal{F}))_n$.

Proposition 3.1 *Let $(R, \mathfrak{m})$ be a Cohen–Macaulay local ring of dimension $d \geq 1$, I an $\mathfrak{m}$-primary ideal and $\mathcal{F} = \{I_n\}$ be an I-admissible filtration of ideals in R. Let $(x_1, \ldots, x_d)$ be a minimal reduction of $\mathcal{F}$. Put $\underline{x}^k = (x_1^k, \ldots, x_d^k)$ for all $k \geq 1$. Then for all $n \in \mathbb{Z}$,*

$$H^d_{G(\mathcal{F})_+}(G(\mathcal{F}))_n = \lim_{\overrightarrow{k}} \frac{I_{dk+n}}{\underline{x}^k I_{(d-1)k+n} + I_{dk+n+1}}.$$

Proof Let $x_i^* = x_i + I_2$ be the image of x_i in $G(\mathcal{F})$. Since $\sqrt{G(\mathcal{F})_+} = \sqrt{(x_1^*, \ldots, x_d^*)}$, by [5, Theorem 5.2.9], $H^d_{G(\mathcal{F})_+}(G(\mathcal{F})) = \lim_{\overrightarrow{k}} H^d((x_1^*)^k, \ldots, (x_d^*)^k, G(\mathcal{F}))$ where $H^d((x_1^*)^k, \ldots, (x_d^*)^k, G(\mathcal{F}))$ is the dth cohomology of the Koszul complex of $G(\mathcal{F})$ with respect to the elements $(x_1^*)^k, \ldots, (x_d^*)^k$. Thus we get the required result. □

Proposition 3.2 *Let $(R, \mathfrak{m})$ be a Cohen–Macaulay local ring of dimension $d \geq 1$, I an $\mathfrak{m}$-primary ideal and $\mathcal{F} = \{I_n\}$ be an I-admissible filtration of ideals in R. Let $(x_1, \ldots, x_d)$ be a minimal reduction of $\mathcal{F}$. Put $\underline{x}^k = (x_1^k, \ldots, x_d^k)$ for all $k \geq 1$. Then for all $n \in \mathbb{Z}$,*

$$H^d_{\mathcal{R}(\mathcal{F})_+}(\mathcal{R}(\mathcal{F}))_n = \lim_{\overrightarrow{k}} \frac{I_{dk+n}}{\underline{x}^k I_{(d-1)k+n}}.$$

Proof Since $\sqrt{\mathcal{R}(\mathcal{F})_+} = \sqrt{(x_1t, \ldots, x_dt)}$, we have $H^d_{\mathcal{R}(\mathcal{F})_+}(\mathcal{R}(\mathcal{F})) = \lim_{\overrightarrow{k}} H^d((x_1t)^k, \ldots, (x_dt)^k, \mathcal{R}(\mathcal{F}))$ by [5, Theorem 5.2.9] where $H^d((x_1t)^k, \ldots, (x_dt)^k, \mathcal{R}(\mathcal{F}))$ is the dth cohomology of the Koszul complex of $\mathcal{R}(\mathcal{F})$ with respect to the elements $(x_1t)^k, \ldots, (x_dt)^k$. Thus we get the required result. □

Lemma 3.3 ([Rees' Lemma] [31, Lemma 1.2] [25, Lemma 2.2]) *Let $(R, \mathfrak{m}, k)$ be a Noetherian local ring of dimension d with infinite residue field k and $I_1, \ldots, I_s$ be $\mathfrak{m}$-primary ideals of R. Let $\mathcal{F} = \{\mathcal{F}(\underline{n})\}_{\underline{n} \in \mathbb{Z}^s}$ be an $\underline{I}$-admissible filtration of ideals in R and S be a finite set of prime ideals of R not containing $I_1 \cdots I_s$. Then for each $i = 1, \ldots, s$, there exists an element $x_i \in I_i$ not contained in any of the prime ideals of S and an integer r_i such that for all $\underline{n} \geq r_i e_i$,*

$$\mathcal{F}(\underline{n}) \cap (x_i) = x_i \mathcal{F}(\underline{n} - e_i).$$

Theorem 3.4 ([31, Theorem 1.3] [25, Theorem 2.3]) *Let $(R, \mathfrak{m})$ be a Noetherian local ring of dimension d with infinite residue field and $I_1, \ldots, I_s$ be $\mathfrak{m}$-primary ideals of R. Let $\mathcal{F} = \{\mathcal{F}(\underline{n})\}_{\underline{n} \in \mathbb{Z}^s}$ be an $\underline{I}$-admissible filtration of ideals in R. Then there exist a set of elements $\{x_{ij} \in I_i : j = 1, \ldots, d; i = 1, \ldots, s\}$ such that $(y_1, \ldots, y_d)\mathcal{F}(\underline{n}) = \mathcal{F}(\underline{n} + e)$ for all large $\underline{n}$ where $y_j = x_{1j} \cdots x_{sj} \in I_1 \cdots I_s$ for all $j = 1, \ldots, d$. Moreover, if the ring is Cohen–Macaulay local then there exist elements $x_{i1} \in I_i$ and integers r_i for all $i = 1, \ldots, s$ such that for all $\underline{n} \geq r_i e_i$, $\mathcal{F}(\underline{n}) \cap (x_{i1}) = x_{i1}\mathcal{F}(\underline{n} - e_i)$ and $y_1 = x_{11} \cdots x_{s1}$.*

Proposition 3.5 ([25, Proposition 3.5]) *Let $(R, \mathfrak{m})$ be a Cohen–Macaulay local ring of dimension two with infinite residue field and $I_1, \ldots, I_s$ be $\mathfrak{m}$-primary ideals in R. Let $\mathcal{F} = \{\mathcal{F}(\underline{n})\}_{\underline{n} \in \mathbb{Z}^s}$ be an $\underline{I}$-admissible filtration of ideals in R. Then for all $\underline{n} \in \mathbb{N}^s$,*

$$[H^1_{\mathcal{R}(\mathcal{F})_{++}}(\mathcal{R}(\mathcal{F}))]_{\underline{n}} \cong \frac{\breve{\mathcal{F}}(\underline{n})}{\mathcal{F}(\underline{n})}.$$

Proof By Lemma 3.3 and Theorem 3.4, there exists a regular sequence $\{y_1, y_2\}$ such that $(y_1, y_2)\mathcal{F}(\underline{n}) = \mathcal{F}(\underline{n} + e)$ for all large $\underline{n}$. For all $k \geq 1$, consider the following complex of $\mathcal{R}(\mathcal{F})$-modules

$$F^{k.} : 0 \longrightarrow \mathcal{R}(\mathcal{F}) \xrightarrow{\alpha_k} \mathcal{R}(\mathcal{F})(ke)^2 \xrightarrow{\beta_k} \mathcal{R}(\mathcal{F})(2ke) \longrightarrow 0,$$

where $\alpha_k(1) = ({y_1}^k \underline{t}^{ke}, {y_2}^k \underline{t}^{ke})$ and $\beta_k(u, v) = {y_2}^k \underline{t}^{ke} u - {y_1}^k \underline{t}^{ke} v$. Since radical of the ideal $(y_1 \underline{t}^e, y_2 \underline{t}^e)\mathcal{R}(\mathcal{F})$ is same as radical of the ideal $\mathcal{R}(\mathcal{F})_{++}$, by [5, Theorem 5.2.9],

$$[H^1_{\mathcal{R}(\mathcal{F})_{++}}(\mathcal{R}(\mathcal{F}))]_{\underline{n}} \cong \varinjlim_{k} \frac{(\ker \beta_k)_{\underline{n}}}{(\operatorname{im} \alpha_k)_{\underline{n}}}.$$

Suppose $(u, v) \in (\ker \beta_k)_{\underline{n}}$ for any $\underline{n} \in \mathbb{N}^s$. Then ${y_2}^k u - {y_1}^k v = 0$. Since $\{y_1, y_2\}$ is a regular sequence, $u = {y_1}^k p$ for some $p \in R$. Thus $v = {y_2}^k p$. Hence $(u, v) = ({y_1}^k p, {y_2}^k p)$. This implies for all $\underline{n} \in \mathbb{N}^s$, $(u, v) \in (\ker \beta_k)_{\underline{n}}$ if and only if $(u, v) = ({y_1}^k p, {y_2}^k p)$ for some $p \in (\mathcal{F}(\underline{n} + ke) : ({y_1}^k, {y_2}^k))$. For $k \gg 0$, by [25, Proposition 3.1], $\breve{\mathcal{F}}(\underline{n}) = (\mathcal{F}(\underline{n} + ke) : ({y_1}^k, {y_2}^k))$ for all $\underline{n} \in \mathbb{N}^s$. Hence for all $\underline{n} \in \mathbb{N}^s$ and $k \gg 0$, $(\ker \beta_k)_{\underline{n}} \cong \breve{\mathcal{F}}(\underline{n})$. Also for all $\underline{n} \in \mathbb{N}^s$,

$$(\operatorname{im} \alpha_k)_{\underline{n}} = \{({y_1}^k p \underline{t}^{ke}, {y_2}^k p \underline{t}^{ke}) : p \in \mathcal{R}(\mathcal{F})_{\underline{n}}\} \cong \mathcal{F}(\underline{n}).$$

Hence $[H^1_{\mathcal{R}(\mathcal{F})_{++}}(\mathcal{R}(\mathcal{F}))]_{\underline{n}} \cong \dfrac{\breve{\mathcal{F}}(\underline{n})}{\mathcal{F}(\underline{n})}$ for all $\underline{n} \in \mathbb{N}^s$. □

Proposition 3.6 *Let $(R, \mathfrak{m})$ be a Cohen–Macaulay local ring of dimension two with infinite residue field, I an $\mathfrak{m}$-primary ideal and $\mathcal{F} = \{I_n\}_{n \in \mathbb{Z}}$ be an I-admissible filtration of ideals in R. Then*

$$[H^1_{\mathcal{R}(\mathcal{F})_+}(\mathcal{R}(\mathcal{F}))]_n = \begin{cases} \breve{I}_n / I_n & \text{if } n \geq 0 \\ R & \text{if } n < 0. \end{cases}$$

Proof By Proposition 3.5, we get $[H^1_{\mathcal{R}(\mathcal{F})_+}(\mathcal{R}(\mathcal{F}))]_n = \breve{I}_n / I_n$ for all $n \geq 0$. Let J be minimal reduction of $\mathcal{F}$ generated by superficial sequence y_1, y_2. For all $k \geq 1$, consider the following complex of $\mathcal{R}(\mathcal{F})$-modules

$$F^{k.} : 0 \longrightarrow \mathcal{R}(\mathcal{F}) \xrightarrow{\alpha_k} \mathcal{R}(\mathcal{F})(k)^2 \xrightarrow{\beta_k} \mathcal{R}(\mathcal{F})(2k) \longrightarrow 0,$$

where $\alpha_k(1) = ({y_1}^k t^k, {y_2}^k t^k)$ and $\beta_k(u, v) = {y_2}^k t^k u - {y_1}^k t^k v$. Since radical of the ideal $(y_1 t, y_2 t)\mathcal{R}(\mathcal{F})$ is same as radical of the ideal $\mathcal{R}(\mathcal{F})_+$, by [5, Theorem 5.2.9],

$$[H^1_{\mathcal{R}(\mathcal{F})_+}(\mathcal{R}(\mathcal{F}))]_n \cong \varinjlim_k \frac{(\ker \beta_k)_n}{(\operatorname{im} \alpha_k)_n}.$$

Now for $n < 0$, $\mathcal{R}(\mathcal{F})_n = 0$. Hence $(\operatorname{im} \alpha_k)_n = 0$.

Suppose $(u, v) \in (\ker \beta_k)_n$ for any $n < 0$. Then ${y_2}^k u - {y_1}^k v = 0$. Since $\{y_1, y_2\}$ is a regular sequence, $u = {y_1}^k p$ for some $p \in R$. Thus $v = {y_2}^k p$. Hence $(u, v) = ({y_1}^k p, {y_2}^k p)$. This implies for all $n < 0$, $(u, v) \in (\ker \beta_k)_n$ if and only if $(u, v) = ({y_1}^k p, {y_2}^k p)$ for some $p \in (\mathcal{F}(n + k) : ({y_1}^k, {y_2}^k)) = R$. □

Proposition 3.7 ([2, Theorem 3.5]) *Let $(R, \mathfrak{m})$ be a Cohen–Macaulay local ring of dimension two with infinite residue field, I an $\mathfrak{m}$-primary ideal and $\mathcal{F} = \{I_n\}_{n\in\mathbb{Z}}$ be an I-admissible filtration of ideals in R. Then*

$$[H^1_{\mathcal{R}(\mathcal{F})_+}(\mathcal{R}'(\mathcal{F}))]_n = \begin{cases} \breve{I}_n/I_n & \text{if } n \geq 0 \\ 0 & \text{if } n < 0. \end{cases}$$

Proof Since $[H^1_{\mathcal{R}(\mathcal{F})_{++}}(\mathcal{R}'(\mathcal{F}))]_n = [H^1_{\mathcal{R}(\mathcal{F})_{++}}(\mathcal{R}(\mathcal{F}))]_n$ for all $n \in \mathbb{N}$ by Proposition 2.4, using Proposition 3.6, we get $[H^1_{\mathcal{R}(\mathcal{F})_{++}}(\mathcal{R}'(\mathcal{F}))]_n = \breve{I}_n/I_n$.

Let J be minimal reduction of $\mathcal{F}$ generated by superficial sequence y_1, y_2. For all $k \geq 1$, consider the following complex of $\mathcal{R}(\mathcal{F})$-modules

$$F^{k.} : 0 \longrightarrow \mathcal{R}'(\mathcal{F}) \xrightarrow{\alpha_k} \mathcal{R}'(\mathcal{F})(k)^2 \xrightarrow{\beta_k} \mathcal{R}'(\mathcal{F})(2k) \longrightarrow 0,$$

where $\alpha_k(1) = ({y_1}^k t^k, {y_2}^k t^k)$ and $\beta_k(u, v) = {y_2}^k t^k u - {y_1}^k t^k v$. Since radical of the ideal $(y_1 \underline{t}, y_2 \underline{t})\mathcal{R}(\mathcal{F})$ is same as radical of the ideal $\mathcal{R}(\mathcal{F})_+$, by [5, Theorem 5.2.9],

$$[H^1_{\mathcal{R}(\mathcal{F})_+}(\mathcal{R}'(\mathcal{F}))]_n \cong \varinjlim_k \frac{(\ker \beta_k)_n}{(\operatorname{im} \alpha_k)_n}.$$

for all $n \in \mathbb{Z}\backslash\mathbb{N}$,

$$(\operatorname{im} \alpha_k)_n = \{({y_1}^k p \underline{t}^{ke}, {y_2}^k p \underline{t}^{ke}) : p \in \mathcal{R}'(\mathcal{F})_n = R\} \cong R.$$

Thus $[H^1_{\mathcal{R}(\mathcal{F})_+}(\mathcal{R}'(\mathcal{F}))]_n = 0$ for all $n \in \mathbb{Z}\backslash\mathbb{N}$. □

Lemma 3.8 ([25, Lemma 2.11]) *Let $I_1, \ldots, I_s$ be $\mathfrak{m}$-primary ideals in a Noetherian local ring $(R, \mathfrak{m})$ of dimension $d \geq 1$ such that grade$(I_1 \cdots I_s) \geq 1$. Let $\mathcal{F} =$*

$\{\mathcal{F}(\underline{n})\}_{\underline{n}\in\mathbb{Z}^s}$ be an $\underline{I}$-admissible filtration of ideals in R. Denote $\mathcal{R}(\underline{I})_{++}$ as $\mathcal{R}_{++}$. Then

$$\lambda_R[H^d_{\mathcal{R}_{++}}(\mathcal{R}'(\mathcal{F}))]_{\underline{n}} \leq \lambda_R[H^d_{\mathcal{R}_{++}}(\mathcal{R}'(\mathcal{F}))]_{\underline{n}-e_i}$$

for all $\underline{n} \in \mathbb{Z}^s$ and $i = 1, \ldots, s$.

Proof By Lemma 3.3 and Theorem 3.4, there exists an ideal $J = (y_1, \ldots, y_d) \subseteq I_1 \cdots I_s$ such that $y_1 = x_{11} \cdots x_{s1}$ is a nonzerodivisor, $x_{i1} \in I_i$ forall $i = 1, \ldots, s$ and $J\mathcal{F}(\underline{n}) = \mathcal{F}(\underline{n}+e)$ for all large $\underline{n}$. Hence $\sqrt{\mathcal{R}(\underline{I})_{++}} = \sqrt{(y_1\underline{t}, \ldots, y_d\underline{t})}$. Consider the short exact sequence of $\mathcal{R}(\underline{I})$-modules,

$$0 \longrightarrow \mathcal{R}'(\mathcal{F})(-e_i) \xrightarrow{x_{i1}t_i} \mathcal{R}'(\mathcal{F}) \longrightarrow \frac{\mathcal{R}'(\mathcal{F})}{x_{i1}t_i\mathcal{R}'(\mathcal{F})} \longrightarrow 0.$$

This gives a long exact sequence of $\underline{n}$-graded components of local cohomology modules,

$$\cdots \longrightarrow [H^d_{\mathcal{R}(\underline{I})_{++}}(\mathcal{R}'(\mathcal{F}))]_{\underline{n}-e_i} \longrightarrow [H^d_{\mathcal{R}(\underline{I})_{++}}(\mathcal{R}'(\mathcal{F}))]_{\underline{n}} \longrightarrow \left[H^d_{\mathcal{R}(\underline{I})_{++}}\left(\frac{\mathcal{R}'(\mathcal{F})}{x_{i1}t_i\mathcal{R}'(\mathcal{F})}\right)\right]_{\underline{n}} \longrightarrow 0.$$

Let $T = \frac{\mathcal{R}(\underline{I})}{x_{i1}t_i\mathcal{R}(\underline{I})}$. Now $\frac{\mathcal{R}'(\mathcal{F})}{x_{i1}t_i\mathcal{R}'(\mathcal{F})}$ is a T-module and $\sqrt{\left(\frac{\mathcal{R}(\underline{I})}{x_{i1}t_i\mathcal{R}(\underline{I})}\right)_{++}} = \sqrt{(y_2\underline{t}, \cdots, y_d\underline{t})T}$. Hence $H^d_{\mathcal{R}(\underline{I})_{++}}\left(\frac{\mathcal{R}'(\mathcal{F})}{x_{i1}t_i\mathcal{R}'(\mathcal{F})}\right) = 0$ which implies the required result. □

Theorem 3.9 ([25, Theorem 3.3]) *Let $(R, \mathfrak{m})$ be a Noetherian local ring of dimension $d \geq 1$ with infinite residue field and $I_1, \ldots, I_s$ be $\mathfrak{m}$-primary ideals in R such that grade$(I_1 \cdots I_s) \geq 1$. Let $\mathcal{F} = \{\mathcal{F}(\underline{n})\}_{\underline{n}\in\mathbb{Z}^s}$ be an $\underline{I}$- admissible filtration of ideals in R. Then for all $\underline{n} \in \mathbb{N}^s$ and $i = 1, \ldots, s$,*

$$[H^0_{G_i(\mathcal{F})_{++}}(G_i(\mathcal{F}))]_{\underline{n}} = \frac{\breve{\mathcal{F}}(\underline{n}+e_i) \cap \mathcal{F}(\underline{n})}{\mathcal{F}(\underline{n}+e_i)}.$$

Proof Let $x \in \mathcal{F}(\underline{n})$ and $x^* = x + \mathcal{F}(\underline{n}+e_i) \in [H^0_{G_i(\mathcal{F})_{++}}(G_i(\mathcal{F}))]_{\underline{n}}$. Then $x^*G_i(\mathcal{F})^k_{++} = 0$ for some $k \geq 1$. Therefore $x\mathcal{F}(e)^k \subseteq \mathcal{F}(\underline{n}+ke+e_i)$. Hence $x \in \breve{\mathcal{F}}(\underline{n}+e_i)$.

Conversely, suppose $x^* \in \breve{\mathcal{F}}(\underline{n}+e_i) \cap \mathcal{F}(\underline{n})/\mathcal{F}(\underline{n}+e_i)$. We show that there exists $m \gg 0$ such that $x^*G_i(\mathcal{F})^m_{++} = 0$. Since $G_i(\mathcal{F})^m_{++} \subseteq \bigoplus_{\underline{p}\geq me} \mathcal{F}(\underline{p})/\mathcal{F}(\underline{p}+e_i)$, it is enough to show that $x^*(\mathcal{F}(\underline{p})/\mathcal{F}(\underline{p}+e_i)) = 0$ for all large $\underline{p}$. By [25, Proposition 3.1], there exists $\underline{m} \in \mathbb{N}^s$ with $\underline{m} \geq e$ such that $\breve{\mathcal{F}}(\underline{r}) = \mathcal{F}(\underline{r})$ for all $\underline{r} \geq \underline{m}$. Thus for all $\underline{r} \geq \underline{m}$,

$$x\mathcal{F}(\underline{r}) \subseteq \breve{\mathcal{F}}(\underline{n}+e_i)\mathcal{F}(\underline{r}) \subseteq \breve{\mathcal{F}}(\underline{n}+\underline{r}+e_i) = \mathcal{F}(\underline{n}+\underline{r}+e_i).$$

Therefore $(x + \mathcal{F}(\underline{n} + e_i))G_i(\mathcal{F})^m_{++} = 0$ for some $m \geq 1$. Hence $(x + \mathcal{F}(\underline{n} + e_i)) \in [H^0_{G_i(\mathcal{F})_{++}}(G_i(\mathcal{F}))]_{\underline{n}}$. □

4 The Postulation Number and the Reduction Number

In [35, Proposition 3] Sally gave a nice relation between the postulation number and the reduction number of the filtration $\{\mathfrak{m}^n\}_{n\in\mathbb{N}}$. In [23, Corollary 3.8] Marley generalised this relation for any I-admissible filtration. In this section, we derive these results using the Grothendieck–Serre formula. We recall few preliminary results about superficial sequences which are useful to apply induction in the study of Hilbert coefficients.

Let $(R, \mathfrak{m})$ be a Noetherian local ring, I be an $\mathfrak{m}$-primary ideal and $\mathcal{F} = \{I_n\}_{n\in\mathbb{Z}}$ be an I-admissible filtration of ideals in R.

Definition 4.1 An element $x \in I_t \backslash I_{t+1}$ is called **superficial element for $\mathcal{F}$ of degree** t if there exists an integer $c \geq 0$ such that $(I_{n+t} : x) \cap I_c = I_n$ for all $n \geq c$.

If the residue field of R is infinite, then there exists a superficial element of degree 1 [32, Proposition 2.3]. If grade $(I_1) \geq 1$ and $x \in I_1$ is superficial for $\mathcal{F}$, Huckaba and Marley [13], showed that x is nonzerodivisor in R and $(I_{n+1} : x) = I_n$ for all large n. If dimension of R is $d \geq 1$, $x \in I_1\backslash I_2$ is superficial element for $\mathcal{F}$ and x is a nonzerodivisor on R then by [23, Lemma A.2.1], $e_i(\mathcal{F}) = e_i(\mathcal{F}')$ for all $0 \leq i < d$ where $R' = R/(x)$ and $\mathcal{F}' = \{I_nR'\}_{n\in\mathbb{Z}}$. The following lemma is due to Blancafort [3, Lemma 3.1.6]. This lemma was first proved by Huckaba [11, Lemma 1.1] for I-adic filtration.

Lemma 4.2 *Let $(R, \mathfrak{m})$ be a Cohen–Macaulay local ring of dimension $d \geq 1$, I an $\mathfrak{m}$-primary ideal and $\mathcal{F} = \{I_n\}_{n\in\mathbb{Z}}$ be an I-admissible filtration of ideals in R. Suppose J is a minimal reduction of $\mathcal{F}$ and there exists an $x \in J\backslash I_2$ such that $x^* = x + I_2$ is a nonzerodivisor in $G(\mathcal{F})$. Let $R' = R/(x)$. Then $r(\mathcal{F}) = r(\mathcal{F}')$ where $\mathcal{F}' = \{I_nR'\}_{n\in\mathbb{Z}}$.*

Proof We denote $r(\mathcal{F})$ and $r(\mathcal{F}')$ by r and s respectively. It is clear that $s \leq r$. We use the notation "$'$" to denote the image in R'. Let $n \geq s$ and $a \in I_{n+1}$. Then $a' \in J'I'_n$. Hence $a = p + xq$ for some $p \in JI_n$ and $q \in R$. Therefore $xq \in I_{n+1}$ which implies $q \in (I_{n+1} : x)$. Since x^* is a nonzerodivisor in $G(\mathcal{F})$, we have $(I_{n+1} : x) = I_n$ for all $n \in \mathbb{Z}$. Hence we get the required result. □

Definition 4.3 If $\underline{x} = x_1, \ldots, x_r \in I_1$, we say $\underline{x}$ is a superficial sequence for $\mathcal{F}$ if for all $0 \leq i < r$, x_{i+1} is superficial for $\mathcal{F}/(x_1, \ldots, x_i)$.

Suppose $(R, \mathfrak{m})$ is Cohen–Macaulay local ring of dimension d, I_1 is an $\mathfrak{m}$-primary ideal and $\mathcal{F} = \{I_n\}$ is an I_1-admissible filtration of ideals in R. Suppose $x_1, \ldots, x_r \in I_1$ and $1 \le r \le d$, then $x_1, \ldots, x_r$ is a superficial sequence for $\mathcal{F}$ if and only if $x_1, \ldots, x_r$ is R-regular sequence and there exists an integer $n_0 \ge 0$ such that for all $1 \le i \le r$,

$$(x_1, \ldots, x_i) \cap I_n = (x_1, \ldots, x_i)I_{n-1} \text{ for all } n \ge n_0.$$

This result was first proved by Valabrega and Valla [39, Corollary 2.7] for I-adic filtration and then by Huckaba and Marley [13] for $\mathbb{Z}$-graded admissible filtrations. Marley [23, Proposition A.2.4] showed that if residue the field is infinite then any minimal reduction of $\mathcal{F}$ can be generated by a superficial sequence for $\mathcal{F}$. The following lemma is due to Huckaba and Marley [13, Lemma 2.1].

Lemma 4.4 ([13, Lemma 2.1]) *Let $(R, \mathfrak{m})$ be a Noetherian local ring of dimension $d \ge 1$, I an $\mathfrak{m}$-primary ideal and $\mathcal{F} = \{I_n\}$ be an I-admissible filtration of ideals in R. Let $x_1, \ldots, x_k$ be a superficial sequence for $\mathcal{F}$. If* grade $G(\mathcal{F})_+ \ge k$ *then $x_1^*, \ldots, x_k^*$ is a regular sequence in $G(\mathcal{F})$ and hence $G(\mathcal{F})/(x_1^*, \ldots, x_k^*) \simeq G(\mathcal{F}/(x_1, \ldots, x_k))$ where x_i^* is image of x_i in $G(\mathcal{F})$.*

Proof By induction it is enough to prove for $k = 1$. Let $(I_{n+1} : x_1) \cap I_c = I_n$ for all $n \ge c$. Let $x^* \in (0 : x_1^*) \cap G(\mathcal{F})_n$ for some $n \in \mathbb{N}$. We show that $x^*(G(\mathcal{F})_+)^{c+1} = 0$. Let $0 \ne z^* \in G(\mathcal{F})_+^{c+1} \cap G(\mathcal{F})_p$. Now $x^* z^* \in G(\mathcal{F})_{n+p}$ and $x_1 x z \in I_{n+p+2}$. Therefore $xz \in (I_{n+p+2} : x_1) \cap I_c = I_{n+p+1}$. Thus $x^* z^* = 0$ in $G(\mathcal{F})$. Hence $x^* \in (0 :_{G(\mathcal{F})} (G(\mathcal{F})_+)^{c+1}) = 0$. □

The next theorem was proved for the $\mathfrak{m}$-adic by Sally [35, Proposition 3]. We have adapted her proof for any admissible filtration.

Theorem 4.5 *Let $(R, \mathfrak{m})$ be a Cohen–Macaulay local ring of dimension $d \ge 1$ with infinite residue field, I an $\mathfrak{m}$-primary ideal and $\mathcal{F} = \{I_n\}$ be an I-admissible filtration of ideals in R. Let $H_R(n) = \lambda(I_n/I_{n+1})$ and $P_R(X) \in \mathbb{Q}[X]$ such that $P_R(n) = H_R(n)$ for all large n. Suppose* grade $G(\mathcal{F})_+ \ge d - 1$. *Then for a minimal reduction $J = (x_1, \ldots, x_d)$ of $\mathcal{F}$, $H_R(r_J(\mathcal{F}) - d) \ne P_R(r_J(\mathcal{F}) - d)$ and $H_R(n) = P_R(n)$ for all $n \ge r_J(\mathcal{F}) - d + 1$.*

Proof We denote $r_J(\mathcal{F})$ by r. We use induction on d. Let $d = 1$. Without loss of generality we assume x_1 is superficial. Then

$$H^0_{G(\mathcal{F})_+}(G(\mathcal{F}))_n = \{z^* \in I_n/I_{n+1} \mid zI_l \in I_{n+l+1} \text{ for all large } l\}.$$

For $n \ge r - 1$, $zx_1^l \in I_{n+l+1} = x_1^l I_{n+1}$ implies $z \in I_{n+1}$. Thus for all $n \ge r - 1$, $H^0_{G(\mathcal{F})_+}(G(\mathcal{F}))_n = 0$.

Now we prove that $H^1_{G(\mathcal{F})_+}(G(\mathcal{F}))_{r-1} \neq 0$ and $H^1_{G(\mathcal{F})_+}(G(\mathcal{F}))_n = 0$ for all $n \geq r$. For each n, consider the following map

$$\frac{I_{k+n}}{x_1^k I_n + I_{k+n+1}} \xrightarrow{\phi_k} \frac{I_{k+n+1}}{x_1^{k+1} I_n + I_{k+n+2}} \text{ where } \phi_k(\overline{z}) = \overline{x_1 z}.$$

For all large k, $I_{k+n+1} = x_1 I_{k+n}$. Hence for all large k, ϕ_k is surjective. Now suppose $\phi_k(\overline{z}) = 0$ for some $\overline{z} \in I_{k+n}/x_1^k I_n + I_{k+n+1}$. Then $x_1 z \in x_1^{k+1} I_n + I_{k+n+2}$. Therefore $x_1 z = x_1^{k+1} a + b$ for some $a \in I_n$ and $b \in I_{k+n+2}$. Thus $b \in (x_1) \cap I_{k+n+2}$. Since x_1 is superficial, for all large k, $b \in x_1 I_{k+n+1}$ and hence $z \in x_1^k I_n + I_{n+k+1}$. Thus for all large k, ϕ_k is injective. Therefore by Proposition 3.1, for all large k,

$$H^1_{G(\mathcal{F})_+}(G(\mathcal{F}))_n \simeq \frac{I_{k+n}}{x_1^k I_n + I_{k+n+1}}.$$

Thus for all $n \geq r$ and for all large k, $I_{k+n} = x_1^k I_n \subseteq x_1^k I_n + I_{k+n+1}$. Hence $H^1_{G(\mathcal{F})_+}(G(\mathcal{F}))_n = 0$ for all $n \geq r$.

Suppose $H^1_{G(\mathcal{F})_+}(G(\mathcal{F}))_{r-1} = 0$. Then for all large k,

$$I_{k+r-1} = x_1^k I_{r-1} + I_{k+r} \subseteq x_1^k I_{r-1}.$$

Let $a \in I_{k+r-2}$. Then $x_1 a \in I_{k+r-1} \subseteq x_1^k I_{r-1}$ implies $a \in x_1^{k-1} I_{r-1}$. Thus $I_{k+r-2} = x_1^{k-1} I_{r-1}$. Using this procedure repeatedly, we get $I_r = x_1 I_{r-1}$ which is a contradiction. Thus $H^1_{G(\mathcal{F})_+}(G(\mathcal{F}))_{r-1} \neq 0$. Therefore by Theorem 2.3, we get the required result.

Suppose $d \geq 2$. Without loss of generality we assume $x_1, \ldots, x_d$ is superficial sequence for $\mathcal{F}$. Since grade $G(\mathcal{F})_+ \geq d-1$, by Lemma 4.4, we have x_1^* is a nonzerodivisor of $G(\mathcal{F})$. By [3, Proposition 3.1.3] $G(\mathcal{F})/(x_1^*) \simeq G(\mathcal{F}/(x_1))$. For all $n \in \mathbb{Z}$, consider the following exact sequence

$$0 \longrightarrow \frac{I_{n-1}}{I_n} \xrightarrow{x_1^*} \frac{I_n}{I_{n+1}} \longrightarrow \frac{I_n}{x_1 I_{n-1} + I_{n+1}} \simeq \frac{I_n + (x_1)}{I_{n+1} + (x_1)} \longrightarrow 0. \quad (4.5.1)$$

Then for all $n \in \mathbb{Z}$,

$$H_{R/(x_1)}(n) = H_R(n) - H_R(n-1) \text{ and hence } P_{R/(x_1)}(n) = P_R(n) - P_R(n-1).$$

Since $\dim R/(x_1) = d-1$ and grade $G(\mathcal{F}/(x_1))_+ \geq d-2$, by induction and Lemma 4.2, we have

$$H_{R/(x_1)}(r-d+1) \neq P_{R/(x_1)}(r-d+1) \text{ and } H_{R/(x_1)}(n) = P_{R/(x_1)}(n) \text{ for all } n \geq r-d+2.$$

Since there exists an integer m, such that for all $n \geq m$, $P_R(n) = H_R(n)$, we have

$$\begin{aligned} P_R(n-1) - H_R(n-1) &= P_R(n) - H_R(n) = \cdots \\ &= P_R(n+m) - H_R(n+m) = 0 \text{ for all } n \geq r-d+2. \end{aligned}$$

Therefore

$$\begin{aligned} 0 &\neq P_{R/(x_1)}(r-d+1) - H_{R/(x_1)}(r-d+1) \\ &= [P_R(r-d+1) - H_R(r-d+1)] - [P_R(r-d) - H_R(r-d)] \\ &= P_R(r-d) - H_R(r-d). \end{aligned}$$

□

The following result is due to Marley [23, Corollary 3.8]. Here we give another proof which follows from Theorem 4.5.

Theorem 4.6 ([23, Corollary 3.8]) *Let $(R, \mathfrak{m})$ be a Cohen–Macaulay local ring of dimension $d \geq 1$ with infinite residue field, I an $\mathfrak{m}$-primary ideal and $\mathcal{F} = \{I_n\}$ be an I-admissible filtration of ideals in R. Let* grade $G(\mathcal{F})_+ \geq d-1$. *Then $r_J(\mathcal{F}) = n(\mathcal{F}) + d$ for any minimal reduction J of $\mathcal{F}$. In particular, $r(\mathcal{F}) = n(\mathcal{F}) + d$.*

Proof Let $H_R(n) = \lambda(I_n/I_{n+1})$ for all n and $P_R(X) \in \mathbb{Q}[X]$ such that $P_R(n) = H_R(n)$ for all large n. Let $d = 1$ and J be any minimal reduction of $\mathcal{F}$. Denote $r_J(\mathcal{F})$ by r. Then degree of the polynomial $P_R(X)$ is zero. Hence $P_R(X) = a$ where a is a nonzero constant. By Theorem 4.5, for all $n \geq r$, $P_R(n) = H_R(n)$. Therefore for all $n \geq r$, we have

$$\lambda\left(\frac{R}{I_n}\right) = (n-r)a + \lambda\left(\frac{R}{I_r}\right) = na + \left(\lambda\left(\frac{R}{I_r}\right) - ra\right).$$

Hence $P_{\mathcal{F}}(n) = H_{\mathcal{F}}(n)$ for all $n \geq r$. Suppose $P_{\mathcal{F}}(r-1) = H_{\mathcal{F}}(r-1)$. Then

$$-a + \lambda\left(\frac{R}{I_r}\right) = \lambda\left(\frac{R}{I_{r-1}}\right).$$

This implies $P_R(r-1) = H_R(r-1)$ which contradicts Theorem 4.5. Thus $r_J(\mathcal{F}) - 1 = n(\mathcal{F})$ for any minimal reduction J of $\mathcal{F}$. Hence we get the result for $d = 1$.

Suppose $d \geq 2$ and $J = (x_1, \ldots, x_d)$ is a minimal reduction of $\mathcal{F}$ consisting of superficial elements. Denote $r_J(\mathcal{F})$ by r. For all $n \in \mathbb{Z}$, we get

$$H_R(n) = H_{\mathcal{F}}(n+1) - H_{\mathcal{F}}(n) \text{ and hence } P_R(n) = P_{\mathcal{F}}(n+1) - P_{\mathcal{F}}(n).$$

Since there exists an integer m, such that for all $n \geq m$, $P_{\mathcal{F}}(n) = H_{\mathcal{F}}(n)$, by Theorem 4.5, we have

$$\begin{aligned}P_{\mathcal{F}}(n)-H_{\mathcal{F}}(n)&=P_{\mathcal{F}}(n+1)-H_{\mathcal{F}}(n+1)\\&=\cdots\\&=P_{\mathcal{F}}(n+m)-H_{\mathcal{F}}(n+m)=0 \text{ for all } n\geq r-d+1.\end{aligned}$$

Again using Theorem 4.5, we get

$$\begin{aligned}0&\neq P_R(r-d)-H_R(r-d)\\&=[P_{\mathcal{F}}(r-d+1)-H_{\mathcal{F}}(r-d+1)]-[P_{\mathcal{F}}(r-d)-H_{\mathcal{F}}(r-d)]\\&=P_{\mathcal{F}}(r-d)-H_{\mathcal{F}}(r-d).\end{aligned}$$

Thus $r_J(\mathcal{F})-d=n(\mathcal{F})$ for any minimal reduction J of $\mathcal{F}$. Hence $r(\mathcal{F})=n(\mathcal{F})+d$. □

5 Nonnegativity and Vanishing of Hilbert Coefficients

In this section, we apply Grothendieck–Serre formula to derive various properties of the Hilbert coefficients. We derive a result of Northcott, Narita, Marley, and Itoh. We also derive a formula for the components of local cohomology modules of Rees algebras in terms of the Hilbert coefficients (Proposition 5.11) which generalises [35, Proposition 5] and [20, Proposition 3.3].

The following theorem is a generalisation of a result due to Northcott [27, Theorem 1].

Theorem 5.1 (Northcott's inequality) *Let $(R,\mathfrak{m})$ be a $d\geq 1$-dimensional Cohen–Macaulay local ring, I an $\mathfrak{m}$-primary ideal and $\mathcal{F}=\{I_n\}_{n\in\mathbb{Z}}$ be an I-admissible filtration of ideals in R. Then*

$$e_1(\mathcal{F})\geq e_0(\mathcal{F})-\lambda\left(\frac{R}{I_1}\right)\geq 0.$$

Proof We use induction on d. Let $d=1$. Since R is Cohen–Macaulay, putting $n=1$ in the Difference Formula (Theorem 2.5) for Rees algebra of $\mathcal{F}$, we have

$$\begin{aligned}e_0(\mathcal{F})-e_1(\mathcal{F})-\lambda\left(\frac{R}{I_1}\right)&=P_{\mathcal{F}}(1)-H_{\mathcal{F}}(1)\\&=\lambda_R[H^0_{\mathcal{R}(\mathcal{F})_+}(\mathcal{R}'(\mathcal{F}))]_1-\lambda_R[H^1_{\mathcal{R}(\mathcal{F})_+}(\mathcal{R}'(\mathcal{F}))]_1\\&=-\lambda_R[H^1_{\mathcal{R}(\mathcal{F})_+}(\mathcal{R}'(\mathcal{F}))]_1\leq 0.\end{aligned}$$

Thus we get the first inequality. Suppose $d\geq 2$ and the result is true for rings with dimension upto $d-1$. Without loss of generality we may assume that the residue field of R is infinite. Let $x\in I_1$ be a superficial element for $\mathcal{F}$. Then $e_0(\mathcal{F})=e_0(\mathcal{F}')$

and $e_1(\mathcal{F}) = e_1(\mathcal{F}')$ where "$'$" denotes the image in $R' = R/(x)$. Since $\lambda(R'/I_1') = \lambda(R/I_1)$, by induction hypothesis we get the first inequality.

For any minimal reduction J of $\mathcal{F}$, J is minimal reduction I_1 by [31, Lemma 1.5]. Hence, we get the second inequality. □

Theorem 5.2 ([25, Theorem 5.6]) *Let $(R, \mathfrak{m})$ be a Cohen–Macaulay local ring of dimension $d \geq 1$ and $I_1, \ldots, I_s$ be $\mathfrak{m}$-primary ideals of R. Let $\mathcal{F} = \{\mathcal{F}(\underline{n})\}_{\underline{n}\in\mathbb{Z}^s}$ be an $\underline{I}$-admissible filtration. Then*

(1) $e_\alpha(\mathcal{F}) \geq 0$ where $\alpha = (\alpha_1, \ldots, \alpha_s) \in \mathbb{N}^s$, $|\alpha| \geq d-1$.
(2) $e_\alpha(\mathcal{F}) \geq 0$ where $\alpha = (\alpha_1, \ldots, \alpha_s) \in \mathbb{N}^s$, $|\alpha| = d-2$ and $d \geq 2$.

Proof (1) For $|\alpha| = d$, the result follows from [31, Theorem 2.4]. Suppose $|\alpha| = d-1$. We use induction on d. Let $d = 1$. Then putting $\underline{n} = \underline{0}$ in the Difference Formula (Theorem 2.5), we get $e_{\underline{0}}(\mathcal{F}) = \lambda_R[H^1_{\mathcal{R}_{++}}(\mathcal{R}'(\mathcal{F}))]_{\underline{0}} \geq 0$. Let $d \geq 2$ and assume the result for rings of dimension $d-1$. Then there exists i such that $\alpha_i \geq 1$. Without loss of generality assume $\alpha_1 \geq 1$. By Lemma 3.3, there exists a nonzerodivisor $x \in I_1$ such that $(x) \cap \mathcal{F}(\underline{n}) = x\mathcal{F}(\underline{n} - e_1)$ for all $\underline{n} \in \mathbb{N}^s$ such that $n_1 \gg 0$. Let $R' = R/(x)$ and $\mathcal{F}' = \{\mathcal{F}(\underline{n})R'\}_{\underline{n}\in\mathbb{Z}^s}$. For all large $\underline{n}$, consider the following short exact sequence

$$0 \longrightarrow \frac{(\mathcal{F}(\underline{n}) : (x))}{\mathcal{F}(\underline{n} - e_1)} \longrightarrow \frac{R}{\mathcal{F}(\underline{n} - e_1)} \xrightarrow{.x} \frac{R}{\mathcal{F}(\underline{n})} \longrightarrow \frac{R}{(x, \mathcal{F}(\underline{n}))} \longrightarrow 0.$$

Since forall large $\underline{n}$, $(\mathcal{F}(\underline{n}) : (x)) = \mathcal{F}(\underline{n} - e_1)$, we get $P_{\mathcal{F}'}(\underline{n}) = P_{\mathcal{F}}(\underline{n}) - P_{\mathcal{F}}(\underline{n} - e_1)$. Hence $(-1)^{d-1-|(\alpha-e_1)|}b_{(\alpha-e_1)}(\mathcal{F}') = (-1)^{d-|\alpha|}e_\alpha(\mathcal{F})$ where

$$P_{\mathcal{F}'}(\underline{n}) = \sum_{\substack{\gamma=(\gamma_1,\ldots,\gamma_s)\in\mathbb{N}^s\\ |\gamma|\leq d-1}} (-1)^{d-1-|\gamma|} b_\gamma(\mathcal{F}') \binom{n_1+\gamma_1-1}{\gamma_1} \cdots \binom{n_s+\gamma_s-1}{\gamma_s}.$$

Since $|(\alpha - e_1)| = d-2 = (d-1)-1$, by induction $b_{(\alpha-e_1)}(\mathcal{F}') \geq 0$. Hence for $|\alpha| = d-1$, $e_\alpha(\mathcal{F}) \geq 0$.

(2) We prove the result using induction on d. For $d = 2$ the result follows from the Difference Formula (Theorem 2.5) for $\underline{n} = \underline{0}$ and Proposition 3.5. The rest is same as for the case $|\alpha| = d-1$. □

As a consequence of this we get the following results which is proved by Marley [23, Propositions 3.19 and 3.23]. The next one is a generalisation of a result due to M. Narita [26, Theorem 1]. Here we give different proof.

Proposition 5.3 *Let $(R, \mathfrak{m})$ be a d-dimensional ($d \geq 2$) Cohen–Macaulay local ring, I an $\mathfrak{m}$-primary ideal and $\mathcal{F} = \{I^n\}_{n\in\mathbb{Z}}$ an admissible I-filtration. Then $e_2(\mathcal{F}) \geq 0$.*

Proof Comparing the expressions of coefficients of Hilbert polynomials for $s = 1$ and $s \geq 2$, by Theorem 5.2, we get the required result. □

It is natural to ask whether $e_i(\mathcal{F})$ are nonnegative for $i \geq 3$ in a Cohen–Macaulay local ring. Narita [26, Theorem 2] and Marley [22, Example 2] gave an example of an ideal in a Cohen–Macaulay local ring with $e_3(I) < 0$.

Example 5.4 [26, Theorem 2] Let Δ be a formal power series $k[[X_1, X_2, X_3, X_4]]$ over a field k and $Q = \Delta/\Delta X_4^3$. Then Q is a Cohen–Macaulay local ring of dimension 3. Let x_1, x_2, x_3, x_4 be the images of X_1, X_2, X_3, X_4 in Q and $I = Qx_1 + Qx_2^2 + Qx_3^2 + Qx_2x_4 + Qx_3x_4$. Then

$$e_3(I) = -\lambda_{Q'}\left(\frac{((IQ')^2 : (x_2Q')^2)}{IQ'}\right) = -\lambda_{Q'}\left(\frac{IQ' + (x_4Q')^2}{IQ'}\right) < 0 \text{ where } Q' = Q/(x_1).$$

Example 5.5 [22, Example 2] Let $I = (X^3, Y^3, Z^3, X^2Y, XY^2, YZ^2, XYZ)$ in the regular local ring $R = k[X, Y, Z]_{(X,Y,Z)}$. Then for all $n \geq 1$,

$$P_I(n) = 27\binom{n+2}{3} - 18\binom{n+1}{2} + 4n + 1.$$

Hence $e_3(I) = -1 < 0$.

However, for $\mathcal{F} = \{\overline{I^n}\}_{n\in\mathbb{Z}}$, Itoh proved that $e_3(\mathcal{F})$ is nonnegative in an analytically unramified Cohen–Macaulay local ring [17, Theorem 3]. In order to prove this, he used an analogue of Theorem 2.3 (see [17, p.114]). In [12, Corollary 3.9], authors gave an alternative proof of this result. We prove this result using the GSF. For this purpose, we recall some results of Itoh about vanishing of graded components of local cohomology modules. See also [10, Theorem 1.2].

Theorem 5.6 ([16, Theorem 2] [17, Proposition 13]) *Let $(R, \mathfrak{m})$ be an analytically unramified Cohen–Macaulay local ring of dimension $d \geq 2$. Let $\mathcal{M} = (t^{-1}, \mathcal{R}(\mathcal{F})_+)$ be the maximal homogeneous ideal of $\mathcal{R}'(\mathcal{F})$. Then the following statements hold true for the filtration $\mathcal{F} = \{\overline{I^n}\}_{n\in\mathbb{Z}}$:*

(1) $H^0_{\mathcal{M}}(\mathcal{R}'(\mathcal{F})) = H^1_{\mathcal{M}}(\mathcal{R}'(\mathcal{F})) = 0$;
(2) $H^2_{\mathcal{M}}(\mathcal{R}'(\mathcal{F})_j = 0$ *for* $j \leq 0$;
(3) $H^i_{\mathcal{M}}(\mathcal{R}'(\mathcal{F})) = H^i_{\mathcal{R}(\mathcal{F})_+}(\mathcal{R}'(\mathcal{F}))$ *for* $i = 0, 1, \ldots, d-1$.

Theorem 5.7 ([17, Theorem 3]) *Let $(R, \mathfrak{m})$ be an analytically unramified Cohen–Macaulay local ring of dimension $d \geq 3$ and I be an $\mathfrak{m}$-primary ideal in R. Then* $\overline{e}_3(I) \geq 0$.

Proof For $\mathcal{F} = \{\overline{I^n}\}_{n\in\mathbb{Z}}$, we set $\overline{\mathcal{R}'}(I) := \mathcal{R}'(\mathcal{F})$. We use induction on d. Let $d = 3$. Then, by the Difference Formula (Theorem 2.5) for Rees algebras and Theorem 5.6, we have

$$\overline{e}_3(I) = h^3_{\overline{\mathcal{R}'}(I)_+}\overline{\mathcal{R}'}(I)_0 \geq 0.$$

Let $d > 3$. We may assume that the residue field of R is infinite. Let $J \subseteq I$ be a reduction of I. Since $\overline{I^n} = \overline{J^n}$ for all n, $\overline{e}_i(I) = \overline{e}_i(J)$ for all $i = 1, \ldots, d$. Therefore it suffices to show that $\overline{e}_3(J) \geq 0$. By [17, Theorem 1 and Corollary 8], there exists a system of generators $x_1, \ldots, x_d$ of J such that, if we put $T = (T_1, \ldots, T_d)$, $R(T) = R[T]_{\mathfrak{m}[T]}$ and $C = R(T)/(\sum_{i=1}^d x_i T_i)$, then C is an analytically unramified Cohen–Macaulay local ring of dimension $d - 1$ and $\overline{e}_3(J) = \overline{e}_3(JC)$. Hence, using induction hypothesis the result follows. □

Itoh [17, p.116] proposed the following conjecture on the vanishing of $\overline{e}_3(I)$ which is still open.

Conjecture 5.8 (*Itoh's Conjecture*) Let $(R, \mathfrak{m})$ be an analytically unramified Gorenstein local ring of dimension $d \geq 3$. Then $\overline{e}_3(I) = 0$ if and only if $\overline{I^{n+2}} = I^n\overline{I^2}$ for every $n \geq 0$.

Itoh proved the "if" part of the Conjecture 5.8 in [16, Proposition 10]. He also proved the "only if" part of the Conjecture 5.8 for $\overline{I} = \mathfrak{m}$ [17, Theorem 3(2)]. By [17, Corollary 8 and Proposition 17], it suffices to prove the Conjecture 5.8 for $d = 3$. Let $d = 3$ and $\overline{e}_3(I) = 0$ for an $\mathfrak{m}$-primary ideal in a Cohen–Macaulay ring R. By [16, Proposition 3] and [17, Corollary 16 and (4.1)], $\overline{I^{n+2}} = I^n\overline{I^2}$ for every $n \geq 0$ if and only if $\overline{\mathcal{R}'}(I)$ is Cohen–Macaulay.

It is not known whether the Itoh's conjecture is true for $\overline{I} = \mathfrak{m}$ in a Cohen–Macaulay local ring R (which need not be Gorenstein). Recently, in [8, Theorem 3.6], the authors proved that the Conjecture 5.8 holds true for $\overline{I} = \mathfrak{m}$ in a Cohen–Macaulay local ring of type at most two. T.T. Phuong [29], showed that if R is an analytically unramified Cohen–Macaulay local ring of dimension $d \geq 2$ then the equality $\overline{e}_1(I) = \overline{e}_0(I) - \lambda(R/\overline{I}) + 1$ leads to the vanishing of $\overline{e}_3(I)$. In [21], authors generalised the result of [8]. They also obtained following result for an arbitrary $\mathfrak{m}$-primary ideal I in an analytically unramified Cohen–Macaulay local ring of dimension 3.

Theorem 5.9 ([21, Theorem 1.1]) *Let $(R, \mathfrak{m})$ be an analytically unramified Cohen–Macaulay local ring of dimension* 3. *Let $\mathcal{M} = (t^{-1}, \overline{\mathcal{R}'}_+)$ and $\overline{\mathcal{R}'} = \bigoplus_{n\in\mathbb{Z}} \overline{I^n}t^n$. Suppose that $\overline{e}_3(I) = 0$. Then*

(1) $H^3_{\mathcal{M}}(\overline{\mathcal{R}'}) = 0,$

(2) *Suppose either that R is equicharacteristic or that $\overline{I} = \mathfrak{m}$, and that I has a reduction generated by x, y, z. If $\overline{\mathcal{R}'}$ is not Cohen–Macaulay, then* $\overline{e}_2(I) - \lambda\left(\dfrac{\overline{I^2}}{(x, y, z)\overline{I}}\right) \geq 3.$

As a consequence they generalised [8, Theorem 3.6].

Corollary 5.10 ([21, Corollary 1.2]) *Let $(R, \mathfrak{m})$ be an analytically unramified Cohen–Macaulay local ring of dimension 3.*

(1) *Suppose $\overline{e}_3(I) = 0$. Then there is an inclusion $H^3_{\overline{\mathcal{R}'}_+}(\overline{\mathcal{R}'})_{-1} \subseteq (0 :_{H^3_{\mathfrak{m}}(R)} \overline{I})$.*
(2) *Suppose $\overline{e}_3(\mathfrak{m}) = 0$. Then $\bar{e}_2(\mathfrak{m}) \leq \operatorname{type}(R)$.*
(3) *$\overline{\mathcal{R}'}(\mathfrak{m})$ is Cohen–Macaulay if $\bar{e}_2(\mathfrak{m}) \leq \operatorname{length}_R(\overline{I^2}/\mathfrak{m}I) + 2$ for any ideal I such that $\overline{I} = \mathfrak{m}$, $\overline{e}_3(\mathfrak{m}) = 0$ and I has a minimal reduction.*

Proof (1): By Theorem 5.9, $H^3_{\mathcal{M}}(\overline{\mathcal{R}'}) = 0$. Hence, by [17, Proposition 13(3)], we get an exact sequence

$$0 \longrightarrow H^3_{\overline{\mathcal{R}'}_+}(\overline{\mathcal{R}'})_{-1} \longrightarrow H^3_{\mathfrak{m}}(R) \longrightarrow H^4_{\mathcal{M}}(\overline{\mathcal{R}'})_{-1} \longrightarrow 0.$$

Thus $H^3_{\overline{\mathcal{R}'}_+}(\overline{\mathcal{R}'})_{-1} \subseteq H^3_{\mathfrak{m}}(R)$. By the Difference Formula (Theorem 2.5) and Theorem 5.6, we get

$$h^3_{\overline{\mathcal{R}'}_+}(\overline{\mathcal{R}'})_0 = \bar{e}_3(I) = 0. \tag{5.10.1}$$

Now consider the exact sequence

$$0 \longrightarrow \overline{\mathcal{R}'}(1) \longrightarrow \overline{\mathcal{R}'} \longrightarrow \overline{G} = \bigoplus_{n \geq 0} \frac{\overline{I^n}}{\overline{I^{n+1}}} \longrightarrow 0$$

which gives the long exact sequence

$$\cdots \longrightarrow H^i_{\overline{\mathcal{R}'}_+}(\overline{\mathcal{R}'})_{n+1} \longrightarrow H^i_{\overline{\mathcal{R}'}_+}(\overline{\mathcal{R}'})_n \longrightarrow H^i_{\overline{G}_+}(\overline{G})_n \longrightarrow \cdots.$$

Using (5.10.1), we get an isomorphism $H^3_{\overline{\mathcal{R}'}_+}(\overline{\mathcal{R}'})_{-1} \simeq H^3_{\overline{G}_+}(\overline{G})_{-1}$. This implies that $H^3_{\overline{\mathcal{R}'}_+}(\overline{\mathcal{R}'})_{-1}$ is an $R/\overline{I}$-module. Therefore $H^3_{\overline{\mathcal{R}'}_+}(\overline{\mathcal{R}'})_{-1} \subseteq (0 :_{H^3_{\mathfrak{m}}(R)} \overline{I})$.

(2) Taking $I = \mathfrak{m}$, by the Difference Formula (Theorem 2.5) and Theorem 5.6, we get $\bar{e}_2(I) = h^3_{\overline{\mathcal{R}'}_+}(\overline{\mathcal{R}'})_{-1}$. Hence by (1) we get the result.

(3) Follows from Theorem 5.9(2). □

The next result was first proved by Sally [35, Proposition 5] for the filtration $\{\mathfrak{m}^n\}_{n \in \mathbb{Z}}$ and then by Johnston and Verma [20, Proposition 3.3] for the filtration

$\{I^n\}_{n\in\mathbb{Z}}$ where I is an $\mathfrak{m}$-primary ideal of R. Here we prove the result for $\mathbb{Z}$-graded admissible filtrations.

Proposition 5.11 *Let $(R, \mathfrak{m})$ be a two-dimensional Cohen–Macaulay local ring, I be any $\mathfrak{m}$-primary ideal of R and $\mathcal{F} = \{I_n\}_{n\in\mathbb{Z}}$ an I-admissible filtration of ideals in R. Then*

(1) $\lambda\left(H^2_{\mathcal{R}(\mathcal{F})_+}(\mathcal{R}(\mathcal{F})_0\right) = e_2(\mathcal{F})$,

(2) $\lambda\left(H^2_{\mathcal{R}(\mathcal{F})_+}(\mathcal{R}(\mathcal{F}))_1\right) = e_0(\mathcal{F}) - e_1(\mathcal{F}) + e_2(\mathcal{F}) - \lambda\left(\dfrac{R}{\breve{I}_1}\right)$,

(3) $\lambda\left(H^2_{\mathcal{R}(\mathcal{F})_+}(\mathcal{R}(\mathcal{F}))_{-1}\right) = e_1(\mathcal{F}) + e_2(\mathcal{F})$.

Proof We have

$$P_{\mathcal{F}}(n) - H_{\mathcal{F}}(n) = \sum_{i\geq 0}(-1)^i h^i_{\mathcal{R}(\mathcal{F})_+}(\mathcal{R}'(\mathcal{F}))_n \text{ for all } n \in \mathbb{Z}. \qquad (5.11.1)$$

(1) Putting $n = 0$ in (5.11.1) and using Propositions 2.4 and 3.7, we get the required result.
(2) Putting $n = 1$ in (5.11.1) and using Propositions 2.4 and 3.7, we get the required result.
(3) Consider the short exact sequence of $\mathcal{R}(\mathcal{F})$-modules

$$0 \longrightarrow \mathcal{R}(\mathcal{F})_+ \longrightarrow \mathcal{R}(\mathcal{F}) \longrightarrow R \cong \mathcal{R}(\mathcal{F})/\mathcal{R}(\mathcal{F})_+ \longrightarrow 0$$

which induces a long exact sequence of local cohomology modules whose nth component is

$$\cdots \longrightarrow H^i_{\mathcal{R}(\mathcal{F})_+}(\mathcal{R}(\mathcal{F})_+)_n \longrightarrow H^i_{\mathcal{R}(\mathcal{F})_+}(\mathcal{R}(\mathcal{F}))_n \longrightarrow H^i_{\mathcal{R}(\mathcal{F})_+}(R)_n \longrightarrow \cdots \text{ for all } i \geq 0.$$

Since R is $\mathcal{R}(\mathcal{F})_+$-torsion, $H^0_{\mathcal{R}(\mathcal{F})_+}(R) = R$ and $H^i_{\mathcal{R}(\mathcal{F})_+}(R) = 0$ for all $i \geq 1$. Hence $H^i_{\mathcal{R}(\mathcal{F})_+}(\mathcal{R}(\mathcal{F})_+) \cong H^i_{\mathcal{R}(\mathcal{F})_+}(\mathcal{R}(\mathcal{F}))$ for all $i \geq 2$ and we have the exact sequence

$$\begin{aligned} 0 \to H^0_{\mathcal{R}(\mathcal{F})_+}(\mathcal{R}(\mathcal{F})_+)_n &\to H^0_{\mathcal{R}(\mathcal{F})_+}(\mathcal{R}(\mathcal{F}))_n \to R \\ \to H^1_{\mathcal{R}(\mathcal{F})_+}(\mathcal{R}(\mathcal{F})_+)_n &\to H^1_{\mathcal{R}(\mathcal{F})_+}(\mathcal{R}(\mathcal{F}))_n \to 0. \end{aligned} \qquad (5.11.2)$$

The short exact sequence of $\mathcal{R}(\mathcal{F})$-modules

$$0 \longrightarrow \mathcal{R}(\mathcal{F})_+(1) \longrightarrow \mathcal{R}(\mathcal{F}) \longrightarrow G(\mathcal{F}) \longrightarrow 0$$

induces the exact sequence

$$\begin{aligned} 0 \longrightarrow H^0_{\mathcal{R}(\mathcal{F})_+}(G(\mathcal{F}))_{-1} &\to H^1_{\mathcal{R}(\mathcal{F})_+}(\mathcal{R}(\mathcal{F})_+)_0 \to H^1_{\mathcal{R}(\mathcal{F})_+}(\mathcal{R}(\mathcal{F}))_{-1} \to H^1_{\mathcal{R}(\mathcal{F})_+}(G(\mathcal{F}))_{-1} \\ &\to H^2_{\mathcal{R}(\mathcal{F})_+}(\mathcal{R}(\mathcal{F}))_0 \to H^2_{\mathcal{R}(\mathcal{F})_+}(\mathcal{R}(\mathcal{F}))_{-1} \to H^2_{\mathcal{R}(\mathcal{F})_+}(G(\mathcal{F}))_{-1} \to 0. \end{aligned} \qquad (5.11.3)$$

Now $H^0_{\mathcal{R}(\mathcal{F})_+}(G(\mathcal{F})) \subseteq G(\mathcal{F})$ are nonzero only in nonnegative degrees. Thus $H^1_{\mathcal{R}(\mathcal{F})_+}(\mathcal{R}(\mathcal{F}))_{-1} \cong R$ and $H^1_{\mathcal{R}(\mathcal{F})_+}(\mathcal{R}(\mathcal{F}))_0 = 0$ by Proposition 3.6. Therefore from the exact sequence (5.11.2), we get the exact sequence

$$0 \to R \to H^1_{\mathcal{R}(\mathcal{F})_+}(\mathcal{R}(\mathcal{F})_+)_0 \to H^1_{\mathcal{R}(\mathcal{F})_+}(\mathcal{R}(\mathcal{F}))_0 = 0.$$

Let f denote the map from $H^1_{\mathcal{R}(\mathcal{F})_+}(\mathcal{R}(\mathcal{F}))_{-1}$ to $H^0_{\mathcal{R}(\mathcal{F})_+}(G(\mathcal{F}))_{-1}$ in the exact sequence (5.11.3). First we prove that f is zero map. From the exact sequence (5.11.3), we get the exact sequence

$$0 \longrightarrow R \xrightarrow{g} R \xrightarrow{f} H^1_{\mathcal{R}(\mathcal{F})_+}(G(\mathcal{F}))_{-1}.$$

Since $R/g(R)$ is contained in $H^1_{\mathcal{R}(\mathcal{F})_+}(G(\mathcal{F}))_{-1}$ and by Proposition 2.2, $H^1_{\mathcal{R}(\mathcal{F})_+}(G(\mathcal{F}))_{-1}$ is of finite length, we have $\lambda_R(R/g(R))$ is finite. Since $g(R)$ is principal ideal in R, we get $R = g(R)$. Therefore f is the zero map. Hence we get the exact sequence

$$0 \to H^1_{\mathcal{R}(\mathcal{F})_+}(G(\mathcal{F}))_{-1} \to H^2_{\mathcal{R}(\mathcal{F})_+}(\mathcal{R}(\mathcal{F}))_0 \to H^2_{\mathcal{R}(\mathcal{F})_+}(\mathcal{R}(\mathcal{F}))_{-1} \to H^2_{\mathcal{R}(\mathcal{F})_+}(G(\mathcal{F}))_{-1} \to 0.$$

Therefore by Theorem 2.3, we get

$$\begin{aligned}[H_{\mathcal{F}}(0) - H_{\mathcal{F}}(-1)] - [P_{\mathcal{F}}(0) - P_{\mathcal{F}}(-1)] &= -\lambda\left(H^1_{\mathcal{R}(\mathcal{F})_+}(G(\mathcal{F}))_{-1}\right) + \lambda\left(H^2_{\mathcal{R}(\mathcal{F})_+}(G(\mathcal{F}))_{-1}\right)\\ &= -\lambda\left(H^2_{\mathcal{R}(\mathcal{F})_+}(\mathcal{R}(\mathcal{F}))_0\right) + \lambda\left(H^2_{\mathcal{R}(\mathcal{F})_+}(\mathcal{R}(\mathcal{F}))_{-1}\right).\end{aligned}$$

Thus by part (1) of the Proposition, we get

$$\lambda\left(H^2_{\mathcal{R}(\mathcal{F})_+}(\mathcal{R}(\mathcal{F}))_{-1}\right) = e_2(\mathcal{F}) - e_2(\mathcal{F}) + e_1(\mathcal{F}) + e_2(\mathcal{F}) = e_1(\mathcal{F}) + e_2(\mathcal{F}).$$

□

6 Huneke–Ooishi Theorem and a Multi-graded Version

In this section we give an application of the GSF to derive a result of Huneke [14] and Ooishi [28] which states that if $(R, \mathfrak{m})$ is a Cohen–Macaulay local ring of dimension $d \geq 1$ and I is an $\mathfrak{m}$-primary ideal then $e_0(I) - e_1(I) = \lambda(R/I)$ if and only if $r(I) \leq 1$. A similar result for admissible filtrations was proved in [3, Theorem 4.3.6] and [13, Corollary 4.9]. In [25, Theorem 5.5], authors gave a partial generalisation of this result for an $\underline{I}$-admissible filtration. First we prove few preliminary results needed.

Lemma 6.1 (Sally machine) [34, Corollary 2.4] [13, Lemma 2.2] *Let $(R, \mathfrak{m})$ be a Noetherian local ring, I_1 an $\mathfrak{m}$-primary ideal in R and $\mathcal{F} = \{I_n\}_{n\in\mathbb{Z}}$ be an I_1-*

admissible filtration of ideals in R. Let $x_1, \ldots, x_r$ be a superficial sequence for $\mathcal{F}$. If $\operatorname{grade} G(\mathcal{F}/(x_1, \ldots, x_r))_+ \geq 1$ *then* $\operatorname{grade} G(\mathcal{F})_+ \geq r+1$.

Proof We use induction on r. Let $r = 1$ and $y \in I_t$ such that image of y in $G(\mathcal{F}/(x_1))_t$ is a nonzerodivisor. Then $(I_{n+tj} : y^j) \subseteq (I_n, x_1)$ for all n, j. Since x_1 is a superficial element for $\mathcal{F}$, there exists integer $c \geq 0$, such that $(I_{n+j} : x_1^j) \cap I_c = I_n$ for all $j \geq 1$ and $n \geq c$. Consider an integer $p > c/t$. For arbitrary n and $j \geq 1$, we prove that

$$y^p(I_{n+j} : x_1^j) \subseteq (I_{n+j+tp} : x_1^j) \cap I_c = I_{n+tp}.$$

Let $a \in (I_{n+j} : x_1^j)$. Then $ay^p x_1^j \in I_{n+j+tp}$. Since $pt > c$, $ay^p \in (I_{n+j+tp} : x_1^j) \cap I_c = I_{n+tp}$. Therefore

$$(I_{n+j} : x_1^j) \subseteq (I_{n+tp} : y^p) \subseteq (I_n, x_1).$$

Thus $(I_{n+j} : x_1^j) = I_n + x_1(I_{n+j} : x_1^{j+1})$ for all n and $j \geq 1$. Iterating this formula n times, we get

$$(I_{n+j} : x_1^j) = I_n + x_1 I_{n-1} + x_1^2 I_{n-2} + \cdots + x_1^n(I_{n+j} : x_1^{j+n}) = I_n.$$

Hence $x_1^* = x_1 + I_2$ is a nonzerodivisor of $G(\mathcal{F})$. Since $G(\mathcal{F})/(x_1^*) \simeq G(\mathcal{F}/(x_1))$, $\operatorname{grade} G(\mathcal{F})_+ \geq 2$.

Now assume $r \geq 2$. Then by $r = 1$ case, we have $\operatorname{grade} G(\mathcal{F}/(x_1, \ldots, x_{r-1}))_+ \geq 2 > 1$. By induction on r, we have $\operatorname{grade} G(\mathcal{F})_+ \geq r$ and since $x_1, \ldots, x_r$ is a superficial sequence for $\mathcal{F}$, by Lemma 4.4, we obtain $x_1^*, \ldots, x_r^*$ is a regular sequence of $G(\mathcal{F})$. Since $G(\mathcal{F})/(x_1^*, \ldots, x_r^*) \simeq G(\mathcal{F}/(x_1, \ldots, x_r))$, $\operatorname{grade} G(\mathcal{F})_+ \geq r+1$. □

The next lemma is due to Marley [23, Lemma 3.14].

Lemma 6.2 *Let $(R, \mathfrak{m})$ be a Cohen–Macaulay local ring of dimension $d \geq 1$, I an $\mathfrak{m}$-primary ideal and $\mathcal{F} = \{I_n\}_{n \in \mathbb{Z}}$ be an I-admissible filtration of ideals in R. Suppose $x \in I_1 \setminus I_2$ such that $x^* = x + I_2$ is a nonzerodivisor in $G(\mathcal{F})$. Let $R' = R/(x)$. Then $n(\mathcal{F}) = n(\mathcal{F}') - 1$ where $\mathcal{F}' = \{I_n R'\}_{n \in \mathbb{Z}}$.*

Proof We use the notation "′" to denote the image in R'. For all n, consider the following short exact sequence of R-modules

$$0 \longrightarrow (I_n : x)/I_n \longrightarrow R/I_n \xrightarrow{.x} R/I_n \longrightarrow R'/I'_n \longrightarrow 0.$$

Therefore $H_{\mathcal{F}'}(n) = \lambda(R'/I'_n) = \lambda((I_n : x)/I_n)$. Since x^* is a nonzerodivisor in $G(\mathcal{F})$, we have $(I_{n+1} : x) = I_n$ for all n. Hence $H_{\mathcal{F}'}(n) = \lambda(I_{n-1}/I_n) = \lambda(R/I_n) - \lambda(R/I_{n-1}) = H_{\mathcal{F}}(n) - H_{\mathcal{F}}(n-1)$ for all n which implies $P_{\mathcal{F}'}(n) = P_{\mathcal{F}}(n) - P_{\mathcal{F}}(n-1)$ for all n. Thus $H_{\mathcal{F}'}(n) = P_{\mathcal{F}'}(n)$ for all $n \geq n(\mathcal{F}) + 2$. Since

$$\begin{aligned} P_{\mathcal{F}'}(n(\mathcal{F})+1) - H_{\mathcal{F}'}(n(\mathcal{F})+1) &= [P_{\mathcal{F}}(n(\mathcal{F})+1) - H_{\mathcal{F}}(n(\mathcal{F})+1)] \\ &\quad -[P_{\mathcal{F}}(n(\mathcal{F})) - H_{\mathcal{F}}(n(\mathcal{F}))] \\ &= -[P_{\mathcal{F}}(n(\mathcal{F})) - H_{\mathcal{F}}(n(\mathcal{F}))] \neq 0, \end{aligned}$$

we get the required result. □

The next theorem is due to Blancafort [3] which is a generalisation of a result of Huneke [14] and Ooishi [28] proved independently. We make use of reduction number and postulation number of admissible filtration of ideals to simplify her proof.

Theorem 6.3 ([3, Theorem 4.3.6]) *Let $(R, \mathfrak{m})$ be a Cohen–Macaulay local ring with infinite residue field of dimension $d \geq 1$, I_1 an $\mathfrak{m}$-primary ideal and $\mathcal{F} = \{I_n\}_{n \in \mathbb{Z}}$ be an I_1-admissible filtration of ideals in R. Then the following are equivalent:*

(1) $e_0(\mathcal{F}) - e_1(\mathcal{F}) = \lambda(R/I_1)$,
(2) $r(\mathcal{F}) \leq 1$.

In this case, $e_2(\mathcal{F}) = \cdots = e_d(\mathcal{F}) = 0$, $G(\mathcal{F})$ is Cohen–Macaulay, $n(\mathcal{F}) \leq 0$, $r(\mathcal{F})$ is independent of the reduction chosen and $\mathcal{F} = \{I_1^n\}$.

Proof (1) ⇒ (2) We use induction on d. Let $d = 1$. For all $n \in \mathbb{Z}$, we have

$$P_{\mathcal{F}}(n) - H_{\mathcal{F}}(n) = -h^1_{\mathcal{R}(\mathcal{F})_+}(\mathcal{R}'(\mathcal{F}))_n.$$

By putting $n = 1$ in this formula, we get $e_0(\mathcal{F}) - e_1(\mathcal{F}) - \lambda(R/I_1) = -h^1_{\mathcal{R}(\mathcal{F})_+}(\mathcal{R}'(\mathcal{F}))_1 = 0$. Therefore by Lemma 3.8, for all $n \geq 1$, $h^1_{\mathcal{R}(\mathcal{F})_+}\mathcal{R}'(\mathcal{F})_n = 0$. Consider the short exact sequence of $\mathcal{R}(\mathcal{F})$-modules,

$$0 \longrightarrow \mathcal{R}'(\mathcal{F})(1) \xrightarrow{t^{-1}} \mathcal{R}'(\mathcal{F}) \longrightarrow G(\mathcal{F}) \longrightarrow 0.$$

This induces a long exact sequence,

$$0 \longrightarrow [H^0_{\mathcal{R}(\mathcal{F})_+}(G(\mathcal{F}))]_n \longrightarrow [H^1_{\mathcal{R}(\mathcal{F})_+}(\mathcal{R}'(\mathcal{F}))]_{n+1} \longrightarrow \cdots.$$

Thus for all $n \in \mathbb{N}$, $[H^0_{\mathcal{R}(\mathcal{F})_+}(G(\mathcal{F}))]_n = 0$. Hence $G(\mathcal{F})$ is Cohen–Macaulay. Let $J = (x)$ be a minimal reduction of $\mathcal{F}$. Without loss of generality x is superficial. For each n, consider the following map

$$\frac{I_{k+n}}{x^k I_n} \xrightarrow{\phi_k} \frac{I_{k+n+1}}{x^{k+1} I_n} \text{ where } \phi_k(\overline{z}) = \overline{xz}.$$

For all large k, $I_{k+n+1} = xI_{k+n}$. Hence for all large k, ϕ_k is surjective. Now suppose $\phi_k(\overline{z}) = 0$ for some $\overline{z} \in I_{k+n}/x^k I_n$. Then $xz \in x^{k+1} I_n$. Therefore $xz = x^{k+1}a$ where

$a \in I_n$, hence $z \in x^k I_n$. Thus for all large k, ϕ_k is injective. Therefore by Proposition 3.2, for all large k,

$$H^1_{\mathcal{R}(\mathcal{F})_+}(\mathcal{R}(\mathcal{F}))_n \simeq \frac{I_{k+n}}{x^k I_n}.$$

By Lemma 3.8 and Proposition 2.4, $H^1_{\mathcal{R}(\mathcal{F})_+}(\mathcal{R}(\mathcal{F}))_n = 0$ for all $n \geq 1$. Then for all large k and $n \geq 1$,

$$I_{k+n} = x^k I_n.$$

Let $a \in I_{k+n-1}$. Then $xa \in I_{k+n} \subseteq x^k I_n$ implies $a \in x^{k-1} I_n$. Thus $I_{k+n-1} = x^{k-1} I_n$. Using this procedure repeatedly we get $I_{n+1} = x I_n$. Thus $r(\mathcal{F}) \leq 1$.

Let $d \geq 2$ and $x \in I_1$ be a superficial element for $\mathcal{F}$. Let $R' = R/(x)$, $\mathcal{F}' = \{I_n R'\}_{n \in \mathbb{Z}}$ and $G' = G(\mathcal{F}')$. Since $e_i(\mathcal{F}) = e_i(\mathcal{F}')$ for all $i < d$, we have

$$e_0(\mathcal{F}') - e_1(\mathcal{F}') = e_0(\mathcal{F}) - e_1(\mathcal{F}) = \lambda\left(\frac{R}{I_1}\right) = \lambda\left(\frac{R'}{I_1 R'}\right).$$

Hence by induction hypothesis, G' is Cohen–Macaulay. Therefore by Sally machine (Lemma 6.1), $G(\mathcal{F})$ is Cohen–Macaulay. This implies that for any minimal reduction J of $\mathcal{F}$, $r_J(\mathcal{F}) = n(\mathcal{F}) + d$ by Theorem 4.6. Thus $r_J(\mathcal{F})$ is independent of the minimal reduction J of I. Let J be a minimal reduction of $\mathcal{F}$ generated by superficial sequence $x_1, \ldots, x_d$. Let $\overline{R} = R/(x_1, \ldots, x_{d-1})$ and $\overline{\mathcal{F}} = \{I_n \overline{R}\}_{n \in \mathbb{Z}}$. Since $G(\mathcal{F})$ is Cohen–Macaulay and $x_1, \ldots, x_d$ is superficial, using Theorem 4.6 and Lemmas 4.4, 4.2 and 6.2, for $d - 1$ times, by induction hypothesis we get

$$r(\mathcal{F}) = n(\mathcal{F}) + d = n(\overline{\mathcal{F}}) + 1 = r(\overline{\mathcal{F}}) \leq 1.$$

(2) $\Rightarrow$ (1) Let J be a minimal reduction of $\mathcal{F}$ such that $r(\mathcal{F}) = r_J(\mathcal{F})$ and J is generated by superficial sequence $x_1, \ldots, x_d$. Let $R' = R/(x_1, \ldots, x_{d-1})$ and $\mathcal{F}' = \{I_n R'\}_{n \in \mathbb{Z}}$. Then $x_d I_n R' = I_{n+1} R'$ for all $n \geq 1$. Since x'_d is nonzerodivisor, $(I_{n+1} R' : x'_d) = I_n R'$ for all $n \geq 1$. Therefore $(x'_d)^*$ (the image of x'_d in $G(\mathcal{F}')$) is nonzerodivisor in $G(\mathcal{F}')$. Hence $G(\mathcal{F}')$ is Cohen–Macaulay. Thus by Lemma 6.1, $G(\mathcal{F})$ is Cohen–Macaulay. Therefore by Theorem 4.6, $n(\mathcal{F}) = r(\mathcal{F}) - d \leq 0$. Hence $P_{\mathcal{F}}(n) = H_{\mathcal{F}}(n)$ for all $n > 0$. By putting $n = 1$ for $d = 1$ case we obtain $e_0(\mathcal{F}) - e_1(\mathcal{F}) = \lambda(R/I_1)$.

Now we prove that if $r(\mathcal{F}) \leq 1$ then $e_2(\mathcal{F}) = \cdots = e_d(\mathcal{F}) = 0$. Without loss of generality assume $d \geq 2$. The condition $r(\mathcal{F}) \leq 1$ implies $G(\mathcal{F})$ is Cohen–Macaulay and $n(\mathcal{F}) = r(\mathcal{F}) - d < 0$. Let $d = 2$. Therefore $e_2(\mathcal{F}) = P_{\mathcal{F}}(0) - H_{\mathcal{F}}(0) = 0$. Now assume $d \geq 3$ and the result is true upto dimension $d - 1$. Let J be minimal reduction of $\mathcal{F}$ generated by superficial sequence $x_1, \ldots, x_d$. Let $R' = R/(x_1, \ldots, x_{d-1})$ and $\mathcal{F}' = \{I_n R'\}_{n \in \mathbb{Z}}$. Then $e_i(\mathcal{F}) = e_i(\mathcal{F}') = 0$ for all $0 \leq i < d$. Since $G(\mathcal{F})$ is Cohen–Macaulay and $n(\mathcal{F}) = r(\mathcal{F}) - d < 0$, we get $(-1)^d e_d(\mathcal{F}) = P_{\mathcal{F}}(0) - H_{\mathcal{F}}(0) = 0$. Therefore $e_0(\mathcal{F}) - e_1(\mathcal{F}) - \lambda(R/I_1) = P_{\mathcal{F}}(1) - H_{\mathcal{F}}(1) = 0$.

Let J be a minimal reduction of $\mathcal{F}$ such that $r(\mathcal{F}) = r_J(\mathcal{F})$ and $r(\mathcal{F}) \leq 1$. Then $I_2 = JI_1 \subseteq I_1^2 \subseteq I_2$. Suppose $I_r = I_1^r$ for all $1 \leq r \leq n$. Then $I_{n+1} = JI_n \subseteq I_1 I_1^n \subseteq I_1^{n+1} \subseteq I_{n+1}$. Thus $\mathcal{F}$ is $\{I_1^n\}_{n \in \mathbb{Z}}$. □

Theorem 6.4 ([25, Theorem 5.5]) *Let $(R, \mathfrak{m})$ be a Cohen–Macaulay local ring of dimension $d \geq 1$ and $I_1, \ldots, I_s$ be $\mathfrak{m}$-primary ideals of R. Let $\mathcal{F} = \{\mathcal{F}(\underline{n})\}_{\underline{n} \in \mathbb{Z}^s}$ be an $\underline{I}$-admissible filtration of ideals in R. Then for all $i = 1, \ldots, s$,*

(1) $e_{(d-1)e_i}(\mathcal{F}) \geq e_1(\mathcal{F}^{(i)})$,
(2) $e(I_i) - e_{(d-1)e_i}(\mathcal{F}) \leq \lambda(R/\mathcal{F}(e_i))$,
(3) $e(I_i) - e_{(d-1)e_i}(\mathcal{F}) = \lambda(R/\mathcal{F}(e_i))$ if and only if $r(\mathcal{F}^{(i)}) \leq 1$ and $e_{(d-1)e_i}(\mathcal{F}) = e_1(\mathcal{F}^{(i)})$, where $\mathcal{F}^{(i)} = \{\mathcal{F}(ne_i)\}_{n \in \mathbb{Z}}$ is an I_i-admissible filtration.

Proof (1) We apply induction on d. Let $d = 1$. Then by Theorem 2.5,

$$P_{\mathcal{F}}(re_i) - \lambda(R/\mathcal{F}(re_i)) = -\lambda_R[H^1_{\mathcal{R}_{++}}(\mathcal{R}'(\mathcal{F}))]_{(re_i)} \text{ for all } r \geq 0.$$

Since $\mathcal{F}^{(i)}$ is I_i-admissible, we have $e(\mathcal{F}^{(i)}) = e(I_i)$. Hence using $P_{\mathcal{F}^{(i)}}(r) = e(I_i)r - e_1(\mathcal{F}^{(i)})$, we get

$$P_{\mathcal{F}^{(i)}}(r) - \lambda(R/\mathcal{F}(re_i)) + [e_1(\mathcal{F}^{(i)}) - e_{\underline{0}}(\mathcal{F})] = -\lambda_R[H^1_{\mathcal{R}_{++}}(\mathcal{R}'(\mathcal{F}))]_{(re_i)} \leq 0.$$

Taking $r \gg 0$, we get $e_{\underline{0}}(\mathcal{F}) \geq e_1(\mathcal{F}^{(i)})$. Let $d \geq 2$. Without loss of generality we may assume that the residue field of R is infinite. By Lemma 3.3, there exists a nonzerodivisor $x_i \in I_i$ such that

$$(x_i) \cap \mathcal{F}(\underline{n}) = x_i \mathcal{F}(\underline{n} - e_i) \text{ for } \underline{n} \in \mathbb{N}^s \text{ where } n_i \gg 0.$$

Let $R' = R/(x_i)$ and $\mathcal{F}' = \{\mathcal{F}(\underline{n})R'\}$ and $\mathcal{F}'^{(i)} = \{\mathcal{F}(ne_i)R'\}$. For all $\underline{n} \in \mathbb{N}^s$ such that $n_i \gg 0$, consider the following exact sequence

$$0 \longrightarrow \frac{(\mathcal{F}(\underline{n}) : (x_i))}{\mathcal{F}(\underline{n} - e_i)} \longrightarrow \frac{R}{\mathcal{F}(\underline{n} - e_i)} \xrightarrow{.x_i} \frac{R}{\mathcal{F}(\underline{n})} \longrightarrow \frac{R}{(x_i, \mathcal{F}(\underline{n}))} \longrightarrow 0.$$

Since for all $\underline{n} \in \mathbb{N}^s$ where $n_i \gg 0$, $(\mathcal{F}(\underline{n}) : (x_i)) = \mathcal{F}(\underline{n} - e_i)$, we get $H_{\mathcal{F}'}(\underline{n}) = H_{\mathcal{F}}(\underline{n}) - H_{\mathcal{F}}(\underline{n} - e_i)$ and hence $P_{\mathcal{F}}(\underline{n}) - P_{\mathcal{F}}(\underline{n} - e_i) = P_{\mathcal{F}'}(\underline{n})$. Therefore $e_{(d-2)e_i}(\mathcal{F}') = e_{(d-1)e_i}(\mathcal{F})$ and $e_1(\mathcal{F}'^{(i)}) = e_1(\mathcal{F}^{(i)})$. Therefore by induction, the result follows.
(2) Using part (1), for all $i = 1, \ldots, s$, we have

$$e(I_i) - e_{(d-1)e_i}(\mathcal{F}) \leq e(I_i) - e_1(\mathcal{F}^{(i)}) \leq \lambda(R/\mathcal{F}(e_i))$$

where the last inequality follows from Theorem 5.1.

(3) Let $e(I_i) - e_{(d-1)e_i}(\mathcal{F}) = \lambda(R/\mathcal{F}(e_i))$. Then by part (1),

$$\lambda(R/\mathcal{F}(e_i)) = e(I_i) - e_{(d-1)e_i}(\mathcal{F}) \leq e(I_i) - e_1(\mathcal{F}^{(i)}) \leq \lambda(R/\mathcal{F}(e_i)),$$

where the last inequality follows by Theorem 5.1. Hence $e_{(d-1)e_i}(\mathcal{F}) = e_1(\mathcal{F}^{(i)})$ and $e(I_i) - e_1(\mathcal{F}^{(i)}) = \lambda(R/\mathcal{F}(e_i))$. Therefore, by Theorem 6.3, $r(\mathcal{F}^{(i)}) \leq 1$.

Conversely, suppose $r(\mathcal{F}^{(i)}) \leq 1$ and $e_{(d-1)e_i}(\mathcal{F}) = e_1(\mathcal{F}^{(i)})$. Again, by Theorem 6.3, $e(I_i) - e_1(\mathcal{F}^{(i)}) = \lambda(R/\mathcal{F}(e_i))$. Hence $e(I_i) - e_{(d-1)e_i}(\mathcal{F}) = \lambda(R/\mathcal{F}(e_i))$. □

Theorem 6.5 ([25, Theorem 5.7]) *Let $(R, \mathfrak{m})$ be a Cohen–Macaulay local ring of dimension two and $I_1, \ldots, I_s$ be $\mathfrak{m}$-primary ideals of R. Let $\mathcal{F} = \{\mathcal{F}(\underline{n})\}_{\underline{n} \in \mathbb{Z}^s}$ be an $\underline{I}$-admissible filtration of ideals in R. Then $e_{\underline{0}}(\mathcal{F}) = 0$ implies $e(I_i) - e_{e_i}(\mathcal{F}) = \lambda\left(\frac{R}{\mathcal{F}(e_i)}\right)$ for all $i = 1, \ldots, s$. Suppose $\breve{\mathcal{F}}$ is $\underline{I}$-admissible filtration, then the converse is also true.*

Proof Let $e_{\underline{0}}(\mathcal{F}) = 0$. By Proposition 3.5, $[H^1_{\mathcal{R}_{++}}(\mathcal{R}'(\mathcal{F}))]_{\underline{0}} = 0$. Hence by Theorem 2.5,

$$\lambda_R[H^2_{\mathcal{R}_{++}}(\mathcal{R}'(\mathcal{F}))]_{\underline{0}} = e_{\underline{0}}(\mathcal{F}) = 0.$$

By Lemma 3.8, $\lambda_R[H^2_{\mathcal{R}_{++}}(\mathcal{R}'(\mathcal{F}))]_{e_i} = 0$ for all $i = 1, \ldots, s$. Then using Theorem 2.5 and Proposition 3.5, $P_{\mathcal{F}}(e_i) - H_{\mathcal{F}}(e_i) = -\lambda\left(\frac{\breve{\mathcal{F}}(e_i)}{\mathcal{F}(e_i)}\right)$ for all $i = 1, \ldots, s$. Hence $e(I_i) - e_{e_i}(\mathcal{F}) = \lambda\left(\frac{R}{\breve{\mathcal{F}}(e_i)}\right)$ for all $i = 1, \ldots, s$.

Suppose $\breve{\mathcal{F}}$ is $\underline{I}$-admissible filtration and $e(I_i) - e_{e_i}(\mathcal{F}) = \lambda\left(\frac{R}{\breve{\mathcal{F}}(e_i)}\right)$ for all $i = 1, \ldots, s$. Then by [25, Proposition 3.1] and Theorem 3.9, for all $\underline{n} \geq \underline{0}$ and $i = 1, \ldots, s$,

$$[H^0_{G_i(\breve{\mathcal{F}})_{++}}(G_i(\breve{\mathcal{F}}))]_{\underline{n}} = \frac{\breve{\breve{\mathcal{F}}}(\underline{n} + e_i) \cap \breve{\mathcal{F}}(\underline{n})}{\breve{\mathcal{F}}(\underline{n} + e_i)} = 0.$$

Since the Hilbert polynomial of $\breve{\mathcal{F}}$ is same as the Hilbert polynomial of $\mathcal{F}$, by [25, Theorem 5.3],

$$P_{\breve{\mathcal{F}}}(\underline{n}) = H_{\breve{\mathcal{F}}}(\underline{n}) \text{ for all } \underline{n} \geq 0. \tag{6.5.1}$$

Thus taking $\underline{n} = \underline{0}$ in the Eq. (6.5.1), we get $e_{\underline{0}}(\mathcal{F}) = e_{\underline{0}}(\breve{\mathcal{F}}) = 0$. □

As a consequence of the above theorem we get a theorem of Huneke [14, Theorem 4.5] for integral closure filtrations. We also obtain a result by Itoh [18, Corollary 5] following from the above theorem.

Corollary 6.6 ([18, Corollary 5], [25, Corollary 5.8]) *Let $(R, \mathfrak{m})$ be a Cohen–Macaulay local ring of dimension two and I be $\mathfrak{m}$-primary ideal of R. Let Q be any minimal reduction of I. Then the following are equivalent.*

(1) $e_1(I) - e_0(I) + \lambda\left(\frac{R}{\breve{I}}\right) = 0.$
(2) $\underbrace{\breve{I}^2} = Q\breve{I}.$
(2′) $\underbrace{I^2} = Q\breve{I}.$
(3) $I^{n+1} = Q^n\breve{I}$ *for all* $n \geq 1.$
(4) $e_2(I) = 0.$

Proof We prove (4) ⇒ (3) ⇒ (2′) ⇒ (2) ⇒ (1) ⇒ (4).

(4) ⇒ (3) : Let $\mathcal{F} = \{I^n\}_{n\in\mathbb{Z}}$. Since $e_2(\mathcal{F}) = e_2(I) = 0$, by Theorem 6.5 and Theorem 6.3, the result follows.
(3) ⇒ (2′) : Put $n = 1$ in (3).
(2′) ⇒ (2) Consider the filtration $\mathcal{F} = \{I^n\}_{n\in\mathbb{Z}}$. Then by [3, Proposition 3.2.3], for all $n \geq 0$, $\overbrace{I^n} = \bigcup_{k\geq 1}(I^{nk+n} : I^{nk})$. It suffices to show that $\breve{I}^2 \subseteq \overbrace{I^2}$. Let $x, y \in \breve{I}$. Then for some large k, $xI^k \subseteq I^{k+1}$ and $yI^k \subseteq I^{k+1}$. Hence $xyI^{2k} \subseteq I^{2k+2}$. This implies that $\breve{I}^2 \subseteq \overbrace{I^2}$.
(2) ⇒ (1) : Follows from [14, Theorem 2.1].
(1) ⇒ (4) : Let $\mathcal{F} = \{I^n\}_{n\in\mathbb{Z}}$. Since $\breve{\mathcal{F}}$ is an I-admissible filtration, the result follows by Theorem 6.5. □

Acknowledgments We thank Professor Markus Brodmann for discussions.

References

1. Bhattacharya, P.B.: The Hilbert function of two ideals. Math. Proc. Camb. Philos. Soc. **53**, 568–575 (1957)
2. Blancafort, C.: On Hilbert functions and cohomology. J. Algebra **192**, 439–459 (1997)
3. Blancafort, C.: Hilbert functions: combinatorial and Homological aspects, Ph.D. Thesis, Universitat De Bercelona (1997)
4. Bruns, W., Herzog, J.: Cohen-Macaulay Rings. Cambridge studies in Advanced Mathematics, vol. 39. Cambridge University Press, Cambridge (1993)
5. Brodmann, M.P., Sharp, R.Y.: Local Cohomology : An Algebraic Introduction with Geometric Applications. Cambridge University Press, Cambridge (1998)
6. Brodmann, M.P.: Private communication (2015)
7. Colomé-Nin, G.: Multigraded structures and the depth of Blow-up algebras, Ph.D. Thesis, Universitat de Barcelona (2008)
8. Corso, A., Polini, C., Rossi, M.E.: Bounds on the normal Hilbert coefficients. arXiv:1410.4233
9. Guerrieri, A., Rossi, M.E.: Hilbert coefficients of Hilbert filtrations. J. Algebra **199**, 40–61 (1998)
10. Hong, J., Ulrich, B.: Specialization and integral closure. J. Lond. Math. Soc. **90**(2), 861–878 (2014)
11. Huckaba, S.: Reduction numbers for ideals of higher analytic spread. Proc. Camb. Phil. Soc. **102**, 49–57 (1987)
12. Huckaba, S., Huneke, C.: Normal ideals in regular rings. J. Reine Angew. Math. **510**, 63–82 (1999)

13. Huckaba, S., Marley, T.: Hilbert coefficients and the depths of associated graded rings. J. Lond. Math. Soc. **56**(2), 64–76 (1997)
14. Huneke, C.: Hilbert functions and symbolic powers. Michigan Math. J. **34**, 293–318 (1987)
15. Herrmann, M., Hyry, E., Ribbe, J., Tang, Z.: Reduction numbers and multiplicities of multigraded structures. J. Algebra **197**, 311–341 (1997)
16. Itoh, S.: Integral closures of ideals generated by regular sequences. J. Algebra **117**, 390–401 (1988)
17. Itoh, S.: Coefficients of normal Hilbert polynomials. J. Algebra **150**, 101–117 (1992)
18. Itoh, S.: Hilbert coefficients of integrally closed ideals. J. Algebra **176**, 638–652 (1995)
19. Jayanthan, A.V., Verma, J.K.: Grothendieck-Serre formula and bigraded Cohen-Macaulay Rees algebras. J. Algebra **254**, 1–20 (2002)
20. Johnston, B., Verma, J.K.: Local cohomology of Rees algebras and Hilbert functions. Proc. Amer. Math. Soc. **123**, 1–10 (1995)
21. Kummini, M., Masuti, S.K.: On Conjectures of Itoh and of Lipman on the cohomlogy of normalized blow-ups (2015). arXiv:1507.03343
22. Marley, T.: The coefficients of the Hilbert polynomial and the reduction number of an ideal. J. Lond. Math. Soc. **40**(2), 1–8 (1989)
23. Marley, T.: Hilbert Functions of Ideals in Cohen-Macaulay Rings, Ph.D. Thesis, Purdue University (1989)
24. Masuti, S.K., Verma, J.K.: Two notes on normal Hilbert polynomial of two ideals. Proc. Indian Natl. Sci. Acad. **80**, 589–595 (2014)
25. Masuti, S.K., Sarkar, P., Verma, J.K.: Hilbert polynomials of multigraded filtrations of ideals. J. Algebra **444**, 527–566 (2015)
26. Narita, M.: A note on the coefficients of Hilbert characteristic functions in semi-regular local rings. Proc. Camb. Philos. Soc. **59**, 269–275 (1963)
27. Northcott, D.G.: A note on the coefficients of the abstract Hilbert function. Lond. Math. Soc. **35**, 209–214 (1960)
28. Ooishi, A.: Δ-genera and sectional genera of commutative rings. Hiroshima Math. J. **17**, 361–372 (1987)
29. Phuong, T.T.: Normal Sally modules of rank one (2015). arXiv:1506.05210
30. Ratliff, L.J., Rush, D.: Two notes on reductions of ideals. Indiana Univ. Math. J **27**, 929–934 (1978)
31. Rees, D.: Generalizations of reductions and mixed multiplicities. J. Lond. Math. Soc. **29**(2), 397–414 (1984)
32. Rhodes, C.: The Hilbert-Samuel polynomial of a filtered module. J. Lond. Math. Soc. **3**, 73–85 (1971)
33. Rossi, M.E., Valla, G.: Hilbert Function of filtered modules. Springer-UMI Lecture Notes in Mathematics, LN 9 (2010)
34. Sally, J.D.: Super-regular sequences. Pacific J. Math. **84**, 465–481 (1979)
35. Sally, J.D.: Reductions, local cohomlogy and Hilbert functions of local rings. Commutative Algebra : Durham 1981 (Durham 1981). London Mathematical Society Lecture Note Series, vol. 72, pp. 231–241. Cambridge Univ. Press, Cambridge (1982)
36. Samuel, P.: La notion de multiplicité en algébre et en géométrie algébrique. J. Math. Pures Appl. **30**, 159–274 (1951)
37. Serre, J.-P.: Faisceaux algébriques cohérents. Ann. of Math. **61**(2), 197–278 (1955)
38. Teissier, B.: Cycles évanscents, sections planes et conditions de Whitney, Astèrisque, 7–8, pp. 285–362. Soc. Math. France, Paris (1973)
39. Valabrega, P., Valla, G.: Form rings and regular sequences. Nagoya Math. J. **72**, 93–101 (1978)

de Rham Cohomology of Local Cohomology Modules

Tony J. Puthenpurakal

Abstract Let K be a field of characteristic zero and let $\mathcal{O}_n$ be the ring $K[[X_1, \ldots, X_n]]$. Let $\mathcal{D}_n = \mathcal{O}_n[\partial_1, \ldots, \partial_n]$ be the ring of K-linear differential operators on $\mathcal{O}_n$. Let M be a holonomic $\mathcal{D}_n$-module. In this paper we prove $H^i(\partial, M) = 0$ for $i < n - \dim M$. Here $\dim M =$ dimension of support of M as an $\mathcal{O}_n$-module. Also let $R = K[X_1, \ldots, X_n]$ and let I be an ideal in R and let $A_n(K) = K < X_1, \ldots, X_n, \partial_1, \ldots, \partial_n >$ be the nth Weyl algebra over K. By a result due to Lyubeznik the local cohomology modules $H^i_I(R)$ are holonomic $A_n(K)$-modules for each $i \geq 0$. In this article we also compute the de Rham cohomology modules $H^*(\partial_1, \ldots, \partial_n; H^*_I(R))$ for certain classes of ideals.

Keywords Local cohomology · Associated primes · D-modules · Koszul homology · de Rham cohomology

1991 Mathematics Subject Classication Primary 13D45 · Secondary 13N10

1 Introduction

Let K be a field of characteristic zero and let $R = K[X_1, \ldots, X_n]$. The ring of K-linear differential operators over R is the nth Weyl algebra $A_n(K) = K < X_1, \ldots, X_n, \partial_1, \ldots, \partial_n >$. Let N be a left $A_n(K)$ module. Now $\partial = \partial_1, \ldots, \partial_n$ are pairwise commuting K-linear maps. So we can consider the de Rham complex $K(\partial; N)$. Notice that the de Rham cohomology modules $H^*(\partial; N)$ are in general only K-vector spaces. By a result due to Bernstein de Rham cohomology modules are finite dimensional if N is holonomic; see [1, Chap. 1, Theorem 6.1]. Note that in [1] holonomic $A_n(K)$ modules are denoted as $\mathcal{B}_n(K)$, the *Bernstein* class of left $A_n(K)$ modules. The main idea is that if N is holonomic $A_n(K)$-module then the kernel and cokernel of ∂_n action on N are holonomic $A_{n-1}(K)$-modules. The finiteness of the

T.J. Puthenpurakal (✉)
Department of Mathematics, Indian Institute of Technology Bombay,
Powai, Mumbai 400076, India
e-mail: tputhen@math.iitb.ac.in

S.T. Rizvi et al. (eds.), *Algebra and its Applications*, Springer Proceedings
in Mathematics & Statistics 174, DOI 10.1007/978-981-10-1651-6_9

de Rham cohomology follows by a routine verification. Although I am unable to find a reference it is known that if M is holonomic then $H^i(\partial, M) = 0$ for $i < n - \dim M$; here $\dim M$ = dimension of support of M as an R-module. However the previously known proofs use sophisticated techniques like derived categories. In this paper we give an elementary proof of this result.

Now consider the case when $\mathcal{O}_n$ be the ring $K[[X_1, \ldots, X_n]]$. Let $\mathcal{D}_n = \mathcal{O}_n[\partial_1, \ldots, \partial_n]$ be the ring of K-linear differential operators on $\mathcal{O}_n$. Let M be a holonomic $\mathcal{D}_n$-module. In this case, the analogue of Bernstein result is due to van den Essen. If M is a holonomic left $\mathcal{D}_n$-module, its de Rham cohomology spaces are again finite-dimensional over K, just as in the polynomial case; in contrast to this case, however, it is not true in general that the cokernel of ∂_n acting on M is a holonomic $\mathcal{D}_{n-1}$-module, which makes the proof *much* more difficult. The kernel of ∂_n is again holonomic, and the cokernel is holonomic whenever M satisfies a certain generic condition called x_n-regularity. It turns out that if M is holonomic, we can always make a linear change of coordinates (which does not affect de Rham cohomology) after which M becomes x_n-regular. The same routine induction argument used by Bernstein is then sufficient to prove finiteness of the de Rham cohomology in the formal power series case as well. For a good exposition of this result see [7]. We may ask whether the result on vanishing of de-Rham cohomology holds in this case too. Our main result is:

Theorem 1 *Let $\mathcal{O}_n$ be the ring $K[[X_1, \ldots, X_n]]$ and let $\mathcal{D}_n = \mathcal{O}_n[\partial_1, \ldots, \partial_n]$ be the ring of K-linear differential operators on $\mathcal{O}_n$. Let M be a holonomic $\mathcal{D}_n$-module. Then $H^i(\partial, M) = 0$ for $i < n - \dim M$.*

Our motivation was not to prove Theorem 1 but rather to give an elementary proof in the case for polynomial rings. It turn's out that surprisingly our proof in polynomial case can be easily generalized to the formal power series case.

Our next motivation was to compute de Rham cohomology for an important class of holonomic modules (for commutative algebraist's) which we now describe: Let I be an ideal in $R = K[X_1, \ldots, X_n]$. For $i \geq 0$ let $H^i_I(R)$ be the ith-local cohomology module of R with respect to I. By a result due to Lyubeznik, see [4], the local cohomology modules $H^i_I(R)$ are *holonomic* $A_n(K)$-modules for each $i \geq 0$. In particular $H^*(\partial; H^*_I(R))$ are finite dimensional K-vector spaces. Analogous results hold in the formal power series rings. In this paper we compute for a few classes of ideals only in the polynomial case. We are not able to do it in the formal power series case.

Throughout let $K \subseteq L$ where L is an algebraically closed field. Let $A^n(L)$ be the affine n-space over L. If I is an ideal in R then

$$V(I)_L = \{\mathbf{a} \in A^n(L) \mid f(\mathbf{a}) = 0; \text{ for all } f \in I\};$$

denotes the variety of I in $A^n(L)$. By Hilbert's Nullstellensatz $V(I)_L$ is always non-empty. We say that an ideal I in R is zero-dimensional if $\ell(R/I)$ is finite and non-zero (here $\ell(-)$ denotes length). This is equivalent to saying that $V(I)_L$ is a finite non-empty set. If S is a finite set then let $\sharp S$ denote the number of elements in S. Our second result is

Theorem 2 *Let $I \subset R$ be a zero-dimensional ideal. Then $H^i(\partial; H^n_I(R)) = 0$ for $i < n$ and*

$$\dim_K H^n(\partial; H^n_I(R)) = \sharp V(I)_L$$

For homogeneous ideals it is best to consider their vanishing set in a projective case. Throughout let $P^{n-1}(L)$ be the projective $n-1$ space over L. We assume $n \geq 2$. Let I be a homogeneous ideal in R. Let

$$V^*(I)_L = \{\mathbf{a} \in P^{n-1}(L) \mid f(\mathbf{a}) = 0; \text{ for all } f \in I\};$$

denote the variety of I in $P^{n-1}(L)$. Note that $V^*(I)_L$ is a non-empty finite set if and only if $\operatorname{ht}(I) = n-1$. We prove

Theorem 3 *Let $I \subset R$ be a height $n-1$ homogeneous ideal. Then*

$$\dim_K H^n(\partial; H^{n-1}_I(R)) = \sharp V^*(I)_L - 1,$$
$$\dim_K H^{n-1}(\partial; H^{n-1}_I(R)) = \sharp V^*(I)_L,$$
$$H^i(\partial; H^{n-1}_I(R)) = 0 \text{ for } i \leq n-2.$$

Let M be a holonomic $A_n(K)$-module. By a result of Lyubeznik the set of associate primes of M as a R-module is finite. Note that the set $\operatorname{Ass}_R(M)$ has a natural partial order given by inclusion. We say P is a *maximal* isolated associate prime of M if P is a maximal ideal of R and also a minimal prime of M. We set $\operatorname{mIso}_R(M)$ to be the set of all maximal isolated associate primes of M. We show

Theorem 4 *Let M be a holonomic $A_n(K)$-module. Then*

$$\dim_K H^n(\partial; M) \geq \sharp \operatorname{mIso}_R(M).$$

We give an application of Theorem 4. Let I be an unmixed ideal of height $\leq n-2$. By Grothendieck vanishing theorem and the Hartshorne–Lichtenbaum vanishing theorem it follows that $H^{n-1}_I(R)$ is supported only at maximal ideals of R. By Theorem 4 we get

$$\sharp \operatorname{Ass}_R H^{n-1}_I(R) \leq \dim_K H^n\left(\partial; H^{n-1}_I(R)\right).$$

We now describe in brief the contents of the paper. In Sect. 2 we discuss a few preliminary results that we need. In Sect. 3 we make a few computations. This is used in Sect. 4 to prove Theorem 2. In Sect. 5 we make some additional computations and use it in Sect. 6 to prove Theorem 3. We prove Theorem 4 in Sect. 7. In Sect. 8 we prove the analogue of Theorem 1 in the polynomial ring case. Finally in Sect. 9 we prove Theorem 1.

2 Preliminaries

In this section we discuss a few preliminary results that we need.

Remark 2.1 Although all the results are stated for de Rham cohomology of a $A_n(K)$-module M, we will actually work with de Rham homology. Note that $H_i(\partial, M) = H^{n-i}(\partial, M)$ for any $A_n(K)$-module. Let $S = K[\partial_1, \dots, \partial_n]$. Consider it as a subring of $A_n(K)$. Then note that $H_i(\partial, M)$ is the ith Koszul homology module of M with respect to ∂.

2.2 Let M be a holonomic $A_n(K)$-module. Then for $i = 0, 1$ the de Rham homology modules $H_i(\partial_n, M)$ are holonomic $A_{n-1}(K)$-modules, see [1, 1.6.2].

The following result is well-known.

Lemma 2.3 *Let $\partial = \partial_r, \partial_{r+1}, \dots, \partial_n$ and $\partial' = \partial_{r+1}, \dots, \partial_n$. Let M be a left $A_n(K)$-module. For each $i \geq 0$ there exist an exact sequence*

$$0 \to H_0(\partial_r; H_i(\partial'; M)) \to H_i(\partial; M) \to H_1(\partial_r; H_{i-1}(\partial'; M)) \to 0.$$

2.4 (linear change of variables) We consider a linear change of variables. Let $U_1, \dots, U_n$ be new variables defined by

$$U_i = d_{i1}X_1 + \dots + d_{in}X_n + c_i \quad \text{for } i = 1, \dots, n$$

where $d_{ij}, c_1, \dots, c_n \in K$ are arbitrary and $D = [d_{ij}]$ is an invertible matrix. We say that the change of variables is homogeneous if $c_i = 0$ for all i.

Let $F = [f_{ij}] = (D^{-1})^{tr}$. Using the chain rule it can be easily shown that

$$\frac{\partial}{\partial U_i} = f_{i1}\frac{\partial}{\partial X_1} + \dots + f_{in}\frac{\partial}{\partial X_n} \quad \text{for } i = 1, \dots, n.$$

In particular we have that for any $A_n(K)$ module M an isomorphism of Koszul homologies

$$H_i\left(\frac{\partial}{\partial U_1}, \dots, \frac{\partial}{\partial U_n}; M\right) \cong H_i\left(\frac{\partial}{\partial X_1}, \dots, \frac{\partial}{\partial X_n}; M\right)$$

for all $i \geq 0$.

2.5 Let I, J be two ideals in R with $J \supset I$ and let M be a R-module. The inclusion $\Gamma_J(-) \subset \Gamma_I(-)$ induces, for each i, an R-module homomorphism

$$\theta^i_{J,I}(M)\colon H^i_J(M) \to H^i_I(M).$$

If $L \supset J$ then we can easily see that

$$\theta^i_{J,I}(M) \circ \theta^i_{L,J}(M) = \theta^i_{L,I}(M). \qquad (\dagger)$$

Lemma 2.6 (with hypotheses as above) *If M is a $A_n(K)$-module then the natural map $\theta^i_{J,I}(M)$ is $A_n(K)$-linear.*

Proof Let $I = (a_1, \ldots, a_s)$. Using (†) we may assume that $J = I + (b)$. Let $C(\mathbf{a}; M)$ be the Čech-complex on M with respect to $\mathbf{a}$. Let $C(\mathbf{a}, b; M)$ be the Čech-complex on M with respect to $\mathbf{a}, b$. Note that we have a natural short exact sequence of complexes of R-modules

$$0 \to C(\mathbf{a}; M)_b[-1] \to C(\mathbf{a}, b; M) \to C(\mathbf{a}; M) \to 0.$$

Since M is an $A_n(K)$-module it is easily seen that the above map is a map of complexes of $A_n(K)$-modules. It follows that the map $H^i(C(\mathbf{a}, b; M)) \to H^i(C(\mathbf{a}; M))$ is $A_n(K)$-linear. It is easy to see that this map is $\theta^i_{J,I}(M)$. □

2.7 Let $\mathfrak{a}, \mathfrak{b}$ be ideals in R and let M be an $A_n(K)$-module. Consider the Mayer–Vietoris sequence is a sequence of R-modules

$$\to H^i_{\mathfrak{a}+\mathfrak{b}}(M) \xrightarrow{\rho^i_{\mathfrak{a},\mathfrak{b}}(M)} H^i_{\mathfrak{a}}(M) \oplus H^i_{\mathfrak{b}}(M) \xrightarrow{\pi^i_{\mathfrak{a},\mathfrak{b}}(M)} H^i_{\mathfrak{a}\cap\mathfrak{b}}(M) \xrightarrow{\delta^i} H^{i+1}_{\mathfrak{a}+\mathfrak{b}}(M) \to ..$$

Then for all $i \geq 0$ the maps $\rho^i_{\mathfrak{a},\mathfrak{b}}(M)$ and $\pi^i_{\mathfrak{a},\mathfrak{b}}(M)$ are $A_n(K)$-linear.

To see this first note that since M is a $A_n(K)$-module all the above local cohomology modules are $A_n(K)$-modules. Further note that, (see [3, 15.1]),

$$\rho^i_{\mathfrak{a},\mathfrak{b}}(M)(z) = \left(\theta^i_{\mathfrak{a}+\mathfrak{b},\mathfrak{a}}(z), \theta^i_{\mathfrak{a}+\mathfrak{b},\mathfrak{b}}(z)\right),$$
$$\pi^i_{\mathfrak{a},\mathfrak{b}}(M)(x, y) = \theta^i_{\mathfrak{a},\mathfrak{a}\cap\mathfrak{b}}(x) - \theta^i_{\mathfrak{b},\mathfrak{a}\cap\mathfrak{b}}(y).$$

Using Lemma 2.6 it follows that $\rho^i_{\mathfrak{a},\mathfrak{b}}(M)$ and $\pi^i_{\mathfrak{a},\mathfrak{b}}(M)$ are $A_n(K)$-linear maps.

Remark 2.8 In fact δ^i is also $A_n(K)$-linear for all $i \geq 0$; [6]. However we will not use this fact in this paper.

2.9 Let $I_1, \ldots, I_n$ be proper ideals in R. Assume that they are pairwise co-maximal i.e., $I_i + I_j = R$ for $i \neq j$. Set $J = I_1 \cdot I_2 \ldots I_n$. Then for any R-module M we have an isomorphism of $A_n(K)$-modules

$$H^i_J(M) \cong \bigoplus_{j=1}^{n} H^i_{I_j}(M) \quad \text{for all } i \geq 0.$$

To prove this result note that I_1 and $I_2 \ldots I_n$ are co-maximal. So it suffices to prove the result for $n = 2$. In this case we use the Mayer–Vietoris sequence of local cohomology, see 2.7, to get an isomorphism of R-modules

$$\pi^i_{I_1,I_2}(R)\colon H^i_{I_1}(R) \oplus H^i_{I_2}(R) \to H^i_{I_1\cap I_2}(R).$$

By 2.7 we also get that $\pi^i_{I_1,I_2}(R)$ is $A_n(K)$-linear.

3 Some Computations

The goal of this section is to compute the Koszul homologies $H_*(\partial_1, \ldots, \partial_n; N)$ when $N = R$ and when $N = E$ the injective hull of $R/(X_1, \ldots, X_n) = K$. It is well-known that

$$E = \bigoplus_{r_1,\ldots,r_n \geq 0} K \frac{1}{X_1 X_2 \ldots X_n X_1^{r_1} X_2^{r_2} \ldots X_n^{r_n}}.$$

Note that E has the obvious structure as a $A_n(K)$-module with

$$X_i \cdot \frac{1}{X_1 \ldots X_n X_1^{r_1} \ldots X_n^{r_n}} = \begin{cases} \frac{1}{X_1 \ldots X_n X_1^{r_1} \ldots X_i^{r_i - 1} \ldots X_n^{r_n}} & \text{if } r_i \geq 1, \\ 0 & \text{otherwise.} \end{cases}$$

and

$$\partial_i \cdot \frac{1}{X_1 \ldots X_n X_1^{r_1} \ldots X_n^{r_n}} = \frac{-r_i - 1}{X_1 \ldots X_n X_1^{r_1} \ldots X_i^{r_i+1} \ldots X_n^{r_n}}$$

It is convenient to introduce the following notation. For $i = 1, \ldots, n$ let $R_i = K[X_1, \ldots, X_i]$, $\mathfrak{m}_i = (X_1, \ldots, X_i)$ and let E_i be the injective hull of $R_i/\mathfrak{m}_i = K$ as a R_i-module. Set $R_0 = E_0 = K$. We prove

Lemma 3.1 *$H_0(\partial_n; E_n) \cong E_{n-1}$ and $H_1(\partial_n; E_n) = 0$ as $A_{n-1}(K)$-modules.*

Proof Since E_n is holonomic $A_n(K)$ module it follows that $H_i(\partial_n; E_n)$ (for $i = 0, 1$) are holonomic $A_{n-1}(K)$-modules [1, Chap. 1, Theorem 6.2]. We first prove $H_1(\partial_n; E_n) = 0$. Let $t \in E_n$ with $\partial_n(t) = 0$. Let

$$t = \sum_{r_1,\ldots,r_n \geq 0} t_r \frac{1}{X_1 \ldots X_n X_1^{r_1} \ldots X_n^{r_n}} \quad \text{with atmost finitely many } t_r \in K \text{ non-zero.}$$

Notice that

$$\partial_n(t) = \sum_{r_1,\ldots,r_n \geq 0} t_r \frac{-r_n - 1}{X_1 \ldots X_{n-1} X_n X_1^{r_1} \ldots X_{n-1}^{r_{n-1}} X_n^{r_n+1}}.$$

Comparing coefficients we get that if $\partial_n(t) = 0$ then $t = 0$.

For computing $H_0(\partial_n; E_n)$ we first note that as K-vector spaces

$$E_n = X \bigoplus Y;$$

where

$$X = \bigoplus_{r_1,\ldots,r_{n-1}\geq 0, r_n=0} K\frac{1}{X_1X_2\ldots X_nX_1^{r_1}X_2^{r_2}\ldots X_{n-1}^{r_{n-1}}}$$

$$Y = \bigoplus_{r_1,\ldots,r_{n-1}\geq 0, r_n\geq 1} K\frac{1}{X_1X_2\ldots X_nX_1^{r_1}X_2^{r_2}\ldots X_n^{r_n}}.$$

For $r_n \geq 1$ note that

$$\partial_n\left(\frac{1}{X_1X_2\ldots X_nX_1^{r_1}X_2^{r_2}\ldots X_n^{r_n-1}}\right) = \frac{-r_n}{X_1X_2\ldots X_nX_1^{r_1}X_2^{r_2}\ldots X_n^{r_n}}.$$

It follows that $E_n/\partial_n E_n = X$. Furthermore notice that $X \cong E_{n-1}$ as $A_{n-1}(K)$-modules. Thus we get $H_0(\partial_n; E_n) \cong E_{n-1}$. □

We now show that

Lemma 3.2 *For* $c = 1, 2, \ldots, n$ *we have,*

$$H_i(\partial_c, \partial_{c+1}, \ldots, \partial_n; E_n) = \begin{cases} 0 & \text{for } i > 0 \\ E_{c-1} & \text{for } i = 0 \end{cases}$$

Proof We prove the result by induction on $t = n - c$. For $t = 0$ it is just the Lemma 3.1. Let $t \geq 1$ and assume the result for $t - 1$. Let $\partial = \partial_c, \partial_{c+1}, \ldots, \partial_n$ and $\partial' = \partial_{c+1}, \ldots, \partial_n$. For each $i \geq 0$ there exist an exact sequence

$$0 \to H_0(\partial_c; H_i(\partial'; E_n)) \to H_i(\partial; E_n) \to H_1(\partial_c; H_{i-1}(\partial'; E_n)) \to 0.$$

By induction hypothesis $H_i(\partial'; E_n) = 0$ for $i \geq 1$. Thus for $i \geq 2$ we have $H_i(\partial; E_n) = 0$. Also note that by induction hypothesis $H_0(\partial'; E_n) = E_c$. So we have

$$H_1(\partial; E_n) = H_1(\partial_c; E_c) = 0 \text{ by Lemma 3.1.}$$

Finally again by Lemma 3.1 we have

$$H_0(\partial; E_n) = H_0(\partial_c; E_c) = E_{c-1}.$$

□

As a corollary to the above result we have

Theorem 3.3 *Let* $\partial = \partial_1, \ldots, \partial_n$. *Then* $H_i(\partial; E_n) = 0$ *for* $i > 0$ *and* $H_0(\partial; E_n) = K$. □

We now compute the de Rham homology $H_*(\partial; R)$. We first prove

Lemma 3.4 $H_0(\partial_n; R_n) = 0$ *and* $H_1(\partial_n; R_n) = R_{n-1}$

Proof This is just calculus. □

The proof of the following result is similar to the proof of Lemma 3.2.

Lemma 3.5 *For $c = 1, 2, \ldots, n$ we have,*

$$H_i(\partial_c, \partial_{c+1}, \ldots, \partial_n; R_n) = \begin{cases} 0 & \text{for } i = 0, 1, \ldots, n-c \\ R_{c-1} & \text{for } i = n-c+1 \end{cases}$$

□

As a corollary to the above result we have

Theorem 3.6 *Let $\partial = \partial_1, \ldots, \partial_n$. Then $H_i(\partial; R_n) = 0$ for $i < n$ and $H_n(\partial; R_n) = K$.* □

We will need the following computation in part 2 of this paper.

Lemma 3.7 *Let f be a non-constant squarefree polynomial in $R = K[X_1, \ldots, X_n]$. Let $\partial = \partial_1, \ldots, \partial_n$. Then $H_n(\partial; R_f) = K$. Furthermore $H_n\left(\partial; H^1_{(f)}(R)\right) = 0$ and*

$$H_i\left(\partial; H^1_{(f)}(R)\right) \cong H_i(\partial; R_f) \quad \text{for } i < n.$$

Proof Note that

$$H_n(\partial; R_f) = \{v \in R_f \mid \partial_i v = 0 \text{ for all } i = 1, \ldots, n\}.$$

Clearly if $v \in R_f$ is a constant then $\partial_i v = 0$ for all $i = 1, \ldots, n$. By a linear change in variables we may assume that $f = X_n^s +$ lower terms in X_n. Note that by 2.4 the de Rham homology does not change.

Suppose if possible there exists a non-constant $v = a/f^r \in H_n(\partial; R_f)$ where f does not divide a if $r \geq 1$. Note that if $r = 0$ then $v \in H_n(\partial; R) = K$. So v is a constant. So assume $r \geq 1$. Since $\partial_n(v) = 0$ we get $f\partial_n(a) = ra\partial_n(f)$.

Since f is squarefree we have $f = f_1 \ldots f_m$ where f_i are distinct irreducible polynomials. As f is monic in X_n we have that f_i is monic in X_n for each i.

Since $f\partial_n(a) = ra\partial_n(f)$ we have that f_i divides $a\partial_n(f)$ for each i. Note that if f_i divides $\partial_n(f)$ then f_i divides $f_1 \ldots f_{i-1}\partial_n(f_i) \cdot f_{i+1} \ldots f_m$. Therefore f_i divides $\partial_n(f_i)$ which is easily seen to be a contradiction since f_i is monic in X_n. Thus f_i divides a for each $i = 1, \ldots, m$. Therefore f divides a, which is a contradiction. Thus $H_n(\partial; R_f)$ only consists of constants.

We have an exact sequence

$$0 \to R \to R_f \to H^1_I(R) \to 0.$$

Notice $H_n(\partial, R) = H_n(\partial; R_f) = K$ and $H_{n-1}(\partial, R) = 0$ (see Theorem 3.6 and Lemma 3.7). So we get $H_n(\partial, H^1_I(R)) = 0$. Also as $H_i(\partial, R) = 0$ for $i < n$ we get

$$H_i\left(\partial; H^1_{(f)}(R)\right) \cong H_i(\partial; R_f) \quad \text{for } i < n.$$ □

4 Proof of Theorem 2

In this section we prove Theorem 2. Throughout $K \subseteq L$ where L is an algebraically closed field. We first prove:

Lemma 4.1 *Let $\mathfrak{m} = (X_1 - a_1, \ldots, X_n - a_n)$, where $a_1, \ldots, a_n \in K$, be a maximal ideal in $R = K[X_1, \ldots, X_n]$. Let $\partial = \partial_1, \ldots, \partial_n$. Then $H_i(\partial; H^n_{\mathfrak{m}}(R)) = 0$ for $i > 0$ and $H_0(\partial; H^n_{\mathfrak{m}}(R)) = K$.*

Proof Let $U_i = X_i - a_i$ for $i = 1, \ldots, n$. Then by 2.4

$$H_i\left(\frac{\partial}{\partial U_1}, \ldots, \frac{\partial}{\partial U_n}; H^n_{\mathfrak{m}}(R)\right) \cong H_i\left(\frac{\partial}{\partial X_1}, \ldots, \frac{\partial}{\partial X_n}; H^n_{\mathfrak{m}}(R)\right)$$

for all $i \geq 0$. Thus we may assume $a_1 = a_2 = \cdots = a_n = 0$. Finally note that $H^n_{\mathfrak{m}}(R) = E$ the injective hull of $R/\mathfrak{m} = K$. So our result follows from Theorem 3.3. □

We now give a proof of Theorem 2.
Proof of Theorem 2 Notice

$$A_n(L) = A_n(K) \otimes_K L$$
$$\text{and } S = L[X_1, \ldots, X_n] = R \otimes_K L.$$

So $A_n(L)$ and S are faithfully flat extensions of $A_n(K)$ and R respectively. It follows that

$$H_i\left(\partial; H^n_{IS}(S)\right) \cong H_i\left(\partial; H^n_I(R)\right) \otimes_K L \quad \text{for all } i \geq 0.$$

Thus we may as well assume that $K = L$ is algebraically closed. Since I is zero-dimensional we have

$$\sqrt{I} = \mathfrak{m}_1 \cap \mathfrak{m}_2 \cap \cdots \cap \mathfrak{m}_r,$$

where $\mathfrak{m}_1, \ldots, \mathfrak{m}_r$ are distinct maximal ideals and $r = \sharp V(I)_L$, the number of points in $V(I)_L$. By 2.9 we have an isomorphism of $A_n(K)$-modules

$$H^j_I(R) \cong \bigoplus_{i=0}^{r} H^j_{\mathfrak{m}_i}(R) \quad \text{for all } j \geq 0.$$

In particular we have that

$$H_j\left(\partial; H^n_I(R)\right) = \bigoplus_{i=0}^{r} H_j\left(\partial; H^n_{\mathfrak{m}_i}(R)\right).$$

Since K is algebraically closed each maximal ideal $\mathfrak{m}$ in R is of the form $(X_1 - a_1, \ldots, X_n - a_n)$. The result follows from Lemma 4.1. □

5 Some Computations-II

Let $R = K[X_1, \ldots, X_n]$ and let $P = (X_1, \ldots, X_{n-1})$. The goal of this section is to compute $H_i(\partial; H_P^{n-1}(R))$ for all $i \geq 0$.

As before it is convenient to introduce the following notation. For $i = 1, \ldots, n$ let $R_i = K[X_1, \ldots, X_i]$, $\mathfrak{m}_i = (X_1, \ldots, X_i)$ and let E_i be the injective hull of $R_i/\mathfrak{m}_i = K$ as a R_i-module.

Notice that $R_{n-1} \subseteq R_n$ is a faithfully flat extension. So

$$R_n \otimes_{R_{n-1}} H^i_{\mathfrak{m}_{n-1}}(R_{n-1}) \cong H^i_{\mathfrak{m}_{n-1}R_n}(R_n) \quad \text{for all } i \geq 0.$$

Thus

$$H^{n-1}_{\mathfrak{m}_{n-1}R_n}(R_n) = E_{n-1}[X_n] = \bigoplus_{j \geq 0} E_{n-1}X_n^j.$$

We first prove the following:

Lemma 5.1 *$H_1(\partial_n; E_{n-1}[X_n]) = E_{n-1}$ and $H_0(\partial_n; E_{n-1}[X_n]) = 0$.*

Proof Let $v \in E_{n-1}[X_n]_j$. So

$$v = \frac{c}{X_1 \ldots X_{n-1} X_1^{r_1} \ldots X_{n-1}^{r_{n-1}}} \cdot X_n^j$$

for some $c \in K$ and $r_1, \ldots, r_{n-1} \geq 0$. Notice that

$$\partial_n(v) = \begin{cases} \frac{cj}{X_1 \ldots X_{n-1} X_1^{r_1} \ldots X_{n-1}^{r_{n-1}}} \cdot X_n^{j-1} & \text{if } j \geq 1, \\ 0 & \text{if } j = 0. \end{cases}$$

It follows that $H_1(\partial_n; E_{n-1}[X_n]) = E_{n-1}$.

Let $v \in E_{n-1}[X_n]_j$ be a homogeneous element. So

$$v = \frac{c}{X_1 \ldots X_{n-1} X_1^{r_1} \ldots X_{n-1}^{r_{n-1}}} \cdot X_n^j$$

for some $c \in K$ and $r_1, \ldots, r_{n-1} \geq 0$. Let

$$u = \frac{c}{j+1} \cdot \frac{1}{X_1 \ldots X_{n-1} X_1^{r_1} \ldots X_{n-1}^{r_{n-1}}} \cdot X_n^{j+1}.$$

Notice that $\partial_n(u) = v$. Thus it follows that $H_0(\partial_n; E_{n-1}[X_n]) = 0$. □

Next we prove

Lemma 5.2 *For $c = 1, 2, \ldots, n$ we have,*

$$H_i(\partial_c, \partial_{c+1}, \ldots, \partial_n; E_{n-1}[X_n]) = \begin{cases} 0 & \textit{for } i \neq 1 \\ E_{c-1} & \textit{for } i = 1. \end{cases}$$

Proof We prove the result by induction on $t = n - c$. For $t = 0$ it is just the Lemma 5.1. Let $t \geq 1$ and assume the result for $t - 1$. Let $\partial = \partial_c, \partial_{c+1}, \ldots, \partial_n$ and $\partial' = \partial_{c+1}, \ldots, \partial_n$. For each $i \geq 0$ we have an exact sequence

$$0 \to H_0(\partial_c; H_i(\partial'; E_{n-1}[X_n])) \to H_i(\partial; E_{n-1}[X_n]) \to H_1(\partial_c; H_{i-1}(\partial'; E_{n-1}[X_n])) \to 0.$$

So $H_i(\partial; E_{n-1}[X_n]) = 0$ for $i \geq 3$ and for $i = 0$. Notice that

$$\begin{aligned} H_2(\partial; E_{n-1}[X_n]) &= H_1(\partial_c; H_1(\partial'; E_{n-1}[X_n])) \\ &= H_1(\partial_c; E_c); \text{ (by induction hypothesis).} \\ &= 0; \text{ by Lemma 3.1.} \end{aligned}$$

Similarly we have

$$\begin{aligned} H_1(\partial; E_{n-1}[X_n]) &= H_0(\partial_c; H_1(\partial'; E_{n-1}[X_n])) \\ &= H_0(\partial_c; E_c); \text{ (by induction hypothesis).} \\ &= E_{c-1}; \text{ by Lemma 3.1.} \end{aligned}$$

□

As a corollary we obtain

Theorem 5.3 *Let $R = K[X_1, \ldots, X_n]$ and let $P = (X_1, \ldots, X_{n-1})$. Let $\partial = \partial_1, \ldots, \partial_n$. Then*

$$H_i(\partial; H_P^{n-1}(R)) = \begin{cases} 0 & \textit{for } i \neq 1 \\ K & \textit{for } i = 1. \end{cases}$$

6 Proof of Theorem 3

In this section we prove Theorem 3. Throughout $K \subseteq L$ where L is an algebraically closed field. We first prove:

Lemma 6.1 *Let $Q = (X_1 - a_1X_n, \ldots, X_{n-1} - a_{n-1}X_n)$, where $a_1, \ldots, a_{n-1} \in K$, be a homogeneous prime ideal in $R = K[X_1, \ldots, X_n]$. Let $\partial = \partial_1, \ldots, \partial_n$. Then $H_i(\partial; H_Q^{n-1}(R)) = 0$ for $i \neq 1$ and $H_1(\partial; H_Q^{n-1}(R)) = K$.*

Proof Let $U_i = X_i - a_iX_n$ for $i = 1, \ldots, n-1$ and let $U_n = X_n$. Then by 2.4

$$H_i\left(\frac{\partial}{\partial U_1}, \ldots, \frac{\partial}{\partial U_n}; H^n_{\mathfrak{m}}(R)\right) \cong H_i\left(\frac{\partial}{\partial X_1}, \ldots, \frac{\partial}{\partial X_n}; H^n_{\mathfrak{m}}(R)\right)$$

for all $i \geq 0$. Thus we may assume $a_1 = a_2 = \cdots = a_{n-1} = 0$. The result follows from Theorem 5.3. □

We now give

Proof of Theorem 3 As shown in the proof of Theorem 2 we may assume that $K = L$ is algebraically closed. We take $X_n = 0$ to be the hyperplane at infinity. After a homogeneous linear change of variables we may assume that there are no zero's of $V(I)$ in the hyperplane $X_n = 0$; see 2.4. Thus

$$\sqrt{I} = Q_1 \cap Q_2 \cap \cdots \cap Q_r$$

where $r = \sharp V(I)$ and $Q_i = (X_1 - a_{i1}X_n, \ldots, X_{n-1} - a_{i,n-1}X_n)$ for $i = 1, \ldots, r$.

We first note that $H^n_I(R) = 0$. This can be easily proved by induction on r and using the Mayer–Vietoris sequence.

We prove the result by induction on r. For $r = 1$ the result follows from Lemma 6.1. So assume $r \geq 2$ and that the result holds for $r - 1$. Set $J = Q_1 \cap \cdots \cap Q_{r-1}$. Then $\sqrt{I} = J \cap Q_r$. Notice that $\sqrt{Q_r + J} = \mathfrak{m} = (X_1, \ldots, X_n)$. By Mayer–Vietoris sequence and the fact that $H^n_{Q_r}(R) = H^n_J(R) = 0$ we get an exact sequence of R-modules

$$0 \to H^{n-1}_J(R) \bigoplus H^{n-1}_{Q_r}(R) \xrightarrow{\alpha} H^{n-1}_I(R) \to H^n_{\mathfrak{m}}(R) \to 0.$$

By 2.7 α is $A_n(K)$ linear. Set $C = \operatorname{coker} \alpha$. So we have an exact sequence of $A_n(K)$-modules

$$0 \to H^{n-1}_J(R) \bigoplus H^{n-1}_{Q_r}(R) \xrightarrow{\alpha} H^{n-1}_I(R) \to C \to 0.$$

Claim: $C \cong H^n_{\mathfrak{m}}(R)$ as $A_n(K)$-modules.

First suppose the claim is true. Then note that the result follows from induction hypothesis and Lemma's 4.1, 6.1.

It remains to prove the claim. Note that $C \cong H^n_{\mathfrak{m}}(R)$ as R-modules. In particular

$$\operatorname{soc}_R(C) = \operatorname{Hom}_R(R/\mathfrak{m}, C) \cong \operatorname{Hom}_R(R/\mathfrak{m}, H^n_{\mathfrak{m}}(R)) \cong K.$$

Let e be a non-zero element of $\operatorname{soc}_R(C)$. Consider the map

$$\phi \colon A_n(K) \to C$$
$$d \mapsto de.$$

Clearly ϕ is $A_n(K)$-linear. Since $\phi(A_n(K)\mathfrak{m}) = 0$ we get an $A_n(K)$-linear map

$$\overline{\phi} \colon \frac{A_n(K)}{A_n(K)\mathfrak{m}} \to C.$$

Note that $A_n(K)/A_n(K)\mathfrak{m} \cong H^n_{\mathfrak{m}}(R)$ as $A_n(K)$-modules.

To prove that $\overline{\phi}$ is an isomorphism, note that $\overline{\phi}$ is R-linear. Since $\overline{\phi}$ induces an isomorphism on socles we get that $\overline{\phi}$ is injective. As $H^n_{\mathfrak{m}}(R)$ is an injective R-module and $\overline{\phi}$ is injective R-linear map we have that $C \cong \text{image}\,\overline{\phi} \oplus \text{coker}\,\overline{\phi}$ as R-modules. Set $N = \text{coker}\,\overline{\phi}$. Note that $\text{soc}_R(N) = 0$. Also note that as R-module C is supported only at $\mathfrak{m}$. So N is supported only at $\mathfrak{m}$. Since $\text{soc}_R(N) = 0$ we get that $N = 0$. So $\overline{\phi}$ is surjective. Thus $\overline{\phi}$ is an $A_n(K)$-linear isomorphism of $A_n(K)$-modules. □

7 Proof of Theorem 4

In this section we prove Theorem 4.

Let A be a Noetherian ring, I an ideal in A and let M be an A-module, not necessarily finitely generated. Set

$$\Gamma_I(M) = \{m \in M \mid I^s m = 0 \text{ for some } s \geq 0\}.$$

The following result is well-known. For lack of a suitable reference we give sketch of a proof here. When M is finitely generated, for a proof of the following result see [2, Proposition 3.13].

Lemma 7.1 (with hypotheses as above)

$$\text{Ass}_A \frac{M}{\Gamma_I(M)} = \{P \in \text{Ass}_A M \mid P \not\supseteq I\}$$

Proof (sketch) Note that if $P \in \text{Ass}_A \Gamma_I(M)$ then $P \supseteq I$. It follows that if $P \in \text{Ass}_A M$ and $P \not\supseteq I$ then $P \in \text{Ass}_A M/\Gamma_I(M)$.

It can be easily verified that if $P \in \text{Ass}_A M/\Gamma_I(M)$ then $P \not\supseteq I$. Also note that if $P \not\supseteq I$ then $\Gamma_I(M)_P = 0$. Thus

$$M_P \cong \left(\frac{M}{\Gamma_I(M)}\right)_P \quad \text{if } P \not\supseteq I.$$

The result follows. □

We now give
Proof of Theorem 4 First consider the case when K is algebraically closed. Set

$$\text{Ass}_A(M) = \text{mIso}_R(M) \sqcup \left(\bigcup_{i=1}^{s} V(P_i) \cap \text{Ass}_A(M) \right).$$

Here $P_1, \ldots, P_s$ are minimal primes of M which are not maximal ideals.

Set $I = P_1 P_2 \ldots P_s$. Note that $\Gamma_I(M)$ is a $A_n(K)$-submodule of M. Set $N = M/\Gamma_I(M)$. By Lemma 7.1 we get that

$$\begin{aligned}\operatorname{Ass}_R N &= \{P \in \operatorname{Ass}_R M \mid P \not\supseteq I\} \\ &= \operatorname{mIso}(M).\end{aligned}$$

Let $\operatorname{mIso}(M) = \{\mathfrak{m}_1, \ldots, \mathfrak{m}_r\}$. Set $J = \mathfrak{m}_1\mathfrak{m}_2 \ldots \mathfrak{m}_r$. Since $\mathfrak{m}_1, \ldots, \mathfrak{m}_r$ are comaximal we get by 2.9 that as $A_n(K)$-modules

$$\Gamma_J(N) = \Gamma_{\mathfrak{m}_1}(N) \oplus \cdots \oplus \Gamma_{\mathfrak{m}_r}(N).$$

Set $E = N/\Gamma_J(N)$. By Lemma 7.1 we get that $\operatorname{Ass}_R E = \emptyset$. So $E = 0$. Thus

$$N = \Gamma_{\mathfrak{m}_1}(N) \oplus \cdots \oplus \Gamma_{\mathfrak{m}_r}(N).$$

Note that

$$\Gamma_{\mathfrak{m}_i}(N) = E_R(R/\mathfrak{m}_i)^{s_i} = H^n_{\mathfrak{m}_i}(R)^{s_i} \quad \text{for some } s_i \geq 1.$$

Since K is algebraically closed we have that for each $i = 1, \ldots, r$ the maximal ideal $\mathfrak{m}_i = (X_1 - a_{i1}, \ldots, X_n - a_{in})$ for some $a_{ij} \in K$. It follows from Lemma 4.1 that

$$H_i(\partial; N) = 0 \text{ for } i \geq 1$$

$$\dim_K H_0(\partial; N) = \sum_{i=1}^{r} s_i.$$

The exact sequence $0 \to \Gamma_I(M) \to M \to N \to 0$ yields an exact sequence of de Rham homologies

$$0 \to H_0(\partial; \Gamma_I(M)) \to H_0(\partial; M) \to H_0(\partial; N) \to 0;$$

since $H_1(\partial; N) = 0$. The result follows. So we have proved the result when K is algebraically closed.

Now consider the case when K is *not* algebraically closed. Let $L = \overline{K}$ the algebraic closure of K. Note that $S = L[X_1, \ldots, X_n] = R \otimes_K L$ and $A_n(L) = A_n(K) \otimes_K L$. Further notice that $M \otimes_K L$ is a holonomic $A_n(L)$-module. Also note that $M \otimes_R S = M \otimes_K L$.

Claim-1: $\sharp \operatorname{mIso}_S(M \otimes_R S) \geq \sharp \operatorname{mIso}_R(M)$.

We assume the claim for the moment. Note that $H_0(\partial, M) \otimes_K L = H_0(\partial, M \otimes_K L)$. So

$$\dim_K H_0(\partial, M) = \dim_L H_0(\partial, M \otimes_K L) \geq \sharp \operatorname{mIso}_S(M \otimes_R S) \geq \sharp \operatorname{mIso}_R(M).$$

The result follows.

It remains to prove Claim-1. By Theorem 23.2(ii) of [5] we have

$$\mathrm{Ass}_S(M \otimes_R S) = \bigcup_{P \in \mathrm{Ass}_R(M)} \mathrm{Ass}_S\left(\frac{S}{PS}\right). \tag{†}$$

Suppose $\mathfrak{m}$ is an isolated maximal prime of M. Notice $S/\mathfrak{m}S$ has finite length. It follows that

$$\sqrt{\mathfrak{m}S} = \mathfrak{m}_1 \cap \mathfrak{m}_2 \cap \cdots \cap \mathfrak{m}_r;$$

for some maximal ideals $\mathfrak{m}_1, \mathfrak{m}_2, \ldots, \mathfrak{m}_r$ of S.

Claim-2: $\mathfrak{m}_1, \mathfrak{m}_2, \ldots, \mathfrak{m}_r \in \mathrm{mIso}_S(M \otimes_R S)$.

Note that Claim-2 implies Claim-1. It remains to prove Claim-2.

Suppose if possible some $\mathfrak{m}_i \notin \mathrm{mIso}_S(M \otimes_R S)$. Then there exist $Q \subsetneq \mathfrak{m}_i$ and $Q \in \mathrm{Ass}_S(M \otimes_R S)$. Note that Q is not a maximal ideal in S. By (†) we have that

$$Q \in \mathrm{Ass}_S\left(\frac{S}{PS}\right) \quad \text{for some } P \in \mathrm{Ass}_R(M).$$

Notice that as Q is not a maximal ideal in S we have that P is not a maximal ideal in R. Also note that by Theorem 23.2(i) of [5] we have

$$P = Q \cap R \subseteq \mathfrak{m}_i \cap R = \mathfrak{m}.$$

Thus $\mathfrak{m}$ is not an isolated maximal prime of M, a contradiction. □

An application of Theorem 4 is the following result:

Corollary 7.3 *Let I be an unmixed ideal of height $\leq n-2$ in R. Then*

$$\sharp\,\mathrm{Ass}_R\, H^{n-1}_I(R) \leq \dim_K H_0\left(\partial, H^{n-1}_I(R)\right).$$

Proof We first show that $M = H^{n-1}_I(R)$ is supported only at maximal ideals of R. As M is I-torsion it follows that any $P \in \mathrm{Supp}(M)$ contains I.

We first show that if $\mathrm{ht}\, P \leq n-2$ then $P \notin \mathrm{Supp}(M)$. Note $M_P = H^{n-1}_{IR_P}(R_P) = 0$ by Grothendieck vanishing theorem as $\dim R_P = \mathrm{ht}\, P \leq n-2$. So $P \notin \mathrm{Supp}(M)$.

Next we prove that if $\mathrm{ht} P = n-1$ then $P \notin \mathrm{Supp}(M)$. Let $\widehat{R_P}$ be the completion of R_P with respect to its maximal ideal. As I is unmixed we have $\dim R_P/I_P > 0$. So $I\widehat{R_P}$ is not $P\widehat{R_P}$-primary. Therefore

$$M_P \otimes_{R_P} \widehat{R_P} = H^{n-1}_{I\widehat{R_P}}(\widehat{R_P}) = 0,$$

by Hartshorne–Lichtenbaum Vanishing theorem. As $\widehat{R_P}$ is a faithfully flat R_P algebra we have $M_P = 0$.

Thus M is supported at only maximal ideals of R. It follows that $\mathrm{Ass}_A(M) = \mathrm{mIso}_R(M)$. The result now follows from Theorem 4. □

8 Elementary Proof of an Analogue of Theorem 1 in the Polynomial Ring Case

In this section we give an elementary proof of the following result:

Theorem 8.1 *Let K be a field of characteristic zero and let $R = K[X_1, \dots, X_n]$. Let $A_n(K)$ be the nth Weyl algebra. Let M be a holonomic $A_n(K)$-module. Then $H^i(\partial, M) = 0$ for $i < n - \dim M$.*

Set $R_{n-1} = K[X_1, \dots, X_{n-1}]$.

We begin by the following result on vanishing (and non-vanishing) of de Rham homology of a simple $A_n(K)$-module. If M is a simple $A_n(K)$-module then it is well-known that $\operatorname{Ass}_R(M)$ is a set with one element.

Theorem 8.2 *Let M be a simple $A_n(K)$-module and assume $\operatorname{Ass}_R(M) = \{P\}$. Set $Q = P \cap R_{n-1}$. Then*

$$H_0(\partial_n; M) = 0 \implies P = QR,$$
$$H_1(\partial_n; M) \neq 0 \implies P = QR.$$

To prove the above theorem we need a criterion for an ideal I to be equal to $(I \cap R_{n-1})R$. This is provided by the following:

Lemma 8.3 *Let I be an ideal in R. Set $J = I \cap R_{n-1}$. Then the following are equivalent:*

(1) $\partial_n(I) \subseteq I$.
(2) $I = JR$.
(3) *Let $\xi \in I$. Let $\xi = \sum_{j=0}^{m} c_j X_n^j$ where $c_j \in R_{n-1}$ for $j = 0, \dots, m$. Then $c_j \in I$ for each j.*

Proof We first prove (1) $\implies$ (3). Let $\xi \in I$. Let $\xi = \sum_{j=0}^{m} c_j X_n^j$ where $c_j \in R_{n-1}$ for $j = 0, \dots, m$. Notice $\partial_n^m(\xi) = m! c_m$. So $c_m \in I$. Thus $\xi - c_m X_n^m \in I$. Iterating we obtain that $c_j \in I$ for all j.

Notice that (3) $\implies$ (1) is trivial. We now show (3) $\implies$ (2). Let $\xi \in I$. Let $\xi = \sum_{j=0}^{m} c_j X_n^j$ where $c_j \in R_{n-1}$ for $j = 0, \dots, m$. By hypothesis $c_j \in I$ for each j. Notice $c_j \in I \cap R_{n-1} = J$. Thus $I \subseteq JR$. The assertion $JR \subseteq I$ is trivial. So $I = JR$.

Finally we prove that (2) $\implies$ (3). If $b \in J$ and $r \in R$ then notice that if $br = \sum_{j=0}^{m} c_j X_n^j$ where $c_j \in R_{n-1}$ for $j = 0, \dots, m$ then each $c_j \in J$. As $I = JR$ each $\xi \in I$ is a finite sum $b_1 r_1 + \cdots + b_s r_s$ where $b_i \in J$ and $r_i \in R$. The assertion follows. □

The following corollary is useful.

Corollary 8.4 *Let P be a prime ideal in R and let I be an ideal in R with $\sqrt{I} = P$. If $\partial_n(I) \subseteq I$ then $P = (P \cap R_{n-1})R$.*

Proof Set $Q = P \cap R_{n-1}$. Let $\xi \in P$. Let $\xi = \sum_{j=0}^{m} c_j X_n^j$ where $c_j \in R_{n-1}$ for $j = 0, \ldots, m$. Notice $\xi^s \in I$ for some $s \geq 1$. Also $\xi^s = c_m^s X_n^{sm} + ..$ lower terms in X_n. By Lemma 8.3 we get that $c_m^s \in I$. It follows that $c_m \in P$. Thus $\xi - c_m X_n^m \in P$. Iterating we obtain that $c_j \in P$ for all j. So by Lemma 8.3 we get that $P = QR$. □

We now give

Proof of Theorem 8.2 First suppose $H_0(\partial_n, M) = 0$. Let $a \in M$ with $P = (0\colon a)$. Say $\partial_n b = a$. Set $I = (0\colon b)$.

We first claim that $I \subseteq P$. Let $\xi \in I^2$. Notice $\partial_n \xi = \xi \partial_n + \partial_n(\xi)$. Also note that $\partial_n(\xi) \in I$. So we have that $\partial_n \xi b = \xi a + \partial_n(\xi) b$. Thus $\xi a = 0$. So $\xi \in P$. Thus $I^2 \subseteq P$. As P is a prime ideal we get that $I \subseteq P$.

Next we claim that $\partial_n(I) \subseteq I$. Let $\xi \in I$. We have $\partial_n \xi b = \xi a + \partial_n(\xi) b$. So $\partial_n(\xi) b = 0$. Thus $\partial_n(\xi) \in I$.

Since M is simple we have that $M = A_n(K)a$. So $b = da$ for some $d \in A_n(K)$. It can be easily verified that there exists $s \geq 1$ with $P^s d \subseteq A_n(K)P$. It follows that $P^s \subseteq I$. Thus $\sqrt{I} = P$. The result follows from Corollary 8.4.

Next suppose $H_1(\partial_n; M) \neq 0$. Say $a \in \ker \partial_n$ is non-zero. Set $J = (0\colon a)$. Let $\xi \in J$. Notice $\partial_n \xi a = \xi \partial_n a + \partial_n(\xi) a$. Thus $\partial_n(\xi) a = 0$. Thus $\partial_n(J) \subseteq J$.

By hypothesis M is simple and $\operatorname{Ass}_R(M) = \{P\}$. Now $\Gamma_P(M)$ is a non-zero $A_n(K)$-submodule of M. As M is simple we have that $M = \Gamma_P(M)$. Thus $P^s a = 0$ for some $s \geq 1$. Thus $P^s \subseteq J$. Also note that for any R-module E the maximal elements in the set $\{(0\colon e) \mid e \neq 0\}$ are associate primes of E. Thus $J = (0\colon a) \subseteq P$. Therefore $\sqrt{J} = P$. The result follows from Corollary 8.4. □

Remark 8.5 Let P be a prime ideal in R. Set $Q = P \cap R_{n-1}$. Then it can be easily seen that

$$\operatorname{ht}_R P - 1 \leq \operatorname{ht}_{R_{n-1}} Q \leq \operatorname{ht}_R P.$$

Furthermore $\operatorname{ht}_{R_{n-1}} Q = \operatorname{ht}_R P$ if and only if $P = QR$.

Remark 8.6 Let M be a holonomic $A_n(K)$-module. Assume M is I-torsion. Set $J = I \cap R_{n-1}$. Then for $i = 0, 1$ the Koszul homology modules $H_i(\partial_n, M)$ are J-torsion holonomic $A_{n-1}(K)$-modules. For holonomicity see 2.2. Also note the sequence

$$0 \to H_1(\partial_n, M) \to M \xrightarrow{\partial_n} M \to H_0(\partial_n, M) \to 0$$

is an exact sequence of $A_{n-1}(K)$-modules. It follows that $H_i(\partial_n, M)$ are J-torsion for $i = 0, 1$.

8.7 Let M be a R-module, not-necessarily finitely generated. By $\dim M$ we mean dimension of support of M. We set $\dim 0 = -\infty$. It can be easily seen that the following are equivalent:

(1) $\dim M \leq n - i$.
(2) $M_P = 0$ for all primes P with $\operatorname{ht} P < i$.

8.8 Let M be a holonomic $A_n(K)$-module. Let $c = \ell_{A_n(K)}(M)$. So we have a composition series

$$0 = V_0 \subsetneq V_1 \subsetneq V_2 \subsetneq \cdots \subsetneq V_c = M.$$

For $i = 1, \ldots, c$, $C_i = V_i/V_{i-1}$ are simple holonomic $A_n(K)$-modules. Let $\operatorname{Ass} C_i = \{P_i\}$. Set $d_i = \operatorname{ht} P_i$ and let $d = \min_i\{d_i\}$. Then

$$\dim M = n - d.$$

To see this let $d_j = d$. Set $P = P_j$. Then $(C_j)_P \neq 0$. So $(V_j)_P \neq 0$. So $M_P \neq 0$. Thus $\dim M \geq n - d$. If $Q \in \operatorname{Spec}(R)$ with $\operatorname{ht} Q < d$ then note that $P_i \nsubseteq Q$ for all i. Therefore $(C_i)_Q = 0$ for all i. It follows that $M_Q = 0$. Therefore $\dim M \leq n - d$ by 8.7. Thus $\dim M = n - d$.

To prove Theorem 8.1 by induction we need the following:

Lemma 8.9 *Let*

$$0 = V_0 \subsetneq V_1 \subsetneq V_2 \subsetneq \cdots \subsetneq V_c = M.$$

be a composition series of a holonomic-module M. For $i = 1, \ldots, c$ set $C_i = V_i/V_{i-1}$. Then

(1) $\dim H_0(\partial_n; M) \leq \max_i\{\dim H_0(\partial_n; C_i)\} \leq \dim M$.
(2) $\dim H_1(\partial_n; M) \leq \max_i\{\dim H_1(\partial_n; C_i)\} \leq \dim M - 1$.

Proof For $i = 1, \ldots, c$ we have an exact sequence

$$\begin{aligned}0 \to H_1(\partial_n; V_{i-1}) \to H_1(\partial_n; V_i) \to H_1(\partial_n; C_i)\\ \to H_0(\partial_n; V_{i-1}) \to H_0(\partial_n; V_i) \to H_0(\partial_n; C_i) \to 0.\end{aligned}$$

Let $\operatorname{Ass} C_i = \{P_i\}$ and $d_i = \operatorname{ht} P_i$. Set $Q_i = P_i \cap R_{n-1}$.

(1) We prove the first inequality. Suppose if possible $H_0(\partial_n; C_i) = 0$ for all i. Then by the above exact sequence we get $H_0(\partial_n; V_i) = 0$ for all i. So $H_0(\partial_n, M) = 0$. Therefore the first inequality holds in this case.

Now suppose $H_0(\partial_n; C_i) \neq 0$ for some i. Set

$$\max_i\{\dim H_0(\partial_n; C_i)\} = n - 1 - c \quad \text{for some } c \geq 0.$$

If $c = 0$ then we have nothing to prove. Now suppose $c > 0$. Let P be a prime in R with $\operatorname{ht} P < c$. Then $H_0(\partial_n; C_i)_P = 0$ for all i. By the above exact sequence we get $H_0(\partial_n; V_i)_P = 0$ for all i. So $H_0(\partial_n, M)_P = 0$. Thus by 8.7 we get $\dim H_0(\partial_n, M) \leq n - 1 - c$.

We now prove that $\dim H_0(\partial_n, C_i) \leq \dim M$ for all i. Set $N_i = H_0(\partial_n, C_i)$. We have nothing to prove if $N_i = 0$. So assume $N_i \neq 0$. By 8.6, N_i is Q_i-torsion. By 8.5 we have $\operatorname{ht} Q_i \geq d_i - 1$. If Q is a prime ideal in R_{n-1} with $\operatorname{ht} Q < d_i - 1$ then $Q \nsupseteq Q_i$. So $(N_i)_Q = 0$. By 8.7

$$\dim N_i \leq n - 1 - (d_i - 1) = n - d_i \leq \dim M.$$

Here the last inequality follows from 8.8.

(2) The proof of the first inequality is same as that in (1). Set $W_i = H_1(\partial_n, C_i)$. We prove $\dim W_i \leq \dim M - 1$ for all i.

If $\dim M = 0$ then note that $d_i = n$ for all i. So P_i is a maximal ideal in R. It follows that $P_i \neq Q_i R$. So by Theorem 8.2 we get $W_i = 0$.

Now assume $\dim M \geq 1$. If $W_i = 0$ then we have nothing to prove. So assume $W_i \neq 0$. Then by Theorem 8.2 we have $P_i = Q_i R$. So by 8.5 $\operatorname{ht} Q_i = \operatorname{ht} P_i = d_i$. By 8.6 W_i is Q_i-torsion. If Q is a prime ideal in R_{n-1} with $\operatorname{ht} Q < d_i$ then $Q \not\supseteq Q_i$. So $(W_i)_Q = 0$. By 8.7

$$\dim W_i \leq n - 1 - d_i \leq \dim M - 1.$$

Here the last inequality follows from 8.8. □

We now give

Proof of Theorem 8.1 We prove by induction on n that $H_i(\partial, M) = 0$ for $i > \dim M$. We first consider the case when $n = 1$. We have nothing to prove when $\dim M = 1$. If $\dim M = 0$ then M is only supported at maximal ideals. Let

$$0 = V_0 \subsetneq V_1 \subsetneq V_2 \subsetneq \cdots \subsetneq V_c = M.$$

be a composition series of M. For $i = 1, \ldots, c$ set $C_i = V_i / V_{i-1}$. Let $P_i = \operatorname{Ass} C_i$. Then P_i is a maximal ideal of R. By 8.2 we have $H_1(\partial_1, C_i) = 0$ for all i. So $H_1(\partial_1, M) = 0$.

Now assume $n \geq 2$. Let $\overline{M} = H_0(\partial_n, M)$ and $M_0 = H_1(\partial_n, M)$. Set $\partial' = \partial_1, \ldots, \partial_{n-1}$. Then we have an exact sequence

$$\cdots \to H_{j+1}(\partial'; \overline{M}) \to H_{j-1}(\partial'; M_0) \to H_j(\partial; M) \to H_j(\partial'; \overline{M}) \to \cdots$$

By Lemma 8.9 we have $\dim \overline{M} \leq \dim M$ and $\dim M_0 \leq \dim M - 1$. So for $j > \dim M$ we have, by induction hypothesis, $H_j(\partial'; \overline{M}) = 0$ and $H_{j-1}(\partial'; M_0) = 0$. So $H_j(\partial; M) = 0$. □

9 Proof of Theorem 1

In this section we prove Theorem 1. The proof of Theorem 1 follows in the same pattern as in proof of Theorem 8.1. Only Lemmas 7.2, 8.3, 7.8 and Remark 7.4 need an explanation.

Remark 9.1 Let M be a holonomic $\mathcal{D}_n$-module. Then $H_1(\partial_n; M)$ is a holonomic $\mathcal{D}_{n-1}$-module; see [8]. However $H_0(\partial_n; M)$ need not be a holonomic $\mathcal{D}_{n-1}$-module; see [9]. Nevertheless there exists a change of variables such that $H_i(\partial_n; M)$ are holonomic $\mathcal{D}_{n-1}$-modules for $i = 0, 1$; see [10].

Iteratively it follows that there exists a change of variables such that $H_i(\partial'; M)$ is finite dimensional K-vector spaces for $i \geq 0$. Note that $H_i(\partial; M) \cong H_i(\partial'; M)$ for all $i \geq 0$ it follows that $H_i(\partial; M)$ are finite dimensional K-vector spaces.

We first generalize Lemma 8.3.

Lemma 9.2 *Let I be an ideal in $\mathcal{O}_n$. Set $J = I \cap \mathcal{O}_{n-1}$. Then the following are equivalent:*

(1) $\partial_n(I) \subseteq I$.
(2) $I = J\mathcal{O}_n$.
(3) Let $\xi \in I$. Let $\xi = \sum_{j=0}^{\infty} c_j X_n^j$ where $c_j \in \mathcal{O}_{n-1}$ for $j \geq 0$. Then $c_j \in I$ for each j.

Proof (1) $\Longrightarrow$ (3) : Let $\xi = \sum_{j=r}^{\infty} c_j X_n^j \in I$ with $c_j \in \mathcal{O}_{n-1}$ for $j \geq r$. Put $v_r = \xi$ and $c_j^{(r)} = c_j$ for $j \geq r$. Put

$$v_{r+1} = v_r - \frac{1}{(r+1)!} X_n^{r+1} \partial_n^{r+1}(v_r) = c_r X_n^r + \sum_{j \geq r+2} c_j^{(r+1)} X_n^j.$$

Here $c_j^{(r+1)} \in \mathcal{O}_{n-1}$ for $j \geq r+2$. By hypothesis $v_{r+1} \in I$.

Now suppose $v_r, v_{r+1}, \ldots, v_{r+s} \in I$ have been constructed where

$$v_{r+s} = c_r X_n^r + \sum_{j \geq r+s+1} c_j^{(r+s)} X_n^j.$$

Put

$$v_{r+s+1} = v_{r+s} - \frac{1}{(r+s+1)!} X_n^{r+s+1} \partial_n^{r+s+1}(v_{r+s}) = c_r X_n^r + \sum_{j \geq r+s+2} c_j^{(r+s+1)} X_n^j.$$

Here $c_j^{(r+s+1)} \in \mathcal{O}_{n-1}$ for $j \geq r+s+2$. By hypothesis $v_{r+s+1} \in I$.

Since $v_{r+s} \in I$ we have that $c_r X_n^r \in I + \mathfrak{m}^{r+s+1}$ for all $s \geq 1$. By Krull's intersection theorem we have $\bigcap_{s \geq 1}(I + \mathfrak{m}^{r+s+1}) = I$. So $c_r X_n^r \in I$. Therefore

$$c_r = \frac{1}{r!} \partial_n^r (c_r X_n^r) \in I$$

Now notice that $\xi - c_r X_n^r = \sum_{j=r+1}^{\infty} c_j X_n^j \in I$. Iteratively one can prove that $c_j \in I$ for all $i \geq r$.

The assertion (3) $\Longrightarrow$ (1) is trivial. We now show (3) $\Longrightarrow$ (2). Let $\xi = \sum_{j=r}^{\infty} c_j X_n^j \in I$ with $c_j \in \mathcal{O}_{n-1}$ for $j \geq r$. Then by hypothesis $c_j \in I$ for $j \geq r$. Set $S = \mathcal{O}_{n-1}[X_n]$. So $\xi_m = \sum_{j=r}^{m} c_j X_n^j \in JS$ for all $m \geq r$. Let $\widehat{\ }$ denote completion with respect to X_n-adic topology. Note $\xi = \lim_m \xi_m \in \widehat{JS} = J\widehat{S} = J\mathcal{O}_n$. It follows that $I \subseteq J\mathcal{O}_n$. The assertion $J\mathcal{O}_n \subseteq I$ is trivial. So $I = J\mathcal{O}_n$.

The proof of (2) $\implies$ (3) is similar to the analogous assertion in Lemma 8.3. $\square$

We now generalize Lemma 7.4.

Corollary 9.3 *Let P be a prime ideal in R and let I be an ideal in R with $\sqrt{I} = P$. If $\partial_n(I) \subseteq I$ then $P = (P \cap R_{n-1})R$.*

Proof Set $Q = P \cap \mathcal{O}_{n-1}$. Let $\xi \in P$. Let $\xi = \sum_{j=r}^{\infty} c_j X_n^j$ where $c_j \in \mathcal{O}_{n-1}$ for $j \geq r$. Notice $\xi^s \in I$ for some $s \geq 1$. Also $\xi^s = c_r^s X_n^{sr} + ..$ higher terms in X_n. By Lemma 9.2 we get that $c_r^s \in I$. It follows that $c_r \in P$. Thus $\xi - c_r X_n^r \in P$. Iterating we obtain that $c_j \in P$ for all $j \geq r$. So by Lemma 9.2 we get that $P = QR$. $\square$

Remark 9.4 Theorem 8.1 generalizes to the case of $\mathcal{D}_n$-modules. The proof is the same.

Remark 9.5 We now generalize Remark 7.4. Let P be a prime ideal in $\mathcal{O}_n$. Set $Q = P \cap \mathcal{O}_{n-1}$. It is elementary that

$$\text{ht}_{\mathcal{O}_{n-1}} Q \leq \text{ht}_{\mathcal{O}_n} P \quad \text{with equality if and only if } P = Q\mathcal{O}_n.$$

However the assertion $\text{ht}\, Q \geq \text{ht}\, P - 1$ requires a proof. I thank J.K. Verma for providing this proof. Note that $\text{ht}\, Q = \text{ht}\, Q\mathcal{O}_n$. Set $A = \mathcal{O}_{n-1}/Q$ and $B = \mathcal{O}_n/Q\mathcal{O}_n = A[[X_n]]$. Set $\mathfrak{n} = P/Q\mathcal{O}_n$. Let S be the non-zero elements of A. Then $\mathfrak{n} \cap S = \emptyset$. So $\text{ht}\, \mathfrak{n} = \text{ht}\, \mathfrak{n} S^{-1} B$. Let $L =$ quotient field of A. Then $S^{-1}B = L[[X_n]]$. It follows that $\text{ht}\, \mathfrak{n} \leq 1$. Therefore $\text{ht}\, P - \text{ht}\, Q \leq 1$. The result follows.

For stating our generalization of Lemma 7.8 we need the following result:

Proposition 9.6 *Let $0 \to N \to M \to L \to 0$ be a short exact sequence of holonomic $\mathcal{D}_n$-modules. The following are equivalent:*

(1) $H_i(\partial_n; M)$ are holonomic $\mathcal{D}_{n-1}$-module for $i = 0, 1$.
(2) $H_i(\partial_n; N)$, $H_i(\partial_n; M)$ are holonomic $\mathcal{D}_{n-1}$-modules for $i = 0, 1$.

Proof Let E be a holonomic $\mathcal{D}_n$-module. Then $H_1(\partial_n; E)$ is a holonomic $\mathcal{D}_{n-1}$-module; see [8]. Note that we have an exact sequence of $\mathcal{D}_{n-1}$-modules

$$H_1(\partial; L) \to H_0(\partial; N) \to H_0(\partial; M) \to H_0(\partial; L) \to 0.$$

(2) $\implies$ (1) : By the above exact sequence $H_0(\partial; M)$ is a holonomic $\mathcal{D}_{n-1}$-module.

We now prove (1) $\implies$ (2). Note that $H_1(\partial; L)$ is holonomic $\mathcal{D}_{n-1}$-module. By the above exact sequence $H_0(\partial; N)$ is a holonomic $\mathcal{D}_{n-1}$-module. Furthermore $H_0(\partial; L)$ is a subquotient of $H_0(\partial; M)$ and so it is holonomic. $\square$

The correct statement which generalizes Lemma 7.8 is the following:

Lemma 9.7 *Let*

$$0 = V_0 \subsetneq V_1 \subsetneq V_2 \subsetneq \cdots \subsetneq V_c = M.$$

be a composition series of a holonomic-module M. For $i = 1, \ldots, c$ set $C_i = V_i / V_{i-1}$. Let $C = \bigoplus_{i=1}^{c} C_i$. Suppose we have a change of variables with $H_i(\partial_n; C)$ holonomic $\mathcal{D}_{n-1}$ module for $i = 0, 1$. Then

(1) $H_i(\partial_n; C_j)$ *are holonomic* $\mathcal{D}_{n-1}$ *module for* $i = 0, 1$ *and* $j = 1, \ldots, c$.
(2) $H_i(\partial_n; M)$ *are holonomic* $\mathcal{D}_{n-1}$*-module for* $i = 0, 1$.
(3) $\dim H_0(\partial_n; M) \leq \max_i\{\dim H_0(\partial_n; C_i)\} \leq \dim M$.
(4) $\dim H_1(\partial_n; M) \leq \max_i\{\dim H_1(\partial_n; C_i)\} \leq \dim M - 1$.

Proof The assertions (1) and (2) follow from Proposition 9.6. The proof of assertions (3) and (4) is similar to that of (1) and (2) in Lemma 7.8. □

We now give
Proof of Theorem 1 Let

$$0 = V_0 \subsetneq V_1 \subsetneq V_2 \subsetneq \cdots \subsetneq V_c = M.$$

be a composition series of a holonomic-module M. For $i = 1, \ldots, c$ set $C_i = V_i / V_{i-1}$. Let $C = \bigoplus_{i=1}^{c} C_i$. Choose a change of variables with $H_i(\partial_n; C)$ holonomic $\mathcal{D}_{n-1}$ module for $i = 0, 1$. Then by Lemma 9.7 we have that $H_i(\partial_n; C_j)$ are holonomic $\mathcal{D}_{n-1}$ module for $i = 0, 1$ and $j = 1, \ldots, c$. Furthermore $H_i(\partial_n; M)$ are holonomic $\mathcal{D}_{n-1}$-module for $i = 0, 1$.

After this choice of variables the proof of Theorem 1 is now identical to proof of Theorem 8.1. □

References

1. Björk, J.-E.: Rings of differential operators. North-Holland Mathematical Library, vol. 21. North-Holland Publishing Co., Amsterdam-New York (1979)
2. Eisenbud, D.: Commutative algebra with a view toward algebraic geometry. Graduate Texts in Mathematics, vol. 150. Springer, New York (1995)
3. Iyengar, S.B., Leuschke, G.J., Leykin, A., Miller, C., Miller, E., Singh, A.K., Walther, U.: Twenty-four hours of local cohomology. Graduate Studies in Mathematics, vol. 87. American Mathematical Society, Providence (2007)
4. Lyubeznik, G.: Finiteness properties of local cohomology modules (an application of D-modules to commutative algebra). Invent. Math. **113**(1), 41–55 (1993)
5. Matsumura, H.: Commutative ring theory. Translated from the Japanese by M. Reid, 2 edn. Cambridge Studies in Advanced Mathematics, vol. 8. Cambridge University Press, Cambridge (1989)
6. Puthenpurakal, T.J.: Injective modules over some rings of differential operators. eprint: arXiv:1301.1176

7. Switala, N., Van den Essen's theorem on the de Rham cohomology of a holonomic D-module over a formal power series ring. eprint: arXiv:1505.01788
8. van den Essen, A.: Le noyau de l'opérateur $\delta/\delta x_n$ agissant sur un D_n-module. C. R. Acad. Sci. Paris Sér. A-B **288**(14), A687–A690 (1979)
9. van den Essen, A.: Un D-module holonome tel que le conoyau de l'opérateur $\delta/\delta x_n$ soit non-holonome. C. R. Acad. Sci. Paris Sér. I Math. **295**(7), 455–457 (1982)
10. van den Essen, A.: Le conoyau de l'opérateur $\delta/\delta x_n$ agissant sur un D_n-module holonome. C. R. Acad. Sci. Paris Sér. I Math. **296**(22), 903–906 (1983)

Central Quotient Versus Commutator Subgroup of Groups

Manoj K. Yadav

Abstract In 1904, Issai Schur proved the following result. If G is an arbitrary group such that $G/\mathrm{Z}(G)$ is finite, where $\mathrm{Z}(G)$ denotes the center of the group G, then the commutator subgroup of G is finite. A partial converse of this result was proved by B.H. Neumann in 1951. He proved that if G is a finitely generated group with finite commutator subgroup, then $G/\mathrm{Z}(G)$ is finite. In this short note, we exhibit few arguments of Neumann, which provide further generalizations of converse of the above mentioned result of Schur. We classify all finite groups G such that $|G/\mathrm{Z}(G)| = |\gamma_2(G)|^d$, where d denotes the number of elements in a minimal generating set for $G/\mathrm{Z}(G)$. Some problems and questions are posed in the sequel.

Keywords Commutator subgroup · Schur's theorem · Class-preserving automorphism

Classifications Primary 20F24 · 20E45

1 Introduction

In 1951, Neumann [18, Theorem 5.3] proved the following result: *If the index of* $\mathrm{Z}(G)$ *in* G *is finite, then* $\gamma_2(G)$ *is finite, where* $\mathrm{Z}(G)$ *and* $\gamma_2(G)$ *denote the center and the commutator subgroup of* G *respectively.* He mentioned [19, End of page 237] that this result can be obtained from an implicit idea of Schur [23], and his proof also used Schur's basic idea. However there is no mention of this fact in [18] in which Schur's paper is also cited. In this note, this result will be termed as 'the Schur's theorem'. Neumann also provided a partial converse of the Schur's theorem [18, Corollary 5.41] as follows: *If* G *is finitely generated by* k *elements and* $\gamma_2(G)$ *is finite, then* $G/\mathrm{Z}(G)$ *is finite, and bounded by* $|G/\mathrm{Z}(G)| \leq |\gamma_2(G)|^k$.

M.K. Yadav (✉)
School of Mathematics, Harish-Chandra Research Institute,
Chhatnag Road, Jhunsi Allahabad 211 019, India
e-mail: myadav@hri.res.in

S.T. Rizvi et al. (eds.), *Algebra and its Applications*, Springer Proceedings
in Mathematics & Statistics 174, DOI 10.1007/978-981-10-1651-6_10

Our first motivation of writing this note is to exhibit an idea of Neumann [18, page 179] which proves much more than what is said above on converse of the Schur's theorem. We quote the text here (with a minor modification in the notations):

"*Let* G *be generated by* $g_1, g_2, \ldots, g_k$. *Then*

$$Z(G) = \cap_{\kappa=1}^{\kappa=k} C_G(g_\kappa);$$

for, an element of G *lies in the center if and only if it (is permutable) commutes with all the generators of* G. *If* G *is an FC-group (group whose all conjugacy classes are of finite length), then* $|G : C_G(g_\kappa)|$ *is finite for* $1 \leq \kappa \leq k$, *and* $Z(G)$, *as intersection of a finite set of subgroups of finite index, also has finite index. The index of the intersection of two subgroups does not exceed the product of the indices of the subgroups: hence in this case one obtains an upper bound for the index of the center, namely*

$$|G : Z(G)| \leq \Pi_{\kappa=1}^{\kappa=k} |G : C_G(g_\kappa)|."$$

Just a soft staring at the quoted text for a moment or two suggests the following. The conclusion does not require the group G to be FC-group. It only requires the finiteness of the conjugacy classes of the generating elements. If a generator of the group G is contained in $Z(G)$, one really does not need to count it. Thus the argument works perfectly well even if G is generated by infinite number of elements, all but finite of them lie in the center of G. Thus the following result holds true.

Theorem A. *Let* G *be an arbitrary group such that* $G/Z(G)$ *is finitely generated by* $x_1 Z(G), x_2 Z(G), \ldots, x_t Z(G)$ *and the conjugacy class of* x_i *in* G *is of finite length for* $1 \leq i \leq t$. *Then* $G/Z(G)$ *is finite. Moreover* $|G/Z(G)| \leq \Pi_{i=1}^{t} |x_i^G|$ *and* $\gamma_2(G)$ *is finite, where* x_i^G *denotes the conjugacy class of* x_i in G.

Neumann's result [18, Corollary 5.41] was reproduced by Hilton [12, Theorem 1]. It seems that Hilton was not aware of Neumann's result. This lead two more publications [21] and [24] dedicated to proving special cases of Theorem A.

Converse of the Schur's theorem is not true in general as shown by infinite extraspecial p-groups, where p is an odd prime. It is interesting to know that example of such a 2-group also exists, which is mentioned on page 238 (second para of Sect. 3) of [19]. It is a central product of infinite copies of quaternion groups of order 8 amalgamated at the center of order 2.

Our second motivation of writing this note is to provide a modification of an innocent looking result of Neumann [20, Lemma 2], which allows us to say little more on converse of the Schur's theorem. A modified version of this lemma is the following.

Lemma 1.1 *Let* G *be an arbitrary group having a normal abelian subgroup* A *such that the index of* $C_G(A)$ *in* G *is finite and* G/A *is finitely generated by* $g_1A, g_2A, \ldots, g_rA$, *where* $|g_i^G| < \infty$ *for* $1 \leq i \leq r$. *Then* $G/Z(G)$ *is finite.*

This lemma helps proving the first three statements of the following result.

Theorem B. *For an arbitrary group G, $G/Z(G)$ is finite if any one of the following holds true:*

(i) $Z_2(G)/Z(Z_2(G))$ *is finitely generated and* $\gamma_2(G)$ *is finite.*
(ii) $G/Z(Z_2(G))$ *is finitely generated and* $G/(Z_2(G)\gamma_2(G))$ *is finite.*
(iii) $\gamma_2(G)$ *is finite and* $Z_2(G) \leq \gamma_2(G)$.
(iv) $\gamma_2(G)$ *is finite and* $G/Z(G)$ *is purely non-abelian.*

Our final motivation is to provide a classification of all groups G upto isoclinism (see Sect. 3 for the definition) such that $|G/Z(G)| = |\gamma_2(G)|^d$ is finite, where d denotes the number of elements in a minimal generating set for $G/Z(G)$, discuss example in various situations and pose some problems. We conclude this section with fixing some notations. For an arbitrary group G, by $Z(G)$, $Z_2(G)$ and $\gamma_2(G)$ we denote the center, the second center and the commutator subgroup of G respectively. For $x \in G$, $[x, G]$ denotes the set $\{[x, g] \mid g \in G\}$. Notice that $|[x, G]| = |x^G|$, where x^G denotes the conjugacy class of x in G. If $[x, G] \subseteq Z(G)$, then $[x, G]$ becomes a subgroup of G. For a subgroup H of G, $C_G(H)$ denotes the centralizer of H in G and for an element $x \in G$, $C_G(x)$ denotes the centralizer of x in G.

2 Proofs

We start with the proof of Lemma 1.1, which is essentially same as the one given by Neumann.

Proof of Lemma 1.1. Let G/A be generated by $g_1A, g_2A, \ldots, g_rA$ for some $g_i \in G$, where $1 \leq i \leq r < \infty$. Let $X := \{g_1, g_2, \ldots, g_r\}$ and A be generated by a set Y. Then $G = \langle X \cup Y\rangle$ and $Z(G) = C_G(X) \cap C_G(Y)$. Notice that $C_G(A) = C_G(Y)$. Since $C_G(A)$ is of finite index, $C_G(Y)$ is also of finite index in G. Also, since $|g_i^G| < \infty$ for $1 \leq i \leq r$, $C_G(X)$ is of finite index in G. Hence the index of $Z(G)$ in G is finite and the proof is complete. □

Proof of Theorem A can be made quite precise by using Lemma 1.1.

Proof of Theorem A. Taking $A = Z(G)$ in Lemma 1.1, it follows that $G/Z(G)$ is finite. Moreover,

$$|G/Z(G)| = |G/\cap_{i=1}^{t} C_G(x_i)| \leq \Pi_{i=1}^{t}|G : C_G(x_i)| = \Pi_{i=1}^{t}|[x_i, G]| = \Pi_{i=1}^{t}|x_i^G|.$$

That $\gamma_2(G)$ is finite now follows from the Schur's theorem. □

For the proof of Theorem B we need the following result of Hall [9] and the subsequent proposition.

Theorem 2.1 *If G is an arbitrary group such that $\gamma_2(G)$ is finite, then $G/Z_2(G)$ is finite.*

Explicit bounds on the order of $G/\mathrm{Z}_2(G)$ were first given by Macdonald [14, Theorem 6.2] and later on improved by Podoski and Szegedy [22] by showing that if $|\gamma_2(G)/(\gamma_2(G)\cap \mathrm{Z}(G))| = n$, then $|G/\mathrm{Z}_2(G)| \leq n^{c\, log_2 n}$ with $c = 2$.

Proposition 2.1 *Let G be an arbitrary group such that $\gamma_2(G)$ is finite and $G/\mathrm{Z}(G)$ is infinite. Then $G/\mathrm{Z}(G)$ has an infinite abelian group as a direct factor.*

Proof Since $\gamma_2(G)$ is finite, by Theorem 2.1 $G/\mathrm{Z}_2(G)$ is finite. Thus $\mathrm{Z}_2(G)/\mathrm{Z}(G)$ is infinite. Again using the finiteness of $\gamma_2(G)$, it follows that the exponent of $\mathrm{Z}_2(G)/\mathrm{Z}(G)$ is finite. Hence by [6, Theorem 17.2] $\mathrm{Z}_2(G)/\mathrm{Z}(G)$ is a direct sum of cyclic groups. Let $G/\mathrm{Z}_2(G)$ be generated by $x_1\,\mathrm{Z}_2(G), \ldots, x_r\,\mathrm{Z}_2(G)$ and $H := \langle x_1, \ldots, x_r\rangle$. Then it follows that modulo $\mathrm{Z}(G)$, $H \cap \mathrm{Z}_2(G)$ is finite. Thus we can write

$$\mathrm{Z}_2(G)/\mathrm{Z}(G) = \langle y_1\,\mathrm{Z}(G)\rangle \times \cdots \times \langle y_s\,\mathrm{Z}(G)\rangle \times \langle y_{s+1}\,\mathrm{Z}(G)\rangle \times \cdots,$$

such that $(H \cap \mathrm{Z}_2(G))\mathrm{Z}(G)/\mathrm{Z}(G) \leq \langle y_1\,\mathrm{Z}(G)\rangle \times \cdots \times \langle y_s\,\mathrm{Z}(G)\rangle$. It now follows that the infinite abelian group $\langle y_{s+1}\,\mathrm{Z}(G)\rangle \times \cdots$ is a direct factor of $G/\mathrm{Z}(G)$, and the proof is complete. □

We are now ready to prove Theorem B.

Proof of Theorem B. Since $\gamma_2(G)$ is finite, it follows from Theorem 2.1 that $G/\mathrm{Z}_2(G)$ is finite. Now using the fact that $\mathrm{Z}_2(G)/\mathrm{Z}(\mathrm{Z}_2(G))$ is finitely generated, it follows that $G/\mathrm{Z}(\mathrm{Z}_2(G))$ is finitely generated. Take $\mathrm{Z}(\mathrm{Z}_2(G)) = A$. Then notice that A is a normal abelian subgroup of G such that the index of $C_G(A)$ in G is finite, since $\mathrm{Z}_2(G) \leq C_G(A)$. Hence by Lemma 1.1, $G/\mathrm{Z}(G)$ is finite, which proves (i).

Again take $\mathrm{Z}(\mathrm{Z}_2(G)) = A$ and notice that $\mathrm{Z}_2(G)\gamma_2(G) \leq \mathrm{C}_G(A)$. (ii) now directly follows from Lemma 1.1. If $\mathrm{Z}_2(G) \leq \gamma_2(G)$, then $\mathrm{Z}_2(G)$ is abelian. Thus (iii) follows from (i). Finally, (iv) follows from Proposition 2.1. This completes the proof of the theorem. □

We conclude this section with an extension of Theorem A in terms of conjugacy class-preserving automorphisms of given group G. An automorphism α of an arbitrary group G is called *(conjugacy) class-preserving* if $\alpha(g) \in g^G$ for all $g \in G$. We denote the group of all class-preserving automorphisms of G by $\mathrm{Aut}_c(G)$. Notice that $\mathrm{Inn}(G)$, the group of all inner automorphisms of G, is a normal subgroup of $\mathrm{Aut}_c(G)$ and $\mathrm{Aut}_c(G)$ acts trivially on the center of G. A detailed survey on class-preserving automorphisms of finite p-groups can be found in [25].

Let G be the group as in the statement of Theorem A. Then G is generated by $x_1, x_2, \ldots, x_t$ along with $\mathrm{Z}(G)$. Since $\mathrm{Aut}_c(G)$ acts trivially on the center of G, it follows that

$$|\,\mathrm{Aut}_c(G)| \leq \Pi_{i=1}^{t}|x_i^G| \tag{2.1}$$

as there are only $|x_i^G|$ choices for the image of each x_i under any class-preserving automorphism. Since $|x_i^G|$ is finite for each x_i, $1 \leq i \leq t$, it follows that $|\,\mathrm{Aut}_c(G)| \leq \Pi_{i=1}^{t}|x_i^G|$ is finite.

We have proved the following result of which Theorem A is a corollary, because $|G/Z(G)| = |\operatorname{Inn}(G)| \leq |\operatorname{Aut}_c(G)|$.

Theorem 2.2 *Let G be an arbitrary group such that $G/Z(G)$ is finitely generated by $x_1 Z(G), x_2 Z(G), \ldots, x_t Z(G)$ and the conjugacy class of x_i in G is of finite length for $1 \leq i \leq t$. Then $\operatorname{Aut}_c(G)$ is finite. Moreover $|\operatorname{Aut}_c(G)| \leq \Pi_{i=1}^{t} |x_i^G|$ and $\gamma_2(G)$ is finite.*

Proof of Theorem A is also reproduced using IA-automorphisms (automorphisms of a group that induce identity on the abelianization) in [7, Theorem 2.3]. Proof goes on the same way as in the case of class-preserving automorphisms.

3 Groups with Maximal Central Quotient

We start with the following concept due to Hall [8]. For a group X, the commutator map $a_X : X/Z(X) \times X/Z(X) \to \gamma_2(X)$ given by $a_X(x_1 Z(X), x_2 Z(X)) = [x_1, x_2]$ is well defined. Two groups K and H are said to be *isoclinic* if there exists an isomorphism ϕ of the factor group $\bar{K} = K/Z(K)$ onto $\bar{H} = H/Z(H)$, and an isomorphism θ of the subgroup $\gamma_2(K)$ onto $\gamma_2(H)$ such that the following diagram is commutative

$$\begin{array}{ccc} \bar{K} \times \bar{K} & \xrightarrow{a_G} & \gamma_2(K) \\ {\scriptstyle \phi\times\phi}\downarrow & & \downarrow{\scriptstyle \theta} \\ \bar{H} \times \bar{H} & \xrightarrow{a_H} & \gamma_2(H). \end{array}$$

The resulting pair (ϕ, θ) is called an *isoclinism* of K onto H. Notice that isoclinism is an equivalence relation among groups.

The following proposition (also see Macdonald's result [14, Lemma 2.1]) is important for the rest of this section.

Proposition 3.1 *Let G be a group such that $G/Z(G)$ is finite. Then there exists a finite group H isoclinic to the group G such that $Z(H) \leq \gamma_2(H)$. Moreover if G is a p-group, then H is also a p-group.*

Proof Let G be the given group. Then by Schur's theorem $\gamma_2(G)$ is finite. Now it follows from a result of Hall [8] that there exists a group H which is isoclinic to G and $Z(H) \leq \gamma_2(H)$. Since $|\gamma_2(G)| = |\gamma_2(H)|$ is finite, $Z(H)$ is finite. Hence, by the definition of isoclinism, H is finite. Now suppose that G is a p-group, then it follows that, $H/Z(H)$ as well as $\gamma_2(H)$ are p-groups. Since $Z(H) \leq \gamma_2(H)$, this implies that H is a p-group. □

For an arbitrary group G with finite $G/Z(G)$, we have

$$|G/Z(G)| \leq |\gamma_2(G)|^d, \tag{3.1}$$

where $d = d(G/Z(G))$. For simplicity we say that a group G has *Property A* if $G/Z(G)$ is finite and equality holds in (3.1) for G. We are now going to classify, upto isoclinism, all groups G having Property A.

Let G be an arbitrary group having Property A. Then by Proposition 3.1 there exists a finite group H isoclinic to G and, by the definition of isoclinism, H has Property A. Thus for classifying all groups G, upto isoclinism, having Property A, it is sufficient to classify all finite group with this property.

Let us first consider non-nilpotent finite groups. For such groups we prove the following result in [5]

Theorem 3.1 *There is no non-nilpotent group G having Property A.*

So we only need to consider finite nilpotent groups. Since a finite nilpotent group is a direct product of its' Sylow p-subgroups, it is sufficient to classify finite p-groups admitting Property A. Obviously, all abelian groups admit Property A. Perhaps the simplest examples of non-abelian groups having Property A are finite extraspecial p-groups. The class of 2-generated finite capable nilpotent groups with cyclic commutator subgroup also admits Property A. A group G is said to be *capable* if there exists a group H such that $G \cong H/Z(H)$. Isaacs [13, Theorem 2] proved: *Let G be finite and capable, and suppose that $\gamma_2(G)$ is cyclic and that all elements of order 4 in $\gamma_2(G)$ are central in G. Then $|G/Z(G)| \le |\gamma_2(G)|^2$, and equality holds if G is nilpotent*. So, if G is a group as in the preceding statement and G is also 2-generated nilpotent, then G admits Property *A*. A complete classification of 2-generated finite capable p-groups of class 2 is given in [15].

Motivated by finite extraspecial p-groups, a more general class of groups G admitting Property A can be constructed as follows. For any positive integer m, let $G_1, G_2, \ldots, G_m$ be 2-generated finite p-groups such that $\gamma_2(G_i) = Z(G_i) \cong X$ (say) is cyclic of order q for $1 \le i \le m$, where q is some power of p. Consider the central product

$$Y = G_1 *_X G_2 *_X \cdots *_X G_m \tag{3.2}$$

of $G_1, G_2, \ldots, G_m$ amalgamated at X (isomorphic to cyclic commutator subgroups $\gamma_2(G_i)$, $1 \le i \le m$). Then $|Y| = q^{2m+1}$ and $|Y/Z(Y)| = q^{2m} = |\gamma_2(Y)|^{d(Y)}$, where $d(Y) = 2m$ is the number of elements in any minimal generating set for Y. Thus Y has Property A. Notice that in all of the above examples, the commutator subgroup is cyclic. Infinite groups having Property A can be easily obtained by taking a direct product of an infinite abelian group with any finite group having Property A.

We now proceed to showing that any finite p-group G of class 2 having Property A is isoclinic to a group Y defined in (3.2).

Let $x \in Z_2(G)$ for a group G. Then, notice that $[x, G]$ is a central subgroup of G. We have the following simple but useful result.

Lemma 3.1 *Let G be an arbitrary group such that $Z_2(G)/Z(G)$ is finitely generated by $x_1 Z(G), x_2 Z(G), \ldots, x_t Z(G)$ such that $exp([x_i, G])$ is finite for $1 \le i \le t$. Then*

$$|Z_2(G)/Z(G)| = \prod_{i=1}^{t} exp([x_i, G]).$$

Proof By the given hypothesis $exp([x_i, G])$ is finite for all i such that $1 \le i \le t$. Suppose that $exp([x_i, G]) = n_i$. Since $[x_i, G] \subseteq Z(G)$, it follows that $[x_i^{n_i}, G] = [x_i, G]^{n_i} = 1$. Thus $x_i^{n_i} \in Z(G)$ and no smaller power of x_i than n_i can lie in $Z(G)$, which implies that the order of $x_i Z(G)$ is n_i. Since $Z_2(G)/Z(G)$ is abelian, we have $|Z_2(G)/Z(G)| = \prod_{i=1}^{t} exp([x_i, G])$. □

Let $\Phi(X)$ denote the Frattini subgroup of a group X. The following result provides some structural information of p-groups of class 2 admitting Property A.

Proposition 3.2 *Let H be a finite p-group of class 2 having Property A and $Z(H) = \gamma_2(H)$. Then*

(i) $\gamma_2(H)$ is cyclic;
(ii) $H/Z(H)$ is homocyclic;
(iii) $[x, H] = \gamma_2(H)$ for all $x \in H - \Phi(H)$;
(iv) H is minimally generated by even number of elements.

Proof Let H be the group given in the statement, which is minimally generated by d elements $x_1, x_2 \dots, x_d$ (say). Since $Z(H) = \gamma_2(H)$, it follows that $H/Z(H)$ is minimally generated by $x_1 Z(H)$, $x_2 Z(H), \dots, x_d Z(H)$. Thus by the identity $|H/Z(H)| = |\gamma_2(H)|^d$, it follows that order of $x_i Z(H)$ is equal to $|\gamma_2(H)|$ for all $1 \le i \le d$. Since the exponent of $H/Z(H)$ is equal to the exponent of $\gamma_2(H)$, we have that $\gamma_2(H)$ is cyclic and $H/Z(H)$ is homocyclic. Now by Lemma 3.1, $|\gamma_2(H)|^d = |H/Z(H)| = \prod_{i=1}^{t} exp([x_i, H])$. Since $[x_i, H] \subseteq \gamma_2(H)$, this implies that $[x_i, H] = \gamma_2(H)$ for each i such that $1 \le i \le d$. Let x be an arbitrary element in $H - \Phi(H)$. Then the set $\{x\}$ can always be extended to a minimal generating set of H. Thus it follows that $[x, H] = \gamma_2(H)$ for all $x \in H - \Phi(H)$. This proves first three assertions.

For the proof of (iv), we consider the group $\bar{H} = H/\Phi(\gamma_2(H))$. Notice that both H as well as $\bar{H}$ are minimally generated by d elements. Since $[x, H] = \gamma_2(H)$ for all $x \in H - \Phi(H)$, it follows that for no $x \in H - \Phi(H)$, $\bar{x} \in Z(\bar{H})$, where $\bar{x} = x\Phi(\gamma_2(H)) \in \bar{H}$. Thus it follows that $Z(\bar{H}) \le \Phi(\bar{H})$. Also, since $\gamma_2(H)$ is cyclic, $\gamma_2(\bar{H})$ is cyclic of order p. Thus it follows that $\bar{H}$ is isoclinic to a finite extraspecial p-group, and therefore it is minimally generated by even number of elements. Hence H is also minimally generated by even number of elements. This completes the proof of the proposition. □

Using the definition of isoclinism, we have

Corollary 3.1 *Let G be a finite p-group of class 2 admitting Property A. Then $\gamma_2(G)$ is cyclic and $G/Z(G)$ is homocyclic.*

We need the following important result.

Theorem 3.2 *([3], Theorem 2.1) Let G be a finite p-group of nilpotency class* 2 *with cyclic center. Then G is a central product either of two generator subgroups with cyclic center or two generator subgroups with cyclic center and a cyclic subgroup.*

Theorem 3.3 *Let G be a finite p-group of class* 2 *having Property A. Then G is isoclinic to the group Y, defined in* (3.2), *for suitable positive integers m and n.*

Proof Let G be a group as in the statement. Then by Proposition 3.1 there exists a finite p-group H isoclinic to G such that $Z(H) = \gamma_2(H)$. Obviously H also satisfies $|H/Z(H)| = |\gamma_2(H)|^d$, where d denotes the number of elements in any minimal generating set of $G/Z(G)$. Then by Proposition 3.2, $\gamma_2(H) = Z(H)$ is cyclic of order $q = p^n$ (say) for some positive integer n, and $H/Z(H)$ is homocyclic of exponent q and is of order q^{2m} for some positive integer m. Since $Z(H) = \gamma_2(H)$ is cyclic, it follows from Theorem 3.2 that H is a central product of 2-generated groups $H_1, H_2, \ldots, H_m$. It is easy to see that $\gamma_2(H_i) = Z(H_i)$ for $1 \le i \le m$ and $|\gamma_2(H)| = q$. This completes the proof of the theorem. □

We would like to remark that Theorem 3.3 is also obtained in [26, Theorem 11.2] as a consequence on study of class-preserving automorphisms of finite p-group. But we have presented a direct proof here.

Now we classify finite p-groups of impotency class larger than 2. Consider the metacyclic groups

$$K := \left\langle x, y \mid x^{p^{r+t}} = 1, y^{p^r} = x^{p^{r+s}}, [x, y] = x^{p^t} \right\rangle, \tag{3.3}$$

where $1 \le t < r$ and $0 \le s \le t$ ($t \ge 2$ if $p = 2$) are nonnegative integers. Notice that the nilpotency class of K is at least 3. Since K is generated by 2 elements, it follows from (2.1) that $|\operatorname{Aut}_c(K)| \le |\gamma_2(K)|^2 = p^{2r}$. It is not so difficult to see that $|\operatorname{Inn}(K)| = |K/Z(K)| = p^{2r}$. Since $\operatorname{Inn}(K) \le \operatorname{Aut}_c(K)$, it follows that $|\operatorname{Aut}_c(K)| = |\operatorname{Inn}(K)| = |\gamma_2(K)|^2 = |\gamma_2(K)|^{d(K)}$ (That $\operatorname{Aut}_c(G) = \operatorname{Inn}(G)$, is, in fact, true for all finite metacylic p-groups). Thus K admits Property A. Furthermore, if H is any 2-generator group isoclinic to K, then it follows that H admits Property A. For a finite p-group having Property A, there always exists a p-group H isoclinic G such that $|H/Z(H)| = |\gamma_2(H)|^d$, where $d = d(H)$. The following theorem now classifies, upto isoclinism, all finite p-groups G of nilpotency class larger than 2 having Property A.

Theorem 3.4 *(Theorem 11.3, [26]) Let G be a finite p-group of nilpotency class at least* 3. *Then the following holds true.*

(i) *If* $|G/Z(G)| = |\gamma_2(G)|^d$, *where* $d = d(G)$, *then* $d(G) = 2$;
(ii) *If* $|\gamma_2(G)/\gamma_3(G)| > 2$, *then* $|G/Z(G)| = |\gamma_2(G)|^d$ *if and only if G is a* 2-*generator group with cyclic commutator subgroup. Furthermore, G is isoclinic to the group K defined in* (3.3) *for suitable parameters;*
(iii) *If* $|\gamma_2(G)/\gamma_3(G)| = 2$, *then* $|G/Z(G)| = |\gamma_2(G)|^d$ *if and only if G is a* 2-*generator* 2-*group of nilpotency class* 3 *with elementary abelian* $\gamma_2(G)$ *of order* 4.

It is clear that the groups G occurring in Theorem 3.4(iii) are isoclinic to certain groups of order 32. Using Magma (or GAP), one can easily show that such groups of order 32 are SmallGroup(32,k) for $k = 6, 7, 8$ in the small group library.

We conclude this section with providing some different type of bounds on the central quotient of a given group. If $|\gamma_2(G)\,Z(G)/Z(G)| = n$ is finite for a group G, then it follows from [22, Theorem 1] that $|G/Z_2(G)| \leq n^{2\,log_2 n}$. Using this and Lemma 3.1 we can also provide an upper bound on the size of $G/Z(G)$ in terms of n, the rank of $Z_2(G)/Z(G)$ and exponents of certain sets of commutators (here these sets are really subgroups of G) of coset representatives of generators of $Z_2(G)/Z(G)$ with the elements of G. This is given in the following theorem.

Theorem 3.5 *Let G be an arbitrary group. Let $|\gamma_2(G)\,Z(G)/Z(G)| = n$ is finite and $Z_2(G)/Z(G)$ is finitely generated by $x_1 Z(G), x_2 Z(G), \ldots, x_t Z(G)$ such that $exp([x_i, G])$ is finite for $1 \leq i \leq t$. Then*

$$|G/Z(G)| \leq n^{2\,log_2 n} \prod_{i=1}^{t} exp([x_i, G]).$$

4 Problems and Examples

Theorem B provides some conditions on a group G under which $G/Z(G)$ becomes finite. It is interesting to solve

Problem 1. Let G be an arbitrary group. Provide a set $\mathcal{C}$ of optimal conditions on G such that $G/Z(G)$ is finite if and only if all conditions in $\mathcal{C}$ hold true.

As we know that there is no finite non-nilpotent group G admitting Property A. Since $\mathrm{Inn}(G) \leq \mathrm{Aut}_c(G)$, it is interesting to consider

Problem 2. Classify all non-nilpotent finite groups G such that $|\mathrm{Aut}_c(G)| = |\gamma_2(G)|^d$, where $d = d(G)$.

A much stronger result than Theorem 3.1 is known in the case when the Frattini subgroup of G is trivial. This is given in the following theorem of Herzog, Kaplan and Lev [10, Theorem A] (the same result is also proved independently by Halasi and Podoski in [11, Theorem 1.1]).

Theorem 4.1 *Let G be any non-abelian group with trivial Frattini subgroup. Then $|G/Z(G)| < |\gamma_2(G)|^2$.*

The following result with the assertion similar to the preceding theorem is due to Isaacs [13].

Theorem 4.2 *If G is a capable finite group with cyclic $\gamma_2(G)$ and all elements of order 4 in $\gamma_2(G)$ are central in G, then $|G : Z(G)| \leq |\gamma_2(G)|^2$. Moreover, equality holds if G is nilpotent.*

So, there do exist nilpotent groups with comparatively small central quotient. A natural problem is the following.

Problem 3. Classify all finite p-groups G such that $|G : Z(G)| \le |\gamma_2(G)|^2$.

Let G be a finite nilpotent group of class 2 minimally generated by d elements. Then it follows from Lemma 3.1 that $|G/Z(G)| \le \prod_{i=1}^{d} exp([x_i, G])$, which in turn implies

$$|G/Z(G)| \le |exp(\gamma_2(G))|^d. \tag{4.1}$$

Problem 4. Classify all finite p-groups G of nilpotency class 2 for which equality holds in (4.1).

Now we discuss some examples of infinite groups with finite central quotient. The most obvious example is the infinite cyclic group. Other obvious examples are the groups $G = H \times \mathcal{Z}$, where H is any finite group and $\mathcal{Z}$ is the infinite cyclic group. Nonobvious examples include finitely generated FC-groups, in which conjugacy class sizes are bounded, and certain Cernikov groups. We provide explicit examples in each case. Let F_n be the free group on n symbols and p be a prime integer. Then the factor group $F_n/(\gamma_2(F_n)^p\gamma_3(F_n))$ is the required group of the first type, where $\gamma_2(F_n)^p = \langle u^p \mid u \in \gamma_2(F_n)\rangle$. Now let $H = \mathcal{Z}(p^\infty) \times A$ be the direct product of quasi-cyclic (Prüfer) group $\mathcal{Z}(p^\infty)$ and the cyclic group $A = \langle a \rangle$ of order p, where p is a prime integer. Now consider the group $G = H \rtimes B$, the semidirect product of H and the cyclic group $B = \langle b \rangle$ of order p with the action by $x^b = x$ for all $x \in \mathcal{Z}(p^\infty)$ and $a^b = ac$, where c is the unique element of order p in $\mathcal{Z}(p^\infty)$. This group is suggested by V. Romankov and Rahul D. Kitture through ResearchGate, and is a Cernikov group.

The following problem was suggested by R. Baer in [1].

Problem 5. Let A and Q be two groups. Obtain necessary and sufficient conditions on A and Q so that there exists a group G with $A \cong (G)$ and $Q \cong G/Z(G)$.

This problem was solved by Baer himself for an arbitrary abelian group A and finitely generated abelian group G. Moskalenko [16] solved this problem for an arbitrary abelian group A and a periodic abelian group G. He [17] also solved this problem for arbitrary abelian group A and a non-periodic abelian group G such that the rank of $G/t(G)$ is 1, where $t(G)$ denotes the tortion subgroup of G. If this rank is more than 1, then he solved the problem when A is a torsion abelian group. For a given group Q, the existence of a group G such that $Q \cong G/Z(G)$ has been studied extensively under the theme 'Capable groups'. However, to the best of our knowledge, Problem 5 has been poorly studied in full generality. Let us restate a special case of this problem in a little different setup. A pair of groups (A, Q), where A is an arbitrary abelian group and Q is an arbitrary group, is said to be a *capable pair* if there exists a group G such that $A \cong Z(G)$ and $Q \cong G/Z(G)$. So, in our situation, the following problem is very interesting.

Problem 6. Classify capable pairs (A, Q), where A is an infinite abelian group and Q is a finite group.

Finally let us get back to the situation when G is a group with finite $\gamma_2(G)$ but infinite $G/Z(G)$. The well-known examples of such type are infinite extraspecial p-groups. Other class of examples can be obtained by taking a central product (amalgamated at their centers) of infinitely many copies of a 2-generated finite p-group of class 2 such that $\gamma_2(H) = Z(H)$ is cyclic of order q, where q is some power of p. Notice that both of these classes consist of groups of nilpotency class 2. Now if we take $G = X \times H$, where X is an arbitrary group with finite $\gamma_2(X)$ and H is with finite $\gamma_2(H)$ and infinite $H/Z(H)$, then $\gamma_2(G)$ is finite but $G/Z(G)$ is infinite. So we can construct nilpotent groups of arbitrary class and even non-nilpotent group with infinite central quotient and finite commutator subgroup.

A non-nilpotent group G is said to be *purely non-nilpotent* if it does not have any nontrivial nilpotent subgroup as a direct factor. With the help of Rahul D. Kitture, we have also been able to construct purely non-nilpotent groups G such that $\gamma_2(G)$ is finite but $G/Z(G)$ is infinite. Let H be an infinite extraspecial p-group. Then we can always find a field $\mathcal{F}_q$, where q is some power of a prime, containing all pth roots of unity. Now let K be the special linear group $sl(p, \mathcal{F}_q)$, which is a non-nilpotent group having a central subgroup of order p. Now consider the group G which is a central product of H and K amalgamated at $Z(H)$. Then G is a purely non-nilpotent group with the required conditions. It will be interesting to see more examples of this type which do not occur as a central product of such infinite groups of nilpotency class 2 with non-nilpotent groups.

By Proposition 2.1 we know that for an arbitrary group G with finite $\gamma_2(G)$ but infinite $G/Z(G)$, the group $G/Z(G)$ has an infinite abelian group as a direct factor. Further structural information is highly welcome.

Problem 7. Provide structural information of the group G such that $\gamma_2(G)$ is finite but $G/Z(G)$ is infinite?

References

1. Baer, R.: Groups with preassigned central and central quotient group. Trans. Am. Math. Soc. **44**, 387–412 (1938)
2. Bosma, W., Cannon, J., Playoust, C.: The Magma algebra system I: the user language. J. Symb. Comput. **24**, 235–265 (1997)
3. Brady, J.M., Bryce, R.A., Cossey, J.: On certain abelian-by-nilpotent varieties. Bull. Aust. Math. Soc. **1**, 403–416 (1969)
4. Cheng, Y.: On finite p-groups with cyclic commutator subgroup. Arch. Math. **39**, 295–298 (1982)
5. Dolfi, S., Pacifici, E., Yadav, M.K.: Bounds for the index of the centre in non-nilpotent groups. preprint
6. Fuchs, L.: Infinite abelian groups Vol. I. Pure and Applied Mathematics, vol. 36, Academic Press, New York (1970)
7. Gumber, D.K., Kalra, H.: On the converse of a theorem of Schur. Arch. Math. (Basel) **101**, 17–20 (2013)
8. Hall, P.: The classification of prime power groups. Journal für die reine und angewandte Mathematik **182**, 130–141 (1940)

9. Hall, P.: Finite-by-nilpotent groups. Proc. Camb. Phil. Soc. **52**, 611–616 (1956)
10. Herzog, M., Kaplan, G., Lev, A.: The size of the commutator subgroup of finite groups. J. Algebr. **320**, 980–986 (2008)
11. Halasi, Z., Podoski, K.: Bounds in groups with trivial Frattini subgroups. J. Algebr. **319**, 893–896 (2008)
12. Hilton, P.: On a theorem of Schur. Int. J. Math. Math. Sci. **28**, 455–460 (2001)
13. Isaacs, I.M.: Derived subgroups and centers of capable groups. Proc. Am. Math. Soc. **129**, 2853–2859 (2001)
14. Macdonald, I.D.: Some explicit bounds in groups with finite derived groups. Proc. London Math. Soc. 3 **11**, 23–56 (1961)
15. Magidin, A., Morse, R.F.: Certain homological functors of 2-generator p-groups of class 2. Contemp. Math. **511**, 127–166 (2010)
16. Moskalenko, A.I.: Central extensions of abelian groups by means of abelian groups. Sibirsk. Mat. Z. **9**, 104–115 (1968)
17. Moskalenko, A.I.: Central extensions of a periodic abelian group by means of an abelian group. Algebra and number theory. Moskov. Gos. Ped. Inst. Ucen. Zap. **375**, 80–84 (1971)
18. Neumann, B.H.: Groups with finite classes of conjugate elements. Proc. London Math. Soc. (3) **1**, 178–187 (1951)
19. Neumann, B.H.: Groups covered by permutable subsets. J. Lond. Math. Soc. **29**, 236–248 (1954)
20. Neumann, B.H.: A problem of Paul Erdös on groups. J. Aust. Math. Soc. **21**(A), 467–472 (1976)
21. Niroomand, P.: The converse of Schur's theorem. Arch. Math. **94**, 401–403 (2010)
22. Podoski, K., Szegedy, B.: Bounds for the index of the centre in capable groups. Proc. Am. Math. Soc. **133**, 3441–3445 (2005)
23. Schur, I.: Über die Darstellung der endlichen Gruppen durch gebrochene lineare Substitutionen. Journal für die reine und angewandte Mathematik **127**, 20–50 (1904)
24. Sury, B.: A generalization of a converse of Schur's theorem. Arch. Math. **95**, 317–318 (2010)
25. LMS: Lect. Note Ser. Bath. **388**(2011), 569–579 (2009)
26. Yadav, M.K.: Class-preserving automorphisms of finite p-groups II. Isr. J. Math. **209**, 355–396 (2015). doi:10.1007/s11856-015-1222-4

Robinson–Schensted Correspondence for the Walled Brauer Algebras and the Walled Signed Brauer Algebras

A. Tamilselvi, A. Vidhya and B. Kethesan

Abstract In this paper, we develop a Robinson–Schensted algorithm for the walled Brauer algebras which gives the bijection between the walled Brauer diagram d and the pairs of standard tri-tableaux of shape $\lambda = (\lambda_1, \lambda_2, \lambda_3)$ with $\lambda_1 = (2^f)$, $\lambda_2 \vdash r - f$ and $\lambda_3 \vdash s - f$, for $0 \leq f \leq \min(r, s)$. As a biproduct, we define a Robinson–Schensted correspondence for the walled signed Brauer algebras which gives the correspondence between the walled signed Brauer diagram d and the pairs of standard signed-tri-tableaux of shape $\lambda = (\lambda_1, \lambda_2, \lambda_3)$ with $\lambda_1 = (2^{2f})$, $\lambda_2 \vdash_b r - f$ and $\lambda_3 \vdash_b s - f$, for $0 \leq f \leq \min(r, s)$. We also derive the Knuth relations and the determinantal formula for the walled Brauer and the walled signed Brauer algebras by using the Robinson–Schensted correspondence.

Keywords Robinson–Schensted correspondence · Walled signed Brauer algebra

Mathematics Subject Classifications 05E10 · 20C30

1 Introduction

In order to characterise invariants of classical groups acting on tensor powers of the vector representations, Brauer [2] introduced a new class of algebras called Brauer algebras. The Brauer algebras used graphs to represent its basis. Hence it can be considered as a class of diagram algebras, that are finite dimensional algebras whose basis consists of diagrams. These basis have interesting combinatorial properties to be studied in their own right.

A. Tamilselvi (✉) · A. Vidhya · B. Kethesan
Ramanujan Institute for Advanced Study in Mathematics, University of Madras,
Chennai 600 005, India
e-mail: tamilselvi.riasm@gmail.com

A. Vidhya
e-mail: vidhyamath@gmail.com

B. Kethesan
e-mail: kethesanb@gmail.com

S.T. Rizvi et al. (eds.), *Algebra and its Applications*, Springer Proceedings
in Mathematics & Statistics 174, DOI 10.1007/978-981-10-1651-6_11

Parvathi and Kamaraj [8] introduced a new class of algebras called signed Brauer algebras $S_f^{(x)}$ which are a generalization of Brauer algebras. Parvathi and Selvaraj [9] studied signed Brauer algebras as a class of centraliser algebras, which are the direct product of orthogonal groups over the field of real numbers $\mathbb{R}$.

The walled Brauer algebras (also known as the rational Brauer algebras) $B_{r,s}(\delta)$ are the subalgebras of the Brauer algebras $B_{r+s}(\delta)$ which arise from a Schur–Weyl duality between $B_{r,s}(\delta)$ and $GL_\delta(\mathbb{C})$ from the actions on the mixed tensor product $V^{\otimes r} \bigotimes (V^*)^{\otimes s}$ of the natural representation (and its dual) for $GL_\delta(\mathbb{C})$. This algebra was studied by Turaev, Koike, Benkart et al., Brundan et al., Cox et al., [1, 3, 4, 7, 18].

Kethesan [6] introduced a new class of algebras called walled signed Brauer algebras $D_{r,s}$ which are subalgebras of signed Brauer algebras D_{r+s} introduced by [8]. He observed that the number of walled Brauer diagrams is the dimension of the regular representation of walled signed Brauer algebras.

Let G be the group of linear transformations on a n-dimensional vector space V. Suppose that G acts diagonally on the k-fold tensor space $V^{\otimes k}$. Then the k-fold tensor space $V^{\otimes k}$ decomposes into irreducible representation of G as centraliser algebras $End_G\left(V^{\otimes k}\right)$. This work was successfully extended to other centraliser algebras, namely Brauer algebras $End_{O(n)}\left(V^{\otimes k}\right)$ where $O(n)$ is the orthogonal group of degree n, Partition algebras $End_{S_n}\left(V^{\otimes k}\right)$ and so on.

A study on complete set of irreducible representations of these algebras have been done by many. For Brauer algebras the irreducible representations have been studied by Brown, Wenzl, etc. The dimension of the regular representation of the walled Brauer algebras is indexed by partition of shape $\lambda = (\lambda_1, \lambda_2, \lambda_3)$ with $\lambda_1 = (2^f)$, $\lambda_2 \vdash r - f$ and $\lambda_3 \vdash s - f$, for $0 \leq f \leq \min(r, s)$. In a similar way, the dimension of the regular representation of the walled signed Brauer algebras is indexed by partition of shape $\lambda = (\lambda_1, \lambda_2, \lambda_3)$ with $\lambda_1 = (2^{2f})$, $\lambda_2 \vdash_b r - f$ and $\lambda_3 \vdash_b s - f$, for $0 \leq f \leq \min(r, s)$.

Using Young diagrams and Young tableaux introduced by Alfred Young in 1900, Robinson gave a bijective correspondence between permutations and pairs of standard Young tableaux of the same shape in an attempt to prove the Littlewood–Richardson rule in [14]. Later in 1961 [16], Schensted gave the simplest description of the correspondence, whose main objective was counting permutations with given lengths of their longest increasing and decreasing subsequences. By using a combination of Robinson–Schensted–Knuth insertion and jeu-de taquin, in [5] the authors provide a bijection between sequences $\{i_1, i_2, \ldots, i_k\}$, $1 \leq i_j \leq n$ and pairs (P_λ, Q_λ) consisting of a standard Young tableau P_λ and a column strict tableau Q_λ of shape λ, thus providing a combinatorial proof of the identity $n^k = \sum_{\substack{\lambda \vdash k \\ |\lambda| \leq n}} f^\lambda d_\lambda$ where f^λ is the number of standard Young tableaux of shape λ and d_λ is the number of column strict tableau of shape λ.

This motivated us to study the Robinson–Schensted correspondence for the walled Brauer algebras and signed walled Brauer algebras.

The Robinson–Schensted correspondence for signed Brauer algebras, G-Brauer algebra and G-vertex colored Partition algebra were studied already by Parvathi and Tamilselvi [11–13].

We construct an explicit bijection between the set of walled Brauer diagrams d and the pairs of standard λ-tri-tableaux of shape $\lambda = (\lambda_1, \lambda_2, \lambda_3)$ with $\lambda_1 = (2^f)$, $\lambda_2 \vdash r - f$ and $\lambda_3 \vdash s - f$, for $0 \le f \le \min(r, s)$. We also derive the Knuth relations and the determinantal formula for the walled Brauer algebras by using the Robinson–Schensted correspondence for the standard tri-tableau of shape $\lambda = (\lambda_1, \lambda_2, \lambda_3)$ with $\lambda_1 = (2^f)$, $\lambda_2 \vdash r - f$ and $\lambda_3 \vdash s - f$, for $0 \le f \le \min(r, s)$.

We construct an explicit bijection between the set of walled signed Brauer diagrams d and the pairs of standard λ-tri-tableaux of shape $\lambda = (\lambda_1, \lambda_2, \lambda_3)$ with $\lambda_1 = (2^{2f})$, $\lambda_2 \vdash_b r - f$ and $\lambda_3 \vdash_b s - f$, for $0 \le f \le \min(r, s)$. We also derive the Knuth relations and the determinantal formula for the walled signed Brauer algebras by using the Robinson–Schensted correspondence for the standard tri-tableau of shape $\lambda = (\lambda_1, \lambda_2, \lambda_3)$ with $\lambda_1 = (2^{2f})$, $\lambda_2 \vdash_b r - f$ and $\lambda_3 \vdash_b s - f$, for $0 \le f \le \min(r, s)$.

2 Preliminaries

2.1 Basic Definitions and Results

We state the basic definitions and some known results which will be used in this paper.

Definition 2.1 ([15]) A sequence of non-negative integers $\lambda = (\lambda_1, \lambda_2, \ldots, \lambda_l)$ is called a partition of n, which is denoted by $\lambda \vdash n$, if

1. $\lambda_i \ge \lambda_{i+1}$, for every $i \ge 1$
2. $\sum_{i=1}^{l} \lambda_i = n$

The λ_i are called the parts of λ.

Definition 2.2 ([15]) Let $\lambda = (\lambda_1, \lambda_2, \ldots, \lambda_l) \vdash n$. The Young diagram of λ is an array of n dots having l left justified rows with row i containing λ_i dots for $1 \le i \le l$.

Example

$$\lambda := \begin{array}{ll} * * \cdot \cdot \cdot * & \lambda_1 \text{ nodes} \\ * * \cdot \cdot * & \lambda_2 \text{ nodes} \\ \vdots \ \vdots & \vdots \\ * * \cdots * & \lambda_l \text{ nodes} \end{array}$$

Definition 2.3 Suppose $\lambda \vdash n$. A Young tableau of shape λ, is an array t obtained by replacing the stars of the Young diagram of λ with the numbers $1, 2, \ldots, n$ bijectively.

Definition 2.4 A tableau t is standard if the entries in the tableau t is increasing along the rows and columns.

Proposition 2.5 ([15]) *If* $\lambda = (\lambda_1, \ldots, \lambda_l) \vdash n$ *then* $f^\lambda = n! \left| \frac{1}{(\lambda_i - i + j)!} \right|_{l \times l}$ *where* f^λ *is the number of standard tableaux of shape* λ.

Definition 2.6 A bipartition λ of n, denoted by $\lambda \vdash_b n$, is an ordered pair of partitions $\left(\lambda^{(1)}, \lambda^{(2)}\right)$ where $\lambda^{(1)} \vdash r$ and $\lambda^{(2)} \vdash s$ such that $r + s = n,\ r, s \geq 0$.

Proposition 2.7 ([10]) *If* $\lambda = (\lambda^{(1)}, \lambda^{(2)}) \vdash_b n$ *then*

$$f^\lambda = n! \left| \frac{1}{(\lambda_i^{(1)} - i + j)!} \right|_{l \times l} \left| \frac{1}{(\lambda_i^{(2)} - i + j)!} \right|_{n-l \times n-l}$$

where $|\lambda^{(1)}| = l$ *and* $|\lambda^{(2)}| = n - l$, f^λ *is the number of standard bitableau of shape* λ.

Definition 2.8 ([15]) A rim hook is a connected skew shape containing no 2×2 square.

Definition 2.9 ([17]) A generalised permutation is a two-line array of integers

$$x = \begin{pmatrix} i_1\ i_2\ \ldots\ i_n \\ x_1\ x_2\ \ldots\ x_n \end{pmatrix}$$

whose column are in lexicographic order, with the top entry taking precedence and $x_l \neq x_m,\ \forall\, l, m$. The set of all generalised permutations denoted by $\mathcal{GP}(n)$

Proposition 2.10 ([17]) *If* $x \in \mathcal{GP}(n)$ *then* $P(x^{-1}) = Q(x)$ *and* $Q(x^{-1}) = P(x)$ *where* $P(x)$, $P(x^{-1})$, $Q(x)$, $Q(x^{-1})$ *be the standard tableaux of shape* $\lambda \vdash n$.

Definition 2.11 ([17]) The generalised permutations x and y differ by a Knuth relation of first kind, denoted by $x \overset{1}{\sim} y$ if

$$x = x_1 \ldots x_{i-1} x_i x_{i+1} \ldots x_n \text{ and}$$
$$y = x_1 \ldots x_{i-1} x_{i+1} x_i \ldots x_n$$

such that $x_i < x_{i-1} < x_{i+1}$.

They differ by a Knuth relation of second kind, denoted by $x \overset{2}{\sim} y$ if

$$x = x_1 \ldots x_i x_{i+1} x_{i-1} \ldots x_n \text{ and}$$
$$y = x_1 \ldots x_{i+1} x_i x_{i-1} \ldots x_n$$

such that $x_i < x_{i-1} < x_{i+1}$.

The two permutations are Knuth equivalent, denoted by $x \overset{K}{\sim} y$ if there is a sequence of permutations such that

$$x = z_1 \overset{i}{\sim} z_2 \overset{j}{\sim} \cdots \overset{l}{\sim} z_k = y \text{ where } i, j, \ldots, l \in \{1, 2\}.$$

Proposition 2.12 ([17]) *If $x, y \in \mathcal{GP}(n)$ then $x \overset{K}{\sim} y \Longleftrightarrow P(x) = P(y)$ where $P(x), P(y)$ are the standard tableaux of shape λ, $\lambda \vdash n$.*

Definition 2.13 ([17]) The generalised permutations x and y differ by a dual Knuth relation of first kind, denoted by $x \overset{1^*}{\sim} y$ if

$$x = x_1 \ldots x_i \ldots x_{i-1} \ldots x_{i+1} \ldots x_n \text{ and}$$
$$y = x_1 \ldots x_{i+1} \ldots x_{i-1} \ldots x_i \ldots x_n$$

such that $x_i < x_{i-1} < x_{i+1}$.

They differ by a dual Knuth relation of second kind, denoted by $x \overset{2^*}{\sim} y$ if

$$x = x_1 \ldots x_{i-1} \ldots x_{i+1} \ldots x_i \ldots x_n \text{ and}$$
$$y = x_1 \ldots x_i \ldots x_{i+1} \ldots x_{i-1} \ldots x_n$$

such that $x_i < x_{i-1} < x_{i+1}$.

The two permutations are dual Knuth equivalent, denoted by $x \overset{K^*}{\sim} y$ if there is a sequence of permutations such that

$$x = z_1 \overset{i^*}{\sim} z_2 \overset{j^*}{\sim} \cdots \overset{l^*}{\sim} z_k = y \text{ where } i, j, \ldots, l \in \{1, 2\}.$$

Lemma 2.14 ([17]) *If $x, y \in \mathcal{GP}(n)$ then $x \overset{K}{\sim} y \Longleftrightarrow x^{-1} \overset{K^*}{\sim} y^{-1}$.*

Proposition 2.15 ([17]) *If $x, y \in \mathcal{GP}(n)$ then $x \overset{K^*}{\sim} y \Longleftrightarrow Q(x) = Q(y)$ where $Q(x), Q(y)$ are the standard tableaux of shape λ, $\lambda \vdash n$.*

Definition 2.16 ([17]) A generalised signed permutation is a two-line array of integers

$$x = \begin{pmatrix} i_1 & i_2 & \ldots & i_n \\ \varepsilon_{x_1} x_1 & \varepsilon_{x_2} x_2 & \ldots & \varepsilon_{x_n} x_n \end{pmatrix}$$

where $\varepsilon_{x_i} \in \{\pm 1\}$, $\forall i$ whose column are in lexicographic order, with the top entry taking precedence and $x_l \neq x_m$, $\forall l, m$. The set of all generalised signed permutations denoted by $\mathcal{GSP}(n)$

Proposition 2.17 ([11]) *If $x \in \mathcal{GSP}(n)$ then $P(x^{-1}) = Q(x)$ and $Q(x^{-1}) = P(x)$ where $P(x)$, $P(x^{-1})$, $Q(x)$, $Q(x^{-1})$ be the standard bi-tableaux of shape $\lambda = (\lambda_1, \lambda_2)$ where $\lambda_1 \vdash k$ and $\lambda_2 \vdash n - k$, $k \geq 0$.*

Definition 2.18 ([17]) The generalised signed permutations x and y differ by a Knuth relation of first kind, denoted by $x \overset{\widetilde{1}}{\sim} y$ if

$$x = \varepsilon_{x_1}x_1 \ldots \varepsilon_{x_{i-1}}x_{i-1}\varepsilon_{x_i}x_i\varepsilon_{x_{i+1}}x_{i+1} \ldots \varepsilon_{x_n}x_n \text{ and}$$
$$y = \varepsilon_{x_1}x_1 \ldots \varepsilon_{x_{i-1}}x_{i-1}\varepsilon_{x_{i+1}}x_{i+1}\varepsilon_{x_i}x_i \ldots \varepsilon_{x_n}x_n$$

such that $x_i < x_{i-1} < x_{i+1}$ and $\varepsilon_{x_{i-1}} = \varepsilon_{x_i} = \varepsilon_{x_{i+1}}$.
They differ by a Knuth relation of second kind, denoted by $x \overset{\widetilde{2}}{\sim} y$ if

$$x = \varepsilon_{x_1}x_1 \ldots \varepsilon_{x_i}x_i\varepsilon_{x_{i+1}}x_{i+1}\varepsilon_{x_{i-1}}x_{i-1} \ldots \varepsilon_{x_n}x_n \text{ and}$$
$$y = \varepsilon_{x_1}x_1 \ldots \varepsilon_{x_{i+1}}x_{i+1}\varepsilon_{x_i}x_i\varepsilon_{x_{i-1}}x_{i-1} \ldots \varepsilon_{x_n}x_n$$

such that $x_i < x_{i-1} < x_{i+1}$ and $\varepsilon_{x_{i-1}} = \varepsilon_{x_i} = \varepsilon_{x_{i+1}}$.
They differ by a Knuth relation of third kind, denoted by $x \overset{\widetilde{3}}{\sim} y$ if

$$x = \varepsilon_{x_1}x_1 \ldots \varepsilon_{x_i}x_i\varepsilon_{x_{i+1}}x_{i+1} \ldots \varepsilon_{x_n}x_n \text{ and}$$
$$y = \varepsilon_{x_1}x_1 \ldots \varepsilon_{x_{i+1}}x_{i+1}\varepsilon_{x_i}x_i \ldots \varepsilon_{x_n}x_n$$

such that $\varepsilon_{x_i} = -\varepsilon_{x_{i+1}}$.

The two permutations are Knuth equivalent, denoted by $x \overset{\widetilde{K}}{\sim} y$ if there is a sequence of permutations such that

$$x = z_1 \overset{i}{\sim} z_2 \overset{j}{\sim} \cdots \overset{l}{\sim} z_k = y \text{ where } i, j, \ldots, l \in \{\widetilde{1}, \widetilde{2}, \widetilde{3}\}.$$

Proposition 2.19 ([11]) *If $x, y \in \mathcal{GSP}(n)$ then $x \overset{\widetilde{K}}{\sim} y \iff P(x) = P(y)$ where $P(x)$, $P(y)$ are the standard bi-tableaux of shape $\lambda = (\lambda_1, \lambda_2)$ where $\lambda_1 \vdash k$ and $\lambda_2 \vdash n-k$, $k \geq 0$.*

Definition 2.20 ([17]) The generalised signed permutations x and y differ by a dual Knuth relation of first kind, denoted by $x \overset{\widetilde{1}^*}{\sim} y$ if

$$x = \varepsilon_{x_1}x_1 \ldots \varepsilon_{x_i}x_i \ldots \varepsilon_{x_{i-1}}x_{i-1} \ldots \varepsilon_{x_{i+1}}x_{i+1} \ldots \varepsilon_{x_n}x_n \text{ and}$$
$$y = \varepsilon_{x_1}x_1 \ldots \varepsilon_{x_{i+1}}x_{i+1} \ldots \varepsilon_{x_{i-1}}x_{i-1} \ldots \varepsilon_{x_i}x_i \ldots \varepsilon_{x_n}x_n$$

such that $x_i < x_{i-1} < x_{i+1}$ and $\varepsilon_{x_{i-1}} = \varepsilon_{x_i} = \varepsilon_{x_{i+1}}$.
They differ by a dual Knuth relation of second kind, denoted by $x \overset{\widetilde{2}^*}{\sim} y$ if

$$x = \varepsilon_{x_1}x_1 \ldots \varepsilon_{x_{i-1}}x_{i-1} \ldots \varepsilon_{x_{i+1}}x_{i+1} \ldots \varepsilon_{x_i}x_i \ldots \varepsilon_{x_n}x_n \text{ and}$$
$$y = \varepsilon_{x_1}x_1 \ldots \varepsilon_{x_i}x_i \ldots \varepsilon_{x_{i+1}}x_{i+1} \ldots \varepsilon_{x_{i-1}}x_{i-1} \ldots \varepsilon_{x_n}x_n$$

such that $x_i < x_{i-1} < x_{i+1}$ and $\varepsilon_{x_{i-1}} = \varepsilon_{x_i} = \varepsilon_{x_{i+1}}$.

They differ by a dual Knuth relation of third kind, denoted by $x \overset{\widetilde{3}^*}{\sim} y$ if

$$x = \varepsilon_{x_1} x_1 \ldots \varepsilon_{x_i} x_i \ldots \varepsilon_{x_{i+1}} x_{i+1} \ldots \varepsilon_{x_n} x_n \text{ and}$$
$$y = \varepsilon_{x_1} x_1 \ldots \varepsilon_{x_i} x_{i+1} \ldots \varepsilon_{x_{i+1}} x_i \ldots \varepsilon_{x_n} x_n$$

such that $\varepsilon_{x_i} = -\varepsilon_{x_{i+1}}$.

The two permutations are dual Knuth equivalent, denoted by $x \overset{\widetilde{K}^*}{\sim} y$ if there is a sequence of permutations such that

$$x = z_1 \overset{i^*}{\sim} z_2 \overset{j^*}{\sim} \cdots \overset{l^*}{\sim} z_k = y \text{ where } i, j, \ldots, l \in \{\widetilde{1}, \widetilde{2}, \widetilde{3}\}.$$

Lemma 2.21 *If* $x, y \in \mathcal{GSP}(n)$ *then* $x \overset{\widetilde{K}}{\sim} y \Longleftrightarrow x^{-1} \overset{\widetilde{K}^*}{\sim} y^{-1}$.

Proposition 2.22 ([11]) *If* $x, y \in \mathcal{GSP}(n)$ *then* $x \overset{\widetilde{K}^*}{\sim} y \Longleftrightarrow Q(x) = Q(y)$ *where* $Q(x), Q(y)$ *are the standard bi-tableaux of shape* $\lambda = (\lambda_1, \lambda_2)$ *where* $\lambda_1 \vdash k$ *and* $\lambda_2 \vdash n - k$, $k \geq 0$.

2.2 Walled Brauer Algebras

Definition 2.23 ([3, 4]) A walled Brauer diagram is a diagram on $2(r+s)$ vertices with $r+s$ edges, vertices being arranged in two rows each row consisting of $r+s$ vertices. Partition the basis diagram of vertices with a wall separating the first r vertices in the upper and lower row from the remaining vertices. Then the walled Brauer algebras $B_{r,s}$ is the subalgebra with basis of those Brauer diagrams such that each vertex must be connected to exactly one other vertex by an edge; edges can cross transversally, no triple intersections. We partition the diagram with a wall. Horizontal edges must cross the wall, i.e., the edges with one vertex in the left side of the wall and other vertex in the right side of the wall in the same row. Whereas, the vertical edges should be either in the left side of the wall or in the right side of the wall whose vertices are in different row.

Example The diagram in $B_{4,3}$ is

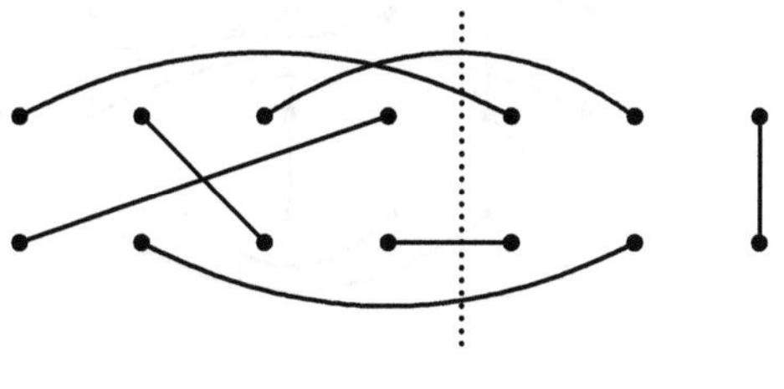

Note that $B_{0,n}(\delta) \cong B_{n,0}(\delta) \cong kS_n$, the group algebras of the symmetric group S_n on n letters. Clearly, the walled Brauer algebras $B_{r,s}(\delta)$ is a subalgebra of the Brauer algebras $B_{r+s}(\delta)$.

Multiplication on Walled Brauer Algebras

Let $B_{r,s}$ denote the set of walled Brauer diagrams on $2(r+s)$ vertices. Let $d_1,\ d_2 \in B_{r,s}$. The multiplication of two diagrams is defined as follows:

1. Place d_1 above d_2.
2. join the ith lower vertex of d_1 with ith upper vertex of d_2
3. Let d_3 be the resulting graph obtain without loops. Then $d_1 d_2 = \delta^r d_3$, where r is the number of loops, and δ is an indeterminate.

Example The diagram in $d_1 d_2$ is

$d_1 =$ 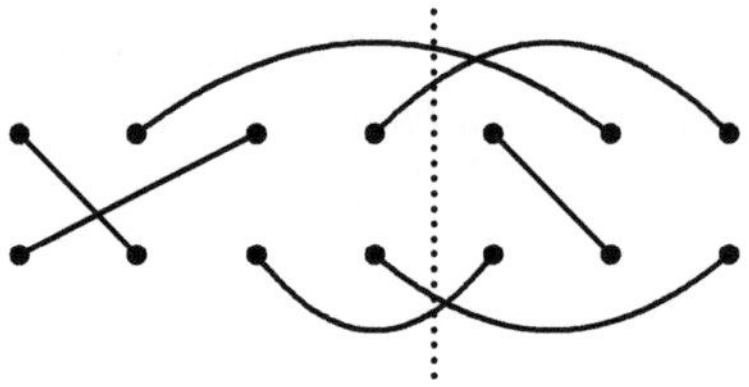

$d_2 =$ 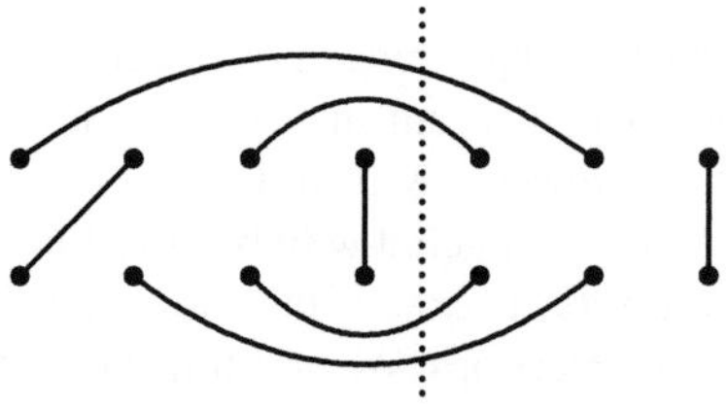

$d_1 d_2 =$ 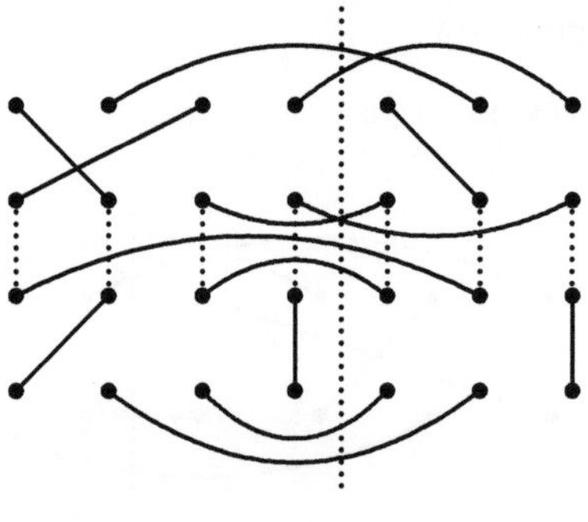

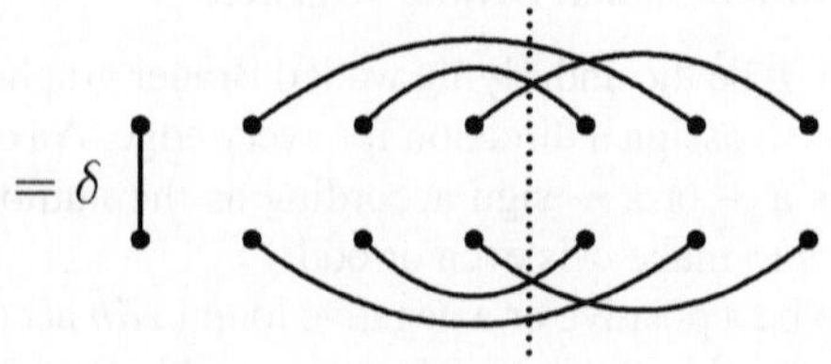

The irreducible representations of the walled Brauer algebras $B_{r,s}(\delta)$ are indexed by tri-partition $\lambda = (\lambda_1, \lambda_2, \lambda_3)$ with $\lambda_1 = (2^f)$, $\lambda_2 \vdash r - f$ and $\lambda_3 \vdash s - f$, for $0 \le f \le \min(r, s)$.

2.3 Walled Signed Brauer Algebras

Definition 2.24 ([6]) A walled signed Brauer diagram is a diagram on $2(r + s)$ vertices with $r + s$ signed edges, the vertices being arranged in two rows each row consisting of $r + s$ vertices. Partition the basis diagram of vertices with a wall separating the first r vertices in the upper and lower row from the remaining vertices. Then the walled signed Brauer algebras $D_{r,s}$ is the subalgebra with basis of those of the signed Brauer diagrams such that each vertex must be connected to exactly one other vertex by a signed edge; signed edges can cross transversally, no triple intersections. We partition the diagram with a wall. Signed horizontal edges must cross the wall, i.e., the signed edges with one vertex in the left side of the wall and other vertex in the right side of the wall in the same row. Whereas, the signed vertical edge should be either in the left side of the wall or in the right side of the wall whose vertices are in different row.

Example The diagram in $D_{4,3}$ is

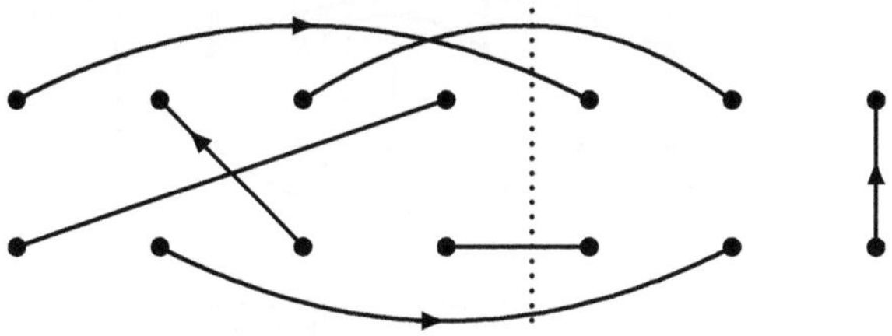

Note that $D_{0,n}(\delta) \cong D_{n,0}(\delta) \cong k\widetilde{S}_n$, the group algebras of the hyperoctahedral group of type of B_n on n letters. Clearly, the walled signed Brauer algebras $D_{r,s}(\delta)$ is a subalgebra of the signed Brauer algebras $D_{r+s}(\delta)$ and also the generalization of walled Brauer algebras.

Multiplication on Walled Signed Brauer Algebras

Let $\overline{a}, \overline{b} \in D_{r,s}$. Let a, b be the underlying walled Brauer graphs, $ab = \delta^d c$, the only thing we have to do is to assign a direction for every edge. An edge α in the product $\overline{a}\overline{b}$ will be labeled as a $+$ or a $-$ sign according as the number of negative edges involved from $\overline{a}$ and $\overline{b}$ to make α is even or odd.

A loop β is said to be a positive or a negative loop in $\overline{a}\overline{b}$ according as the number of negative edges involved in the loop β is even or odd. Then $\overline{a}\overline{b} = \delta^{2d_1+d_2}\overline{c}$, where d_1 is the number of positive loops, d_2 is the number of negative loops and $\overline{c}$ is the walled signed Brauer diagram obtained by multiplication of $\overline{a}$ and $\overline{b}$.

Example The diagram in $\overline{a}\overline{b}$ is

$\overline{a} =$

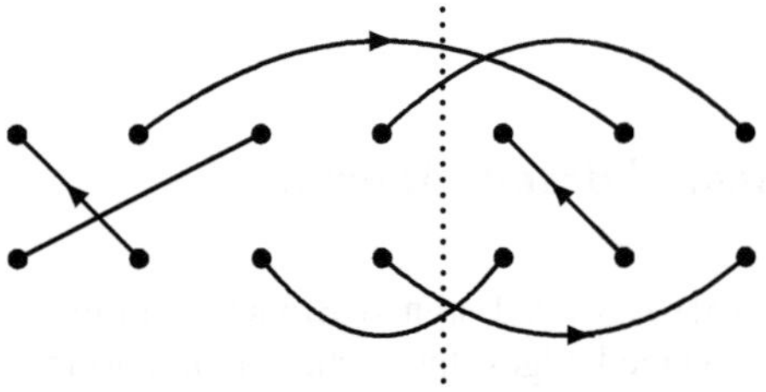

$\overline{b} =$

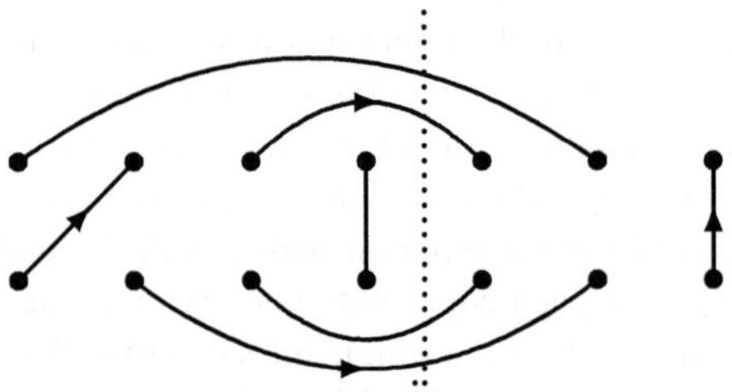

$\overline{a}\overline{b} =$

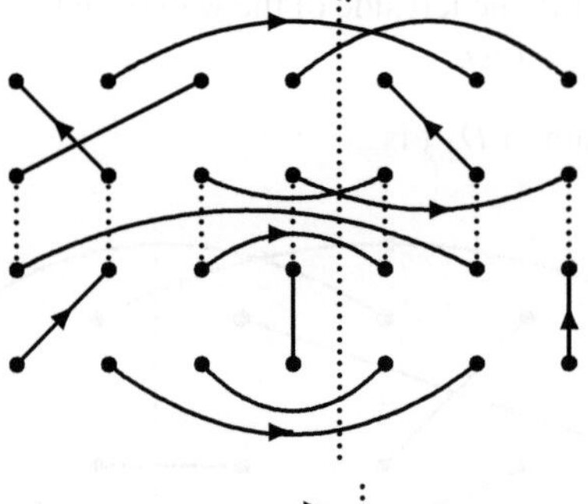

$= \delta$

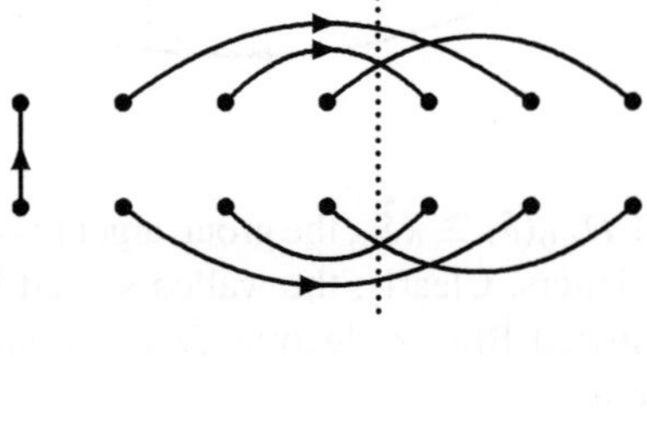

The irreducible representations of the walled signed Brauer algebras $D_{r,s}(\delta)$ are indexed by signed tri-partition of shape $\lambda = (\lambda_1, \lambda_2, \lambda_3)$ with $\lambda_1 = (2^{2f})$, $\lambda_2 \vdash_b r - f$ and $\lambda_3 \vdash_b s - f$, for $0 \leq f \leq \min(r, s)$.

3 The Robinson–Schensted Correspondence for the Walled Brauer Algebras

3.1 The Robinson–Schensted Correspondence

In this section, we define a Robinson–Schensted algorithm for the walled Brauer algebras which gives the correspondence between the walled Brauer diagram d and the pairs of standard tri-tableaux of shape $\lambda = (\lambda_1, \lambda_2, \lambda_3)$ with $\lambda_1 = (2^f)$, $\lambda_2 \vdash r - f$ and $\lambda_3 \vdash s - f$, for $0 \leq f \leq \min(r, s)$.

Definition 3.1 A tripartition ν of n will be an ordered triple of partitions (ν_1, ν_2, ν_3) where $\nu_1 = (2^f)$, $\nu_2 \vdash r - f$ and $\nu_3 \vdash s - f$, for $0 \leq f \leq \min(r, s)$, for $r + s = n$.

Definition 3.2 A standard block is defined as the block consisting of two nodes $d^{(1)}, d^{(2)}$ adjacent to each other such that $d^{(1)} < d^{(2)}$. i.e. $\boxed{d^{(1)}\ d^{(2)}}$. We call $d^{(1)}$ as the first node of the block and $d^{(2)}$ as the second node of the block.

Definition 3.3 A block tableau of shape 2^f is a tableau consisting of standard blocks one below the other.

Definition 3.4 A column standard block tableau of shape 2^f is a block tableau of shape 2^f if the first nodes of each block are increasing read from top to bottom.

Definition 3.5 A tri-tableau is a triple (t_1, t_2, t_3) where t_1, t_2, t_3 are any tableau.

Definition 3.6 A standard tri-tableau (t_1, t_2, t_3) is a tri-tableau (t_1, t_2, t_3) where t_1 is a column standard block tableau and t_2, t_3 are standard tableaux.

Definition 3.7 Given a walled Brauer diagram $d \in B_{r,s}$, we may associate a quadruple $[d_1, d_2, w_1, w_2]$ such that

$$\begin{aligned}
d_1 &= \{\ (i, d_1(i))|\ \text{the edge joining the vertices } i \text{ and } d_1(i) \text{ in the first row}\} \\
&= \{(i_1, d_1(i_1)), (i_2, d_1(i_2)), \ldots, (i_f, d_1(i_f))\} \\
d_2 &= \{\ (j, d_2(j))|\ \text{the edge joining the vertices } j \text{ and } d_2(j) \text{ in the second} \\
&\qquad \text{row}\ \} \\
&= \{(i_1, d_2(i_1)), (i_2, d_2(i_2)), \ldots, (i_f, d_2(i_f))\} \\
w_1 &= \{\ (k, w_1(k))|\ \text{the edge joining the vertex } k \text{ left to the wall in the first} \\
&\qquad \text{row and the vertex } w_1(k) \text{ left to the wall in the second row}\}
\end{aligned}$$

$$= \{(i_1, w_1(i_1)), \ldots, (i_r, w_1(i_{r-f}))\}$$

$$w_2 = \{\ (l, w_2(l)) |\ \text{the edge joining the vertex } l \text{ right to the wall in the first row and the vertex } w_2(l) \text{ right to the wall in the second row}\}$$

$$= \{(i_1, w_2(i_1)), \ldots, (i_s, w_2(i_{s-f}))\}$$

such that $i_1 < i_2 < \ldots < i_{r-f}$ and $j_1 < j_2 < \ldots < j_{s-f}$ where f is the number of horizontal edges in a row of d, $r - f$ is the number of vertical edges left to the wall of d and $s - f$ is the number of vertical edges right to the wall of d.

Theorem 3.8 *The map*

$$d \overset{R-S}{\underset{B_{r,s}}{\longleftrightarrow}} [(P_1(d), P_2(d), P_3(d)), (Q_1(d), Q_2(d), Q_3(d))]$$

provides a bijection between the set of walled Brauer diagrams d and the pairs of standard λ*-tri-tableaux of shape* $\lambda = (\lambda_1, \lambda_2, \lambda_3)$ *with* $\lambda_1 = (2^f)$, $\lambda_2 \vdash r - f$ *and* $\lambda_3 \vdash s - f$, *for* $0 \le f \le \min(r, s)$.

Proof We first descibe the map that given a diagram $d \in B_{r,s}$, produces a pair of tri-tableaux. "$d \overset{R-S}{\underset{B_{r,s}}{\longleftrightarrow}} [(P_1(d), P_2(d), P_3(d)), (Q_1(d), Q_2(d), Q_3(d))]$"

We construct a sequence of tableaux

$$\emptyset = P_1^0, P_1^1, \ldots, P_1^f$$
$$\emptyset = Q_1^0, Q_1^1, \ldots, Q_1^f$$
$$\emptyset = P_2^0, P_2^1, \ldots, P_2^{r-f}$$
$$\emptyset = Q_2^0, Q_2^1, \ldots, Q_2^{r-f}$$
$$\emptyset = P_3^0, P_3^1, \ldots, P_3^{s-f}$$
$$\emptyset = Q_3^0, Q_3^1, \ldots, Q_3^{s-f}$$

where f is the number of horizontal edges of d, $r - f$ is the number of vertical edges left to the wall of d and $s - f$ is the number of vertical edges right to the wall of d. The edges joining the vertices (x_1, x_2) are inserted into $P_1(d)$, $P_2(d)$, $P_3(d)$, $Q_1(d)$ and placed in $Q_2(d)$, $Q_3(d)$ so that shP_1^i =shQ_1^i, for all i, shP_1^j =shQ_1^j, for all j and shP_2^k =shQ_2^k, for all k.

Begin with the tableau $P_1^0 = P_2^0 = P_3^0 = Q_1^0 = Q_2^0 = Q_3^0 = \emptyset$. Then, recursively define the standard tableau by the following:

If $(l', m') \in d_2$ then P_1^k = insertion of (l, m) in P_1^{k-1}.

If $(l, m) \in d_1$ then Q_1^k = insertion of (l, m) in Q_1^{k-1}.

If $(l, m') \in w_1$ then P_2^k = insertion of m in P_2^{k-1} and place l in Q_2^{k-1} where the insertion terminates in P_2^{k-1} when m is inserted.

If $(l, m') \in w_2$ then P_3^k = insertion of m in P_3^{k-1} and place l in Q_3^{k-1} where the insertion terminates in P_3^{k-1} when m is inserted.

The operations of insertion and placement will now be described.

First, we give the insertion on $P_1(d)$. Let $(i_k, d_2(i_k)) \in d_2$ and i_k, $d_2(i_k)$ be the elements not in $P_1(d)$. To insert i_k, $d_2(i_k)$ into $P_1(d)$, we proceed as follows. Place the block containing i_k, $d_2(i_k)$ below the block containing i_{k-1}, $d_2(i_{k-1})$. Insertion on $Q_1(d)$ is the same as in $P_1(d)$.

Now we give the insertion on $P_2(d)$. Let $(i_k, w_1(i_k)) \in w_1$ and $w_1(i_k)$ be the element not in $P_2(d)$. To insert $w_1(i_k)$ into $P_2(d)$, we proceed as follows.

RS1 Set $R :=$ the first row of $P_2(d)$.
RS2 While $w_1(i_k)$ is less than some element of row R, do

RSa Let y be the smallest element of R greater than $w_1(i_k)$ and replace y by $w_1(i_k)$ in R.
RSb Set $w_1(i_k) := y$ and $R :=$ the next row down.

RS3 Now $w_1(i_k)$ is greater than every element of R, so place $w_1(i_k)$ at the end of the row R and stop.

The placement of i_k in $Q_2(d)$ is even easier than insertion. Suppose that $Q_2(d)$ is a partial tableau of shape μ and if i_k is greater than every element of $Q_2(d)$, then place i_k in $Q_2(d)$ along the cell where the insertion in $P_2(d)$ terminates.

Let $(i_k, w_2(i_k)) \in w_2$. The insertion of $w_2(i_k)$ in $P_3(d)$ and placement of i_k in $Q_3(d)$ are the same as in $P_2(d)$ and $Q_3(d)$, respectively.

Hence $d \xrightarrow[B_{r,s}]{R-S} [(P_1(d), P_2(d), P_3(d)), (Q_1(d), Q_2(d), Q_3(d))]$.

To prove "$[(P_1(d), P_2(d), P_3(d)), (Q_1(d), Q_2(d), Q_3(d))] \xleftarrow[B_{r,s}]{R-S} d$". We merely reverse the preceding algorithm step by step. We begin by defining

$$\left(P_1^f, P_2^{r-f}, P_3^{s-f}\right) = (P_1(d), P_2(d), P_3(d))$$
$$\left(Q_1^f, Q_2^{r-f}, Q_3^{s-f}\right) = (Q_1(d), Q_2(d), Q_3(d))$$

where f is the number of horizontal edges of d, $r - f$ is the number of vertical edges left to the wall of d and $s - f$ is the number of vertical edges right to the wall of d.

To recover all the elements of w_1 using the following rules.

Assuming that P_2^k, Q_2^k has been constructed, we will find $w_1(i_k)$ (the kth element of w_1) and P_2^{k-1}, Q_2^{k-1}. We write the (i, j) entry of P_2^k as $P_2^{i,j}$.

Find the cell (i, j) containing i_k in Q_2^k. Since this is the largest element in Q_2^k, $P_2^{i,j}$ must have been the last element to be displaced in the construction of P_2^k. We can now use the following procedure to delete $P_2^{i,j}$ from $P_2(d)$. For convenience, we assume the existence of an empty zeroth row above the first row of P_2^k.

SR1 Set $x := P_2^{i,j}$ and erase $P_2^{i,j}$.
Set $R :=$ the $(i-1)$ row of P_2^k.
SR2 While R is not the zeroth row of P_2^k, do

SRa Let y be the largest element of R smaller than x and replace y by x in R.
SRb Set $x := y$ and $R :=$ the next row down up.

SR3 Now x has been removed from the first row, so set $w_1(i_k) := x$.

It is easy to see that P_2^{k-1} is P_2^k after the deletion process just described is complete and Q_2^{k-1} is Q_2^k with k erased. Continuing in this way, we eventually recover all the elements of w_1 in reverse order. We can recover all the elements of w_2 as in w_1.

We are yet to find the elements in d_1, d_2.

We may recover the elements of d_2 such that the pair $(x_k, d_2(x_k))$ is the block in the cells $((2k, 1), (2k, 2))$ of $P_1(d)$, for every k.

Similarly, we may recover the elements of d_1 such that the pair $(x_k, d_1(x_k))$ is the element in the cells $((2k, 1), (2k, 2))$ of $Q_1(d)$, for every k.

Thus we recover the quadruple $[d_1, d_2, w_1, w_2]$ from the pair of tri-tableaux $[(P_1(d), P_2(d), P_3(d)), (Q_1(d), Q_2(d), Q_3(d))]$.

Hence $[(P_1(d), P_2(d), P_3(d)), (Q_1(d), Q_2(d), Q_3(d))] \overset{R-S}{\underset{B_{r,s}}{\longleftarrow}} d$ which completes the proof. □

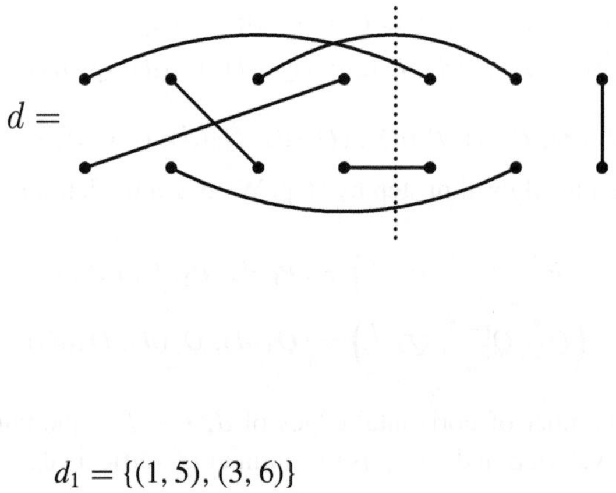

Example Let

$$d_1 = \{(1, 5), (3, 6)\}$$
$$d_2 = \{(2', 6'), (4', 5')\}$$
$$w_1 = \{(2, 3'), (4, 1')\}$$
$$w_2 = \{(7, 7')\}$$

$$P_1^0 = \emptyset \qquad P_1^1 = \boxed{2\,6} \qquad P_1^2 = \begin{array}{|c|}\hline 2\,6 \\ \hline 4\,5 \\ \hline\end{array}$$

$$Q_1^0 = \emptyset \qquad Q_1^1 = \boxed{1\,5} \qquad Q_1^2 = \begin{array}{|c|}\hline 1\,5 \\ \hline 3\,6 \\ \hline\end{array}$$

$$P_2^0 = \emptyset \qquad P_2^1 = \boxed{3} \qquad P_2^2 = \begin{array}{|c|}\hline 1 \\ \hline 3 \\ \hline\end{array}$$

$$Q_2^0 = \emptyset \qquad Q_2^1 = \boxed{2} \qquad Q_2^2 = \begin{array}{|c|}\hline 2 \\ \hline 4 \\ \hline\end{array}$$

$$P_3^0 = \emptyset \qquad P_3^1 = \boxed{7}$$

$$Q_3^0 = \emptyset \qquad Q_3^1 = \boxed{7}$$

Thus

$$d \overset{R-S}{\underset{B_{r,s}}{\longleftrightarrow}} \left[\left(\begin{array}{|c|}\hline 2\,6 \\ \hline 4\,5 \\ \hline\end{array}, \begin{array}{|c|}\hline 1 \\ \hline 3 \\ \hline\end{array}, \boxed{7}\right), \left(\begin{array}{|c|}\hline 1\,5 \\ \hline 3\,6 \\ \hline\end{array}, \begin{array}{|c|}\hline 2 \\ \hline 4 \\ \hline\end{array}, \boxed{7}\right)\right]$$

Definition 3.9 The flip of any walled Brauer diagram d is the diagram of d reflected over its horizontal axis, which is denoted by flip(d).

Proposition 3.10 *Let $d \in B_{r,s}$. If*

$$d \overset{R-S}{\underset{B_{r,s}}{\longrightarrow}} [(P_1(d), P_2(d), P_3(d)), (Q_1(d), Q_2(d), Q_3(d))].$$

Then flip(d) $\overset{R-S}{\underset{B_{r,s}}{\longrightarrow}} [(Q_1(d), Q_2(d), Q_3(d)), (P_1(d), P_2(d), P_3(d))]$ where $P_1(d)$, $Q_1(d)$ are the column standard block tableaux and $P_2(d)$, $P_3(d)$, $Q_2(d)$, $Q_3(d)$ are the standard tableaux constructed by the above insertion.

Proof Suppose $d \in B_{r,s}$, then we can recover the triple $[d_1, d_2, w_1, w_2]$ by the Definition 3.7. By the definition flip (d) has the triple $[d_2, d_1, w_1^{-1}, w_2^{-1}]$. Hence the proof follows by Proposition 2.10. □

3.2 The Knuth relations

In this section, we derive the Knuth relations for the walled Brauer algebras by using the Robinson-Schensted correspondence for the standard λ-tri-tableaux of shape $\lambda = (\lambda_1, \lambda_2, \lambda_3)$ with $\lambda_1 = (2^f)$, $\lambda_2 \vdash r - f$ and $\lambda_3 \vdash s - f$, for $0 \leq f \leq \min(r, s)$.

Definition 3.11 Let $d, d' \in B_{r,s}$. Then d and d' are Knuth equivalent, denoted by $d \overset{K_{B_{r,s}}}{\sim} d'$ if the following condition holds.

1. $d_2 = d_2'$.
2. $w_1 \overset{K}{\sim} w_1'$ where $w_1, w_1' \in \mathcal{GP}(a)$, a is the number of vertical edges left to the wall of both d and d'.
3. $w_2 \overset{K}{\sim} w_2'$ where $w_2, w_2' \in \mathcal{GP}(b)$, b is the number of vertical edges right to the wall of both d and d'.

Proposition 3.12 *Let $d, d' \in B_{r,s}$. Then*

$$d \overset{K_{B_{r,s}}}{\sim} d' \iff (P_1(d), P_2(d), P_3(d)) = (P_1(d'), P_2(d'), P_3(d'))$$

where $P_1(d), P_1(d')$ are the column standard block tableaux of shape $\lambda_1 = (2^f)$, $P_2(d), P_2(d')$ are the standard tableaux of shape λ_2, $\lambda_2 \vdash r - f$, $P_3(d), P_3(d')$ are the standard tableaux of shape λ_3, $\lambda_3 \vdash s - f$ and $0 \leq f \leq \min(r, s)$.

Proof The proof follows from the Definition 3.11 and by the Proposition 2.19 □

Definition 3.13 Let $d, d' \in B_{r,s}$. Then d and d' are dual Knuth equivalent, denoted by $d \overset{K^*_{B_{r,s}}}{\sim} d'$ if the following condition holds.

1. $d_1 = d_1'$.
2. $w_1 \overset{K^*}{\sim} w_1'$ where $w_1, w_1' \in \mathcal{GP}(a)$, a is the number of vertical edges left to the wall of both d and d'.
3. $w_2 \overset{K^*}{\sim} w_2'$ where $w_2, w_2' \in \mathcal{GP}(b)$, b is the number of vertical edges right to the wall of both d and d'.

Proposition 3.14 *Let $d, d' \in B_{r,s}$. Then*

$$d \overset{K^*_{B_{r,s}}}{\sim} d' \iff (Q_1(d), Q_2(d), Q_3(d)) = (Q_1(d'), Q_2(d'), Q_3(d'))$$

where $Q_1(d), Q_1(d')$ are the column standard block tableaux of shape $\lambda_1 = (2^f)$, $Q_2(d), Q_2(d')$ are the standard tableaux of shape λ, $\lambda \vdash r - f$, $Q_3(d), Q_3(d')$ are the standard tableaux of shape μ, $\mu \vdash s - f$ and $0 \leq f \leq \min(r, s)$.

Proof The proof follows from the Definition 3.13 and by the Proposition 2.22 □

3.3 The Determinantal Formula

In this section, we constructed the determinantal formula, which gives the dimensions of the irreducible representations of walled Brauer algebras.

Set $\frac{1}{r!} = 0$ if $r < 0$.

Lemma 3.15 $\begin{vmatrix} \frac{1}{1!} & \frac{1}{2!} & \frac{1}{3!} & \frac{1}{4!} & \cdots & \frac{1}{(f-1)!} & \frac{1}{f!} \\ \frac{1}{0!} & \frac{1}{1!} & \frac{1}{2!} & \frac{1}{3!} & \cdots & \frac{1}{(f-2)!} & \frac{1}{(f-1)!} \\ 0 & \frac{1}{0!} & \frac{1}{1!} & \frac{1}{2!} & \cdots & \frac{1}{(f-3)!} & \frac{1}{(f-2)!} \\ 0 & 0 & \frac{1}{0!} & \frac{1}{1!} & \cdots & \frac{1}{(f-4)!} & \frac{1}{(f-3)!} \\ \vdots & \vdots & \vdots & \vdots & \vdots & \vdots & \vdots \\ 0 & 0 & 0 & 0 & \cdots & \frac{1}{0!} & \frac{1}{1!} \end{vmatrix} = \frac{1}{f!}$, *where f is a positive integer.*

Proof Let $A_f = \begin{pmatrix} \frac{1}{1!} & \frac{1}{2!} & \frac{1}{3!} & \frac{1}{4!} & \cdots & \frac{1}{(f-1)!} & \frac{1}{f!} \\ \frac{1}{0!} & \frac{1}{1!} & \frac{1}{2!} & \frac{1}{3!} & \cdots & \frac{1}{(f-2)!} & \frac{1}{(f-1)!} \\ 0 & \frac{1}{0!} & \frac{1}{1!} & \frac{1}{2!} & \cdots & \frac{1}{(f-3)!} & \frac{1}{(f-2)!} \\ 0 & 0 & \frac{1}{0!} & \frac{1}{1!} & \cdots & \frac{1}{(f-4)!} & \frac{1}{(f-3)!} \\ \vdots & \vdots & \vdots & \vdots & \vdots & \vdots & \vdots \\ 0 & 0 & 0 & 0 & \cdots & \frac{1}{0!} & \frac{1}{1!} \end{pmatrix}$.

We prove the lemma by induction.

From the definition of A_f, it is clear that $|A_1| = \frac{1}{1!}$ and $|A_2| = \frac{1}{2!}$.

Assume the lemma is true for $f - 1$. That is, $|A_{f-1}| = \frac{1}{(f-1)!}$.

To prove $|A_f| = \frac{1}{f!}$.

Consider

$$|A_f| = \begin{vmatrix} \frac{1}{1!} & \frac{1}{2!} & \frac{1}{3!} & \frac{1}{4!} & \cdots & \frac{1}{(f-1)!} & \frac{1}{f!} \\ \frac{1}{0!} & \frac{1}{1!} & \frac{1}{2!} & \frac{1}{3!} & \cdots & \frac{1}{(f-2)!} & \frac{1}{(f-1)!} \\ 0 & \frac{1}{0!} & \frac{1}{1!} & \frac{1}{2!} & \cdots & \frac{1}{(f-3)!} & \frac{1}{(f-2)!} \\ 0 & 0 & \frac{1}{0!} & \frac{1}{1!} & \cdots & \frac{1}{(f-4)!} & \frac{1}{(f-3)!} \\ \vdots & \vdots & \vdots & \vdots & \vdots & \vdots & \vdots \\ 0 & 0 & 0 & 0 & \ldots & \frac{1}{0!} & \frac{1}{1!} \end{vmatrix}$$

$$= \frac{1}{1!}|A_{f-1}| - \frac{1}{0!} \begin{vmatrix} \frac{1}{2!} & \frac{1}{3!} & \frac{1}{4!} & \cdots & \frac{1}{(f-1)!} & \frac{1}{f!} \\ \frac{1}{0!} & \frac{1}{1!} & \frac{1}{2!} & \cdots & \frac{1}{(f-3)!} & \frac{1}{(f-2)!} \\ 0 & \frac{1}{0!} & \frac{1}{1!} & \cdots & \frac{1}{(f-4)!} & \frac{1}{(f-3)!} \\ \vdots & \vdots & \vdots & \vdots & \vdots & \vdots \\ 0 & 0 & 0 & \ldots & \frac{1}{0!} & \frac{1}{1!} \end{vmatrix}$$

$$= \frac{1}{1!}|A_{f-1}| - \frac{1}{2!}|A_{f-2}| + \frac{1}{0!} \begin{vmatrix} \frac{1}{3!} & \frac{1}{4!} & \cdots & \frac{1}{(f-1)!} & \frac{1}{f!} \\ \frac{1}{0!} & \frac{1}{1!} & \cdots & \frac{1}{(f-4)!} & \frac{1}{(f-3)!} \\ \vdots & \vdots & \vdots & \vdots & \vdots \\ 0 & 0 & \ldots & \frac{1}{0!} & \frac{1}{1!} \end{vmatrix}$$

$$= \frac{1}{1!}|A_{f-1}| - \frac{1}{2!}|A_{f-2}| + \frac{1}{3!}|A_{f-3}| - \cdots + (-1)^{f-1}\frac{1}{0!} \begin{vmatrix} \frac{1}{(f-1)!} & \frac{1}{f!} \\ 0 & \frac{1}{1!} \end{vmatrix}$$

$$= \frac{1}{1!}|A_{f-1}| - \frac{1}{2!}|A_{f-2}| + \frac{1}{3!}|A_{f-3}| - \cdots + (-1)^{f-1}\frac{1}{(f-1)! \cdot 1!}$$
$$+ (-1)^f \frac{1}{(f)! \cdot 0!}$$

$$= \frac{1}{1! \cdot (f-1)!} - \frac{1}{2! \cdot (f-2)!} + \frac{1}{3! \cdot (f-3)!} - \cdots + (-1)^{f-1}\frac{1}{(f-1)! \cdot 1!}$$
$$+ (-1)^f \frac{1}{(f)! \cdot 0!} \text{ (by induction)}$$

$$= \frac{1}{f!}\left(fC_1 - fC_2 + fC_3 - \cdots + (-1)^{f-1} fC_{f-1} + (-1)^f fC_f\right)$$
$$= \frac{1}{f!}.$$

Hence the proof. □

Theorem 3.16 (Determinantal Formula) *If* $\rho = (\lambda, \mu, \nu)$ *with* $\lambda = (2^f)$, $\mu \vdash r - f$, $\nu \vdash s - f$ *for fixed* $r \geq 0,\ s \geq 0$ *where* $0 \leq f \leq \min(r, s)$ *then*

$$h^\rho = r!s! \left| \frac{1}{(\lambda_i - i + j - 1)!} \right|_{f \times f} \left| \frac{1}{(\mu_i - i + j)!} \right|_{r-f \times r-f} \left| \frac{1}{(\nu_i - i + j)!} \right|_{s-f \times s-f}$$

where h^ρ *is the number of standard tri-tableaux of shape* ρ.

Proof Number of ways of choosing f horizontal edges such that each edge contains one vertex from left wall of d and another from the right wall of d is $f! \cdot rC_f \cdot sC_f$. i.e. the number of column standard block tableau of shape 2^f is

$$f! \cdot rC_f \cdot sC_f = \frac{r!s!}{f!(r-f)!(s-f)!}. \tag{1}$$

By Proposition 2.5, the number of standard tableaux of shape $\mu \vdash r - f$, is

$$(r-f)! \left| \frac{1}{(\mu_i - i + j)!} \right|_{r-f \times r-f}. \tag{2}$$

By Proposition 2.5, the number of standard tableaux of shape $\nu \vdash s - f$, is

$$(s-f)! \left| \frac{1}{(\mu_i - i + j)!} \right|_{s-f \times s-f}. \tag{3}$$

It suffices to prove $\frac{1}{f!} = \left| \frac{1}{(\lambda_i - i + j - 1)!} \right|_{f \times f}$.

Consider

$$\left| \frac{1}{(\lambda_i - i + j - 1)!} \right|_{f \times f} = \begin{vmatrix} \frac{1}{1!} & \frac{1}{2!} & \frac{1}{3!} & \frac{1}{4!} & \cdots & \frac{1}{(f-1)!} & \frac{1}{f!} \\ \frac{1}{0!} & \frac{1}{1!} & \frac{1}{2!} & \frac{1}{3!} & \cdots & \frac{1}{(f-2)!} & \frac{1}{(f-1)!} \\ 0 & \frac{1}{0!} & \frac{1}{1!} & \frac{1}{2!} & \cdots & \frac{1}{(f-3)!} & \frac{1}{(f-2)!} \\ 0 & 0 & \frac{1}{0!} & \frac{1}{1!} & \cdots & \frac{1}{(f-4)!} & \frac{1}{(f-3)!} \\ \vdots & \vdots & \vdots & \vdots & \vdots & \vdots & \vdots \\ 0 & 0 & 0 & 0 & \cdots & \frac{1}{0!} & \frac{1}{1!} \end{vmatrix}$$

$$= \frac{1}{f!} \text{ by Lemma 3.15.} \tag{4}$$

Substitute 4 in 1 we get

$$f! \cdot rC_f \cdot sC_f = \frac{r!s!}{(r-f)!(s-f)!} \left| \frac{1}{(\lambda_i - i + j - 1)!} \right|_{f \times f} \tag{5}$$

By Eqs. 2, 3 and 5, we get the number of standard tri-tableaux of shape ρ, is

$$h^{\rho} = r!s! \left| \frac{1}{(\lambda_i - i + j - 1)!} \right|_{f \times f} \left| \frac{1}{(\mu_i - i + j)!} \right|_{r-f \times r-f} \left| \frac{1}{(\nu_i - i + j)!} \right|_{s-f \times s-f}.$$

Hence the proof. □

4 The Robinson–Schensted Correspondence for the Walled Signed Brauer Algebras

4.1 The Robinson–Schensted Correspondence

In this section, we define a Robinson–Schensted correspondence for the walled signed Brauer algebras which gives the correspondence between the walled signed Brauer diagram d and the pairs of standard signed-tri-tableaux of shape $\lambda = (\lambda_1, \lambda_2, \lambda_3)$ with $\lambda_1 = (2^{2f})$, $\lambda_2 \vdash_b r - f$ and $\lambda_3 \vdash_b s - f$, for $0 \leq f \leq \min(r, s)$.

Definition 4.1 Each 2×1 and 1×2 rectangular boxes consisting of two nodes is called as a 2-domino.

Definition 4.2 A 2-domino in which all the nodes are filled with the same number from the set $\{1, 2, \ldots, n\}$ is defined as a 2-tablet.

Definition 4.3 A signed tripartition ν of n will be an ordered triple of partitions (ν_1, ν_2, ν_3) where $\nu_1 = (2^{2f})$, $\nu_2 \vdash_b r - f$ and $\nu_3 \vdash_b s - f$, for $0 \leq f \leq \min(r, s)$, $r + s = n$.

Definition 4.4 A standard horizontal block is defined as the block consisting of two horizontal 2-tablets $d^{(1)}, d^{(2)}$ one above the other such that $d^{(1)} < d^{(2)}$, i.e., $\begin{array}{|cc|}\hline d^{(1)} & d^{(1)} \\ \hline d^{(2)} & d^{(2)} \\ \hline\end{array}$.

We call $d^{(1)}$ as the first 2-tablet of the horizontal block and $d^{(2)}$ as the second 2-tablet of the horizontal block. We call horizontal block as positive block.

Definition 4.5 A standard vertical block is defined as the block consisting of two vertical 2-tablets $d^{(1)}, d^{(2)}$ adjacent to each other such that $d^{(1)} < d^{(2)}$. i.e. $\begin{array}{|c|c|}\hline d^{(1)} & d^{(2)} \\ d^{(1)} & d^{(2)} \\ \hline\end{array}$.

We call $d^{(1)}$ as the first 2-tablet of the vertical block and $d^{(2)}$ as the second 2-tablet of the vertical block. We call vertical block as negative block.

Definition 4.6 A block domino tableau of shape 2^{2f} is a tableau consisting either of the standard horizontal block or standard vertical block.

Definition 4.7 A column standard block domino tableau of shape 2^{2f} is a block domino tableau of shape 2^{2f} if the first tablets of each block are increasing read from top to bottom.

Definition 4.8 A standard signed tri-tableau (t_1, t_2, t_3) is a tri-tableau (t_1, t_2, t_3) where t_1 is a column standard block domino tableau and t_2, t_3 are standard bitableaux.

Definition 4.9 Given a walled signed Brauer diagram $d \in D_{r,s}$, we may associate a quadruple $[d_1, d_2, w_1, w_2]$ such that

$$\begin{aligned}
d_1 &= \{\ (i, d_1(i), c(d_1(i)))|\ \text{the edge joining the vertices } i \text{ and } d_1(i) \text{ in the first} \\
&\quad\ \text{row with sign } c(d_1(i))\} \\
&= \{(i_1, d_1(i_1), c(d_1(i_1))), (i_2, d_1(i_2), c(d_1(i_2))), \ldots, (i_f, d_1(i_f), c(d_1(i_f)))\} \\
d_2 &= \{\ (j, d_2(j), c(d_2(j)))|\ \text{the edge joining the vertices } j \text{ and } d_2(j) \text{ in the} \\
&\quad\ \text{second row with sign } c(d_2(j))\} \\
&= \{(i_1, d_2(i_1), c(d_2(i_1))), (i_2, d_2(i_2), c(d_2(i_2))), \ldots, (i_f, d_2(i_f), c(d_2(i_f)))\} \\
w_1 &= \{\ (k, w_1(k), c(w_1(k)))|\ \text{the edge joining the vertex } k \text{ left to the wall in} \\
&\quad\ \text{the first row and the vertex } w_1(k) \text{ left to the wall in the second row} \\
&\quad\ \text{with sign } c(w_1(k))\} \\
&= \{(i_1, w_1(i_1), c(w_1(i_1))), \ldots, (i_{r-f}, w_1(i_{r-f}), c(w_1(i_{r-f})))\} \\
w_2 &= \{\ (l, w_2(l), c(w_2(l)))|\ \text{the edge joining the vertex } l \text{ right to the wall in} \\
&\quad\ \text{the first row and the vertex } w_2(l) \text{ right to the wall in the second row} \\
&\quad\ \text{with sign } c(w_2(l))\} \\
&= \{(i_1, w_2(i_1), c(w_2(i_1))), \ldots, (i_{s-f}, w_2(i_{s-f}), c(w_2(i_{s-f})))\}
\end{aligned}$$

such that $i_1 < i_2 < \cdots < i_m$ for any $m > 1$, where f is the number of signed horizontal edges in a row of d, $r - f$ is the number of signed vertical edges left to the wall of d and $s - f$ is the number of signed vertical edges right to the wall of d.

Theorem 4.10 *The map*

$$d \underset{D_{r,s}}{\overset{R-S}{\longleftrightarrow}} [(P_1(d), P_2(d), P_3(d)), (Q_1(d), Q_2(d), Q_3(d))]$$

where $P_i(d) = \left(P_i^{(1)}(d), P_i^{(2)}(d)\right)$ *and* $Q_i(d) = \left(Q_i^{(1)}(d), Q_i^{(2)}(d)\right)$, *for* $i = 2, 3$ *provides a bijection between the set of walled signed Brauer diagrams d and the pairs of standard* λ*-signed-tri-tableaux of shape* $\lambda = (\lambda_1, \lambda_2, \lambda_3)$ *with* $\lambda_1 = (2^{2f})$, $\lambda_2 \vdash_b r - f$ *and* $\lambda_3 \vdash_b s - f$, *for* $0 \leq f \leq \min(r, s)$.

Proof We first describe the map that, given a diagram $d \in D_{r,s}$, produces a pair of signed-tri-tableaux.

$$''d \underset{D_{r,s}}{\overset{R-S}{\longleftrightarrow}} [(P_1(d), P_2(d), P_3(d)), (Q_1(d), Q_2(d), Q_3(d))]$$

where $P_i(d) = \left(P_i^{(1)}(d), P_i^{(2)}(d)\right)$ and $Q_i(d) = \left(Q_i^{(1)}(d), Q_i^{(2)}(d)\right)$, for $i = 2, 3.''$

We construct a sequence of tableaux

$$\begin{aligned}
\emptyset &= P_1^0, P_1^1, \ldots, P_1^f \\
\emptyset &= Q_1^0, Q_1^1, \ldots, Q_1^f \\
\emptyset &= \left(P_2^{(1)}, P_2^{(2)}\right)^0, \left(P_2^{(1)}, P_2^{(2)}\right)^1, \ldots, \left(P_2^{(1)}, P_2^{(2)}\right)^{r-f} \\
\emptyset &= \left(Q_2^{(1)}, Q_2^{(2)}\right)^0, \left(Q_2^{(1)}, Q_2^{(2)}\right)^1, \ldots, \left(Q_2^{(1)}, Q_2^{(2)}\right)^{r-f} \\
\emptyset &= \left(P_3^{(1)}, P_3^{(2)}\right)^0, \left(P_3^{(1)}, P_3^{(2)}\right)^1, \ldots, \left(P_3^{(1)}, P_3^{(2)}\right)^{s-f} \\
\emptyset &= \left(Q_3^{(1)}, Q_3^{(2)}\right)^0, \left(Q_3^{(1)}, Q_3^{(2)}\right)^1, \ldots, \left(Q_3^{(1)}, Q_3^{(2)}\right)^{s-f}
\end{aligned}$$

where f is the number of signed horizontal edges of d, $r - f$ is the number of signed vertical edges left to the wall of d and $s - f$ is the number of signed vertical edges right to the wall of d. The edges joining the vertices (x_1, x_2) with sign c are inserted into $P_1(d), P_2(d), P_3(d), Q_1(d)$ and placed in $Q_2(d), Q_3(d)$ so that $\mathrm{sh}P_1^i = \mathrm{sh}Q_1^i$, for all i, $\mathrm{sh}P_1^j = \mathrm{sh}Q_1^j$, for all j and $\mathrm{sh}P_2^k = \mathrm{sh}Q_2^k$, for all k.

Begin with the tableau $P_1^0 = P_2^0 = P_3^0 = Q_1^0 = Q_2^0 = Q_3^0 = \emptyset$. Then recursively define the standard tableau by the following.

If $(l', m', c) \in d_2$ then $P_1^k =$ insertion of (l, m, c) in P_1^{k-1}.

If $(l, m, c) \in d_1$ then $Q_1^k =$ insertion of (l, m, c) in Q_1^{k-1}.

If $(l, m', c) \in w_1$ then $P_2^k =$ insertion of m with sign c in P_2^{k-1} and place l in Q_2^{k-1} where the insertion terminates in P_2^{k-1} when m is inserted.

If $(l, m', c) \in w_2$ then $P_3^k =$ insertion of m with sign c in P_3^{k-1} and place l in Q_3^{k-1} where the insertion terminates in P_3^{k-1} when m is inserted.

The operations of insertion and placement will now be described.

First we give the insertion on $P_1(d)$. Let $(i_k, d_2(i_k), c(d_2(i_k))) \in d_2$ and $i_k, d_2(i_k)$ be the elements not in $P_1(d)$. To insert $i_k, d_2(i_k)$ with sign $c(d_2(i_k))$ into $P_1(d)$, we proceed as follows.

If $c(d_2(i_k)) = 1$ then the positive block i.e. $\boxed{\begin{matrix} i_k & i_k \\ d_2(i_k) & d_2(i_k) \end{matrix}}$ is to be inserted into $P_1(d)$ along the cells $(i, j), (i, j+1), (i+1, j), (i+1, j+1)$.

If $c(d_2(i_k)) = -1$ then the negative block i.e. $\beta_x = \begin{array}{|c|c|} \hline i_k & d_2(i_k) \\ \hline i_k & d_2(i_k) \\ \hline \end{array}$ is to be inserted into $P_1(d)$ along the cells $(i, j), (i+1, j), (i, j+1), (i+1, j+1)$.

Now place the block containing $i_k, d_2(i_k)$ below the block containing i_{k-1}, $d_2(i_{k-1})$. Insertion on $Q_1(d)$ is the same as in $P_1(d)$.

Now we give the insertion on $P_2(d)$. The insertion is just as in Robinson–Schensted correspondence for the symmetric group, we give it here for the sake of completion. Let $(i_k, w_1(i_k), c(w_1(i_k))) \in w_1$ and $w_1(i_k)$ be the element not in $P_2(d)$. To insert $w_1(i_k)$ with sign $c(w_1(i_k))$ into $P_2(d)$, we proceed as follows.

If $c(w_1(i_k)) = 1$, then insert $w_1(i_k)$ in $P_2^{(1)}(d)$.

If $c(w_1(i_k)) = -1$, then insert $w_1(i_k)$ in $P_2^{(2)}(d)$.

RS1 Set $R :=$ the first row of $P_2^{(i)}(d)$.

RS2 While $w_1(i_k)$ is less than some element of row R, do

RSa Let y be the smallest element of R greater than $w_1(i_k)$ and replace y by $w_1(i_k)$ in R.

RSb Set $w_1(i_k) := y$ and $R :=$ the next row down.

RS3 Now $w_1(i_k)$ is greater than every element of R, so place $w_1(i_k)$ at the end of the row R and stop.

The placement of i_k in $Q_2^{(i)}(d)$ is even easier than insertion. Suppose that $Q_2^{(i)}(d)$ is a partial tableau of shape μ and if i_k is greater than every element of $Q_2^{(i)}(d)$, then place i_k in $Q_2^{(i)}(d)$ along the cell where the insertion in $P_2^{(i)}(d)$ terminates.

Let $(i_k, w_2(i_k), c(w_2(i_k))) \in w_2$. The insertion of $w_2(i_k)$ with sign $c(w_2(i_k))$ in $P_3(d)$ and placement of i_k in $Q_3(d)$ are the same as in $P_2(d)$ and $Q_3(d)$, respectively.

Hence

$$d \xrightarrow[D_{r,s}]{R-S} [(P_1(d), P_2(d), P_3(d)), (Q_1(d), Q_2(d), Q_3(d))]$$

where $P_i(d) = \left(P_i^{(1)}(d), P_i^{(2)}(d)\right)$ and $Q_i(d) = \left(Q_i^{(1)}(d), Q_i^{(2)}(d)\right)$, for $i = 2, 3$.

To prove

$$''[(P_1(d), P_2(d), P_3(d)), (Q_1(d), Q_2(d), Q_3(d))] \xleftarrow[D_{r,s}]{R-S} d$$

where $P_i(d) = \left(P_i^{(1)}(d), P_i^{(2)}(d)\right)$ and $Q_i(d) = \left(Q_i^{(1)}(d), Q_i^{(2)}(d)\right)$, for $i = 2, 3''$.

We merely reverse the preceding algorithm step by step. We begin by defining

$$\left(P_1^f, P_2^{r-f}, P_3^{s-f}\right) = (P_1(d), P_2(d), P_3(d))$$

$$\left(Q_1^f, Q_2^{r-f}, Q_3^{s-f}\right) = (Q_1(d), Q_2(d), Q_3(d))$$

where f is the number of signed horizontal edges of d, $r - f$ is the number of signed vertical edges left to the wall of d and $s - f$ is the number of signed vertical edges right to the wall of d.

To recover all the elements of w_1 using the following rules. Recovering the elements is just as in reverse Robinson–Schensted correspondence for the symmetric group, we give it here for the sake of completion.

Assuming that P_2^k, Q_2^k has been constructed, we will find $w_1(i_k)$ (the kth element of w_1) and P_2^{k-1}, Q_2^{k-1}. We write the (i, j) entry of P_2^k as $P_2^{i,j}$.

Find the cell (i, j) containing i_k in $\left(Q_2^l\right)^k$. Since this is the largest element in $\left(Q_2^l\right)^k$, $\left(P_2^l\right)^{i,j}$ must have been the last element to be displaced in the construction of $\left(P_2^l\right)^k$. We can now use the following procedure to delete $\left(P_2^l\right)^{i,j}$ from $\left(P_2^l\right)(d)$. For convenience, we assume the existence of an empty zeroth row above the first row of $\left(P_2^l\right)^k$.

SR1 Set $x := P_2^{i,j}$ and erase $P_2^{i,j}$.
Set $R :=$ the $(i-1)$ row of P_2^k.
SR2 While R is not the zeroth row of P_2^k, do

SRa Let y be the largest element of R smaller than x and replace y by x in R.
SRb Set $x := y$ and $R :=$ the next row down up.

SR3 Now x has been removed from the first row, so set $w_1(i_k) := x$.

It is easy to see that P_2^{k-1} is P_2^k after the deletion process just described is complete and Q_2^{k-1} is Q_2^k with k erased. Continuing in this way, we eventually recover all the elements of w_1 in reverse order. We can recover all the elements of w_2 as in w_1.

We are yet to find the elements in d_1, d_2.

We may recover the elements of d_2 such that the pair $(x_k, d_2(x_k), c(d_2(x_k)))$ is the block in the cells $((2k-1, 1), (2k-1, 2), (2k, 1), (2k, 2))$ of $P_1(d)$, for every k and the $c(d_2(x_k)) = 1$ $(c(d_2(x_k)) = -1)$ if the block is positive block (negative block).

Similarly, we may recover the elements of d_1 such that the pair $(x_k, d_1(x_k), c(d_1(x_k)))$ is the element in the cells $((2k-1, 1), (2k-1, 2), (2k, 1), (2k, 2))$ of $Q_1(d)$, for every k and the $c(d_1(x_k)) = 1$ $(c(d_1(x_k)) = -1)$ if the block is positive block (negative block).

Thus we recover the quadruple $[d_1, d_2, w_1, w_2]$ from the pair of tri-tableaux $[(P_1(d), P_2(d), P_3(d)), (Q_1(d), Q_2(d), Q_3(d))]$.

Hence $[(P_1(d), P_2(d), P_3(d)), (Q_1(d), Q_2(d), Q_3(d))] \xleftarrow[D_{r,s}]{R-S} d$ which completes the proof. □

Example Let

$d =$ 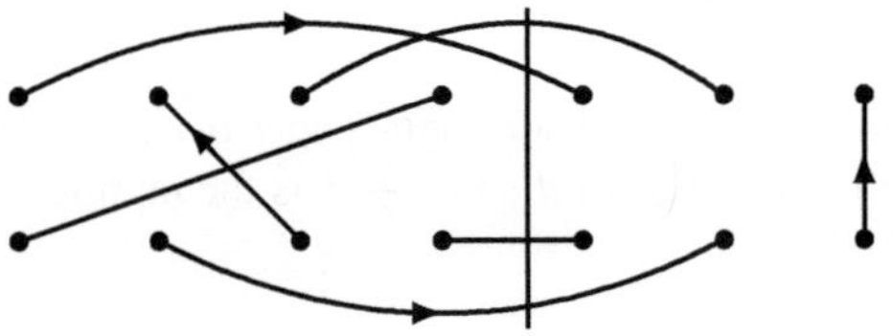

Then

$$\begin{aligned} d_1 &= \{(1, 5, -1), (3, 6, 1)\} \\ d_2 &= \{(2', 6', -1), (4', 5', 1)\} \\ w_1 &= \{(2, 3', -1), (4, 1', 1)\} \\ w_2 &= \{(7, 7', -1)\} \end{aligned}$$

$$P_1^0 = \emptyset \qquad P_1^1 = \begin{array}{|c|c|}\hline 2 & 6 \\ \hline 2 & 6 \\ \hline \end{array} \qquad P_1^2 = \begin{array}{|c|c|}\hline 2 & 6 \\ \hline 2 & 6 \\ \hline 4 & 4 \\ \hline 5 & 5 \\ \hline \end{array}$$

$$Q_1^0 = \emptyset \qquad Q_1^1 = \begin{array}{|c|c|}\hline 1 & 5 \\ \hline 1 & 5 \\ \hline \end{array} \qquad Q_1^2 = \begin{array}{|c|c|}\hline 1 & 5 \\ \hline 1 & 5 \\ \hline 3 & 3 \\ \hline 6 & 6 \\ \hline \end{array}$$

$$P_2^0 = (\emptyset, \emptyset) \qquad P_2^1 = (\emptyset, \boxed{3}) \qquad P_2^2 = (\boxed{1}, \boxed{3})$$

$$Q_2^0 = (\emptyset, \emptyset) \qquad Q_2^1 = (\emptyset, \boxed{2}) \qquad Q_2^2 = (\boxed{4}, \boxed{2})$$

$$P_3^0 = (\emptyset, \emptyset) \qquad P_3^1 = (\emptyset, \boxed{7})$$

$$Q_3^0 = (\emptyset, \emptyset) \qquad Q_3^1 = (\emptyset, \boxed{7})$$

Thus $d \overset{R-S}{\underset{D_{r,s}}{\longleftrightarrow}} (P, Q)$ where

$$P = \left(\begin{array}{|c|c|}\hline 2 & 6 \\ \hline 2 & 6 \\ \hline 4 & 4 \\ \hline 5 & 5 \\ \hline \end{array}, (\boxed{1}, \boxed{3}), (\emptyset, \boxed{7}) \right)$$

$$Q = \left(\begin{array}{|c|c|}\hline 1 & 5 \\ \hline 1 & 5 \\ \hline 3 & 3 \\ \hline 6 & 6 \\ \hline \end{array}, (\boxed{4}, \boxed{2}), (\emptyset, \boxed{7}) \right)$$

Definition 4.11 The flip of any walled signed Brauer diagram d is the diagram of d reflected over its horizontal axis, which is denoted by flip(d).

Proposition 4.12 *Let $d \in D_{r,s}$. If*

$$d \xrightarrow[D_{r,s}]{R-S} [(P_1(d), P_2(d), P_3(d)), (Q_1(d), Q_2(d), Q_3(d))].$$

Then flip$(d) \xrightarrow[D_{r,s}]{R-S} [(Q_1(d), Q_2(d), Q_3(d)), (P_1(d), P_2(d), P_3(d))]$ where $P_1(d)$, $Q_1(d)$ are the column standard block domino tableaux and $P_2(d)$, $P_3(d)$, $Q_2(d)$, $Q_3(d)$ are the standard bitableaux constructed by the above insertion.

Proof Suppose $d \in D_{r,s}$, then we can recover the quadruple $[d_1, d_2, w_1, w_2]$ by the Definition 4.9. By the Definition 4.11, flip(d) has the quadruple $[d_2, d_1, w_1^{-1}, w_2^{-1}]$. Hence the proof follows by Proposition 2.17. □

4.2 The Knuth relations

In this section, we derive the Knuth relations for the walled signed Brauer algebras by using the Robinson-Schensted correspondence for the standard λ-signed-tri-tableaux of shape $\lambda = (\lambda_1, \lambda_2, \lambda_3)$ with $\lambda_1 = (2^{2f})$, $\lambda_2 \vdash_b r - f$ and $\lambda_3 \vdash_b s - f$, for $0 \leq f \leq \min(r, s)$.

Definition 4.13 Let $d, d' \in D_{r,s}$. Then d and d' are Knuth equivalent, denoted by $d \overset{K_{D_{r,s}}}{\sim} d'$ if the following condition holds.

1. $d_2 = d_2'$.
2. $w_1 \overset{K}{\sim} w_1'$ where $w_1, w_1' \in \mathcal{GSP}(a)$, a is the number of signed vertical edges left to the wall of both d and d'.
3. $w_2 \overset{K}{\sim} w_2'$ where $w_2, w_2' \in \mathcal{GSP}(b)$, b is the number of signed vertical edges right to the wall of both d and d'.

Proposition 4.14 *Let $d, d' \in D_{r,s}$. Then*

$$d \overset{K_{D_{r,s}}}{\sim} d' \iff (P_1(d), P_2(d), P_3(d)) = (P_1(d'), P_2(d'), P_3(d'))$$

where $P_1(d)$, $P_1(d')$ are the column standard block domino tableaux of shape $\lambda_1 = (2^{2f})$, $P_2(d)$, $P_2(d')$ are the standard bitableaux of shape λ_2, $\lambda_2 \vdash_b r - f$, $P_3(d)$, $P_3(d')$ are the standard bitableaux of shape λ_3, $\lambda_3 \vdash_b s - f$ and $0 \leq f \leq \min(r, s)$.

Proof The proof follows from the Definition 4.13 and by the Proposition 2.19 □

Definition 4.15 Let $d, d' \in D_{r,s}$. Then d and d' are dual Knuth equivalent, denoted by $d \overset{K^*_{D_{r,s}}}{\sim} d'$ if the following condition holds.

1. $d_1 = d_1'$.
2. $w_1 \overset{K^*}{\sim} w_1'$ where $w_1, w_1' \in \mathcal{GSP}(a)$, a is the number of signed vertical edges left to the wall of both d and d'.
3. $w_2 \overset{K^*}{\sim} w_2'$ where $w_2, w_2' \in \mathcal{GSP}(b)$, b is the number of signed vertical edges right to the wall of both d and d'.

Proposition 4.16 *Let $d, d' \in D_{r,s}$. Then*

$$d \overset{K^*_{D_{r,s}}}{\sim} d' \Longleftrightarrow (Q_1(d), Q_2(d), Q_3(d)) = (Q_1(d'), Q_2(d'), Q_3(d'))$$

where $Q_1(d), Q_1(d')$ are the column standard block domino tableaux of shape $\lambda_1 = (2^{2f})$, $Q_2(d), Q_2(d')$ are the standard bitableaux of shape λ_2, $\lambda_2 \vdash_b r - f$, $Q_3(d), Q_3(d')$ are the standard bitableaux of shape λ_3, $\lambda_3 \vdash_b s - f$ and $0 \leq f \leq \min(r, s)$.

Proof The proof follows from the Definition 4.15 and by the Proposition 2.22. □

4.3 The Determinantal Formula

In this section, we constructed the determinantal formula, which gives the dimensions of the irreducible representations of walled signed Brauer algebras.

Set $\dfrac{1}{r!} = 0$ if $r < 0$.

Theorem 4.17 (Determinantal Formula) *If $\rho = (\lambda, \mu, \nu)$ with $\lambda = (2^{2f})$, $\mu \vdash_b r - f$, $\nu \vdash_b s - f$ for fixed $r \geq 0$, $s \geq 0$ where $0 \leq f \leq \min(r, s)$ then*

$$h^\rho = 2^f r! s! \left| \frac{1}{(\lambda_i - i + j - 1)!} \right|_{f \times f} \left| \frac{1}{(\mu_i^{(1)} - i + j)!} \right|_{l \times l}$$
$$\left| \frac{1}{(\mu_i^{(2)} - i + j)!} \right|_{r-f-l \times r-f-l} \left| \frac{1}{(\nu_i^{(1)} - i + j)!} \right|_{m \times m}$$
$$\left| \frac{1}{(\nu_i^{(2)} - i + j)!} \right|_{s-f-m \times s-f-m}$$

where h^ρ is the number of standard signed tri-tableaux of shape ρ.

Proof Number of ways of choosing f signed horizontal edges such that each edge contains one vertex from left wall of d and another from the right wall of d is $2^f f! \cdot rC_f \cdot sC_f$. i.e. the number of column standard block domino tableau of shape 2^{2f} is

$$2^f f! \cdot rC_f \cdot sC_f = \frac{2^f r! s!}{f!(r-f)!(s-f)!}. \tag{6}$$

By Proposition 2.7, the number of standard bitableaux of shape $\mu \vdash_b r - f$, is

$$(r-f)!\left|\frac{1}{(\mu_i^{(1)}-i+j)!}\right|_{l\times l}\left|\frac{1}{(\mu_i^{(2)}-i+j)!}\right|_{r-f-l\times r-f-l} \tag{7}$$

where $|\mu^{(1)}| = l$ and $|\mu^{(2)}| = r - f - l$.

By Proposition 2.7, the number of standard bitableaux of shape $\nu \vdash_b s - f$, is

$$(s-f)!\left|\frac{1}{(\nu_i^{(1)}-i+j)!}\right|_{m\times m}\left|\frac{1}{(\nu_i^{(2)}-i+j)!}\right|_{s-f-m\times s-f-m} \tag{8}$$

where $|\nu^{(1)}| = m$ and $|\nu^{(2)}| = s - f - m$.

It suffices to prove $\frac{1}{f!} = \left|\frac{1}{(\lambda_i - i + j - 1)!}\right|_{f\times f}$.

Consider

$$\left|\frac{1}{(\lambda_i-i+j-1)!}\right|_{f\times f} = \begin{vmatrix} \frac{1}{1!} & \frac{1}{2!} & \frac{1}{3!} & \frac{1}{4!} & \cdots & \frac{1}{(f-1)!} & \frac{1}{f!} \\ \frac{1}{0!} & \frac{1}{1!} & \frac{1}{2!} & \frac{1}{3!} & \cdots & \frac{1}{(f-2)!} & \frac{1}{(f-1)!} \\ 0 & \frac{1}{0!} & \frac{1}{1!} & \frac{1}{2!} & \cdots & \frac{1}{(f-3)!} & \frac{1}{(f-2)!} \\ 0 & 0 & \frac{1}{0!} & \frac{1}{1!} & \cdots & \frac{1}{(f-4)!} & \frac{1}{(f-3)!} \\ \vdots & \vdots & \vdots & \vdots & \vdots & \vdots & \vdots \\ 0 & 0 & 0 & 0 & \cdots & \frac{1}{0!} & \frac{1}{1!} \end{vmatrix}$$

$$= \frac{1}{f!} \text{ by Lemma 3.15.} \tag{9}$$

Substitute Eq. 9 in 6 we get

$$2^f f! \cdot rC_f \cdot sC_f = \frac{2^f r!s!}{(r-f)!(s-f)!}\left|\frac{1}{(\lambda_i-i+j-1)!}\right|_{f\times f} \tag{10}$$

By Eqs. 7, 8, and 10, we get the number of standard tri-tableaux of shape ρ, is

$$h^\rho = 2^f r!s!\left|\frac{1}{(\lambda_i-i+j-1)!}\right|_{f\times f}\left|\frac{1}{(\mu_i^{(1)}-i+j)!}\right|_{l\times l}$$
$$\left|\frac{1}{(\mu_i^{(2)}-i+j)!}\right|_{r-f-l\times r-f-l}\left|\frac{1}{(\nu_i^{(1)}-i+j)!}\right|_{m\times m}$$
$$\left|\frac{1}{(\nu_i^{(2)}-i+j)!}\right|_{s-f-m\times s-f-m}$$

where $|\mu^{(1)}| = l$, $|\mu^{(2)}| = r - f - l$, $|\nu^{(1)}| = m$ and $|\nu^{(2)}| = s - f - m$. Hence the proof. □

References

1. Benkart, G., Chakrabarti, M., Halverson, T., Leduc, R., Lee, C., Stroomer, J.: Tensor product representations of general linear groups and their connections with Brauer algebras. J. Algebra **166**, 529–567 (1994)
2. Brauer, R.: On algebras which are connected with the semisimple continuous groups. Ann. Math. **38**, 857–872 (1937)
3. Brundan, J., Stroppel, C.: Gradings on walled Brauer algebras and Khovanov's arc algebra. Adv. Math. **231**, 709–773 (2012)
4. Cox, A.G., De Visscher, M., Doty, S., Martin, P.P.: On the blocks of the walled Brauer algebra. J. Algebra **320**, 169–212 (2008)
5. Halverson, T., Lewandowski, T.: RSK insertion for set partitions and diagram algebras. Electron. J. Comb. **11**(2), R24 (2004–2005)
6. Kethesan, B.: The Structure of Walled Signed Brauer Algebra (preprint)
7. Koike, K.: On the decomposition of tensor products of the representations of classical groups: by means of universal characters. Adv. Math. **74**, 57–86 (1989)
8. Parvathi, M., Kamaraj, M.: Signed Brauer's algebras. Commun. Algebra **26**(3), 839–855 (1998)
9. Parvathi, M., Selvaraj, C.: Signed Brauer's algebras as centralizer algebras. Commun. Algebras **27**(12), 5985–5998 (1999)
10. Parvathi, M., Sivakumar, B., Tamilselvi, A.: R-S correspondence for the hyperoctahedral group of type B_n - a different approach. J. Algebra Discret. Math. **1**, 87–108 (2007)
11. Parvathi, M., Tamilselvi, A.: Robinson Schensted correspondence for the signed Brauer algebras. Electron. J. Comb. **14**, # R49 (2007)
12. Parvathi, M., Tamilselvi, A.: Robinson Schensted correspondence for the G-Brauer algebras. Contemporary Mathematics Proceedings **456** (2008)
13. Parvathi, M., Tamilselvi, A.: Robinson Schensted correspondence for the G-vertex colored partition algebras. Asian-Eur. J. Math. **3**(2), 369–385 (2010)
14. de Robinson, G.B.: On the representations of the symmetric group. Am. J. Math **60**, 745–760 (1938)
15. Sagan, B.E.: Graduate Texts in Mathematics. The symmetric group. Representations, combinatorial algorithms, and symmetric functions, vol. 203, 2nd edn. Springer, New York (2001)
16. Schensted, C.: Longest increasing and decreasing subsequences. Canad. J. Math. **13**, 179–191 (1961)
17. Schützenberger, M.P.: Quelques remarques sur une construction de Schensted. Math. Scand. **12**, 117–128 (1963)
18. Turaev, V.: Operator invariants of tangles and R-matrices. Izv. Akad. Nauk SSSR Ser. Math. **53**, 1073–1107 (1989). (in Russian)

Γ-Semigroups: A Survey

M.K. Sen and S. Chattopadhyay

Abstract The concept of Γ-semigroup is a generalization of semigroup. Let S and Γ be two nonempty sets. S is called Γ-semigroup if there exists a mapping $S \times \Gamma \times S \longrightarrow S$, written as $(a, \alpha, b) \longrightarrow a\alpha b$, satisfying the identity $(a\alpha b)\beta c = a\alpha(b\beta c)$ for all $a, b, c \in S$ and $\alpha, \beta \in \Gamma$. This article is a survey of some works published by different authors on Γ-semigroups.

Keywords Γ-semigroup · Regular Γ-semigroup · Orthodox semigroup · Orthodox Γ-semigroup · Right(left) inverse semigroup · Right(left) inverse Γ-semigroup · Γ-group · Semidirect product · E-inversive Γ-semigroup

AMS Mathematics Subject Classification (2010) 20M17

1 Introduction

In 1964, N. Nobusawa published a paper [24] entitled "On a generalisation of ring theory." In that paper [24] Nobusawa introduced a new type of algebraic system which is known as Γ-ring. The class of Γ-rings contains not only all rings but also Hestenes ternary rings [17]. Many fundamental results of ring theory were extended to Γ-rings. There is a large literature dealing with Γ-rings, some of them are in [1–4]. Following this in 1981, M. K. Sen [29] first introduced the notion of Γ-semigroup as follows:

Let S and Γ be two nonempty sets. S is called Γ-semigroup if there exist mappings $S \times \Gamma \times S \longrightarrow S$, written as $(a, \alpha, b) \longrightarrow a\alpha b$, and $\Gamma \times S \times \Gamma \longrightarrow \Gamma$, written as

---

M.K. Sen (✉)
Department of Pure Mathematics, University of Calcutta, Calcutta, India
e-mail: senmk6@yahoo.com

S. Chattopadhyay
Sovarani Memorial College, Jagathballavpur, India
e-mail: chatterjees04@yahoo.co.in

S.T. Rizvi et al. (eds.), *Algebra and its Applications*, Springer Proceedings in Mathematics & Statistics 174, DOI 10.1007/978-981-10-1651-6_12

$(\alpha, a, \beta) \longrightarrow \alpha a \beta$, satisfying the identities $a\alpha(b\beta c) = a(\alpha b \beta)c = (a\alpha b)\beta c$ for all $a, b, c \in S$ and $\alpha, \beta \in \Gamma$.

In 1986, M. K. Sen and N. K. Saha [35] weakened the defining conditions of Γ-semigroup and redefined Γ-semigroup as follows:

Definition 1.1 Let $S = \{a, b, c, \dots\}$ and $\Gamma = \{\alpha, \beta, \gamma, \dots\}$ be two nonempty sets. S is called a Γ-semigroup if there exists a mapping $S \times \Gamma \times S \longrightarrow S$, written as $(a, \alpha, b) \longrightarrow a\alpha b \in S$ satisfying $(a\alpha b)\beta c = a\alpha(b\beta c)$, for all $a, b, c \in S$ and for all $\alpha, \beta \in \Gamma$.

Let S be an arbitrary semigroup. Let 1 be a symbol not representing any element of S. Let us extend the binary operation defined on S to $S \cup \{1\}$ by defining $11 = 1$ and $1a = a1 = a$ for all $a \in S$. It can be shown that $S \cup \{1\}$ is a semigroup with identity element 1. Let $\Gamma = \{1\}$. If we take $ab = a1b$, it can be shown that the semigroup S is a Γ-semigroup where $\Gamma = \{1\}$. Thus a semigroup can be considered to be a Γ-semigroup.

Let S be a Γ-semigroup and α be a fixed element of Γ. Define $a.b = a\alpha b$ for all $a, b \in S$. It can be shown that $(S, .)$ is a semigroup and denote this semigroup by S_α.

In [35] Sen and Saha proved that in a Γ-semigroup S if S_α is a group for some $\alpha \in \Gamma$, then S_α is a group for all $\alpha \in \Gamma$.

Definition 1.2 A Γ-semigroup S is called Γ-group if S_α is a group for some (hence for all) $\alpha \in \Gamma$.

Dutta and Adhikari described in [12] that operator semigroups to be a very effective tool in studying Γ-semigroups. In the paper [27] Sardar,Gupta, and Shum established that there is a close connection between the Morita equivalence of monoids and Γ-semigroups. Idempotent elements play an important role in semigroup theory. The notion of idempotent element in a Γ-semigroup was defined in [35] as follows.

Definition 1.3 Let S be a Γ-semigroup and $\alpha \in \Gamma$. Then $e \in S$ is said to be an α-idempotent if $e\alpha e = e$. The set of all α-idempotents is denoted by E_α. We denote $\bigcup_{\alpha \in \Gamma} E_\alpha$ by $E(S)$. The elements of $E(S)$ are called idempotent elements of S. If $S = E(S)$ then S is called an idempotent Γ-semigroup [40].

Many classical notions and results of the theory of semigroups have been extended and generalized to Γ-semigroups. Now we have many papers on Γ-semigroup covering diverse aspects of this topic. So there is enough material to be surveyed, rather it is not possible to cover arbitrarily many directions. In this survey, we cover only some examples of Γ-semigroups, and results on Regular Γ-semigroups, Right and left-orthodox Γ semigroups, Green's relations in Γ-semigroups, Γ-semigroup $T(A, B)$, Congruences on Γ-semigroups, Semidirect Product of a Monoid and a Γ-Semigroup, Semidirect Product of a Semigroup and a Γ-Semigroup and E-inversive Γ-semigroups.

2 Examples

Example 2.1 [35] Let A be a nonempty set and S be the set of all mappings from A to A. Then S is a semigroup with respect to the usual composition of mappings, which is known as full transformation semigroup on A. But this result does not happen if S to be the set of all mappings from a nonempty set A to another nonempty set B. Now, if Γ is the set of all mappings from B to A and $a\alpha b$, $\alpha a\beta$ denote the usual product of mappings, where $a, b \in S$ and $\alpha, \beta \in \Gamma$ then $a\alpha b \in S$ and $\alpha a\beta \in \Gamma$. Moreover, $(a\alpha b)\beta c = a(\alpha b\beta)c = a\alpha(b\beta c)$ for all $a, b, c \in S$ and for all $\alpha, \beta \in \Gamma$.

Example 2.2 [35] Let S be the set of all $m \times n$ matrices and Γ be the set of all $n \times m$ matrices over a field. Then for $A_{m,n}, B_{m,n} \in S$, the usual matrix multiplication $A_{m,n}B_{m,n}$ cannot be defined, i.e., S is not a semigroup under the usual matrix multiplication when $m \neq n$. But for all $A_{m,n}, B_{m,n}, C_{m,n} \in S$ and $P_{n,m}, Q_{n,m} \in \Gamma$, $A_{m,n}P_{n,m}B_{m,n}$ is defined and an element of S. Also we notice that $(A_{m,n}P_{n,m}B_{m,n})$ $Q_{n,m}C_{m,n} = A_{m,n}P_{n,m}(B_{m,n}Q_{n,m}C_{m,n})$. Hence S is a Γ-semigroup.

Example 2.3 [6] Let $A = \{1, 2, 3\}$ and $B = \{4, 5\}$. S denotes the set of all mappings from A to B. Here members of S will be described by the images of the elements 1, 2, 3. For example, the map $1 \to 4$, $2 \to 5$, $3 \to 4$ will be written as $(4, 5, 4)$. A map from B to A will be described in the same fashion. For example $(1, 2)$ denotes $4 \to 1$, $5 \to 2$. Now $S = \Big\{(4, 4, 4), (4, 4, 5), (4, 5, 4), (4, 5, 5), (5, 5, 5), (5, 4, 5), (5, 4, 4), (5, 5, 4)\Big\}$ and let $\Gamma = \{(1, 1), (1, 2), (2, 3), (3, 1)\}$. Let $f, g \in S$ and $\alpha \in \Gamma$. We define $f\alpha g$ by $(f\alpha g)(a) = f\alpha\Big(g(a)\Big)$ for all $a \in A$. So $f\alpha g$ is a mapping from A to B and hence $f\alpha g \in S$ and we can show that $(f\alpha g)\beta h = f\alpha(g\beta h)$ for all $f, g, h \in S$ and $\alpha, \beta \in \Gamma$. So S is a Γ-semigroup

In the Example 2.3 we see that Γ does not contain all the mappings from B to A.

3 Regular Γ-Semigroup

The notion of regularity of a Γ-semigroup was introduced in [35] by Sen and Saha.

Definition 3.1 [35] Let S be a Γ-semigroup. An element $a \in S$ is said to be regular in the Γ-semigroup S if $a \in a\Gamma S\Gamma a$ where $a\Gamma S\Gamma a = \{a\alpha b\beta a : b \in S, \alpha, \beta \in \Gamma\}$. S is said to be regular if every element of S is regular.

Example 3.2 [35] Let M be the set of all 3×2 matrices and Γ be the set of all 2×3 matrices over the field of rational numbers. Then M is a Γ-semigroup[see Example 2.2]. Let $A \in M$ where $A = \begin{pmatrix} a & b \\ c & d \\ e & f \end{pmatrix}$. Then, the $B \in \Gamma$ is taken according to the following cases such that $(ABA)BA = ABA = A$.

Case (1): When the submatrix $\begin{pmatrix} a & b \\ c & d \end{pmatrix}$ is nonsingular. Then $ad - bc \neq 0$ and e, f may be both zero or one of them is zero or both of them is nonzero. Then $B = \begin{pmatrix} \frac{d}{ad-bc} & \frac{-b}{ad-bc} & 0 \\ \frac{-c}{ad-bc} & \frac{a}{ad-bc} & 0 \end{pmatrix} \in \Gamma$ and $ABA = A$.

Case (2) and Case (3) are considered by taking $af - be \neq 0$, $cf - de \neq 0$, respectively and the corresponding B is taken as $\begin{pmatrix} \frac{f}{af-be} & 0 & \frac{-b}{af-be} \\ \frac{-e}{af-be} & 0 & \frac{-a}{af-be} \end{pmatrix}$ and $\begin{pmatrix} 0 & \frac{f}{cf-de} & \frac{-d}{cf-de} \\ 0 & \frac{-e}{cf-de} & \frac{c}{cf-de} \end{pmatrix}$, respectively, such that $ABA = A$.

Case (4): When the submatrices are singular then either $ad - bc = 0$ and $cf - de = 0$ or $ad - bc = 0$ and $af - de = 0$. If all the elements of A are zero then the case is trivial. Next consider at least one of the elements of A is nonzero say $a_{ij} \neq 0, i = 1, 2, 3, j = 1, 2$. Then the element b_{ji} of B can be taken as $(a_{ij})^{-1}$ and other elements of B are zero and then $ABA = A$. Thus A is regular and hence M is regular.

Example 3.3 Let S be the set of all positive integers of the form $4n + 1$ and Γ be the set of all positive integers of the form $4n + 3$. If $a\alpha b$ is $a + \alpha + b$ for all $a, b \in S$ and $\alpha \in \Gamma$ then S is a Γ-semigroup. Since for the element $2 \in S$, there do not exist any $a \in S$ and $\alpha, \beta \in \Gamma$ such that $2 + \alpha + a + \beta + 2 = 2$, 2 is not a regular element in S. Hence S is not a regular Γ -semigroup.

Let S be a Γ-semigroup and α be a fixed element of Γ. It is evident that for a Γ-semigroup S, if S_α is a regular semigroup for some $\alpha \in \Gamma$ then S is a regular Γ-semigroup but

(i) S_α may not be a regular semigroup for some $\alpha \in \Gamma$ yet S may be a regular Γ-semigroup.

Example 3.4 Let $S = \{(a, 0) : a \in R\} \cup \{(0, b) : b \in R\}$ where R denotes the field of real numbers. Let $\Gamma = \{(0, 5), (0, 1), (3, 0), (1, 0)\}$. Defining $S \times \Gamma \times S \to S$ by $(a, b)(\alpha, \beta)(c, d) = (a\alpha c, b\beta d)$ for all $(a, b), (c, d) \in S$ and $(\alpha, \beta) \in \Gamma$, we can show that S is a Γ-semigroup. S_α is not a regular semigroup for any $\alpha \in \Gamma$. Let $(a, 0) \in S$. If $a = 0$, then $(a, 0)$ is regular. Suppose $a \neq 0$, then $(a, 0)(3, 0)(\frac{1}{3a}, 0)$ $(1, 0)(a, 0) = (a, 0)$. Similarly, we can show that $(0, b)$ is also regular for all $b \in R$. Hence S is a regular Γ-semigroup.

(ii) In a Γ-semigroup S, if S_α is a regular semigroup for some $\alpha \in \Gamma$ then there may exist a $\beta \in \Gamma$ such that S_β is not a regular semigroup.

Example 3.5 Let $S = \{(a, b) : a, b \in R$, the field of real numbers$\}$ and $\Gamma = \{(9, 7), (0, 3)\}$. Defining $(a, b)(\alpha, \beta)(c, d)$ by $(a, b)(\alpha, \beta)(c, d) = (a\alpha c, b\beta d)$ for (a, b), $(c, d) \in S$ and $(\alpha, \beta) \in \Gamma$ we find that S is a Γ-semigroup. In this Γ-semigroup $S_{(0,3)}$ cannot be a regular semigroup but $S_{(9,7)}$ is a regular semigroup.

In a regular semigroup S an element $b \in S$ is said to be an inverse of an element a of S if $a = aba$ and $b = bab$. This was generalized by Sen and Saha in [35].

Definition 3.6 Let $a \in S$ and $\alpha, \beta \in \Gamma$. An element $b \in S$ is called (α, β)- inverse of a if $a = a\alpha b\beta a$ and $b = b\beta a\alpha b$. In this case we write $b \in V_\alpha^\beta(a)$.

In the paper [21], Braja described some interesting characterizations of regular Γ-semigroups. Also in [42], Xhilari and Braja studied completely regular Γ-semigroups by quasi-ideals and established a necessary and sufficient condition that an element be completely regular.

4 Green's Relations in Γ-Semigroups

Green's relations play a fundamental role in semigroup theory and it is natural to consider them in the context of Γ-semigroup. These notions were discussed in [26] by N.K.Saha, in [13] by T. K. Dutta and T. K. Chatterjee and in [39] by A. Seth.

Definition 4.1 [26] Let S be a Γ-semigroup. For $a, b \in S$, The binary relations $\mathcal{L}, \mathcal{R}, \mathcal{H}, \mathcal{D}$ and $\mathcal{J}$ are given by
$a\mathcal{L}b$ if $S\Gamma a \cup \{a\} = S\Gamma b \cup \{b\}$;
$a\mathcal{R}b$ if $a\Gamma S \cup \{a\} = b\Gamma S \cup \{b\}$;
$a\mathcal{H}b$ if $a\mathcal{L}b$ and $a\mathcal{R}b$;
$a\mathcal{D}b$ if $a\mathcal{L}c$ and $c\mathcal{R}b$ for some $c \in S$;
$a\mathcal{J}b$ if $a\Gamma S \cup S\Gamma a \cup S\Gamma a\Gamma S \cup \{a\} = b\Gamma S \cup S\Gamma b \cup S\Gamma b\Gamma S \cup \{b\}$.

Theorem 4.2 *Let S be a regular Γ-semigroup. Then*

(i) $a\mathcal{L}b$ *if and only if there exist* $\alpha, \beta, \delta \in \Gamma$, *and* $a' \in V_\alpha^\beta(a)$, $b' \in V_\alpha^\delta(b)$ *such that* $a'\beta a = b'\delta b$.
(ii) $a\mathcal{R}b$ *if and only if there exist* $\alpha, \beta, \delta \in \Gamma$, *and* $a' \in V_\alpha^\beta(a)$, $b' \in V_\gamma^\beta(b)$ *such that* $a\alpha a' = b\gamma b'$.
(iii) $a\mathcal{H}b$ *if and only if there exist* $\gamma, \delta \in \Gamma$, *and* $a' \in V_\gamma^\delta(a)$, $b' \in V_\gamma^\delta(b)$ *such that* $a\gamma a' = b\gamma b'$ *and* $a'\delta a = b'\delta b$.

In semigroups, $\mathcal{D}$-class of the Green's relation has some interesting property. If one element of a D-class is regular then every element of that D-class is regular. Here, we state the extended result to Γ-semigroups.

Theorem 4.3 *[26] Let S be a Γ-semigroup and* $a \in S$. *Let* D_a *denote the* $\mathcal{D}$*-class of S containing a. If a is regular, then every element of* D_a *is regular.*

Let us refer some papers [7], [8], [10], [11], [13], [18], [19], [22] and [26] in which authors studied Green's relations in Γ semigroups.

5 Γ-Semigroup $T(A, B)$

Let A and B be two nonempty sets, $S = T(A, B)$ be the set of all mappings from the set A to the set B and $\Gamma = T(B, A)$ be the set of all mappings from B to A. S becomes a Γ-semigroup with respect to usual mapping product $f\alpha g$ for all $f, g \in S$ and $\alpha \in \Gamma$.

In [36] Seth studied the Green's equivalences in $T(A, B)$ and described a method for systematic computation of regular $\mathcal{D}$-classes of a Γ-subsemigroup of $T(A, B)$. Some of important definitions and results from[38] are given below.

Let $f \in T(A, B)$. Then the kernel of f which is defined by $Kerf = \{(a, b) \in A \times A : f(a) = f(b)\}$ is an equivalence relation on A and so there is a bijection from $A/Kerf$ to Imf. He defined common cardinal number of $A/Kerf$ and Imf as rank of f and denoted it by $rankf$.

Theorem 5.1 *Given $f, g \in T(A, B)$, $Imf \subseteq Img(Kerg \subseteq Kerf)$ if and only if there exist $h \in T(A, B), \alpha \in T(B, A)$ such that $f = h\alpha g(f = g\alpha h)$. Moreover, $rankf \leq rankg$ if and only if there exist $h, j \in T(A, B), \alpha, \beta \in T(B, A)$ such that $f = j\alpha g\beta h$.*

Definition 5.2 Let A and B be two finite sets. $S \subseteq T(A, B)$ and $\Gamma \subseteq T(B, A)$ be such that $S\Gamma S \subseteq S$. Minimal rank of S is defined as $min\{rankf : f \in S\}$

Theorem 5.3 *Let A, B be two finite sets. Let $S \subseteq T(A, B)$ and $\Gamma \subseteq T(B, A)$ be such that $S\Gamma S \subseteq S$. Let r be the minimal rank of S. The minimal ideal I of S coincides with the set of all $f \in S$ such that $rankf = r$. For $f \in I$, the minimal left (right) ideal containing f consists of all $g \in S$ such that $Imf = Img(Kerf = Kerg)$.*

Theorem 5.4 *Two elements of $T(A, B)$ are $\mathcal{D}$-equivalent if and only if they have same rank.*

Theorem 5.5 *For two nonempty sets A and B and $\Gamma = T(B, A)$, let $R = \{r : r$ is a cardinal number $\leq min(\mid A \mid, \mid B \mid)\}$. Then (i) There is a one-to-one correspondence between the set of all $\mathcal{D}$ -classes of $T(A, B)$ and the set R such that the $\mathcal{D}$-class D_r corresponding to $r \in R$ consists of all elements of $T(A, B)$ of rank r.*

(ii) Let $r \in R$. Then there is a one-to-one correspondence between the set of all $\mathcal{L}$-classes contained in D_r and the set of all subsets B' of B of cardinal r such that the $\mathcal{L}$- class corresponding to B' consists of all elements of $T(A, B)$ having range B'.

(iii) Let $r \in R$. Then there is a one -to-one correspondence between the set of all $\mathcal{R}$-classes contained in D_r and the set of all equivalence relation π on A for which $\mid A/\pi \mid = r$ such that $\mathcal{R}$-class corresponding to π consists of all elements of $T(A, B)$ having kernel π.

(iv) Let $r \in R$. Then there is a one-to-one correspondence between the set of all $\mathcal{H}$-classes contained in D_r and the set of all pairs (π, B') where π is an equivalence relation on A and B' is a subset of B and $\mid A/\pi \mid = \mid B' \mid = r$ such that $\mathcal{H}$-class corresponding to (π, B') consists of all elements of $T(A, B)$ having Kernel π and range B'.

Theorem 5.6 *$T(A, B)$ is a regular Γ-semigroup.*

In the paper [16] Heidari and Amooshai associated left and right transformation semigroups to a Γ-semigroup and established relationships between the ideals of a Γ-semigroup and ideals of its left and right transformation semigroups.

6 Congruences on Γ-Semigroups

The notion of congruence on a Γ-semigroup was introduced by Dutta and Chatterjee in [13].

Definition 6.1 Let S be a Γ-semigroup. An equivalence relation ρ on S is said to be a right (left) congruence on S if $(a, b) \in \rho$ implies $(a\alpha c, b\alpha c) \in \rho$, (resp. $(c\alpha a, c\alpha b) \in \rho$) for all $a, b, c \in S$ and for all $\alpha \in \Gamma$. An equivalence relation ρ on S, which is both left and right congruence is called a congruence relation on S.

Let S be a Γ-semigroup and ρ be a congruence relation on S. Let S/ρ be the set of all equivalence classes of S. If $a\rho, b\rho$ be any two elements of S/ρ and $\alpha \in \Gamma$ then, we define $(a\rho)\alpha(b\rho) = (a\alpha b)\rho$. It can be shown that S/ρ is a Γ-semigroup.

In [14] Dutta and Chattopadhyay studied Rees congruence on a Γ -semigroup.

Definition 6.2 [37] Let S be a Γ-semigroup. A congruence ρ on S is called a Γ-group congruence if S/ρ is a Γ-group.

In [37] Seth defined a kernel normal system of a Γ-group and proved that the kernel of a congruence on a Γ-group is a kernel normal system. Again it was shown that a kernel normal system of a Γ-group determines a congruence.

Definition 6.3 [37] The set $\mathcal{A} = \{A_\alpha : \alpha \in T \subseteq \Gamma\}$ is defined to be a kernel normal system of a Γ -group S if and only if for all $\alpha, \beta \in T$, the following hold : (i) Each A_α is a normal subgroup of the group S_α. (ii) $A_\alpha \cap A_\beta = \phi$ if $\alpha \neq \beta$. (iii) Each γ-idempotent in $S(\gamma \in \Gamma)$ is contained in some element of $\mathcal{A}$. (iv) $a\alpha b_\alpha^{-1} \in A_\alpha$ implies $a\alpha b_\beta^{-1} \in A_\beta$ for all $\alpha, \beta \in \Gamma$.

Theorem 6.4 *Let S be a Γ-group and ρ be a congruence on S. Then the kernel of ρ is a kernel normal system of S.*

Theorem 6.5 *Let $\mathcal{A} = \{A_\alpha : \alpha \in T \subseteq \Gamma\}$ be a kernel normal system of a Γ-group S. Then $\rho_{\mathcal{A}} = \{(a, b) \in S \times S : a\alpha b_\alpha^{-1} \in A_\alpha$ for some $\alpha \in T\}$ is a congruence on S.*

In [38] Seth investigated least Γ-group congruences on a regular Γ-semigroup.

A family $\{K_\alpha : \alpha \in \Gamma\}$ of subsets of S is said to be normal family if the following conditions hold.

(i) $E_\alpha \subseteq K_\alpha$ for $\alpha \in \Gamma$,
(ii) $a \in K_\alpha$ and $b \in K_\beta \Rightarrow a\alpha b \in K_\beta, a\beta b \in K_\alpha$,

(iii) $a' \in V_{\alpha}^{\beta}(a)$ and $c \in K_{\gamma} \Rightarrow a\alpha c\gamma a', a\gamma c\alpha a' \in K_{\beta}$.

Let $\mathcal{N} = \bigcup_{i \in \Lambda} K_i$ be the collection of all normal families of subsets of S where $K_i = \{K_{i\alpha} : \alpha \in \Gamma\}$. Let $U_{\alpha} = \bigcap_{i \in \Lambda} K_{i\alpha}$ and $U = \{U_{\alpha} : \alpha \in \Gamma\}$. Then U is a normal family of subsets of S. Moreover U is the least member in $\mathcal{N}$ if we define a partial order in $\mathcal{N}$ by $K_i \leq K_j$.

Let $K = \{K_{\alpha} : \alpha \in \Gamma\}$ be a normal family of subsets of S. The family $KW = \{(KW)_{\gamma} : \gamma \in \Gamma\}$ where $(KW)_{\gamma} = \{x \in S : e\alpha x \in K_{\gamma}$ for some $\alpha \in \Gamma$ and $e \in K_{\alpha}\}$ is called the closure of K. K is said to be closed if $K = KW$. Let $\overline{N}$ denote the set of all closed families in $\mathcal{N}$. We now state the following theorems from [38].

Theorem 6.6 *The mapping* $K \to \rho_K = \{(a, b) \in S \times S : a\gamma b' \in K_{\delta}$ *for some* $b' \in V_{\gamma}^{\delta}(b)\}$ *is a one-to-one order preserving mapping of* $\overline{N}$ *onto the set of* Γ*-group congruences on S.*

Theorem 6.7 *The least* Γ*-group congruence* σ *on a* Γ *-semigroup S is given by* $\sigma = \rho_U$ *and* $Ker\sigma = UW$.

7 Some Classes of Regular Γ-Semigroups

It is known that the notion of inverse semigroup is the most natural generalization of the notion of groups. This notion was generalized in the theory of Γ-semigroup. In 1987, Seth and Saha [25] introduced inverse Γ-semigroup.

Definition 7.1 [25] A regular Γ-semigroup S is called an inverse Γ-semigroup if $|V_{\alpha}^{\beta}(a)| = 1$ for all $a \in S$ and for all $\alpha, \beta \in \Gamma$, whenever $V_{\alpha}^{\beta}(a) \neq \emptyset$. That is every element $a \in S$ has a unique (α, β) -inverse whenever (α, β)-inverse of a exists.

The following theorem gives a useful necessary and sufficient condition for a regular Γ-semigroup to be an inverse Γ-semigroup.

Theorem 7.2 *[25] Let S be a* Γ*-semigroup. S is an inverse* Γ*-semigroup if and only if (i) S is regular and (ii) if e and f be any two* α*-idempotents of S then* $e\alpha f = f\alpha e$, *where* $\alpha \in \Gamma$.

There are several results proved by Seth and Saha which are given below.

Theorem 7.3 *[25] Let S be an inverse* Γ*-semigroup. The minimum* Γ*-group congruence on S is given by* $\sigma = \{(a, b) \in S \times S : e\alpha a = f\beta b$ *for some* α*-idempotent e and for some* β*-idempotent f of S* $\}$.

Definition 7.4 Let S be a Γ-semigroup. A congruence ρ on S is said to be idempotent separating if for any two α-idempotents e and f of S, $(e, f) \in \rho$ implies $e = f$.

Theorem 7.5 *[25] Let S be an inverse Γ-semigroup. Define a relation μ on S by $\mu = \{(a, b) \in S \times S$: there exist $\gamma, \delta \in \Gamma$, $a' \in V_\gamma^\delta(a)$, $b' \in V_\gamma^\delta(b)$, satisfying $a\alpha e\gamma a' = b\alpha e\gamma b'$ for every α-idempotent $e \in S$, where α is any element of Γ$\}$. Then, μ is the maximum-idempotent separating congruence on S.*

In 2001, Chattopadhyay [5] defined right inverse Γ-semigroup and studied it. The main results are given below.

Definition 7.6 A regular Γ-semigroup *S* is called a right(resp. left) inverse Γ-semigroup if for any α-idempotent *e* and and β-idempotent f, $e\alpha f\beta e = f\beta e$(resp. $e\beta f\alpha e = e\beta f$).

Theorem 7.7 *The following conditions on a regular Γ-semigroup S are equivalent:*

(i) $e\Gamma S \bigcap f\Gamma S = e\alpha f\Gamma S (= f\beta e\Gamma S)$ for any α-idempotent e and β-idempotent f;
(ii) S is a right inverse Γ-semigroup;
(iii) if a is an element of S and $a' \in V_\alpha^{\beta_1}(a)$, $a'' \in V_\alpha^{\beta_2}(a)$ then $a'\beta_1 a = a''\beta_2 a$;
(iv) for any α-idempotent e of S, the elements of the set $F_\beta(e) = \{x \in S : x \in V_\alpha^\beta(e)\}$ are (A) β-idempotents and satisfy (B) $x\beta y = y$ for all $x, y \in F_\beta(e)$;
(v) for any $x \in S$, $e \in E_\alpha$, $\beta \in \Gamma$ and $x' \in V_\alpha^\beta(x)$, if $x \in S\Gamma e$ then $x' \in e\Gamma S$;
(vi) for any two α-idempotents e and f, $S\Gamma e = S\Gamma f$ implies $e = f$.

Theorem 7.8 *Let S be a right inverse Γ-semigroup. Then the binary relation δ on S defined by $\delta = \{(a, b) \in S \times S : (x\alpha a, x\alpha b) \in \mu$ for all $x \in S$ and for all $\alpha \in \Gamma\}$ is the maximum-idempotent separating congruence on S where $\mu = \{(a, b) \in S \times S$: there exist $\gamma, \delta \in \Gamma$, $a' \in V_\gamma^\delta(a)$ and $b' \in V_\gamma^\delta(b)$ satisfying $a'\delta e\alpha a = b'\delta e\alpha b$ for any α-idempotent e of S$\}$.*

Orthodox semigroups were first studied by Hall [15] and Yamada in [43] and [44]. A regular semigroup is said to be an orthodox semigroup if the set of all idempotents of the semigroup forms a subsemigroup. In 1990, Sen and Saha [30] generalized this notion in Γ-semigroups.

Definition 7.9 A regular Γ-semigroup *S* is called an orthodox Γ -semigroup if for an α-idempotent *e* and a β-idempotent f of *S*, $e\alpha f$ and $f\alpha e$ are β-idempotents in *S*.

Example 7.10 [30] Let Q^* denote the set of all nonzero rational numbers. Let Γ be the set of all positive integers. Let $a \in Q^*$, $\alpha \in \Gamma$ and $b \in Q^*$. Define $a\alpha b$ by $|a|\alpha b$. For this operation Q^* is a Γ-semigroup. Let $\frac{p}{q} \in Q^*$. Now, $|\frac{p}{q}|q|\frac{1}{p}|1\frac{p}{q} = \frac{p}{q}$. Hence this is a regular Γ-semigroup. Here $\frac{1}{q} (q \in \Gamma)$ is a q-idempotent. These are the only idempotents of Q^*. Now $|\frac{1}{q}|q\frac{1}{p}$ is a p-idempotent. Hence Q^* is an orthodox Γ-semigroup.

It is clear that any inverse Γ-semigroup is an orthodox Γ -semigroup. Some results of orthodox Γ-semigroup studied by Sen and Saha [30] are given below,

Theorem 7.11 *A regular Γ-semigroup S is an orthodox Γ-semigroup if and only if for any α-idempotent e, each of the elements $V_\alpha^\beta(e)$ and $V_\beta^\alpha(e)$ are β-idempotent where $V_\alpha^\beta(e) \neq \phi$ and $V_\beta^\alpha(e) \neq \phi$*

Theorem 7.12 *A regular Γ-semigroup S is an orthodox Γ-semigroup if and only if for $a, b \in S$, $\alpha_1, \alpha_2, \beta_1, \beta_2 \in \Gamma$, $a' \in V_{\alpha_1}^{\alpha_2}(a)$ and $b' \in V_{\beta_1}^{\beta_2}(b)$, we have $b'\beta_2 a' \in V_{\beta_1}^{\alpha_2}(a\alpha_1 b)$ and $b'\alpha_1 a' \in V_{\beta_1}^{\alpha_2}(a\beta_2 b)$.*

Theorem 7.13 *A regular Γ-semigroup S is an orthodox Γ-semigroup if and only if for any α-idempotent e and any γ-idempotent f with $V_\alpha^\beta(e) \bigcap V_\gamma^\beta(f) \neq \phi$, $V_\alpha^\beta(e) = V_\gamma^\beta(f)$.*

Theorem 7.14 *A regular Γ-semigroup S is an orthodox Γ-semigroup if and only if for $a, b \in S$, $V_\alpha^\beta(a) \cap V_\alpha^\beta(b) \neq \emptyset$ for some $\alpha, \beta \in \Gamma$ implies that $V_\gamma^\delta(a) = V_\gamma^\delta(b)$ for all $\gamma, \delta \in \Gamma$.*

Theorem 7.15 *Let S be a Γ-semigroup. Define a relation μ on S by $\mu = \{(a, b) \in S \times S :$ there exist $\gamma, \delta \in \Gamma, a' \in V_\gamma^\delta(a), b' \in V_\gamma^\delta(b)$ satisfying $a\alpha e\gamma a' = b\alpha e\gamma b'$ and $a'\delta e\alpha a = b'\delta e\alpha b$ for every α-idempotent $e \in S\}$. Then μ is the maximum-idempotent separating congruence on S.*

Theorem 7.16 *Let S be an orthodox Γ-semigroup. Then minimum Γ-group congruence on S is given by $\sigma = \{(a, b) \in S \times S : e\alpha a = b\beta f$ for some α-idempotent e and for some β -idempotent f of $S.\}$*

Theorem 7.17 *Let S be an orthodox Γ-semigroup. Then the relation ρ on S is defined by $\rho = \{(a, b) \in S \times S : V_\alpha^\beta(a) = V_\alpha^\beta(b)$, for all $\alpha, \beta \in \Gamma.\}$ is the minimum inverse Γ-semigroup congruence on S.*

Later in 2005, Chattopadhyay [6] introduced right orthodox Γ semigroups as follows.

Definition 7.18 A regular Γ-semigroup S is called a right(resp. left) orthodox Γ-semigroup if for any α-idempotent e and β -idempotent f of S, $e\alpha f$ (resp. $f\alpha e$) is a β -idempotent.

Clearly every orthodox Γ-semigroup is a right orthodox Γ -semigroup as well as left orthodox Γ-semigroup. An important result of right orthodox Γ-semigroups is given below.

Theorem 7.19 *In a regular Γ-semigroup S, the following are equivalent:*

(i) S is a right orthodox Γ-semigroup;
(ii) for any α-idempotent e and β-idempotent f, $V_\beta^\delta(e\alpha f) = V_\alpha^\delta(f\beta e)$ for all $\delta \in \Gamma$;
(iii) for any α-idempotent e and β-idempotent f, if $e\mathcal{R}f$ then $V_\alpha^\delta(e) = V_\beta^\delta(f)$ for all $\delta \in \Gamma$.

In [40] Sheng, Zhao, Zhang introduced band Γ-semigroup. In this paper the band Γ-semigroups and its general structure were discussed. It was determined that a band Γ-semigroup is a semilattice of rectangular band Γ-semigroups. Then, a general structure theorem for band Γ-semigroups was discussed, which generalises the structure theorem for bands due to Petrich.

8 Semidirect Product of a Semigroup and a Γ-Semigroup

In the paper [33] semidirect product of a monoid and a Γ-semigroup and in [32], semidirect product of a semigroup and a Γ-semigroup were introduced by Sen and Chattopadhyay. The authors determined the necessary and sufficient conditions for the semidirect product to be right orthodox, to be left inverse and to be right inverse Γ-semigroup respectively in both the cases. The definition of semidirect product was introduced as follows.

Definition 8.1 [33] Let S be a monoid and T be a Γ-semigroup. Let $End(T)$ denote the set of all endomorphisms on T, i.e., the set of all mappings $f : T \to T$ satisfying $f(a\alpha b) = f(a)\alpha f(b)$ for all $a, b \in T, \alpha \in \Gamma$. Clearly $End(T)$ is a semigroup. Let $\Phi : S \twoheadrightarrow End(T)$ be a given 1-preserving antimorphism, i.e., $\Phi(sr) = \Phi(r)\Phi(s)$ for all $r, s \in S$ and $\Phi(1)$ is the identity mapping from T to T. If $s \in S$ and $t \in T$, write t^s for $(\Phi(s))(t)$ and $T^s = \{t^s : t \in T\}$. Let $S \times_\Phi T = \{(s, t) : s \in S, t \in T\}$. Define $(s_1, t_1)\alpha(s_2, t_2) = (s_1 s_2, t_1^{s_2}\alpha t_2)$ for all $(s_i, t_i) \in S \times_\Phi T, I = 1, 2$ and $\alpha \in \Gamma$. Then $S \times_\Phi T$ is a Γ-semigroup. This Γ-semigroup $S \times_\Phi T$ is called the semidirect product of the monoid S and the Γ-semigroup T.

Theorem 8.2 *[33] Let S be a monoid and T be a Γ-semigroup. Let $\Phi : S \twoheadrightarrow End(T)$ be a given 1-preserving antimorphism. Then the semidirect product $S \times_\Phi T$ is a right(resp. left) orthodox Γ-semigroup if and only if*

(i) S is an orthodox semigroup and T is a right(resp. left) orthodox Γ-semigroup,
(ii) for every $e \in E(S)$ and every $t \in T$, $t \in t^e\Gamma T$ and
(iii) if t^e is an α-idempotent, then t^{ge} is an α -idempotent for every $g \in E(S)$, where $e \in E(S), t \in T$.

Theorem 8.3 *[33] Let S be a monoid, T be a Γ-semigroup and $\Phi : S \twoheadrightarrow End(T)$ be a given 1-preserving antimorphism. Then the semidirect product $S \times_\Phi T$ is a right inverse Γ-semigroup if and only if*

(i) S is a right inverse semigroup and T is a right inverse Γ -semigroup and
(ii) for every $e \in E(S)$ and every $t \in T$, $t \in t^e\Gamma T$.

Theorem 8.4 *[33] Let S be a monoid, T be a Γ-semigroup and $\Phi : S \twoheadrightarrow End(T)$ be a given 1-preserving antimorphism. Then the semidirect product $S \times_\Phi T$ is a left inverse Γ-semigroup if and only if*

(i) S *is a left inverse semigroup and* T *is a left inverse* Γ *-semigroup and*
(ii) *for every* $e \in E(S)$ *and every* $t \in T$, $t = t^e$.

In [32] the authors studied the semidirect product of a semigroup and a Γ-semigroup. Authors took S as a semigroup and T as a Γ-semigroup. $\Phi : S \nrightarrow End(T)$ be a given antimorphism. They defined the semidirect product of S and T in same fashion of previous definition. If $S \times_\Phi T$ is a regular Γ-semigroup and if S has no identity element then T may not be regular. So the absence of the identity element in S may be the reason of the failure of the results described above. In this paper they studied the the results in case of absence of the identity element in S.

Theorem 8.5 *[32] Let* S *be a semigroup and* T *be a* Γ*-semigroup. Let* $\Phi : S \nrightarrow End(T)$ *be a given antimorphism. Then the semidirect product* $S \times_\Phi T$ *is a right(resp. left) orthodox* Γ*-semigroup if and only if*

(i) S *is an orthodox semigroup and* T^e *is a right(resp. left) orthodox* Γ*-semigroup for every* $e \in E(S)$,
(ii) *for every* $e \in E(S)$ *and every* $t \in T$, $t \in t^e \Gamma T$ *and*
(iii) *for every* α*-idempotent* t^e, t^{ge} *is an* α *-idempotent, where* $e, g \in E(S), t \in T$.

Theorem 8.6 *[32] Let* S *be a semigroup,* T *be a* Γ*-semigroup and* $\Phi : S \nrightarrow End(T)$ *be a given antimorphism. Then the semidirect product* $S \times_\Phi T$ *is a right inverse* Γ*-semigroup if and only if*

(i) S *is a right inverse semigroup and* T^e *is a right inverse* Γ*-semigroup for every* $e \in E(S)$ *and*
(i) *for every* $e \in E(S)$ *and every* $t \in T$, $t \in t^e \Gamma T$.

Theorem 8.7 *[32] Let* S *be a semigroup,* T *be a* Γ*-semigroup and* $\Phi : S \nrightarrow End(T)$ *be a given antimorphism. Then the semidirect product* $S \times_\Phi T$ *is a left inverse* Γ*-semigroup if and only if*

(i) S *is a left inverse semigroup and* T^e *is a left inverse* Γ*-semigroup for every* $e \in E(S)$ *and*
(ii) *for every* $e \in E(S)$ *and every* $t \in T$, $t = t^e$.

In [32] Sen and Chattaopadhyay also introduced the notion of Wreath product of a semigroup and a Γ-semigroup and investigated some interesting properties of this product.

9 E-Inversive Γ-Semigroups

Generalizing the regular Γ-semigroup in [34] Sen and Chattaopadhyay introduced the notion E-inversive Γ-semigroup. The main results are given below.

Definition 9.1 Let S be a Γ-semigroup. An element $a \in S$ is called E-inversive if there exist $x \in S$, $\alpha, \beta \in \Gamma$ such that $a\alpha x \in E_\beta$. S is called E-inversive Γ-semigroup if every $a \in S$ is E-inversive.

Definition 9.2 Let S be a Γ-semigroup with zero. A nonzero element $a \in S$ is called E^*-inversive if there exist $x \in S$, $\alpha, \beta \in \Gamma$ such that $0 \neq a\alpha x \in E_\beta$. S is called E^*-inversive Γ-semigroup if every nonzero element $a \in S$ is E^*-inversive.

Definition 9.3 For a Γ-semigroup S, $a \in S$ and $\alpha, \beta \in \Gamma$ the set $W_\alpha^\beta(a)$ is defined by $W_\alpha^\beta(a) = \{x \in S : x\beta a\alpha x = x\}$.

Theorem 9.4 *An element a of a Γ-semigroup S is E-inversive if and only if $W_\alpha^\beta(a) \neq \emptyset$ for some $\alpha, \beta \in \Gamma$.*

Theorem 9.5 *In an E-inversive Γ-semigroup S, $E_\mu \neq \emptyset$ for all $\mu \in \Gamma$.*

Theorem 9.6 *In a Γ-semigroup S, the following conditions are equivalent:*
(i) for two E-inversive elements $a, b \in S$, $a\alpha b$ is an E-inversive element for some $\alpha \in \Gamma$;
(ii) for $e, f \in E(S)$, $e\alpha_1 f$ is an E-inversive element of S for some $\alpha_1 \in \Gamma$.

Theorem 9.7 *Let S be an E-inversive Γ-semigroup. If for every $a \in S$ and $\alpha \in \Gamma$ there exists only one $x \in S$ such that $a\alpha x \in E_\alpha$ then S is a Γ-group.*

Theorem 9.8 *Let $S \times_\Phi T$ be a semidirect product of a semigroup S and a Γ-semigroup T. Then $S \times_\Phi T$ is E-inversive if and only if for all $s \in S, t \in T$ there exists $s' \in W(s)$ such that $t^{s's}$ is an E-inversive element of the Γ-semigroup $T^{s's} = \{t^{s's} : t \in T\}$. If S is an E-inversive semigroup and T is an E-inversive Γ-semigroup, then every semidirect product of S and T is an E-inversive Γ-semigroup.*

In the paper [28] Sattayaporn characterized some properties of E-inversive Γ-semigroup. Moveover, the author also introduced a Γ-group congruence on any E-inversive Γ-semigroup.

Acknowledgments The authors would like to thank the learned referee for his valuable suggestions that improved the praper.

References

1. Booth, G.L.: Operator rings of a Γ -ring. Math. Jpn. **31**(2), 175–183 (1986)
2. Booth, G.L., Groenwald, N.J.: On uniformly strongly prime gamma rings. Bull. Austral. Math. Soc. **37**, 437–445 (1988)
3. Booth, G.L.: The strongly prime radical of gamma ring. Quaest. Math. **11**, 339–359 (1988)
4. Booth, G. L.: Characterising the nil radical of a gamma ring. Quaest. Math. **9**(1–4), 55-67 (1986)
5. Chattopadhyay, S.: Right Inverse Γ-*semigroup*. Bull. Cal. Math. Soc. **93**, 435–442 (2001)
6. Chattopadhyay, S.: Right Orthodox Γ-*semigroup*. Southeast Asian Bull. Math.**29**, 23-30 (2005)

7. Chinram, R., Siammai, P.: On Green's relations for Γ-semigroups and reductiveΓ-semigroups. Int. J. Algebr. **2**(4), 187–195 (2008)
8. Chinram, R., Siammai, P.: On Green's Lemma and Green's Theorem for Γ-semigroups. Lovachevskii J. Math. **30**(3), 208–213 (2009)
9. Clifford, A. H., Preston, G. B.: The Algebraic Theory Of Semigroups, Vol-1, Am. Math. Soc. (1961)
10. Dine, J., Hila, K.: On Green's equivalences on Γ- groupoid, Manuscript
11. Dine, J., Hila, K.: On Green's relations, 2^0 regularity and quasi-ideals in Γ- semigroups, Manuscript
12. Dutta, T.K., Adhikari, N.: C : On Γ-semigroup with right and left unities. Soochow J. Math. **19**(4), 461–474 (1993)
13. Dutta, T.K., Chatterjee, T.K.: Green's equivalences on Γ-semigroup. Bull. Calcutta Math. Soc. **80**, 30–35 (1988)
14. Dutta, T. K., Chattopadhyay, S.: On uniformly strongly prime Γ-semigroup, Analele Stiintifice Ale Universitatii "AL.I.CUZA" IASI Tomul LII,s.I, Mathematica, f.2, 325–335 (2006)
15. Hall. T. E.: On regular semigroups whose idempotents form a subsemigroup. Bull. Aust. Math. Soc. I 195–208 (1969)
16. Heidari, D., Amooshahi, M.: Transformation semigroups associated to Γ- semigroups. Discuss. Math. Gen. Algebr. Appl. **33**, 249–259 (2013)
17. Hestenes, M.R.: On ternary algebra. Scripta. Mathematica. **29**, 253–272 (1953)
18. Hila, K.: On regular, semiprime and quasi relexive Γ- semigroups and minimal quasi ideals. Lovachevskii J. Math. **29**(3), 141–152 (2008)
19. Hila, K., Dine, J.: Study on the structure of periodic Γ- semigroups, Math. Rep. **13**(63) 3, 271–284 (2011)
20. Howie, J. M.: An Introduction to Semigroup Theory. Academic Press, New York (1976)
21. Braja, I.: Characterizations of regular Gamma semi-groups using quazi-ideals. Int. J. Math. Anal. **3**(36), 1789–1794, (2009)
22. Kehayopulu, N.: Green's relations and the relation N in the Γ− semigroups. Quasigroups Relat. Syst. **22**, 89–96 (2014)
23. Liu, G. X.: Orthodox congruences on regular Γ-semigroups, Pure Appl. Math.(Xian), **14**(1), 80–84 (1998)
24. Nobusawa, N.: On a generalisation of ring theory. Osaka J. Math. **1**, 81–89 (1964)
25. Saha, N. K., Seth. A.: Inverse Γ -semigroup. J. Pure Math. **6**, 1–8 (1987-88)
26. Saha, N.: K : On Γ-semigroup II. Bull. cal. Math. Soc. **79**, 331–335 (1987)
27. Sardar, S., Gupta, S., Shum, K.: Gamma semigroups with unities and Morita equivalence for Monoids. Eur. J. Pure Appl. Math. **6**(1), 1–10 (2013)
28. Sattayaporn, S.: Γ- group congruences on E- inversive Γ-semigroup. Thai J. Math. **7**(2), 405–415 (2009)
29. Sen, M. K.: On Γ-semigroup. In: Proceedings of International Conference On Algebra And It's Application, Decker Publication, New York, pp. 301–308 (1981)
30. Sen, M.K., Saha, N.K.: Orthodox Γ -semigroup. Int. J. Mat. Math. Soc. **13**(3), 527–534 (1990)
31. Sen, M.K., Seth, A.: The maximum idempotent-separating congruence in a regular Γ-semigroup. Bull. Calcutta Math. Soc. **82**(2), 131–137 (1990)
32. Sen M. K., Chattopadhyay, S.: Wreath product of a semigroup and a Γ-semigroup. Discuss. Math. Gen. Algebr. Appl. **28**, 161–178 (2008)
33. Sen, M. K., Chattopadhyay, S.: Semidirect product of a semigroup and a Γ-semigroup. East-West J. Math. **6**(2), 131–138(2004)
34. Sen, M.K., Chattopadhyay, S.: E-inversive Γ-semigroups. Kyungpook Math. J. **49**, 457–471 (2009)
35. Sen, M.K., Saha, N.: K : On Γ-semigroup I. Bull. cal. Math. Soc. **78**, 181–186 (1986)
36. Seth, A.: On Γ-semigroup T(A, B). Bull. Cal. Math. Soc. **80**, 424–434 (1988)
37. Seth, A.: Congruences on Γ-group. Soochow Jr. Math. **17**, 233–237 (1991)
38. Seth, A.: Γ-group congruences on regular Γ-semigroups. Int. J. Math. Math. Sci. **15**(1), 103–106 (1992)

39. Seth, A.: Rees's theorem for Γ -semigroup. Bull. Cal. Math. Soc. **81**, 217–226 (1989)
40. Sheng, D.C., Zhao, X. Z., Zhang, W. T.: Band Γ-semigroup. J. Northwest Univ. **30**(4), 279–283 (2000)
41. Sheng, D. C., Liu, G. X.: Γ -semigroups generated by idempotent elements. Pure Appl. Math. **14**(4), 35–38 (1998)
42. Xhillari, T., Braja, I.: Completely Regular Γ-semigroups. Int. J. Algebr. **4**(20), 995–1020 (2010)
43. Yamada, M.: Regular semigroups whose idempotents satisfy permutation identities. Pacific J. Math. **21**(2), 371–392 (1967)
44. Yamada, M.: On regular semigroup in which the idempotents form a band. Pacific J. Math. **33**, 261–272 (1970)

Comparability Axioms in Orthomodular Lattices and Rings with Involution

N.K. Thakare, B.N. Waphare and Avinash Patil

Abstract In this article, a Schröder–Bernstein type theorem is proved for orthomodular lattices. Various comparability axioms available in Baer $*$-rings are introduced in orthomodular lattices. Some applications to complete orthomodular lattices are given. The related classical results in Baer $*$-rings are generalized to $*$-rings.

Keywords Orthomodular lattice · Psuedocomplemented lattice · Relatively semi-orthocomplemented lattice · Parallelogram law

MSC(2010) Primary: 06C15 · Secondary: 06D15

1 Introduction

We assume that the reader is familiar with basics of lattice theory. A bounded lattice is an algebra $(L, (\wedge, \vee, 0, 1))$ where $(L, \wedge, \vee)$ is a lattice with 0 and 1. Two elements a and b of a lattice L are said to form a *modular pair*, denoted by $(a, b)M$, when $(c \vee a) \wedge b = c \vee (a \wedge b)$ holds for all $c \leq b$. An element z of a lattice L with 0 and 1 is called a *central element* when there exist two lattices L_1 and L_2 and an isomorphism between L and the direct product of L_1, L_2 such that z corresponds to the element $[1_1, 0_2] \in L_1 \times L_2$. The set of all central elements of L is called the *center* of L and it is denoted by $Z(L)$. An *orthocomplementation* on a bounded lattice

N.K. Thakare
13, Nimbapushpa, Jawahar Nagar, O Sakri Road, Dhule 424001, India
e-mail: nkthakare@gmail.com

B.N. Waphare (✉)
Center for Advanced Study in Mathematics, Savitribai Phule Pune University, Pune 411007, India
e-mail: bnwaph@yahoo.com; bnwaph@math.unipune.ernet.in

A. Patil
Garware College of Commerce, Karve Road, Pune 411004, India
e-mail: avipmj@gmail.com

S.T. Rizvi et al. (eds.), *Algebra and its Applications*, Springer Proceedings in Mathematics & Statistics 174, DOI 10.1007/978-981-10-1651-6_13

is a unary operation satisfying $a \vee a^\perp = 1$, $a \wedge a^\perp = 0$, $a \le b$ implies $b^\perp \le a^\perp$, $(a^\perp)^\perp = a$. An easy consequence of this are DeMorgan laws $(a \vee b)^\perp = a^\perp \wedge b^\perp$, $(a \wedge b)^\perp = a^\perp \vee b^\perp$.

An *ortholattice* is an algebra $(L, (\wedge, \vee, ^\perp, 0, 1))$ where $(L, (\wedge, \vee, 0, 1))$ is a bounded lattice and $\perp$ is an orthocomplementation on it. An *orthomodular lattice* (abbreviated: OML) is an ortholattice satisfying the orthomodular law: 'If $a \le b$, then $a \vee (a^\perp \wedge b) = b$'. This law can be again replaced by the equation '$a \vee (a^\perp \wedge (a \vee b)) = a \vee b$', see Kalmbach [4] and Stern [11]. Two elements a and b of an OML are said to be *strong perspective* if they have a common complement in $[0, a \vee b]$. The *relative center property* holds in an OML L, if the center of any interval $[0, a]$ of L is the set $\{a \wedge c \mid c \in Z(L)\}$.

In the second section, we consider a relatively semi-orthocomplemented lattice L with 0 and 1 and an equivalence relation on L satisfying some conditions. A Schröder–Bernstein type theorem is proved for OMLs. Similar results were proved in [7, 9]. Here we release the assumptions namely, orthogonal additivity and completeness in OMLs.

We introduce comparability axioms and finiteness in OMLs. In Baer $*$-rings, several comparability axioms such as parallelogram law, generalized comparability, partial comparability, finiteness, etc., are well studied. *Baer $*$-rings* are rings with involution in which right annihilator of any subset is generated by a projection. Berberian [1], Kaplansky [5] carry out detailed investigation of comparability axioms in Baer $*$-rings. There are several deep and interesting open problems mentioned in Kaplansky [5], Berberian [1], Thakare [16] some of which our group succeed in solving; see Thakare and Baliga [13], Waphare [17]. In [14–16] Thakare and Waphare gave nice interplay among these axioms in Baer $*$-rings.

In the third and fourth sections, Psuedocomplementedness is used to obtain some important results involving comparability axioms in OML. In the final section, we provide applications of the results to comparability axioms in general $*$-rings and to OML to have relative center property by considering the strong perspectivity of elements in OMLs.

2 Schröder–Bernstein Type Theorem for OML

In this section, let L be a relatively semi-orthocomplemented lattice with 0 and 1; Z be its center and $\sim$ be an equivalence relation satisfying the following axiom:

"If $a \sim b$, then there is a lattice isomorphism ϕ of $(a]$ onto $(b]$ such that $\phi(x) \sim x$ for every $x \in (a]$ and that $x \perp y \Leftrightarrow \phi(x) \perp \phi(y)$, for $x, y \in (a]$". Here $(a]$ is the principal ideal in L generated by $a \in L$.

Let $\{a_i\}_{i \in I}$ and $\{b_i\}_{i \in I}$ be orthogonal families of the elements of L indexed by the same indexing set I, let $a = \sup a_i$, $b = \sup b_i$ and suppose that $a_i \sim b_i$, for all $i \in I$. Then the obvious question is

Does it follow that $a \sim b$?

If the answer to this question is always affirmative, we say that equivalence $\sim$ is additive in L (or completely additive). If it is affirmative whenever $|I| \leq \aleph$, we say that $\sim$ is $\aleph$-additive; if it is affirmative whenever $a \perp b$, we say that equivalence $\sim$ is orthogonally additive. The term orthogonally $\aleph$-additive is self-explanatory. If I is finite, then we say that equivalence $\sim$ is finitely additive.

For $a, b \in L$ we say that a is dominated by b and write $a \lesssim b$ if there exists $b_1 \leq b$ such that $a \sim b_1 \leq b$. The property $a \lesssim b, b \lesssim a$ implies $a \sim b$ is studied by several mathematicians in different situations. Murray and von Neumann [10] studied the property in rings of operators. Maeda [7] proved that the above property holds in a completely additive, relatively semi-orthocomplemented complete lattice. Maeda [9], himself released the condition of complete additivity and proved the above property in a finitely additive and orthogonally additive relatively semi-complemented complete lattice. Here we succeed to release some of these conditions in OMLs. In the first step, we release the condition of orthogonal additivity to $\aleph_0$ orthogonal additivity, and completeness to σ-completeness. Here by σ-completeness of lattice we mean a lattice in which every countable subset has supremum as well as infimum.

Here is our stipulated Schröder–Bernstein type theorem for OML.

Theorem 2.1 *Let L be a complete OML with finitely additive equivalence relation $\sim$. Then $a \lesssim b$, $b \lesssim a$ implies $a \sim b$.*

Proof Let $a' \leq a, b' \leq b$ such that $a \sim b' \leq b$ and $b \sim a' \leq a$. Assume that $\phi : (a] \to (b]$ and $\psi : (b] \to (a]$ are corresponding isomorphisms. The plan of the proof is to construct an order preserving mapping $\phi_0 : (b] \to (b]$ to which the fixed point theorem is applied. The mapping ϕ_0 is taken to be the composite of four mappings ϕ_1, ϕ_2, ϕ_3 and ϕ_4 defined as follows.

Define $\phi_1 : (b] \to (a]$ by the mapping $g \to \psi(g)$. It is clear that ϕ_1 is an order preserving mapping. Define $\phi_2 : (a] \to (a]$ by $\phi_2(g) = a \wedge g^{\perp}$; Thus ϕ_2 is order reversing. Similarly define $\phi_3 : (a] \to (b]$ by $\phi_3(g) = \phi(g)$ and $\phi_4 : (b] \to (b]$ by $\phi_4(g) = b \wedge g^{\perp}$. Finally define ϕ_0 to be the composite $\phi_0 = \phi_4 \circ \phi_3 \circ \phi_2 \circ \phi_1$. Thus ϕ_0 is order preserving, explicitly, $\phi_0(g) = b \wedge (\phi(a \wedge (\psi(g))^{\perp})^{\perp}$. Since $(b]$ is complete, the fixed point theorem yields an element $g_0 \leq b$ such that $\phi_0(g_0) = g_0$. That is $b \wedge (\phi(a \wedge (\psi(g_0))^{\perp}))^{\perp} = g_0$ implies $b^{\perp} \vee \phi(a \wedge (\psi(g_0))^{\perp}) = g_0^{\perp}$ gives $b \wedge (b^{\perp} \vee \phi(a \wedge (\psi(g_0))^{\perp})) = b \wedge g_0^{\perp}$. Since $(b^{\perp}, b)M$ and $\phi(a \wedge (\psi(g_0))^{\perp}) \leq b$, we have $b \wedge g_0^{\perp} = \phi(a \wedge (\psi(g_0))^{\perp})$. Since $a \wedge (\psi(g_0))^{\perp} \sim \phi(a \wedge (\psi(g_0))^{\perp}) = b \wedge g_0^{\perp}$, $g_0 \sim \psi(g_0)$ and $g_0 \perp b \wedge g_0^{\perp}$, $\psi(g_0) \perp a \wedge \psi(g_0)^{\perp}$, we have $\psi(g_0) \vee ((\psi(g_0))^{\perp} \wedge a) \sim (b \wedge g_0^{\perp}) \vee g_0$. By orthomodularity we get $a \sim b$. □

As a corollary of this theorem we provide lattice theoretic proof of the Schröder–Bernstein theorem of set theory.

Corollary 2.2 *Let X and Y be two sets such that X is numerically equivalent to a subset of Y and Y is numerically equivalent to a subset of X. Then X is numerically equivalent to Y.*

We proceed further to release some more conditions by proving some required lemmas.

Lemma 2.3 *Let L be an OML with finitely additive, orthogonally $\aleph$-additive equivalence relation. Suppose that supremum of every orthogonal family with cardinality $\leq \aleph$ exists.*

Let $\{a_i\}_{i \in I}$ be an orthogonal family of mutually equivalent elements with $|I| \leq \aleph$, and let $J \subset I$ with $|J| = |I|$. Define $a = \sup\{a_i \mid i \in I\}$, $b = \sup\{a_i \mid i \in J\}$. Then $a \sim b$.

Proof Write $J = J' \cup J''$, where $J' \cap J'' = \emptyset$ and $|J'| = |J''| = |J|(= |I|)$. Define $b' = \sup\{a_i \mid i \in J'\}, b'' = \sup\{a_i \mid i \in J''\}, g = \sup\{a_i \mid i \in J'' \cup (I - J)\}$. Since a_i's are orthogonal, we have $b' \perp g$. As $J' \cup (J'' \cup (I - J)) = I$, we get that $a = b' \vee g$ by associativity of suprema. Similarly $b = b' \vee b''$. Since J' and $J'' \cup (I - J)$ have the same cardinality and since $b' \perp g$, we have $b' \sim g$ by the hypothesis. Similarly $b'' \sim b'$. Also we have $J' \cap J'' = \emptyset$ therefore $b' \perp b''$. By the assumed finite additivity, $b' \vee b'' \sim g \vee b'$. Thus $a \sim b$ as required. □

Lemma 2.4 *Let L be an OML in which every orthogonal sequence has supremum. Then every decreasing sequence $\{e_n\}_{n=1}^{\infty}$ has the infimum. Explicitly,* $\inf e_n = e_1 \wedge g^{\perp}$ *where* $g = \sup\{e_n \wedge e_{n+1}^{\perp} \mid n = 1, 2, \ldots\}$.

Proof Clearly the family $\{e_n \wedge e_{n+1}^{\perp} \mid n = 1, 2, \ldots\}$ is orthogonal and $e_1 \wedge g^{\perp} \leq e_1$. By associativity of supremum we have, $g = (e_1 \wedge e_2^{\perp}) \vee g_0$, where $g_0 = \sup\{e_2 \wedge e_3^{\perp}, \ldots\}$. Therefore $g^{\perp} = (e_1 \wedge e_2^{\perp})^{\perp} \wedge g_0^{\perp} = (e_1^{\perp} \vee e_2) \wedge g_0^{\perp}$ implies $e_1 \wedge g^{\perp} = [e_1 \wedge (e_1^{\perp} \vee e_2)] \wedge g_0^{\perp}$. Since $e_2 \leq e_1$ and $(e_1^{\perp}, e_1)M$, we have $e_1 \wedge g^{\perp} = e_2 \wedge g_0^{\perp} \leq e_2$. Similarly, $e_1 \wedge g^{\perp} \leq e_3, e_4$ and so on.

Now, let h be an element in L such that $h \leq e_n$ for $n = 1, 2, \ldots$. Consider $h \vee (e_n \wedge e_{n+1}^{\perp})^{\perp} = h \vee (e_n^{\perp} \vee e_{n+1}) = e_n^{\perp} \vee e_{n+1} = (e_n \wedge e_{n+1}^{\perp})^{\perp}$. Thus $h \leq (e_n \wedge e_{n+1}^{\perp})^{\perp}$, for $n = 1, 2, \ldots$ implies $h^{\perp} \geq e_n \wedge e_{n+1}^{\perp}$, for $n = 1, 2, \ldots$ implies $h^{\perp} \geq g$ implies $g^{\perp} \geq h$ gives $g^{\perp} \wedge e_1 \geq e_1 \wedge h = h$. Hence $e_1 \wedge g^{\perp} = \inf e_n$ □

Now we prove the stipulated result.

Theorem 2.5 *Let L be an OML with finitely additive, orthogonally $\aleph_0$-additive equivalence relation $\sim$. Also assume that supremum of every orthogonal sequence exists. Then $e \lesssim f$, $f \lesssim e$ implies $e \sim f$.*

Proof Assuming $e \sim f' \leq f$, $f \sim e' \leq e$ and $\phi_\circ, \phi_1$ are corresponding orthoisomorphisms, it is to be shown that $e \sim f$.

Put $\phi_1(f') = e''$. Then we have the following situation: $e'' \leq e' \leq e$ and $e'' = \phi_1(f') \sim f' \sim e$. If we prove that $e' \sim e$, then $f \sim e' \sim e$, which is required to prove. Therefore it is sufficient to prove $e' \sim e$. Let $\phi = \phi_1 \circ \phi_\circ$. Since ϕ is an order preserving bijection of $(e]$ onto $(e'']$, we may define a sequence $e_0, e_2, e_4, \ldots$ as $e_0 = e$, $e_2 = \phi(e_0) = \phi(e) = \phi_1(f') = e''$, $e_4 = \phi(e_2), \ldots$ In general $e_{2n} = \phi(e_{2(n-1)})$. Define another sequence $e_1, e_3, e_5, \ldots$ by the same technique, starting with e'; put $e_1 = e'$, $e_3 = \phi(e_1)$, $e_5 = \phi(e_3), \ldots . e_{2n+1} = \phi(e_{2n-1})$, $(n = 1, 2, \ldots)$. Observe that $e_0 \geq e_1 \geq e_2 \geq \ldots$. We now look at the "gaps" in the decreasing sequence $e_0 \geq e_1 \geq e_2 \geq \ldots$, i.e., the set $\{e_0 \wedge e_1^{\perp}, e_1 \wedge e_2^{\perp}, e_2 \wedge e_3^{\perp}, \ldots\}$. It is easy to prove that the above sequence is orthogonal and

$$e_n \wedge e_{n+1}^{\perp} \sim e_{n+2} \wedge e_{n+3}^{\perp} \tag{2.1}$$

By Lemma 2.4, we define $e_\infty = \inf \{e_n \mid n = 0, 1, 2, \ldots\}$. Obviously any truncation of the sequence e_n has the same infimum, in particular, $e_\infty = \inf \{e_n \mid n = 1, 2, \ldots\}$.

Consider the following two sequences of orthogonal elements,

$$e_\infty, e_0 \wedge e_1^{\perp}, e_1 \wedge e_2^{\perp}, e_2 \wedge e_3^{\perp}, \ldots \tag{2.2}$$

$$e_\infty, e_1 \wedge e_2^{\perp}, e_2 \wedge e_3^{\perp}, e_3 \wedge e_4^{\perp}, \ldots \tag{2.3}$$

(the second sequence merely omits the second term of the first sequence).

Let $k = \sup \{e_0 \wedge e_1^{\perp}, e_1 \wedge e_2^{\perp}, e_2 \wedge e_3^{\perp}, \ldots\}$. By Lemma 2.4, we have $e_\infty = e_0 \wedge k^{\perp}$. Therefore $k \vee e_\infty = k \vee (e_0 \wedge k^{\perp}) = e_0$. Thus by the associativity of suprema the sup(2.2)$= e_\infty \vee k = e_0$. Similarly, sup(2.3)$= e_1 = e'$. Now we define $g = \sup\ \{e_0 \wedge e_1^{\perp},\ e_2 \wedge e_3^{\perp},\ e_4 \wedge e_5^{\perp}, \ldots\}$, $g' = \sup\ \{e_2 \wedge e_3^{\perp},\ e_4 \wedge e_5^{\perp}, \ldots\}$, $h = e_\infty \vee \sup \{e_1 \wedge e_2^{\perp}, e_3 \wedge e_4^{\perp}, e_5 \wedge e_6^{\perp}, \ldots\}$. Again by associativity of suprema, $g \vee h$, coincides with the sup(2.2), thus $e = g \vee h$. Similarly, $g' \vee h$ is the sup(2.3) and $e' = g' \vee h$. It follows from the equivalence (2.1), the definitions of g and g', and Lemma 2.3 that $g \sim g'$. Observe that $g \perp h$, $g' \perp h$. Therefore we have $e \sim e'$. □

3 Parallelogram Law and Comparability Axioms in OML

There are several comparability axioms available in Baer $*$-rings. The set of projections of a Baer $*$-ring forms an orthomodular lattice under the partial order, '$e \leq f$ if and only if $e = ef = fe$'. In this section we extend the concepts of generalized comparability, partial comparability to OMLs.

Also S. Maeda [7–9] developed a dimension theory on relatively semi-orthocomplemented complete lattices. We also succeed here to release the condition of completeness upto some extent. We start with necessary definitions and axioms on relatively semi-orthocomplemented lattice. Hereafter in this section, let L be a relatively semi-orthocomplemented lattice with 0 and 1; Z be its center and $\sim$ be an equivalence relation as stated in previous section.

Definition 3.1 (a) The set $Z_0 = \{z \in Z \mid a \lesssim z \text{ implies } a \leq z\}$ is called the relative center with respect to the given equivalence relation.

(b) Relative central cover of an element $a \in L$ (notation $e(a)$) is an element $z \in Z_0$ such that $a \leq z$ and whenever $a \leq z_0$, $z_0 \in Z_0$, we have $z \leq z_0$. It is clear that if L is a complete lattice then relative central cover, i.e.,, $e(a)$ exists for every element $a \in L$ and $a \sim b$ implies $e(a) = e(b)$.

Following result is obvious; also see Maeda [8, Lemma 2.1].

Lemma 3.2 *The relative center Z_0 of L has the following properties:*

(i) *Z_0 is a Boolean sublattice of L.*
(ii) *If $a \sim b$ and $z \in Z_0$, then $z \wedge a \sim z \wedge b$.*

We introduce the concept of generalized comparability and very orthogonality in relatively semi-orthocomplemented lattices with 0 and 1.

Definition 3.3 Elements e, f in L are said to be generalized comparable if there exists $h \in Z_0$ such that $h \wedge e \lesssim h \wedge f$ and $h' \wedge f \lesssim h' \wedge e$, where h' is complement of h. We say that L has generalized comparability (briefly, L has GC) if every pair of elements is generalized comparable.

Definition 3.4 Elements e, f in L are said to be very orthogonal if there exists $h \in Z_0$ such that $e \leq h$ and $f \leq h'$.

Observe that the definitions of generalized comparability and orthogonality are symmetric.

Lemma 3.5 *Following statements are equivalent in a lattice in which relative central cover of elements involved exists.*

(i) *a, b are very orthogonal.*
(ii) *$e(a) \wedge e(b) = 0$.*

Proof $(i) \implies (ii)$: Let $h \in Z_0$ such that $a \leq h$ and $b \leq h'$. By definition of relative central cover we get that $e(a) \leq h$ and $e(b) \leq h'$. Therefore $e(a) \wedge e(b) \leq h \wedge h' = 0$, thus $e(a) \wedge e(b) = 0$.
$(ii) \implies (i)$: Let $e(a) \wedge e(b) = 0$. Put $h = e(a)$. It is immediate that $a \leq h$ and $h \wedge e(b) = 0$. Thus $a \leq h$ and $e(b) \leq h'$. That is, $a \leq h$ and $b \leq e(b) \leq h'$ as required. □

We give here a nice characterization of generalized comparability for relatively semi-orthocomplemented lattice L.

Theorem 3.6 *Let the equivalence relation be finitely additive in L. Then the following statements are equivalent.*

(a) *e, f are generalized comparable.*
(b) *There exists orthogonal decompositions $e = e_1 \vee e_2$, $f = f_1 \vee f_2$ with $e_1 \sim f_1$ and e_2, f_2 are very orthogonal.*

Proof (*a*) $\Longrightarrow$ (*b*): Let $h \in Z_0$ such that $h \wedge e \sim f_1^1 \leq h \wedge f$ and $h' \wedge f \sim e_1^2 \leq h' \wedge e$. Put $e_1^1 = h \wedge e$ and $f_1^2 = h' \wedge f$; we have $e_1^1 \sim f_1^1$ and $f_1^2 \sim e_1^2$. As $(e_1^1)^{\perp} = h' \vee e^{\perp} \geq h' \geq h' \wedge e \geq e_1^2$, we have $e_1^1 \perp e_1^2$. Similarly $f_1^1 \perp f_1^2$. Let $e_1 = e_1^1 \vee e_1^2$, $f_1 = f_1^1 \vee f_1^2$. By finite additivity we get that $e_1 \sim f_1$. Clearly $e_1 \leq e$ and $f_1 \leq f$. Let e_2 be the relative orthocomplement of e_1 in e and f_2 be the relative orthocomplement of f_1 in f. Thus $e = e_1 \vee e_2$ and $f = f_1 \vee f_2$ and $e_1 \sim f_1$. It is sufficient to show that e_2, f_2 are very orthogonal. Since $h \wedge e_2 = h \wedge e \wedge e_2 = e_1^1 \wedge e_2 \leq e_1 \wedge e_2 = 0$ and $h' \wedge f_2 = h' \wedge f \wedge f_2 = f_1^2 \wedge f_2 \leq f_1 \wedge f_2 = 0$, we have $h \wedge e_2 = 0$ and $h' \wedge f_2 = 0$, i.e.,, $e_2 \leq h'$ and $f_2 \leq h$. Thus e_2 and f_2 are very orthogonal.
(*b*) $\Longrightarrow$ (*a*) Let there be the required orthogonal decomposition, say $e = e_1 \vee e_2$, $f = f_1 \vee f_2$ such that $e_1 \sim f_1$ and e_2, f_2 are very orthogonal. Suppose $h \in Z_0$, with the property that $e_2 \leq h$ and $f_2 \leq h'$. Using Lemma 3.2 we get that $h' \wedge e = h' \wedge e_1 \sim h' \wedge f_1 \leq h' \wedge f$ and $h \wedge f = h \wedge f_1 \sim h' \wedge e_1 \leq h \wedge e$. Hence e, f are generalized comparable. □

We introduce the concept of partial comparability and give its connection with generalized comparability in lattices under consideration.

Definition 3.7 Elements e, f in L are said to be partially comparable if there exists nonzero elements e_0, f_0 such that $e_0 \leq e$, $f_0 \leq f$ and $e_0 \sim f_0$. We say that L has partial comparability (briefly L has PC) if every pair e, f in L is either partially comparable or very orthogonal.

Proposition 3.8 *Let the equivalence relation be finitely additive in L. Then GC is stronger than PC.*

Proof Assuming e, f are not partially comparable, it is to be shown that e, f are very orthogonal. By Theorem 3.6, we have orthogonal decompositions $e = e_1 \vee e_2$, $f = f_1 \vee f_2$, $e_1 \sim f_1$ and e_2, f_2 are very orthogonal. By the assumption we have $e_1 = f_1 = 0$. Hence $e = e_2$, $f = f_2$ are very orthogonal. □

Berberian [1, p. 83] raises the question " If a Baer $*$-ring has PC, does it follow that it has GC?"

It is proved by Maeda [8, Lemma 4.1] that with the orthogonal additivity, PC implies orthogonal GC (i.e., GC for orthogonal pairs). It is also proved by Maeda [7, p. 222] with complete additivity, PC implies GC. We give here the statement and its proof as we are using rather different language and different methods.

Theorem 3.9 *Let L be a complete lattice and equivalence relation be completely additive. Then L has PC if and only if L has GC.*

Proof It is clear that if e, f are very orthogonal then they are generalized comparable. Assuming e, f are not very orthogonal, let $\{e_i\}_{i \in I}, \{f_i\}_{i \in I}$ be a maximal pair of orthogonal families of nonzero elements such that $e_i \leq e$, $f_i \leq f$ and $e_i \sim f_i$, $\forall i \in I$. (an application of PC starts the Zorn's Lemma argument). Set $e' = \sup e_i$, $f' = \sup f_i$. Let e'', f'' be relative complements of e', f' in e, f respectively. On one hand $e' \sim f'$, by complete additivity. On the other hand e'', f'' are very orthogonal (if not, an application of PC would contradict the maximality). Thus in view of Theorem 3.6 the orthogonal decompositions $e = e' \vee e''$, $f = f' \vee f''$ shows that e, f are generalized comparable. □

The parallelogram law P is defined in OMLs as follows; see Kalmbach [4].

Definition 3.10 An OML, L is said to satisfy the parallelogram law P if for any pair e, f in L, we have $e \wedge (e^{\perp} \vee f) \sim f \wedge (f^{\perp} \vee e)$.

We see a simple application of parallelogram law.

Proposition 3.11 *Let L be an OML satisfying the parallelogram law P. Then for any pair e, f in L there exist orthogonal decompositions $e = e_1 \vee e_2$, $f = f_1 \vee f_2$ with $e_1 \sim f_1$, $e_2 \perp f$ and $f_2 \perp e$.*

Proof Let $e_1 = e \wedge (e^{\perp} \vee f)$, $f_1 = f \wedge (f^{\perp} \vee e)$. Then by parallelogram law P we have $e_1 \sim f_1$. Set $e_2 = e \wedge e_1^{\perp}$, $f_2 = f \wedge f_1^{\perp}$. Clearly $e = e_1 \vee e_2$, $f = f_1 \vee f_2$. It remains to prove that $e_2 \perp f$ and $f_2 \perp e$. Since $e_2^{\perp} = e^{\perp} \vee e_1 = e^{\perp} \vee (e \wedge (e^{\perp} \vee f)) = e^{\perp} \vee f$. Thus $f \leq e^{\perp} \vee f = e_2^{\perp}$. Therefore $e_2 \perp f$. Similarly $e \perp f_2$. □

With the help of parallelogram law, we reduce the condition of complete additivity in Theorem 3.9 to orthogonal additivity.

Theorem 3.12 *Let L be an orthomodular complete lattice satisfying the parallelogram law P and with an orthogonally additive equivalence relation. Then L has PC if and only if L has GC.*

Proof First observe that if e, f are orthogonal then e, f are generalized comparable; see Maeda [8, Lemma 4.1]. Let e, f be any pair in L. Since L satisfies parallelogram law P, we have the orthogonal decompositions $e = e' \vee e''$, $f = f' \vee f''$ with $e' \sim f'$, $e'' \perp f$ and $f'' \perp e$. It is clear that $e'' \perp f''$ and hence they are generalized comparable. Therefore by Theorem 3.6 we have $e'' = e_1 \vee e_2$, $f'' = f_1 \vee f_2$ with $e_1 \sim f_1$ and e_2, f_2 are very orthogonal. Thus we have $e = e' \vee e_1 \vee e_2$, $f = f' \vee f_1 \vee f_2$. By finite additivity we get that $e' \vee e_1 \sim f' \vee f_1$. Also observe that $e_2 \perp (e' \vee e_1)$ and $f_2 \perp (f' \vee f_1)$. Again by applying Theorem 3.6, we have e, f are generalized comparable. □

We prove here a nice result as a combined effect of GC and P.

Theorem 3.13 *Let L be an OML with GC, P and finitely additive equivalence relation. Then for any pair e, f in L there exists an element $h \in Z_0$ such that $h \wedge e \lesssim h \wedge f$ and $h' \wedge e^{\perp} \lesssim h' \wedge f^{\perp}$.*

Proof Applying GC to the pair $e \wedge f^{\perp}$ and $e^{\perp} \wedge f$, we get an element $h \in Z_0$ such that

$$h \wedge (e \wedge f^{\perp}) \lesssim h \wedge (e^{\perp} \wedge f) \tag{3.1}$$

and

$$h' \wedge (e^{\perp} \wedge f) \lesssim h' \wedge (e \wedge f^{\perp}) \tag{3.2}$$

It follows from P that $e \wedge (e^{\perp} \vee f) \sim f \wedge (f^{\perp} \vee e)$ and $e^{\perp} \wedge (e \vee f^{\perp}) \sim f^{\perp} \wedge (e^{\perp} \vee f)$. Therefore by Lemma 3.2, we have

$$h \wedge e \wedge (e^{\perp} \vee f) \sim h \wedge f \wedge (f^{\perp} \vee e) \tag{3.3}$$

and

$$h' \wedge e^{\perp} \wedge (e \vee f^{\perp}) \sim h' \wedge f^{\perp} \wedge (e^{\perp} \vee f) \tag{3.4}$$

Consider the conditions (3.1) and (3.3). It is clear that $h \wedge (e \wedge f^{\perp}) \perp (h \wedge e) \wedge (e^{\perp} \vee f)$ and $h \wedge (e^{\perp} \wedge f) \perp (h \wedge f) \wedge (e \vee f^{\perp})$.

We get by finite additivity that $[h \wedge (e \wedge f^{\perp})] \vee [(h \wedge e) \wedge (e^{\perp} \vee f)] \lesssim [h \wedge (e^{\perp} \wedge f)] \vee [(h \wedge f) \wedge (e \vee f^{\perp})]$.

Therefore $h \wedge \{(e \wedge f^{\perp}) \vee [e \wedge (e^{\perp} \vee f)]\} \lesssim h \wedge \{(e^{\perp} \wedge f) \vee [f \wedge (e \vee f^{\perp})]\}$.

Since $(e \wedge f^{\perp})^{\perp} = e^{\perp} \vee f$ and $(e^{\perp} \wedge f)^{\perp} = e \vee f^{\perp}$, we get by Maeda and Maeda [6, Theorem 29.13, p. 132] that $h \wedge e \lesssim h \wedge f$. Similarly, by conditions (3.2) and (3.4), we get that $h' \wedge e^{\perp} \lesssim h' \wedge f^{\perp}$. □

We close the section by adding one more result in the process.

Theorem 3.14 *Let L be a lattice in which every element has a relative central cover and having PC. Suppose $\{e_i\}_{i \in I}$ is a family in L with the following property:*

For every nonzero element $h \in Z_0$ the set of indices $\{i \in I \mid h \wedge e_i \neq 0\}$ is infinite. Given any positive integer n there exists n−indices $i_1, i_2, \ldots, i_n$ and nonzero elements $g_k \leq e_{i_k} (k = 1, 2, \ldots, n)$ such that $g_1 \sim g_2 \sim \cdots \sim g_n$.

Proof The proof is by induction on n. The case $n = 1$ is trivial as the set $\{i \mid 1 \wedge e_i \neq 0\}$ is infinite, and any of its member will serve as i, with $g_1 = e_{i_1}$.

Assume inductively that all is well with $n - 1$, and consider n. By assumption, there exist distinct indices $i_1, i_2, \ldots, i_{n-1}$ and nonzero elements $f_1, f_2, \ldots, f_{n-1}$ such that $f_j \leq e_{i_j} (j = 1, 2, \ldots, n-1)$ such that $f_1 \sim f_2 \sim \cdots \sim f_{n-1}$. Since the relative central cover $e(f_1) \neq 0$, it is clear by the hypothesis that there exists an index i_n, distinct from $i_1, i_2, \ldots, i_{n-1}$ such that $e(f_1) \wedge e_{i_n} \neq 0$. Then $e(f_1) \wedge e(e_{i_n}) \neq 0$. Thus f_1 and e_{i_n} are very orthogonal. Citing PC there exist nonzero elements g_1, g_n such that $g_1 \leq f_1, g_n \leq e_{i_n}$ and $g_1 \sim g_n$. For $j = 2, \ldots, n-1$ the equivalence $f_1 \sim f_j$ transforms g_1 into $g_j \leq f_j$ with $g_1 \sim g_j$, thus $g_n \sim g_1 \sim g_j$ $(j = 2, \ldots, n-1)$. □

4 Finiteness in OMLs

In this section, we consider L as an OML with an equivalence relation as in previous section. We start with the definition of finite element and related aspects.

Definition 4.1 An element e in L is said to be finite if $e \sim f \leq e$ implies $e = f$. An OML, L is said to be finite if every element in L is finite, it is said to be infinite if it is not finite. L is said to be properly infinite if 0 is the only finite element in Z_0.

Following result is easy and proved by Maeda [9, p.219].

Lemma 4.2 *Let L be a lattice with finitely additive equivalence relation. Then $f \lesssim e$, e is finite together imply that f is finite.*

We hasten to add a basic result about infinite lattices.

Proposition 4.3 *Let L be an orthomodular σ-complete lattice with finite additivity and orthogonal $\aleph_0$-additivity. Then L is infinite if and only if there exists a sequence $\{f_n\}$ of orthogonal, mutually equivalent nonzero elements.*

Proof By hypothesis, there exists an element $e \neq 1$ such that $e \sim 1$ and $\phi : L \to (e]$ be an orthoisomorphism. In particular ϕ is an order preserving bijection. Define $e_1 = 1$ and inductively $e_{n+1} = \phi(e_n)$, for $n = 1, 2, \ldots$ In particular $e_2 = \phi(1) = e$. Since $e_1 \geq e_2$ and $e_1 \neq e_2$, an application of ϕ to the inequality $e_1 \geq e_2$ yields $e_2 \geq e_3$, $e_2 \neq e_3$. Continuing inductively, we see that the sequence $\{e_n\}$ is strictly decreasing. Defining, $f_n = e_n \wedge e_{n+1}^{\perp}$ $(n = 1, 2, \ldots)$. We have an orthogonal sequence of nonzero elements; moreover, $\phi(f_n) = \phi(e_n) \wedge \phi(e_{n+1})^{\perp} = e_{n+1} \wedge e_{n+2}^{\perp} = f_{n+1}$ shows that $f_n \sim f_{n+1}$.

Conversely, suppose that $\{f_n\}$ is an orthogonal sequence of nonzero elements such that $f_1 \sim f_2 \sim \cdots$ By hypothesis, we may define $e = \sup\{f_n \mid n \geq 1\}$ and $f = \sup\ \{f_n \mid n \geq 2\}$. Then $f_1 \perp f$ and $e = f_1 \vee f$. We have $e \sim f$, by Lemma 2.3, where $f \leq e$ and $e \wedge f^{\perp} = (f_1 \vee f) \wedge f^{\perp} = f_1 \neq 0$. It follows that L is not finite. □

It is useful to have terminology to describe orthogonal families such as those occurring in the above proposition.

Definition 4.4 (a) An orthogonal family of nonzero elements $\{e_i\}$ is called a partition with terms e_i, if $e = \sup\ \{e_i\}$ exists, it is called a partition of e.
(b) Two equipotent partitions $\{e_i\}_{i \in I}$, $\{f_i\}_{i \in I}$ are equivalent if $e_i \sim f_i$, for all $i \in I$.
(c) A partition $\{e_i\}$ is homogeneous if its terms are mutually equivalent.
(d) A homogeneous partition $\{e_i\}$ is called maximal if it cannot be enlarged; that is, there does not exist an element e such that $e \sim e_i$ and $e \perp e_i$, for all i.

Remark 1 If $e \sim f$ and $\{e_i\}$ is a partition of e, then there exists a partition $\{f_i\}$ of f that is equivalent to $\{e_i\}$. If in addition $\{e_i\}$ is homogeneous then so is $\{f_i\}$.

Remark 2 Every homogeneous partition can be enlarged to a maximal one (a routine application of Zorn's Lemma).

We proceed further to obtain a homogeneous partition of 1 in a properly infinite lattice L, in the presence of orthogonal GC. In this direction we take a first step by proving the following result.

Proposition 4.5 *Let L be a complete lattice with orthogonal GC, finitely additive equivalence relation and having infinitely many terms, then there exists a nonzero element $h \in Z_0$ and a homogeneous partition $\{f_i\}$ of h that is equivalent to $h \wedge e_i$.*

Proof Fix an index i_0; for simplicity, write $i_0 = 1$. Set $e = \sup e_i$. Since $e^\perp$ and e_1 are orthogonal, by hypothesis there exists an element $h \in Z_0$ such that

$$h \wedge e^\perp \lesssim h \wedge e_1 \qquad \text{and} \qquad h' \wedge e_1 \lesssim h' \wedge e^\perp \tag{4.1}$$

Necessarily h is not orthogonal to e_1; for $h \perp e_1$ implies $e_1 = h' \wedge e_1 \lesssim h' \wedge e^\perp \leq e^\perp$, contrary to the maximality of $\{e_i\}$. Since $\{h \wedge e_i\}$ is a homogeneous partition of $h \wedge e$, it will suffice by Remark 1 above, to show that $h \wedge e \sim h$. In view of Theorem 2.1, it is sufficient to prove that $h \lesssim h \wedge e$. Let $f = \sup \{e_i \mid i \neq 1\}$; thus $e = f \vee e_1$ and $f \perp e_1$. Since the family $\{e_i\}$ is infinite, we have $e \sim f$, therefore

$$h \wedge e \sim h \wedge f \tag{4.2}$$

From (4.1) and (4.2), we have $(h \wedge e) \vee (h \wedge e^\perp) \lesssim (h \wedge f) \vee (h \wedge e_1)$ implies $h \lesssim h \wedge (f \vee e_1) = h \wedge e$. □

Here, we obtain a sequence which is a homogeneous partition of 1 in a properly infinite lattice with orthogonal GC.

Proposition 4.6 *If L is a properly infinite complete lattice with orthogonal GC and orthogonally additive equivalence relation, then there exists a sequence e_n, that is a homogeneous partition of* 1.

Proof By Proposition 4.3 there exists a homogeneous partition with infinitely many terms, which we can suppose to be maximal. Invoking Proposition 4.5, there exists a nonzero element $h \in Z_0$ (the relative center of L) that possesses a homogeneous partition $\{f_i\}_{i \in I}$ with infinitely many terms. Since $\aleph_0 |I| = |I|$, the index set I can be written as the union of a disjoint sequence of equipotent sets I_n, $I = I_1 \cup I_2 \cup I_3 \cup \cdots$

Defining $f_n = \sup \{f_i \mid i \in I_n\}$ $(n = 1, 2, \ldots)$ we have $f_m \sim f_n$, for all m, n by orthogonal additivity and $\sup f_n = h$. Summarizing, there exists a nonzero element $h \in Z_0$ and a sequence $\{f_n\}$ that is a homogeneous partition of h.

Let $\{h_\alpha\}_{\alpha \in \Lambda}$ be a maximal family of orthogonal, nonzero elements in Z_0 such that for each $\alpha \in \Lambda$, there exists a sequence $\{e_{\alpha_n}\}$ that is a homogeneous partition of h_α. Defining $e_n = \sup \{e_{\alpha_n} \mid \alpha \in \Lambda\}$ $(n = 1, 2, \ldots)$

We have $e_m \sim e_n$ for all m, n and $\sup e_n = \sup h_\alpha$. It will suffice to show that $\sup h_\alpha = 1$. Assume to the contrary that $\sup h_\alpha \neq 1$, i.e.,, $(\sup h_\alpha)^\perp \neq 0$. Then $(\sup h_\alpha)^\perp$ is infinite (because L is properly infinite); by the first part of the proof, it contains a nonzero element $h \in Z_0$ and a sequence $\{f_n\}$ that is a homogeneous partition of h. This contradicts the maximality of the family $\{h_\alpha\}_{\alpha \in \Lambda}$. □

In the conclusion of above proposition, the e_n's are mutually equivalent, this can be achieved by making them equivalent to 1.

Theorem 4.7 *Let L be a properly infinite complete lattice with orthogonal GC and orthogonally additive equivalence relation. Then*

(1) There exists an orthogonal sequence of elements f_n such that $\sup f_n = 1$ and $f_n \sim 1, \forall n$.
(2) For each positive integer m, there exists orthogonal elements $g_1, g_2, \ldots, g_m$ such that $g_1 \vee g_2 \vee \cdots \vee g_m = 1$ and $g_i \sim 1, \forall i$.

Proof By Proposition 4.6, there exists a sequence e_n that is homogeneous partition of 1.

(1) Write the index set $I = \{1, 2, \ldots\}$ as the union of disjoint sequences of infinite subsets, $I = I_1 \cup I_2 \cup \cdots \cup I_n \cup \cdots$ and define $f_n = \sup \{e_i \mid i \in I_n\}$. The f_n's are mutually orthogonal, $\sup f_n = 1$ and $f_n \sim 1$.
(2) The proof is similar, based on a partition of I into infinite subsets $I_1, I_2, \ldots, I_m$. □

We also prove the following interesting result.

Theorem 4.8 *Let L be a finite lattice with GC and finitely additive equivalence relation. Then $e \sim f$ implies $e^\perp \sim f^\perp$.*

Proof Apply GC to the pair $e^\perp$, $f^\perp$; we have an element $h \in Z_0$ such that

$$h \wedge e^\perp \sim f_1 \leq h \wedge f^\perp \tag{4.3}$$

$$h' \wedge f^\perp \sim e_1 \leq h' \wedge e^\perp \tag{4.4}$$

Since $h \wedge e \sim h \wedge f$, join with (4.3) yields $h = (h \wedge e) \vee (h \wedge e^\perp) \sim (h \wedge f) \vee f_1 \leq h$.

Therefore by finiteness we have $(h \wedge f) \vee f_1 = h$ implies $[(h \wedge f) \vee f_1] \wedge f^\perp = h \wedge f^\perp$ implies $f_1 \vee [(h \wedge f) \wedge f^\perp] = h \wedge f^\perp$ gives $f_1 = h \wedge f^\perp$. Thus (4.3) becomes $h \wedge e^\perp \sim h \wedge f^\perp$. Similarly (4.4) becomes $h' \wedge e^\perp \sim h' \wedge f^\perp$. Adding these equivalences we have $e^\perp \sim f^\perp$. □

As applications of pseudocomplementedness we prove two important results. In the first result, we prove that supremum of very orthogonal family of finite elements is finite and in the second result we prove that the finite elements form a lattice.

Theorem 4.9 *Let L be a pseudocomplemented complete lattice. If $\{e_i\}_{i \in I}$ is a very orthogonal family of finite elements and if $e_0 = \sup e_i$, then e_0 is also finite.*

Proof Write $h_i = e(e_i)$, $h = \sup h_i$. First, we prove that $h_i \wedge e_0 = e_i$, for every i. Set $x = h_i \wedge e_0 \wedge e_i^\perp$ for fixed i, implies $x \wedge e_i = 0$. Also $x \wedge e_j = h_i \wedge e_0 \wedge e_i^\perp \wedge e_j \wedge h_j = (h_i \wedge h_j) \wedge e_0 \wedge e_i^\perp \wedge e_j = 0$; by assumed very orthogonality. Thus $x \wedge e_0 = 0$; as L is pseudocomplemented. This gives us $x = 0$ implies $x \vee e_i = e_i$ implies $e_i \vee (h_i \wedge e_0 \wedge e_i^\perp) = e_i$ implies $h_i \wedge e_0 = e_i$; since L is orthomodular.

Now suppose e_i's are finite and $e_0 \sim f \leq e_0$. We have to prove that $e_0 = f$. Since $h_i \wedge e_0 \sim h_i \wedge f \leq h_i \wedge e_0$ we have $e_i \sim h_i \wedge f \leq e_i$ implies $h_i \wedge f = e_i$, $\forall i$, implies $h_i \wedge f = h_i \wedge e_0$, $\forall i$ gives $h_i \wedge f \wedge f^{\perp} = h_i \wedge e_0 \wedge f^{\perp}$ implies $h_i \wedge e_0 \wedge f^{\perp} = 0, \forall i \in I$, then $h \wedge e_0 \wedge f^{\perp} = 0$, since L is pseudocomplemented. As $e_0 \leq h$ we get that $e_0 \wedge f^{\perp} = 0$. Now the orthomodularity together with $f \leq e_0$ yields that $e_0 = f \vee (e_0 \wedge f^{\perp}) = f$ as required. □

Before proving the next result, it should be mentioned here that Maeda [8, p.387] has proved the same result but under the assumption of completeness and orthogonal additivity. Here we succeed to release the condition of completeness up to σ-completeness and orthogonal additivity upto $\aleph_0$-orthogonal additivity with the help of pseudocomplementedness.

Theorem 4.10 *Let L be a σ-complete pseudocomplemented lattice with GC, P, finitely and orthogonally $\aleph_0$-additive equivalence relation. If e, f are finite elements then $e \vee f$ is also finite.*

Proof Note that the lattice $(e \vee f]$ also satisfies all the axioms of the theorem except P in which e, f are finite elements, dropping down to it, we can suppose that $e \vee f = 1$. Citing P we have, $e^{\perp} = (e \vee f) \wedge e^{\perp} \sim f \wedge (e \wedge f)^{\perp} \leq f$. Thus $e^{\perp} \lesssim f$. Since f is finite so is $e^{\perp}$. Thus $1 = e \vee e^{\perp}$ is the sum of orthogonal finite elements. Changing the notation we can suppose that $e \perp f$ and $e \vee f = 1$.

Assume to the contrary that L is not finite and let $\{g_n\}$ be a sequence of orthogonal equivalent, nonzero elements. Define $g = \sup \{g_n \mid n = 1, 2, \ldots\}$, $g' = \sup \{g_n \mid n$ is odd $\}$, $g'' = \sup \{g_n \mid n$ is even $\}$. Then $g' \perp g''$, $g = g' \vee g''$ and $g' \sim g \sim g''$. Applying GC to the pair $g' \wedge e$, $g'' \wedge f$ there exists an element $h \in Z_0$ such that

(i) $h \wedge (g' \wedge e) \lesssim h \wedge (g'' \wedge f)$
(ii) $h' \wedge (g'' \wedge f) \lesssim h' \wedge (g' \wedge e)$

Citing parallelogram law P we have

(iii) $h \wedge [g' \wedge (g' \wedge e)^{\perp}] \sim h \wedge [(g' \vee e) \wedge e^{\perp}]$.

The left-hand sides if (i) and (iii) are obviously orthogonal; prior to gluing them we check that right-hand sides are orthogonal too. Since $g'' \wedge f \perp g'$ also $g'' \wedge f \perp e$, hence $g'' \wedge f \perp g' \vee e$. Combining (i) and (iii) we have $h \wedge g' = [h \wedge (g' \wedge e)] \vee [h \wedge g' \wedge (g' \wedge e)^{\perp}] \lesssim [h \wedge (g'' \wedge f)] \vee [h \wedge (g' \vee e) \wedge e^{\perp}]$. Since $g'' \wedge f \leq f$ and $(g' \vee e) \wedge e^{\perp} \leq e^{\perp} = f$, we have $h \wedge g' \lesssim f$. Further by $h \wedge g \sim h \wedge g'$, we have $(*)$ $h \wedge g \lesssim f$. Again citing P,

(iv) $h' \wedge [g'' \wedge (g'' \wedge f)^{\perp}] \sim h' \wedge [(g'' \vee f) \wedge f^{\perp}]$, by similar arguments (ii) and (iv) gives, $(**)$ $h' \wedge g \lesssim e$.

From $(*)$ and $(**)$ we see that $h \wedge g$, $h' \wedge g$ are finite. But $h \wedge g$ is the supremum of the sequence $\{h \wedge g_n\}$ of orthogonal, equivalent elements. It follows By Lemma 2.3 that $h \wedge g_n = 0, \forall n$. Thus $h \wedge g = 0$, as L is pseudocomplemented Similarly $h' \wedge g = 0$. Thus $g = 0$, a contradiction. □

We provide an example of a lattice which is incomplete but satisfies all the axioms of Theorem 4.10.

Example 4.11 Let $L = \{X \mid X \subseteq \mathbb{R}$, and X is either countable or complement of a countable set$\}$. It is clear that L is a lattice under inclusion. In fact it is a ring of sets. Put orthocomplementation $\perp$ as $A^{\perp} = A^{c}$, a set theoretic complement. Let numerical equivalence relation be the equivalence relation $\sim$.

It follows that L is not a complete lattice but it satisfies all the axioms of Theorem 4.10. Hence the set of all finite elements in L forms a sublattice. It is interesting to note that an element of L is finite if and only if it is a finite subset of $\mathbb{R}$.

5 Applications

Let A be a $*$-ring in which the set of all projections $\tilde{A}$ forms a bounded lattices. Then it is easy to see that the lattice $\tilde{A}$ is orthocomplemented with orthocomplementation defined by $e \to e^{\perp} = 1 - e$. In fact it is an orthomodular lattice.

Lemma 5.1 *Let A be a $*$-rings with 1 in which $\tilde{A}$ forms a lattice. Then $\tilde{A}$ is an orthomodular lattice.*

Proof Let $e \leq f$. Therefore $(1-e)f = f - ef = f(1-e)$, i.e.,, $1-e$ and f are commuting projections, hence $(1-e) \wedge f = (1-e)f$. Also e and $(1-e)f$ are orthogonal projections. Thus $e \vee [(1-e) \wedge f] = e + (1-e)f = e + f - ef = e + f - e = f$. It follows by Maeda and Maeda [6, Theorem 29.13, p.132] that $\tilde{A}$ is an orthomodular lattice. □

Let A be a $*$-ring and $x \in A$. We say that x possesses a central cover if there exists a smallest central projection h such that $hx = x$. If such a projection h exists, then it is unique, it is called the *central cover of* x, denoted by $h = c(x)$. A projection e is said to be *dominated* by the projection f, denoted by $e \precsim f$, if $e \sim g \leq f$, for some projection g in A. Two projections e and f are said to be *generalized comparable* if there exists a central projection h such that $he \precsim hf$ and $(1-h)f \precsim (1-h)e$. A $*$-ring is said to satisfy the *generalized comparability*(GC) if any two projections are generalized comparable. Two projections e and f are said to be *partially comparable* if there exist nonzero projections e_0, f_0 in R such that $e_0 \leq e$, $f_0 \leq f$ and $e_0 \sim f_0$. If for any pair of projections in A, $eAf \neq 0$ implies e and f are partially comparable, then A is said to satisfy *partial comparability*(PC). More about comparability axioms on set of projection can be found in Berberian [1].

Before proving the next result we mention here the restrictions on a $*$-ring A.

(#) $\tilde{A}$ forms a complete lattice.

(##) The relative center Z_0 with respect to the equivalence relation $\sim$ in $\tilde{A}$ coincides with the set of all central projections in A. (It follows that for $f \in \tilde{A}$, $c(f)$, $e(f)$ exists and $c(f) = e(f)$).

Theorem 5.2 *Let A be a $*$-ring in which equivalence is completely additive and A satisfies conditions (#) and (##) mentioned above. Then A has PC implies that A has GC.*

Proof First, we prove that Ã has PC. Let $c(e) \wedge c(f) \neq 0$ implies $eAf \neq 0$. Since A has PC, we have nonzero sub-projections $e_0 \leq e$, $f_0 \leq f$ such that $e_0 \sim f_0$. Hence Ã has PC. Now observe that all condition of Theorem 3.12 are satisfied, hence we have GC in Ã. We prove that A has GC. Let $e, f \in$ Ã; since Ã has GC, we have an element $h \in Z_0$ such that $h \wedge e \lesssim h \wedge f$ and $h' \wedge f \lesssim h' \wedge e$. By assumption, we have a central projection h such that $he = h\wedge \lesssim h \wedge f = hf$ and $(1-h)f \lesssim (1-h)e$. □

We provide an example of a $*$-ring which is not a Baer $*$-ring but satisfies all the axioms of Theorem 5.2.

Example 5.3 Consider the ring $A = \mathbb{Z}_4 = \{0, 1, 2, 3\}$ with identity mapping as an involution. Then Ã={0, 1}= the set of projections in A. It is clear that A satisfies all the conditions of Theorem 5.2; but it is not a Baer $*$-ring.

Now, we see another open problem raised by Berberian [1, p. 110].
Open Problem: If A is a Baer $*$-ring with GC and e, f are finite projections in A, is $e \vee f$ finite?

Let us see the most general partial answer provided by Berberian [1, p.102].

Theorem 5.4 *Let A be a Rickart $*$-ring with GC, satisfying the parallelogram law P, such that every sequence of orthogonal projections in A has a supremum. If e, f are finite projections in A, then $e \vee f$ is also finite.*

We prove the above theorem by dropping the condition that the underlying $*$-ring be a Rickart $*$-ring.

Theorem 5.5 *Let A be a $*$-ring with GC, P and in which the set of projections forms a pseudocomplemented, σ-complete lattice. If A satisfies the above condition (##) and moreover $\sim$ is orthogonally $\aleph_0$-additive, then $e, f \in \tilde{A}$ are finite imply that $e \vee f$ is finite.*

Proof Observe that Ã satisfies all the conditions of Theorem 4.10. Hence $e \vee f$ is finite. □

We provide an example of a $*$-ring which satisfies all the axioms of Theorem 5.4 but it is not a Rickart $*$-ring.

Example 5.6 Consider the ring of sets L as in Example 4.11. It is a ring with intersection as the multiplication and symmetric difference as the addition. Therefore L is a $*$-ring with identity involution. Put $A = \mathbb{Z}_4 \oplus L$. As $\mathbb{Z}_4$ is not a Rickart $*$ring, therefore A is not a Rickart $*$-ring; but A clearly satisfies all the axioms of Theorem 5.5. In this $*$-ring we see that $e \vee f$ in finite whenever e and f are finite.

For the remaining part of the section $\sim$ denotes the relation "strong perspectivity," which is reflexive and symmetric relation. Chevalier defines [2, Definition 3.1]:

Definition 5.7 An OML, L is said to satisfy the axiom of comparability (abbreviated $A.C.$), if $x, y \in L$, there exists a central element h such that: $x \wedge h \lesssim y \wedge h$ and $y \wedge h^{\perp} \lesssim x \wedge h^{\perp}$, for all $x, y \in L$.

Observe that, the axiom $A.C.$ defined by Chevalier is the modified GC by taking $h \in Z(L)$ instead of taking in Z_0 and an analog of Theorem 3.6 is also proved [2, Proposition 2].

In this section, we will use GC with modification, i.e.,, taking $h \in Z(L)$ in Definition 3.3. Observe that the definitions of GC, PC and Theorem 3.6 are valid if we use strong perspectivity (as transitivity of $\sim$ is not required). The advantage of considering $\sim$ as strong perspectivity is that every OML satisfies the parallelogram law P (as $\phi_a(b) = a \wedge (a^{\perp} \vee b)$ is strongly perspective to $\phi_b(a)$).

Recall [8, Definition 3.1], two elements a and b of a lattice L are said to be unrelated if $a_1 \leq a$, $b_1 \leq b$ and $a_1 \sim b_1$ together imply $a_1 = b_1 = 0$. Using this, we have the following result.

Theorem 5.8 *Every OML with relative center property has PC under strong perspectivity.*

Proof It is enough to show that two unrelated elements are very orthogonal. Let a and b be unrelated. Then by parallelogram law, $a \wedge (a^{\perp} \vee b) \sim b \wedge (b^{\perp} \vee a)$. As a and b are unrelated, we have $a \wedge (a^{\perp} \vee b) = 0$ and $b \wedge (b^{\perp} \vee a) = 0$. Now $a \wedge (a^{\perp} \vee b) = 0$ gives $a^{\perp} \vee b \leq a^{\perp}$ implies $b \leq a^{\perp}$. Similarly $b \wedge (b^{\perp} \vee a) = 0$ gives $a \leq b^{\perp}$. Now by [4, Lemma 9, p.108], there exists a central element $h \in L$ such that $a \leq h$ and $b \leq h^{\perp}$, i.e.,, a and b are very orthogonal. Hence L has PC. □

Corollary 5.9 *Let L be a complete OML with finitely additive strong perspectivity. Then L has relative center property if and only if it has PC under strong perspectivity.*

Proof If L has relative center property then by Theorem 5.8, L has PC. Conversely if L has PC then by Theorem 3.9 L has GC and by [3, Proposition 7], L has relative center property. □

References

1. Berberian, S.K.: Baer ∗-Rings. Springer, Berlin (1972)
2. Chevalier, G.: Around the relative center property in orthomodular lattices. Proc. Am. Math. Soc. **112**(4), 935–948 (1991)
3. Chevalier, G.: Commutators and decompositions of orthomodular lattices. Order **6**, 181–194 (1989)
4. Kalmbach, G.: Orthomodul. Lattices. Academic Press. Inc., London (1983)
5. Kaplansky, I.: Rings of Operators. Benjamin, New York (1968)
6. Maeda, F., Maeda, S.: Theory of Symmetric Lattices. Springer, Berlin (1970)
7. Maeda, S.: Dimension function of certain lattices. J. Sci. Hiroshima Univ. Ser. A **22**, 211–237 (1955)
8. Maeda, S.: Dimension theory on relatively semiorthocomplemented complete lattices. J. Sci. Hiroshima Univ. Ser. A **25**, 369–404 (1961)
9. Maeda, S.: On the lattice of projections of a Baer ∗-ring. J. Sci. Hiroshima Univ. Ser. A **22**, 76–88 (1958)
10. Murry, J., von Neumann, J.: On rings of operators. Ann. Math. **37**(2), 116–229 (1936)

11. Manfred, S.: Semimodular Lattices - Theory and Applications, Encyclopedia of Mathematics and its Applications, vol. 73. Cambridge University Press, Cambridge (1999). xiv+370 pp
12. Thakare, N.K.: Encounters with problems from algebra, discrete mathematics and soft as well as hard analysis. Math. Stud. **68**, 219–241 (1999)
13. Thakare, N.K., Baliga, A.: A Unitary as a product of symmetries. Proc. Am. Math. Soc. **123**(4), 1005–1008 (1995)
14. Thakare, N.K., Waphare, B.N.: Partial solutions to the open problem of unitification of a weakly Rickart $*$-ring. Indian J. Pure Appl. Math. **28**, 189–195 (1997)
15. Thakare, N.K., Waphare, B.N.: Baer $*$-rings with finitely many elements. J. Comb. Math. Comb. Comput. **26**, 161–164 (1998)
16. Thakare, N.K., Waphare, B.N.: Problems, some open. In: Tariq Rizvi, S., Zaidi, S.M.A. (eds.) Baer *-rings: A Survey, Trends in Theory of Rings and Modules. Anamaya Publishers, New Delhi (2005)
17. Waphare, B.N.: Solutions of Some Open Problems From The Theory of Baer *-rings, M. Phil thesis accepted by University of Poona (1988)
18. Waphare, B.N.: Symmetricity in Posets, Comparability Axioms in Orthomodular Lattices with applications, Ph.D. thesis accepted by University of Poona (1992)

Structure Theory of Regular Semigroups Using Categories

A.R. Rajan

Abstract Structure theory of regular semigroups has been using theory of categories to a great extent. Structure theory of regular semigroups developed by K.S.S. Nambooripad using inductive groupoids, structure of combinatorial regular semigroups developed by A.R. Rajan and several other structure theories have made extensive use of categories. The theory of cross connections developed by K.S.S. Nambooripad has provided an abstract description of the category of left ideals of a regular semigroup which he called normal category. The first appearance of categories in structure theory can be traced to Schein's structure theory of inverse semigroups which uses groupoids as a basic object where groupoids are categories in which all morphisms are isomorphisms. Schein described the category of isomorphisms between order ideals of the set of idempotents of an inverse semigroup and called them inductive groupoids. Some instances of appearance of categories in structure theory of certain classes of regular semigroups are presented here.

Keywords Ordered groupoid · Reflective subcategory · Normal category

1 Introduction

Structure theory of semigroups and especially that of regular semigroups uses theory of categories to a great extent. Structure theory of regular semigroups developed by K.S.S. Nambooripad [5] using inductive groupoids, structure of combinatorial regular semigroups given by A.R. Rajan [7]) etc. makes extensive use of categories. A more detailed use of categories can be found in the structure theory for regular semigroups developed by K.S.S. Nambooripad using cross connections. The first appearance of categories in structure theory can be traced to Schein's structure theory of inverse semigroups [10], which use groupoids as a basic object which is a specialised category. Another recent reference to the use of categories in structure theory is Lawson [3].

A.R. Rajan (✉)
Department of Mathematics, University of Kerala, Kariavattom, India
e-mail: arrunivker@yahoo.com

S.T. Rizvi et al. (eds.), *Algebra and its Applications*, Springer Proceedings in Mathematics & Statistics 174, DOI 10.1007/978-981-10-1651-6_14

2 Categories

A category $\mathcal{C}$ is usually defined as a structure consisting of two components, a class $v\mathcal{C}$ called the class of objects of the category $\mathcal{C}$ and a class $\mathcal{M}(\mathcal{C})$ called the class of morphisms. A category is said to be a small category if the class of objects is a set. We give here a description for small categories which can be realised as a generalization of the concept of semigroup.

In the following, we consider a small category as the set of all morphims of the category and describe the vertex set as the set of all identity morphisms so that $v\mathcal{C}$ can be considered as a subset of $\mathcal{C}$. Further the domain and the codomain of the morphisms are realised as given by two mappings d and r from $\mathcal{C}$ to $v\mathcal{C}$.

Definition 2.1 A small category is a 5-tuple $(\mathcal{C}, v\mathcal{C}, d, r, \circ)$ where $\mathcal{C}$ is a set, $v\mathcal{C}$ is a subset of $\mathcal{C}, d, r : \mathcal{C} \to v\mathcal{C}$ are surjective mappings and $\circ$ is a partial binary operation on $\mathcal{C}$ such that the following conditions hold:

1. For $a \in v\mathcal{C},\ d(a) = a = r(a)$
2. The domain of the partial binary operation $\circ$ is

$$\{(f, g) \in \mathcal{C} \times \mathcal{C} : r(f) = d(g)\}$$

For $a, b \in \mathcal{C}$ we write

$$\mathcal{C}(a, b) = \{f \in \mathcal{C} : d(f) = a \text{ and } r(f) = b\}$$

and is called the set of morphisms from a to b.
3. If $f \in \mathcal{C}(a, b)$ and $g \in \mathcal{C}(b, c)$ then $f \circ g \in \mathcal{C}(a, c)$
4. $\circ$ is associative in the sense that

$$f \circ (g \circ h) = (f \circ g) \circ h$$

whenever both sides are defined.
5. If $f \in \mathcal{C}(a, b)$ then $a \circ f = f = f \circ b$
6. For $a, b, a', b' \in v\mathcal{C},\ \mathcal{C}(a, b) \cap \mathcal{C}(a', b') = \emptyset$ if $a \neq a'$ or $b \neq b'$.

Remark 2.2 By this description small categories can be considered as sets with partial binary operation which is associative and admits unique left and right identity for each element. In particular groups and union of groups are small categories.

We consider several instances of category theoretic descriptions of the structure of semigroups. For a regular semigroup S one of the much used categories associated with it is the groupoid $G(S)$ defined as follows. A groupoid is a category in which every morphism is invertible.

$$G(S) = \{(x, x') : x' \text{ is an inverse of } x\}$$

with product defined by

$$(x, x')(y, y') = \begin{cases} (xy, y'x') \text{ if } x'x = yy' \\ \text{undefined otherwise} \end{cases}$$

It may be observed that for $(x, x') \in G(S)$ the left identity of (x, x') is (xx', xx') and the right identity is $(x'x, x'x)$. For each idempotent e of S we may identify $(e, e) \in G(S)$ with e and consider $E(S)$ as the vertex set of $G(S)$.

Groupoids which arise as $G(S)$ of a regular semigroup S have been characterised by Nambooripad [5] and such groupoids are called inductive groupoids. This is a generalisation of the concept of inductive groupoid introduced by Schein [10] for describing inverse semigroups. Schein's inductive groupoids are the inductive groupoids $G(S)$ arising from inverse semigroups. In the case when S is an inverse semigroup the groupoid $G(S)$ can be completely described in terms of the associated partial order relation. Such groupoids called ordered groupoids are defined as follows:

Definition 2.3 A groupoid G with a partial order $\leq$ defined on it is called an **ordered groupoid** if the following are satisfied:

(OG1) If $u \leq x$, $v \leq y$ in G and if the products uv, xy exist in G then $uv \leq xy$
(OG2) If $u \leq x$ then $u^{-1} \leq x^{-1}$
(OG3) If $x \in G$ and if $e \leq xx^{-1}$ with $e \in vG$ then there exists a unique $e * x \in G$ such that

$$e * x \leq x \text{ and } (e * x)(e * x)^{-1} = e.$$

$e * x$ is called the restriction of x to e.

A special class of ordered groupoids called inductive groupoids are defined below.

Definition 2.4 An ordered groupoid $(G, \leq)$ is said to be an inductive groupoid if the set vG of identities of G form a semilattice under the induced partial order of G.

The following theorem characterises ordered groupoids arising from inverse semigroups as inductive groupoids.

Theorem 2.5 *A groupoid G is isomorphic to the inductive groupoid $G(S)$ of an inverse semigroup S if and only if there is a partial order on G making it an ordered groupoid such that the partially ordered subset of identities is a semilattice.*

There is a class of groupoids associated with a regular semigroup which arises in the structure theory called the class of Rees groupoids. These are subgroupoids $G(D)$ of $G(S)$ determined by the Green's $\mathcal{D}$-classes. For a $\mathcal{D}$-class D of S

$$G(D) = \{(x, x') \in G(S) : x \in D\}.$$

A Rees groupoid may be regarded as the inductive groupoid $G(S)^*$ of non zero elements of a completely 0-simple semigroup S.

It can be seen that an inductive groupoid G as described by Nambooripad [5] is a Rees groupoid if and only if it is connected in the sense that for each pair e, f of identities in G there exists $x \in G$ such that e is left identity of x and f is right identity of x.

3 Combinatorial Semigroups

There are special classes of semigroups where the induced groupoids have simple descriptions. One example is the class of combinatorial regular semigroups. A regular semigroup S is said to be combinatorial if all its maximal subgroups are trivial [7]. The Rees groupoids arising from these semigroups are called combinatorial Rees groupoids.

Theorem 3.1 *A groupoid G is a combinatorial Rees groupoid if and only if for each pair e, f of identities of G there exists a unique $x \in G$ such that e is left identity of x and f is right identity of x.*

If G is a groupoid satisfying the condition above with set of identities E. Then $S = E \times E \cup \{0\}$ becomes a combinatorial completely 0-simple semigroup by defining product by

$$(e, f)(g, h) = \begin{cases} (e, g) \text{ if } f = g \\ 0 \text{ otherwise} \end{cases}$$

and $(e, f)0 = 0(e, f) = 0$. Further in this case $G(S)$ is isomorphic to G and so G is a combinatorial Rees groupoid.

Structure theorem for combinatorial locally inverse semigroups (A.R. Rajan [8]) gives purely category theoretic description of the structure of these semigroups.

A locally inverse semigroup is a regular semigroup S such that eSe is an inverse subsemigroup of S for every idempotent e of S. These semigroups are also known as pseudo inverse semigroups. The set of idempotents of these semigroups are characterised as pseudo semilattices or local semilattices.

A pseudo semilattice can be regarded as a set E with a binary operation $\wedge$ satisfying certain axioms which are weaker than those for a semilattice.

The following theorem given in [8] describes pseudo semilattices. The following terminology is used here. For a category $\mathcal{C}$ and $a \in v\mathcal{C}$ the star of a is the full subcategory of $\mathcal{C}$ with vertex set

$$v\text{Star}(a) = \{b \in v\mathcal{C} : \text{ there is a morphism } x : b \to a \text{ in } \mathcal{C}\}.$$

A functor $F : \mathcal{C} \to \mathcal{D}$ is said to be a star isomorphism if

$$F|\,\mathrm{Star}(a) : \mathrm{Star}(a) \to \mathrm{Star}(F(a))$$

is an isomorphism.

A **preorder** $\mathcal{P}$ is a category $\mathcal{P}$ in which for all $a, b \in v\mathcal{P}$ the morphism set $\mathcal{P}(a, b)$ contains at most one element.

$\mathcal{P}$ is said to be strict preorder if

$$\mathcal{P}(a, b) \cup \mathcal{P}(b, a)$$

contains at most one element.

In this case $v\mathcal{P}$ becomes a partially ordered set by defining partial order as follows.

$$a \leq b \text{ if } \mathcal{P}(a, b) \text{ is nonempty.}$$

In this case we say $\mathcal{P}$ is a partial order.

Another category concept that is used here is that of adjoints. Adjoint relation between two categories provides a pair of functors giving an equivalence of the categories. A special case is adjoint for inclusion functor.

A subcategory $\mathcal{D}$ of a category $\mathcal{C}$ is said to be a **reflective** subcategory of $\mathcal{C}$, if the inclusion functor from $\mathcal{D}$ to $\mathcal{C}$ has a left adjoint.

Theorem 3.2 *Let I and Λ be strict preorders and Δ be a reflective subcategory of the preorder $I \times \Lambda$. Let $F : I \times \Lambda \to \Delta$ be the left adjoint of the inclusion. Define product $\wedge$ on Δ by*

$$(i, \lambda) \wedge (j, \mu) = F(j, \lambda).$$

Then $(\Delta, \wedge)$ is a pseudo semilattice if and only if the projections p_1 and p_2 from Δ to I and Λ respectively are star bijections.

This theorem can be extended to provide a characterization of combinatorial locally inverse semigroups.

4 Normal Categories

Another category used in structure theory is the category of principal left or right ideals. These categories arise in the theory of cross connections developed by Nambooripad [6] and Grillet [1]. Grillet used the partially ordered sets of $\mathcal{L}$-classes and $\mathcal{R}$-classes, where $\mathcal{L}$ and $\mathcal{R}$ are the Green's equivalences relating to left and right ideals of the semigroup. He associated certain mappings called normal mappings with these partially ordered sets in his cross connection theory. This theory could provide the structure for fundamental regular semigroups. The general case was provided by Nambooripad using the category of principal left ideals and principal right ideals. These categories are abstractly described as normal categories.

The normal category $\mathcal{L}(S)$ of principal left ideals of a regular semigroup S is described as follows:

$$v\mathcal{L}(S) = \{Se : e \in E(S)\}$$

and for $Se, Sf \in v\mathcal{L}(S)$, a morphism from Se to Sf is a right translation

$$\rho(e, u, f) : Se \to Sf \text{ which maps } x \in Se \text{ to } xu \in Sf$$

for $u \in eSf$.

When S is a combinatorial regular semigroup the principal left[right] ideals can be represented by order ideals of the partially ordered sets $S/\mathcal{L}$ and $S/\mathcal{R}$. Morphisms of $\mathcal{L}(S)$ in this case can be replaced by restrictions of normal mappings to the principal order ideals of $S/\mathcal{L}$. The normality properties of Grillet and Nambooripad also coincide in this case.

Characterisation of normal categories associated with various special classes of semigroups can be carried out. For example in the case of inverse semigroups, the normal category has been characterised by the uniqueness on normal factorizations of morphisms (A.R. Rajan [9]).

Acknowledgments The author thanks the Kerala State Council for Science Technology and Environment (KSCSTE) for the support under Emeritus Scientist Scheme which facilitated the participation in the conference.

References

1. Grillet, P.A.: Structure of regular semigroups I a representation, II cross connections, III the reduced case. Semigroup Forum 8(1974), 177–183, 254–265
2. Howie, J.M.: Fundamentals of Semigroup Theory. Academic press, New York (1995)
3. Lawson, M.V., Wallis, A.R.: Categorical and semigroup theoretic descriptions of Bass - Serre theory (2014, preprint)
4. Munn, W.D.: pseudo-iverses in semigroups. Proc. Camb. Phil. Soc. **51**, 396–399 (1955)
5. Nambooripad, K.S.S.: Structure of regular semigroups I. Mem. Amer. Math. Soc. **22**(224) (1979)
6. Nambooripad, K.S.S.: Theory of cross connections, Pub. No. 38, Centre for Mathematical Sciences, Trivandrum (1984)
7. Nambooripad, K.S.S., Rajan, A.R.: Structure of combinatorial regular semigroups. Quart. J. Math. (Oxford) **29**, 489–504 (1978)
8. Rajan, A.R.: A Structutre theorem for combinatorial pseudo inverse semigroups, Colloq. Janos Bolyai Inst., 39, Semigroups, Szeged 325–346 (1981)
9. Rajan, A.R.: Normal categories of Inverse semigroups. East - West J. Math. Bangkok **16**(2), 122–130 (2014)
10. Schein, B.M.: On the theory of generalised groups and generalised heaps, Izv. dat. Saratov Univ., saratov University, Saratov, 286-324 (1966) (in Russian), English Translation: Amer. Math. Soc. (2) 89-122, (1979)

Biorder Ideals and Regular Rings

P.G. Romeo and R. Akhila

Abstract In [4] (Structure of regular semigroups, 1979) K.S.S. Nambooripad introduced biordered sets as a partial algebra (E, ω^r, ω^l) where ω^r and ω^l are two quasiorders on the set E satisfying biorder axioms; to study the structure of a regular semigroup. John von Neumann (Continuous Geometry, 1960 in [5]) described the complemented modular lattice of principle ideals of a regular ring. In this paper, we introduced the biorder ideals of a regular ring and showed that these ideals form a complemented modular lattice.

Keywords Biordered set · Sandwitch set

1991 Mathematics Subject Classification 20M10

In many algebraic systems like semigroups, rings, algebras, the idempotent elements are important structural objects and can be used effectively in analyzing the structure of the algebraic system under consideration. The concept of biordered set was originally introduced by Nambooripad [1972, 1979] to describe the structure of the set of idempotents of a semigroup. He identified a partial binary operation on the set of idempotents $E(S)$ of a semigroup S arising from the binary operation in S. The resulting structure on $E(S)$ involving two quasiorders is abstracted as a biordered set. In this paper, we propose to extend biordered set approach to rings to study the structure of regular rings.

1 Preliminaries

First, we recall some basic definitions regarding semigroups, biordered sets, and rings needed in the sequel. A set S in which for every pair of elements $a, b \in S$ there is an element $a \cdot b \in S$ which is called the product of a by b is called a groupoid.

P.G. Romeo (✉) · R. Akhila
Department of Mathematics, Cochin University of Science and Technology, Kochi, Kerala, India
e-mail: romeo_parackal@yahoo.com

R. Akhila
e-mail: akhila.ravikumar@gmail.com

S.T. Rizvi et al. (eds.), *Algebra and its Applications*, Springer Proceedings in Mathematics & Statistics 174, DOI 10.1007/978-981-10-1651-6_15

A groupoid S is a semigroup if the binary operation on S is associative. An element $a \in S$ is called regular if there exists an element $a' \in S$ such that $aa'a = a$, if every element of S is regular then S is a regular semigroup. An element $e \in S$ such that $e \cdot e = e$ is called an idempotent and the set of all idempotents in S will be denoted by $E(S)$.

1.1 Biordered Sets

By a partial algebra E, we mean a set together with a partial binary operation on E. Then $(e, f) \in D_E$ if and only if the product ef exists in the partial algebra E. If E is a partial algebra, we shall often denote the underlying set by E itself; and the domain of the partial binary operation on E will then be denoted by D_E. Also, for brevity, we write $ef = g$, to mean $(e, f) \in D_E$ and $ef = g$. The dual of a statement T about a partial algebra E is the statement T^* obtained by replacing all products ef by its left–right dual fe. When D_E is symmetric, T^* is meaningful whenever T is. On E we define

$$\omega^r = \{(e, f) : fe = e\} \, \omega^l = \{(e, f) : ef = e\}$$

and $\mathcal{R} = \omega^r \cap (\omega^r)^{-1}$, $\mathcal{L} = \omega^l \cap (\omega^l)^{-1}$, *and* $\omega = \omega^r \cap \omega^l$. We will refer ω^r and ω^l as the right and the left quasiorder of E.

Definition 1 Let E be a partial algebra. Then E is a biordered set if the following axioms and their duals hold

(1) ω^r and ω^l are quasi orders on E and

$$D_E = (\omega^r \cup \omega^l) \cup (\omega^r \cup \omega^l)^{-1}$$

(2) $f \in \omega^r(e) \Rightarrow f \mathcal{R} fe\omega e$
(3) $g\omega^l f$ *and* $f, g \in \omega^r(e) \Rightarrow ge\omega^l fe.$
(4) $g\omega^r f\omega^r e \Rightarrow gf = (ge)f$
(5) $g\omega^l f$ and $f, g \in \omega^r(e) \Rightarrow (fg)e = (fe)(ge).$

We shall often write $E =< E, \omega^l, \omega^r >$ to mean that E is a biordered set with quasiorders ω^l, ω^r. The relation ω defined is a partial order and

$$\omega \cap (\omega)^{-1} \subset \omega^r \cap (\omega^l)^{-1} = 1_E.$$

Definition 2 Let $\mathcal{M}(e, f)$ denote the quasiordered set $(\omega^l(e) \cap \omega^r(f), <)$ where $<$ is defined by $g < h \Leftrightarrow eg\omega^r eh$, and $gf\omega^l hf$. Then the set

$$S(e, f) = \{h \in M(e, f) : g < h \text{ for all } g \in M(e, f)\}$$

is called the sandwich set of e and f.

(1) $f, g \in \omega^r(e) \Rightarrow S(f, g)e = S(fe, ge)$

The biordered set E is said to be regular if $S(e, f) \neq \emptyset \ \forall e, f \in E$. The following theorem shows that if S is a regular semigroup, then $E(S)$ is a regular biordered set.

Theorem 1 ([4], Theorem 1.1) *Let S be a semigroup such that $E(S) \neq \phi$.*

(1) The partial algebra $E(S)$ is a biordered set.
(2) For $e, f \in E(S)$ define

$$S_1(e, f) = \{h \in M(e, f) : ehf = ef\}$$

Then $S_1(e, f) \subset S(e, f)$.
(3) If $e, f \in E(S)$ then ef is a regular element of S if and only if $S_1(e, f) = S(e, f) \neq \phi$.
(4) If S is regular, then $E(S)$ is a regular biordered set.

Remark 1 For $e \in E$, $\omega^r(e)$ $\left[\omega^l(e)\right]$ are principal right [left] ideals and $\omega(e)$ is a principal two sided ideal and are called biorder ideals generated by e.

1.2 Lattices

A lattice is a partially ordered set in which each pair of elements has a least upper bound and a greatest lower bound. If a and b are elements of a lattice, we denote their greatest lower bound (meet) and least upper bound (join) by $a \wedge b$ and $a \vee b$, respectively. It is easy to see that $a \vee b$ and $a \wedge b$ are unique. The notations $a \wedge b$ and $a \vee b$ are analogous to the notations for the intersection and union of two sets. However some properties of union and intersection of sets do not carry over to lattices, for instance, the distributive laws are false in some lattices. But many of the well-known lattices posses the modularity property which is a weak form of distributive property.

Definition 3 A lattice is called modular (or a Dedekind lattice) if

$$(a \vee b) \wedge c = a \vee (b \wedge c) \text{ for all } a \leq c.$$

A lattice is bounded if it has both a maximum element and a minimum element, we use the symbols 0 and 1 to denote the minimum element and maximum element of a lattice. A bounded lattice L is said to be complemented if for each element a of L, there exists at least one element b such that $a \vee b = 1$ and $a \wedge b = 0$. The element b is referred to as a complement of a. It is quite possible for an element of a complemented lattice to have many different complements. An element x is called a complement of a in b if $a \vee x = b$ and $a \wedge x = 0$.

Definition 4 Two elements a and b of a lattice L are said to be perspective (in symbols $a \sim b$) if there exists x in L such that $a \vee x = b \vee x, a \wedge x = b \wedge x = 0$ Such an element x is called an axis of perspective.

1.3 Principal Ideals of Regular Ring

A ring is a set R together with two binary operations '+', '·' with the following properties.

(1) The set $(R, +)$ is an abelian group.
(2) The set $(R, \cdot)$ is a semigroup.
(3) The operation $\cdot$ is distributive over $+$.

A ring $(R, +, \cdot)$ is regular if for every $a \in R$ there exists an element a' such that $aa'a = a$, i.e., the ring is regular if the multiplicative semigroup is a regular semigroup.

Definition 5 A subset a of a ring $\mathcal{R}$ is called right ideal in case

$$x + y \in a, \quad xz \in a$$

for each $x, y \in a$ and $z \in \mathcal{R}$.

Similarly, we a can define the left ideal and a is called an ideal if it is both a right and a left ideal. The set of all right (left) ideals of $\mathcal{R}$ is denoted by $R_{\mathcal{R}}(L_{\mathcal{R}})$. The intersection of any class of right(left) ideals is again a right (left) ideal and also for any $a \subset \mathcal{R}$ there is a unique least extension a_r, (a_l) which is a right (left) ideal.

Proposition 1 *If $R \subset R_{\mathcal{R}}$ is any class of right ideals, there exists both a smallest right ideal (least upper bound of R) containing every element of R and a greatest right ideal (greatest lower bound of R) contained in every element of R. Thus $R_{\mathcal{R}}$ is a continuous lattice with $\subset$ and the operations thus defined. The zero element of $R_{\mathcal{R}}$ is $(0)_r = 0$ and the unit element is $(1)_r = \mathcal{R}$.*

Definition 6 A principal right [left] ideal is one of the from $(a)_r[(a)_l]$. The class of all principal right [left] ideals will be denoted by $\bar{R}_{\mathcal{R}}$ $[\bar{L}_{\mathcal{R}}]$.

In [5] John von Neumann describes the structure of principal ideals of a regular ring. Here we recall some of those results.

Lemma 1 *Let $\mathcal{R}$ be a ring, $e \in \mathcal{R}$, then*

- *e is idempotent if and only if $(1 - e)$ is idempotent.*
- *$\langle e \rangle_r$ if the set of all x such that $x = ex$ is a principal right ideal.*
- *$\langle e \rangle_r$ and $\langle 1 - e \rangle_r$ are mutual inverses.*
- *If $\langle e \rangle_r = \langle f \rangle_r$ and If $\langle 1 - e \rangle_r = \langle 1 - f \rangle_r$ where e and f are idempotents, then $e = f$.*

Theorem 2 *Two right ideals a and b are inverses if and only if there exists an idempotent e such that $a = \langle e \rangle_r$ and $b = \langle 1 - e \rangle_r$.*

Proof Let a and b be inverse right ideals, then there exists elements x, y with $x + y = 1, x \in a, y \in b$. If $z \in a$ then $xz + yz = x$. Since $z, xz \in a, yz \in a$. But $yz \in b$, hence $yz = 0$. Thus $z = xz \in (x)_r$ for every $z \in a$ and $a \subset (x)_r$. Bust $x \in a$, hence $a = (x)_r$. Similarly $b = (y)_r = (1 - x)_r$, since $x + y = 1$. Finally, since $z = xz$ for every $z \in a$ this holds for $z = x$ and x is idempotent. □

Theorem 3 *The following statements are equivalent*

(1) Every principal right ideal $\langle a \rangle_r$ has an inverse right ideal.
(2) For every a there exists an idempotent e such that $\langle a \rangle_r = \langle e \rangle_r$.
(3) For every a there exists an idempotent x such that $axa = a$.
(4) For every a there exists an idempotent f such that $\langle a \rangle_l = \langle f \rangle_l$.
(5) Every principal ideal $\langle a \rangle_l$ has an inverse left ideal.

Definition 7 A ring $\mathcal{R}$ is said to be regular if $\mathcal{R}$ possesses anyone of the equivalent properties of the above Theorem.

Theorem 4 *The set $\bar{R}_{\mathcal{R}}$ is a complemented modular lattice partially ordered by $\subset$, the meet being $\cap$ and join $\cup$, its zero is $(0)_r$ and its unit is $(1)_r$.*

2 Biorder Ideals of Regular Rings

Analogous to von Neumann's construction of the principal ideals of a regular ring, we proceed to describe the structure of the biorder ideals of regular rings.

Proposition 2 *Let e and f be idempotents in a regular ring R. Then the following are equivalent:*

(1) $ef = 0$
(2) $e\omega^l(1 - f)$
(3) $f\omega^r(1 - e)$

Proof Suppose $ef = 0$. Then

$$e(1 - f) = e - ef = e.$$

Conversely, $e\omega^l(1 - f)$ then $e(1 - f) = e - ef = e$ and hence $ef = 0$. Proof (3) is similar. □

Proposition 3 *Let e and f be idempotents in a regular ring R. Then the following holds.*

(1) $e\omega^l f$ *if and only if* $(1 - f)\omega^r(1 - e)$
(2) $e\omega^r f$ *if and only if* $(1 - f)\omega^l(1 - e)$

Proof Let $e\omega^l f$. Then,

$$(1-e)(1-f) = 1-e-f+ef = 1-e-f+e = 1-f$$

Conversely, suppose $(1-e)(1-f) = (1-f)$, then

$$1-e-f+ef = 1-f$$

hence $e\omega^l f$. Proof (2) is similar. □

Corollary 1 *Let e and f be idempotents in the ring R. Then the following old.*

(1) $\omega^l(e) = \omega^l(f)$ if and only if $\omega^r(1-e) = \omega^r(1-f)$
(2) $\omega^r(e) = \omega^r(f)$ if and only if $\omega^l(1-e) = \omega^l(1-f)$

Proposition 4 *Let e and f be idempotents in the ring R, if $\omega^r(e) = \omega^r(f), \omega^r(1-e) = \omega^r(1-f)$, where e, f are idempotents, then $e = f$*

Proof Since $\omega^r(e) = \omega^r(f)$, $ef = f$. Therefore, $(1-e)f = 0$. Since $\omega^r(1-e) = \omega^r(1-f)$, by replacing e and f by $(1-e)$ and $(1-f)$ respectively, we get $e(1-f) = 0$. That is, $ef = e$ and so $e = f$. □

Lemma 2 *Let $e, f, g \in E_R$ with $ef = fe = 0$. Then $e+f$ is an idempotent and the following hold.*

(1) $e\omega(e+f)$ and $f\omega(e+f)$
(2) If $e\omega^l g$ and $f\omega^l g$, then $(e+f)\omega^l g$
(3) If $e\omega^r g$ and $f\omega^r g$, then $(e+f)\omega^r g$

Proof Given $e, f \in E_R$ with $ef = fe = 0$, then

$$(e+f)^2 = e^2 + ef + fe + f^2 = e + f.$$

(1) $e(e+f) = e^2 + ef = e + ef = e$, and $(e+f)e = e^2 + fe = e + fe = e$. Thus $e\omega(e+f)$. Similarly, we can prove $f\omega(e+f)$.
(2) Given $e\omega^l g$ and $f\omega^l g$. Therefore, $(e+f)g = eg + fg = e + f$, i.e., $(e+f)\omega^l g$. Proof (3) is similar. □

Lemma 3 *Let $\omega^r(e) \cup \omega^r(f) = \omega^r(e+f'')$ where $f''\mathcal{R}f'$ and $f' = (1-e)f$.*

Proof Define

$$\begin{aligned}\omega^r(e) \cup \omega^r(f) &= \{eg + fh : g, h \in E_R; gh = hg = 0\}\\ &= \{eg = efh + (1-e)fh : g, h \in E_R; gh = hg = 0\}\\ &= \{e(g+fh) + (1-e)fh : g, h \in E_R; gh = hg = 0\}\end{aligned}$$

Let $f' = (1-e)f$. Then $f' \in S(f, 1-e)$ so that $f' \in E_R$, $ff' = f'$ and $(1-e)f' = f'$. So $ef' = 0$ and $\omega^r(e) \cup \omega^r(f) = \omega^r(e) \cup \omega^r(f')$. Define

$f'' = f'(1-e).$, then $f'f'' = f'f'(1-e) = f'(1-e) = f''$ and $f''f' = f'(1-e)f' = f'f' = f'$. Further f'' is an idempotent, $\omega^r(f') = \omega^r(f'')$ and $\omega^r(e) \cup \omega^r(f) = \omega^r(e) \cup \omega^r(f'')$. Now, $ef'' = ef'(1-e) = 0$ and $f''e = f'(1-e)e = 0$ so, by Lemma above $(e + f'')$ is an idempotent.

Next we proceed to prove that $\omega^r(e) \cup \omega^r(f) = \omega^r(e + f'')$. For, consider $e + f''$, then

$$e + f'' = e^2 + (f'')^2 = e \cdot e + f'' \cdot f'' \in \omega^r(e) \cup \omega^r(f'') \text{ where } ef'' = 0.$$

So, $\omega^r(e + f'') \subseteq \omega^r(e) \cup \omega^r(f'')$ and $e\omega^r(e + f'')$ and $f''\omega^r(e + f'')$. That is

$$\omega^r(e) \subseteq \omega^r(e + f'') \text{ and } \omega^r(f'') \subseteq \omega^r(e + f'')$$

thus $\omega^r(e) \cup \omega^r(f'') \subseteq \omega^r(e + f'')$, hence $\omega^r(e) \cup \omega^r(f) = \omega^r(e + f'')$. □

Denote by Ω_R the class of all principal ω^r−ideals and by Ω_L the class of all principal ω^l-ideals. In the light of the above lemma we have the following theorem:

Theorem 5 *Ω_R is closed with respect to the operation $\cup$ defined in Ω_R.*

Next we introduce the notion of annihilators in principal ω^r and ω^l-ideals.

Definition 8 For every ω^r-ideal we define

$$(\omega^r(e))^L = \{y : yz = 0 \text{ for every } z \in \omega^r(e)\}$$

and for every ω^l-ideal,

$$(\omega^l(e))^R = \{y : zy = 0 \text{ for every } z \in \omega^l(e)\}$$

then $(\omega^r(e))^L$ is a left ideal and $(\omega^l(e))^R$ is a right ideal.

Proposition 5 *For $e \in E_R$, $(\omega^l(e))^R$ is a principal ω^r-ideal and $(\omega^r(e))^L$ is a principal ω^l-ideal. In fact, $(\omega^l(e))^R = \omega^r(1-e)$ and $(\omega^r(e))^L = \omega^l(1-e)$.*

Proof

$$\begin{aligned}
\omega^r(e) &= \{g : e.g. = g\} \\
&= \{g : (1-e)g = 0\} \\
&= \{g : u(1-e) = 0; \text{ for every } u \in E_R\} \\
&= \{g : \text{ for every } h \in \omega^l(e), hg = 0\}
\end{aligned}$$

where $h = u(1-e)$. Since $h(1-e) = u(1-e)(1-e) = u(1-e) = h$ we have $h \in \omega^l(1-e)$. Thus $\omega^r(e) = (\omega^l(1-e))^R$. □

Lemma 4 *Let $e, f \in E_R$ and $\omega^r(e)$ and $\omega^r(f)$ are ideals generated by e and f, then*

(1) $\omega^r(e) \subset \omega^r(f) \Rightarrow (\omega^r(e))^L \supset (\omega^r(f))^L$
(2) $\omega^r(e) = (\omega^r(e))^{LR}$ *and* $(\omega^r(e))^L = (\omega^r(e))^{LRL}$

Proof (1) Let $g \in (\omega^r(f))^L$, then $gh = 0$ for every $h \in \omega^r(f)$. If $\omega^r(e) \subset \omega^r(f)$ then for $h \in \omega^r(e)$, $gh = 0$ for every $h \in \omega^r(e)$. Thus $g \in (\omega^r(e))^L$ and so

$$(\omega^r(f))^L \subset (\omega^r(e))^L.$$

(2) Let $g \in \omega^r(e)$. Consider $h \in (\omega^r(e))^L$, for $z \in \omega^r(e)$, $hz = 0$. Hence $hg = 0$ so $g \in (\omega^r(e))^{LR}$ and $\omega^r(e) \subset (\omega^r(e)^{LR}$. Now by (1) we have

$$\omega^r(e) \subseteq (\omega^r(e))^{LR}; (\omega^r(e))^L \supseteq (\omega^r(e))^{LRL}$$

Replace $\omega^r(e)$ by $(\omega^r(e))^L$ we get $(\omega^r(e))^L \subseteq (\omega^r(e))^{LR}$. Hence $(\omega^r(e))^L = (\omega^r(e))^{LRL}$. But $\omega^r(e) = (\omega^l(1-e))^{RLR} = (\omega^l(1-e))^R = \omega^r(e)$, thus $\omega^r(e) = (\omega^r(e))^L$. □

In the following proposition, we establish the relation between Ω_L and Ω_R by using the relation between principal ω-ideals and annihilators.

Proposition 6 *Let R be a regular ring and E_R the set of idempotents on R. Let Ω_L and Ω_R denote the lattice of principal ω^l-ideals and principal ω^r-ideals of E_R. Define ϕ and ψ on Ω_L and Ω_R by*

$$\phi(\omega^l(e)) = (\omega^l(e))^R \text{ and } \psi(\omega^r(e)) = (\omega^r(e))^L$$

then ϕ and ψ are mutually inverse anti-isomorphisms.

Proof Let $I \in \Omega_L$. Therefore, there exists an idempotent, e such that $I = \omega^l(e)$ and

$$\phi(I) = \phi(\omega^l(e)) = (\omega^l(e))^R = \omega^r(1-e)$$

Thus ϕ maps Ω_L to Ω_R. Also ϕ reverses the order, for let $I, J \in \Omega_L$ with $I \subseteq J$, then there exists idempotents $e, f \in E_R$ such that $\omega^l(e) \subseteq \omega^l(f)$. But if $\omega^l(e) \subseteq \omega^l(f)$ then $(\omega^l(f)^R \subseteq (\omega^l(e))^R$ and $\phi(J) \subseteq \phi(I)$. Similarly ψ is an order reserving map from Ω_R to Ω_L. Moreover for $I \in \Omega_L$ and $I = \omega^l(e)$ then

$$(\psi\phi(I)) = \psi(\phi(\omega^l(e))) = \psi(\omega^r(1-e)) = (\omega^r(1-e))^L = \omega^l(1-(1-e)) = \omega^l(e) = I.$$

For I in Ω_R, $(\phi\psi)(I) = I$. Hence ϕ and ψ are mutually inverse anti-isomorphisms between Ω_L and Ω_R. □

Lemma 5 *Let $\omega^r(e)$ and $\omega^r(f)$ be principal right ω ideals generated by e and f. Then $(\omega^r(e) \cup \omega^r(f))^L = (\omega^r(e))^L \cap (\omega^r(f))^L$.*

Proof

$$\begin{aligned}(\omega^r(e))^L \cap (\omega^r(f))^L &= \{g : gh = 0 \; \forall \; h \in \omega^r(e) \text{ and } g : gh = 0 \; \forall \; h \in \omega^r(f)\} \\ &= \{g : gh = 0 \; \forall \; h \in \omega^r(e) \cup \omega^r(f)\} \\ &= \{g : gh = 0 \; \forall \; h \in (\omega^r(e) \cup \omega^r(f)\}\end{aligned}$$

Hence $(\omega^r(e))^L \cap (\omega^r(f))^L = (\omega^r(e) \cup \omega^r(f))^L$. □

Lemma 6 *For two principal ω^r-ideals, $\omega^r(e)$ and $\omega^r(f)$ their intersection is also a principal ω^r-ideal.*

Proof By the above Lemma

$$\begin{aligned}\omega^r(e) \cap \omega^r(f) &= (\omega^r(e))^{LR} \cap (\omega^r(f))^{LR} \\ &= ((\omega^r(e))^L \cup (\omega^r(f))^L)^R\end{aligned}$$

But $(\omega^r(e))^L$ and $(\omega^r(f))^L$ are principal ω^l-ideals, and so $(\omega^r(e))^L \cup (\omega^r(f))^L$ is also a principal ω^l-ideal. Hence $\omega^r(e) \cap \omega^r(f)$ is a principal ω^r-ideal. □

For any idempotent $e \in E_R$, $(1 - e) \in E_R$ and $\omega^r(e) \cup \omega^r(1 - e) = \omega^r(e + 1 - e) = \omega^r(1) = E_R$ and $\omega^r(e) \cap \omega^r(1 - e) = \{0\}$. Thus $\omega^r(e)$ and $\omega^r(1 - e)$ are complements of each other in the lattice of principal right ω-ideals of E_R. Similarly, $\omega^l(e)$ and $\omega^l(1 - e)$ are complements of each other in the lattice of all principal left ω-ideals of E_R.

Thus we have the following theorem:

Theorem 6 *Let R be a ring then the set of all principal ω^l-ideals Ω_L and the set of all principal ω^r-ideals Ω_R of E_R are complemented modular lattices ordered by the relation $\subset$, the meet being $\cap$ and the join $\cup$; its zero is 0, and its unit is $\omega^l(1)[\omega^r(1)]$.*

References

1. Clifford, A.H., Preston, G.B.: The Algebraic Theory of Semigroups, vol. 1, p. 7. Math. Surveys of the American. Math. Soc., Providence (1964)
2. Easdown, D.: Biordered sets comes from Semigroups. J. Algebra, **96**, 581–591, 87d:06020 (1985)
3. Howie, J.M.: An Introduction To Semigroup Theory, Academic Press Inc. (London) (1976). ISBN: 75-46333
4. Nambooripad, K.S.S.: Structure of Regular Semigroups (MEMOIRS, No.224), American Mathematical Society (1979). ISBN-13: 978-0821 82224
5. von Neumann, J.: Continuous Geometry. Princeton University Press, London (1960)

Products of Generalized Semiderivations of Prime Near Rings

Asma Ali and Farhat Ali

Abstract Let N be a near ring. An additive mapping $F : N \longrightarrow N$ is said to be a generalized semiderivation on N if there exists a semiderivation $d : N \longrightarrow N$ associated with a function $g : N \longrightarrow N$ such that $F(xy) = F(x)y + g(x)d(y) = d(x)g(y) + xF(y)$ and $F(g(x)) = g(F(x))$ for all $x, y \in N$. The purpose of the present paper is to prove some theorems in the setting of semigroup ideal of a 3-prime near ring admitting a pair of suitably-constrained generalized semiderivations, thereby extending some known results on derivations and generalized derivations. We show that if N is 2-torsion free and F_1 and F_2 are generalized semiderivations such that $F_1F_2 = 0$, then $F_1 = 0$ or $F_2 = 0$; we prove other theorems asserting triviality of F_1 or F_2; and we also prove some commutativity theorems.

Keywords 3-prime near-rings · Semiderivations · Generalized semiderivations

2010 Mathematics Subject Classification 16N60 · 16W25 · 16Y30

1 Introduction

Throughout the paper, N denotes a zero-symmetric left near ring with multiplicative centre Z; and for any pair of elements $x, y \in N$, $[x, y]$ denotes the commutator $xy - yx$. A near ring N is called zero-symmetric if $0x = 0$, for all $x \in N$ (recall that left distributivity yields that $x0 = 0$). The near ring N is said to be 3-prime if $xNy = \{0\}$ for $x, y \in N$ implies that $x = 0$ or $y = 0$. A near ring N is called 2-torsion free if $(N, +)$ has no element of order 2. A nonempty subset U of N is

---

First author is supported by a grant from Science and Engineering Research Board (SERB), DST, New Delhi, India. Grant No. SR/S4/MS:852/13.

A. Ali (✉) · F. Ali
Department of Mathematics, Aligarh Muslim University, Aligarh 202002, India
e-mail: asma_ali2@rediffmail.com

F. Ali
e-mail: 04farhatamu@gmail.com

S.T. Rizvi et al. (eds.), *Algebra and its Applications*, Springer Proceedings in Mathematics & Statistics 174, DOI 10.1007/978-981-10-1651-6_16

called a semigroup right (resp. semigroup left) ideal if $UN \subseteq U$ (resp. $NU \subseteq U$); and if U is both a semigroup right ideal and a semigroup left ideal, it is called a semigroup ideal. An additive mapping $f : N \longrightarrow N$ is said to be a right (resp. left) generalized derivation with associated derivation D if $f(xy) = f(x)y + xD(y)$ (resp. $f(xy) = D(x)y + xf(y)$), for all $x, y \in N$, and f is said to be a generalized derivation with associated derivation D on N if it is both a right generalized derivation and a left generalized derivation on N with associated derivation D. Motivated by a definition given by Bergen [5] for rings, we define an additive mapping $d : N \longrightarrow N$ is said to be a semiderivation on a near ring N if there exists a function $g : N \longrightarrow N$ such that $(i)\, d(xy) = d(x)g(y) + xd(y) = d(x)y + g(x)d(y)$ and $(ii)\, d(g(x)) = g(d(x))$, for all $x, y \in N$. In case g is the identity map on N, d is of course just a derivation on N, so the notion of semiderivation generalizes that of derivation. But the generalization is not trivial for example take $N = N_1 \oplus N_2$, where N_1 is a zero symmetric near ring and N_2 is a ring. Then the map $d : N \longrightarrow N$ defined by $d((x, y)) = (0, y)$ is a semiderivation associated with function $g : N \longrightarrow N$ such that $g(x, y) = (x, 0)$. However d is not a derivation on N. An additive mapping $F : N \to N$ is said to be a generalized semiderivation of N if there exists a semiderivation $d : N \longrightarrow N$ associated with a map $g : N \longrightarrow N$ such that $(i)\ F(xy) = F(x)y + g(x)d(y) = d(x)g(y) + xF(y)$ and $(ii)\ F(g(x)) = g(F(x))$ for all $x, y \in N$. All semiderivations are generalized semiderivations. If g is the identity map on N, then all generalized semiderivations are merely generalized derivations, again the notion of generalized semiderivation generalizes that of generalized derivation. Moreover, the generalization is not trivial as the following example shows:

Example 1.1 Let S be a 2-torsion free left near ring and let

$$N = \left\{ \begin{pmatrix} 0 & x & y \\ 0 & 0 & 0 \\ 0 & 0 & z \end{pmatrix} \mid x, y, z \in S \right\}.$$

Define maps $F, d, g : N \to N$ by

$$F\begin{pmatrix} 0 & x & y \\ 0 & 0 & 0 \\ 0 & 0 & z \end{pmatrix} = \begin{pmatrix} 0 & xy & 0 \\ 0 & 0 & 0 \\ 0 & 0 & 0 \end{pmatrix};\quad d\begin{pmatrix} 0 & x & y \\ 0 & 0 & 0 \\ 0 & 0 & z \end{pmatrix} = \begin{pmatrix} 0 & 0 & y \\ 0 & 0 & 0 \\ 0 & 0 & z \end{pmatrix}$$

and

$$g\begin{pmatrix} 0 & x & y \\ 0 & 0 & 0 \\ 0 & 0 & z \end{pmatrix} = \begin{pmatrix} 0 & x & 0 \\ 0 & 0 & 0 \\ 0 & 0 & 0 \end{pmatrix}.$$

It can be verified that N is a left near ring and F is a generalized semiderivation with associated semiderivation d and a map g associated with d. However F is not a generalized derivation on N.

2 Preliminary Results

We begin with several lemmas, most of which have been proved elsewhere.

Lemma 2.1 ([2, Lemmas 1.2 and 1.3]) *Let N be a 3-prime near ring.*

(i) If $z \in Z\backslash\{0\}$, then z is not a zero divisor.
(ii) If $Z\backslash\{0\}$ and x is an element of N for which $xz \in Z$, then $x \in Z$.
(iii) If x is an element of N which centralizes some nonzero semigroup right ideal, then $x \in Z$.
(iv) If $Z\backslash\{0\}$ contains an element z for which $z + z \in Z$, then $(N, +)$ is abelian.

Lemma 2.2 ([2, Lemmas 1.3 and 1.4]) *Let N be a 3-prime near ring and U be a nonzero semigroup ideal of N.*

(i) If $x \in N$ and $xU = \{0\}$, or $Ux = \{0\}$, then $x = 0$.
(ii) If $x, y \in N$ and $xUy = \{0\}$, then $x = 0$ or $y = 0$.

Lemma 2.3 ([2, Lemma 1.5]) *If N is a 3-prime near ring and Z contains a nonzero semigroup left ideal or a nonzero semigroup right ideal, then N is a commutative ring.*

Lemma 2.4 ([4, Lemma 2.4]) *Let N be an arbitrary near ring. Let S and T be non empty subsets of N such that $st = -ts$ for all $s \in S$ and $t \in T$. If $a, b \in S$ and c is an element of T for which $-c \in T$, then $(ab)c = c(ab)$.*

Lemma 2.5 *Let N be a 3-prime near ring and U be a nonzero semigroup ideal of N. If N admits a nonzero semiderivation d of N associated with a map g, then $d \neq 0$ on U.*

Proof Let $d(u) = 0$, for all $u \in U$. Replacing u by xu, we get $d(xu) = 0$, for all $x \in N$ and $u \in U$. Thus $d(x)g(u) + xd(u) = 0$, for all $x \in N$ and $u \in U$, i.e., $d(x)g(u) = 0$. The result follows by Lemma 2.2(i).

Lemma 2.6 *Let N be a 3-prime near ring admitting a nonzero semiderivation d with a map g such that $g(xy) = g(x)g(y)$ for all $x, y \in N$. Then N satisfies the following partial distributive law:*

$$(d(x)y + g(x)d(y))z = d(x)yz + g(x)d(y)z \text{ for all } x, y, z \in N.$$

Proof Let $x, y, z \in N$, by defining d we have

$$\begin{aligned} d(xyz) &= d(xy)z + g(xy)d(z) \\ &= (d(x)y + g(x)d(y))z + g(x)g(y)d(z). \end{aligned} \tag{2.1}$$

On the other hand,

$$\begin{aligned} d(xyz) &= d(x)yz + g(x)d(yz) \\ &= d(x)yz + g(x)(d(y)z + g(y)d(z)) \\ &= d(x)yz + g(x)d(y)z + g(x)g(y)d(z). \end{aligned} \tag{2.2}$$

Combining (2.1) and (2.2), we obtain

$$\begin{aligned} &(d(x)y + g(x)d(y))z + g(x)g(y)d(z) \\ &\quad = d(x)yz + g(x)d(y)z + g(x)g(y)d(z) \quad \text{for all} \quad x, y, z \in N. \end{aligned}$$

$$(d(x)y + g(x)d(y))z = d(x)yz + g(x)d(y)z \quad \text{for all} \quad x, y, z \in N.$$

Lemma 2.7 *Let N be a 2-torsion free 3-prime near ring and U be a nonzero semigroup ideal of N. If d is a nonzero semiderivation of N associated with a map g such that $g(uv) = g(u)g(v)$, for all $u, v \in U$. If $a \in N$ and $ad(U) = \{0\}$ (or $d(U)a = \{0\}$), then $a = 0$.*

Proof Let $ad(u) = 0$, for all $u \in U$. Replacing u by uv, $a(d(u)g(v) + ud(v)) = 0$, for all $u, v \in U$. Thus $ad(u)g(v) + aud(v) = 0$, for all $u, v \in U$ or $aud(v) = 0$, for all $u, v \in U$. Choosing v such that $d(v) \neq 0$ and applying Lemma 2.2(ii), we get $a = 0$.

Lemma 2.8 *Let N be a 2-torsion free 3-prime near ring and U be a nonzero semigroup ideal of N. Suppose that d is a semiderivation on N associated with a map g such that $g(U) = U$. If $d^2(U) = \{0\}$, then $d = 0$.*

Proof Suppose $d^2(U) = \{0\}$. Then for $u, v \in U$ exploit the definition of d in different ways to obtain

$$0 = d^2(uv) = d(d(uv)) = d(d(u)v + g(u)d(v)) \text{ for all } u, v \in U,$$

$$= d^2(u)v + g(d(u))d(v) + d(g(u))d(v) + g(u)d^2(v),$$

$$= d(g(u))d(v) + d(g(u))d(v).$$

Note that $g(d(u)) = d(g(u))$ and $g(U) = U$, we get

$$2d(u)d(v) = 0 \text{ for all } u, v \in U.$$

Since N is a 2-torsion free, we get

$$d(u)d(v) = 0 \text{ for all } u, v \in U.$$

Replacing v by wv in the above relation, we get

$$d(u)d(wv) = 0 \text{ for all } u, v, w \in U.$$

$$d(u)(d(w)v + g(w)d(v)) = 0 \text{ for all } u, v, w \in U.$$

$$d(u)d(w)v + d(u)g(w)d(v) = 0 \text{ for all } u, v, w \in U.$$

This implies that

$$d(u)g(w)d(v) = 0 \text{ for all } u, v, w \in U.$$

$$d(u)wd(v) = 0 \text{ for all } u, v, w \in U.$$

$$d(U)Ud(U) = \{0\}.$$

Thus we obtain that $d = 0$ on U by Lemma 2.2(ii).

Lemma 2.9 *Let N be a 3-prime near ring and U be a nonzero semigroup ideal of N. Suppose d is a nonzero semiderivation of N associated with a map g such that $g(uv) = g(u)g(v)$, for all $u, v \in U$. If $d(U) \subseteq Z$, then N is a commutative ring.*

Proof We begin by showing that $(N, +)$ is abelian, which by Lemma 2.1(iv) is accomplished by producing $z \in Z\backslash\{0\}$ such that $z + z \in Z$. Let a be an element of U such that $d(a) \neq 0$. Then for all $x \in N$, $ax \in U$ and $ax + ax = a(x + x) \in U$, so that $d(ax) \in Z$ and $d(ax) + d(ax) \in Z$; hence we need only show that there exists $x \in N$ such that $d(ax) \neq 0$. Suppose this is not the case, so that $d((ax)a) = 0 = d(ax)g(a) + axd(a) = axd(a)$ for all $x \in N$. Since $d(a)$ is not zero divisor by Lemma 2.1(i), we get $aN = \{0\}$, so that $a = 0$—a contradiction. Therefore $(N, +)$ is abelian as required.

We are given that $[d(u), x] = 0$ for all $u \in U$ and $x \in N$. Replacing u by uv, we get $[d(uv), x] = 0$, which yields $[d(u)v + g(u)d(v), x] = 0$ for all $u, v \in U$ and $x \in N$. Since $(N, +)$ is abelian and $d(U) \subseteq Z$, we have

$$d(u)[v, x] + d(v)[x, g(u)] = 0 \text{ for all } u, v \in U \text{ and } x \in N. \tag{2.3}$$

Replacing x by $g(u)$, we obtain $d(u)[v, g(u)] = 0$ for all $u, v \in U$; and choosing $u \in U$ such that $d(u) \neq 0$ and applying Lemma 2.1(iii), we get $g(u) \in Z$. It then follows from (2.3) that $d(u)[v, x] = 0$ for all $v \in U$ and $x \in N$; therefore $U \subseteq Z$ and Lemma 2.3 completes the proof.

Lemma 2.10 *Let N be a 2-torsion free 3-prime near ring and U be a nonzero semigroup ideal of N. Suppose that d is a nonzero semiderivation of N associated with a map g such that $g(U) = U$ and $g(uv) = g(u)g(v)$ for all $u, v \in U$. If $[d(U), d(U)] = \{0\}$, then N is a commutative ring.*

Proof By hypothesis $[d(U), d(U)] = \{0\}$. Thus $d(u)d(vd(w)) = d(vd(w))d(u)$, for all $u, v, w \in U$, i.e., $d(u)(d(v)g(d(w)) + vd^2(w)) = (d(v)g(d(w)) + vd^2(w))$ $d(u)$, for all $u, v, w \in U$. Then by Lemma 2.6, we get $d(u)d(v)g(d(w)) + d(u)$ $vd^2(w) = d(v)g(d(w))d(u) + vd^2(w)d(u)$. This implies that $d(u)d(v)d(g(w)) +$ $d(u)vd^2(w) = d(v)d(g(w))d(u) + vd^2(w)d(u)$ i.e., $d(u)d(v)d(w) + d(u)vd^2$ $(w) = d(v)d(w)d(u) + vd^2(w)d(u)$ for all $u, v, w \in U$ and since $[d(U), d(U)] =$ $\{0\}$, we obtain

$$d(u)vd^2(w) = vd^2(w)d(u) \text{ for all } u, v, w \in U. \tag{2.4}$$

Replace v by xv, to get

$$d(u)xvd^2(w) = xvd^2(w)d(u) \text{ for all } u, v, w \in U \text{ and } x \in N.$$

Using (2.4), the above relation yields that $d(u)xvd^2(w) = xd(u)vd^2(w)$, for all $u, v, w \in U$ and $x \in N$, i.e., $[d(u), x]vd^2(w) = 0$, for all $u, v, w \in U$ and $x \in N$ by Lemma 2.6. Thus $[d(u), x]Ud^2(w) = 0$, for all $u, w \in U$ and $x \in N$. Since $d^2(U) \neq 0$ by Lemma 2.8, Lemma 2.2(ii) gives $d(U) \subseteq Z$, and the result follows by Lemma 2.9.

Lemma 2.11 *Let N be a 3-prime near ring and U be a nonzero semigroup ideal of N. If F is a nonzero generalized semiderivation of N with associated semiderivation d and a map g associated with d such that $g(U) = U$, then $F \neq 0$ on U.*

Proof Let $F(u) = 0$ for all $u \in U$. Replacing u by ux, we get $F(ux) = 0$ for all $u \in U$ and $x \in N$. Thus

$$F(u)x + g(u)d(x) = 0 = Ud(x) \text{ for all } x \in N$$

and it follows by Lemma 2.2(i) that $d = 0$. Therefore, we have

$$F(xu) = F(x)u = 0 \text{ for all } u \in U \text{ for all } x \in N$$

and another appeal to Lemma 2.2(i) gives $F = 0$, which is a contradiction.

Lemma 2.12 *Let N be a 3-prime near ring admitting a generalized semiderivation F associated with a semiderivation d. If g is an onto map associated with d such that $g(xy) = g(x)g(y)$ for all $x, y \in N$, then N satisfies the following partial distributive laws:*

(i) $(F(x)y + g(x)d(y))z = F(x)yz + g(x)d(y)z$ *for all* $x, y, z \in N$.
(ii) $(d(x)g(y) + xF(y))z = d(x)g(y)z + xF(y)z$ *for all* $x, y, z \in N$.

Proof (i) Let $x, y, z \in N$,

$$\begin{aligned} F(xyz) &= F(xy)z + g(xy)d(z) \\ &= (F(x)y + g(x)d(y))z + g(x)g(y)d(z). \end{aligned}$$

On the other hand,

$$\begin{aligned} F(xyz) &= F(x)yz + g(x)d(yz) \\ &= F(x)yz + g(x)(d(y)z + g(y)d(z)) \\ &= F(x)yz + g(x)d(y)z + g(x)g(y)d(z). \end{aligned}$$

Combining both expressions of $F(xyz)$, we obtain

$$(F(x)y + g(x)d(y))z = F(x)yz + g(x)d(y)z \text{ for all } x, y, z \in N.$$

(ii) For all $x, y, z \in N$ we have $F((xy)z) = F(xy)z + g(xy)d(z) = (d(x)g(y) + xF(y))z + g(x)g(y)d(z)$ and $F(x(yz)) = d(x)g(yz) + xF(yz) = d(x)g(y)g(z) + x(F(y)z + g(y)d(z)) = d(x)g(y)z + xF(y)z + g(x)g(y)d(z)$. Comparing the two expression, we get the required result.

Lemma 2.13 *Let N be a 3-prime near ring and U be a nonzero semigroup ideal of N. Suppose that F is a nonzero generalized semiderivation of N with associated semiderivation d and a map g associated with d such that $g(U) = U$ and $g(uv) = g(u)g(v)$ for all $u, v \in U$. If $a \in N$ and $aF(U) = 0$ (or $F(U)a = 0$), then $a = 0$.*

Proof Suppose that $aF(U) = \{0\}$. Then for $u, v \in U$

$$aF(uv) = aF(u)v + ag(u)d(v) = aud(v) = 0 \text{ for all } u, v \in U \text{ and } a \in N.$$

So by Lemma 2.2(ii), $a = 0$ or $d(U) = \{0\}$. If $d(U) = \{0\}$, then

$$ad(u)g(v) + auF(v) = 0 = auF(v) \text{ for all } u, v \in U;$$

and since $F(U) \neq \{0\}$ by Lemma 2.11, $a = 0$.

Lemma 2.14 *Let N be a 3-prime near ring admitting a generalized semiderivation F associated with a semiderivation d and an additive map g associated with d. Then N satisfies the following laws:*

(i) $d(x)y + g(x)d(y) = g(x)d(y) + d(x)y$ *for all* $x, y \in N$.
(ii) $d(x)g(y) + xd(y) = xd(y) + d(x)g(y)$ *for all* $x, y \in N$.
(iii) $F(x)y + g(x)d(y) = g(x)d(y) + F(x)y$ *for all* $x, y \in N$.
(iv) $d(x)g(y) + xF(y) = xF(y) + d(x)g(y)$ *for all* $x, y \in N$.

Proof (i) $d(x(y + y)) = d(x)(y + y) + g(x)d(y + y) = d(x)y + d(x)y + g(x)d(y) + g(x)d(y)$, and $d(xy + xy) = d(xy) + d(xy) = d(x)y + g(x)d(y) + d(x)y + g(x)d(y)$. Comparing these two equations, we get the desired result.
(ii) Again, calculate $d((x + x)y)$ and $d(xy + xy)$ and compare.

(iii) $F(x(y + y)) = F(x)(y + y) + g(x)d(y + y) = F(x)y + F(x)y + g(x) d(y) + g(x)d(y)$, and $F(xy + xy) = F(x)y + g(x)d(y) + F(x)y + g(x)d(y)$. Comparing these two equations, we get the desired result.
(iv) Again, calculate $F((x + x)y)$ and $F(xy + xy)$ and compare.

Lemma 2.15 *Let N be a 3-prime near ring and U be a nonzero semigroup ideal of N. Suppose that N admits nonzero semiderivations d_1, d_2 associated with a map g such that $g(uv) = g(u)g(v)$ for all $u, v \in U$. If $d_1(x)d_2(y) + d_2(y)d_1(x) \in Z$ for all $x, y \in U$ and at least one of $d_1(U) \cap Z$ and $d_2(U) \cap Z$ is nonzero, then N is a commutative ring.*

Proof Assume that $d_1(U) \cap Z \neq \{0\}$. Let $x \in U$ such that $d_1(x) \in Z \backslash \{0\}$, and $y \in U$. Then $d_1(x)d_2(y) + d_2(y)d_1(x) = d_1(x)(2d_2(y)) = d_1(x)(d_2(2y)) \in Z$. Therefore, $d_2(2U) \subseteq Z$. Since $2U$ is nonzero semigroup left ideal, our conclusion follows by Lemma 2.9, then N is commutative ring.

Lemma 2.16 *Let N be a 2-torsion free 3-prime near ring. If U is a nonzero semigroup ideal of N, then $2U \neq \{0\}$ and $d(2U) \neq \{0\}$ for any nonzero semiderivation d associated with a map g such that $g(U) = U$.*

Proof Let $x \in N$ with $x + x \neq 0$. Then for every $u \in U, u(x + x) = ux + ux \in 2U$; and by Lemma 2.2(i), we get $\{0\} \neq U(x + x) \subseteq 2U$. Since $2U$ is a semigroup left ideal, it follows by Lemma 2.5 that $d(2U) \neq \{0\}$.

Lemma 2.17 *Let N be a 3-prime near ring. If F is a generalized semiderivation with associated semiderivation d and a map g associated with d such that $g(U) = U$, then $F(Z) \subseteq Z$.*

Proof Let $z \in Z$ and $x \in N$. Then F(zx) = F(xz); that is $F(z)x + g(z)d(x) = d(x)g(z) + xF(z)$. Applying Lemma 2.14(iii), we get $g(z)d(x) + F(z)x = d(x)g(z) + xF(z)$; $zd(x) + F(z)x = d(x)z + xF(z)$. It follows that $F(z)x = xF(z)$ for all $x \in N$, so $F(Z) \subseteq Z$.

Lemma 2.18 *Let N be a 3-prime near ring and U be a nonzero semigroup ideal of N. Suppose that N admits a semiderivation d associated with a map g such that $g(U) = U$ and $g(uv) = g(u)g(v)$ for all $u, v \in U$. If $d^2(U) \neq \{0\}$ and $a \in N$ such that $[a, d(U)] = \{0\}$, then $a \in Z$.*

Proof Let $C(a) = \{x \in N | ax = xa\}$. Note that $d(U) \subseteq C(a)$. Thus, if $y \in C(a)$ and $u \in U$, both $d(yu)$ and $d(u)$ are in $C(a)$; hence $(d(y)g(u) + yd(u))a = a(d(y)g(u) + yd(u))$ and $d(y)g(u)a + yd(u)a = ad(y)g(u) + ayd(u)$; $d(y)ua + yd(u)a = ad(y)u + ayd(u)$. Since $yd(u) \in C(a)$, we conclude that $d(y)ua = ad(y)u$. Thus

$$d(C(a))U \subseteq C(a). \tag{2.5}$$

Choosing $z \in U$ such that $d^2(z) \neq 0$, and let $y = d(z)$. Then $y \in C(a)$; and by (2.5), $d(y)u \in C(a)$ and $d(y)uv \in C(a)$ for all $u, v \in U$. Thus, $0 = [a, d(y)uv] = ad(y)uv - d(y)uva = d(y)uav - d(y)uva = d(y)u(av - va)$. Thus $d(y)U(av - va) = 0$ for all $v \in U$; and by Lemma 2.2(ii), a centralizes U. By Lemma 2.1(iii), $a \in Z$.

Lemma 2.19 *Let N be a 3-prime near ring and F be a generalized semiderivation of N with associated nonzero semiderivation d and a map g associated with d such that $g(U) = U$ and $g(uv) = g(u)g(v)$ for all $u, v \in U$. If $d(F(N)) = \{0\}$, then $d^2(x)d(y) + d(x)d^2(y) = 0$ for all $x, y \in N$ and $F(d(N)) = \{0\}$.*

Proof Assume that $d(F(x)) = 0$ for all $x \in N$. It follows that $d(F(xy)) = d(F(x)y) + d(g(x)d(y)) = d(F(x)y) + d(xd(y)) = 0$ for all $x, y \in N$, that is,

$$d(F(x))g(y) + F(x)d(y) + d(x)g(d(y)) + xd^2(y) = 0 \text{ for all } x, y \in N.$$

This implies that

$$F(x)d(y) + d(x)d(g(y)) + xd^2(y) = 0.$$

$$F(x)d(y) + d(x)d(y) + xd^2(y) = 0 \text{ for all } x, y \in N. \quad (2.6)$$

Applying d again, we get

$$F(x)d^2(y) + d^2(x)d(y) + d(x)d^2(y) + d(x)d^2(y) + xd^3(y) = 0 \text{ for all } x, y \in N. \quad (2.7)$$

Taking $d(y)$ instead of y in (2.6) gives $F(x)d^2(y) + d(x)d^2(y) + xd^3(y) = 0$, hence (2.7) yields

$$d^2(x)d(y) + d(x)d^2(y) = 0 \text{ for all } x, y \in N. \quad (2.8)$$

Now, substitute $d(x)$ for x in (2.6), to obtain $F(d(x))d(y) + d^2(x)d(y) + d(x)d^2(y) = 0$; and use (2.8) to conclude that $F(d(x))d(y) = 0$ for all $x, y \in N$. Thus, by Lemma 2.7, $F(d(x)) = 0$ for all $x \in N$.

Lemma 2.20 *Let N be a 2-torsion free 3-prime near ring and F be a nonzero generalized semiderivation of N with associated semiderivation d and a map g associated with d such that $g(U) = U$; $g(uv) = g(u)g(v)$ for all $u, v \in U$ and $F(V) \subseteq U$ for some nonzero semigroup ideal V contained in U. If $a \in N$ and $[a, F(U)] = \{0\}$, then $a \in Z$.*

Proof If $d = 0$, then for all $x \in U$ and $y \in N$, $aF(x)y = F(x)ya$; hence $F(U)[a, y] = \{0\}$ and $a \in Z$ by Lemma 2.13. Therefore, we may assume $d \neq 0$. Let $C(a)$ denotes the centralizer of a, and let $y \in C(a)$ for all $u \in U$, $F(yu) \in C(a)$ -i.e. $(d(y)g(u) + yF(u))a = a(d(y)g(u) + yF(u))$ and by Lemma 2.12(ii) $d(y)g(u)a + yF(u)a = ad(y)g(u) + ayF(u); d(y)ua + yF(u)a = ad(y)u + ayF(u)$. Now $yF(u)a = ayF(u)$, and it follows that $d(y)u \in C(a)$; therefore $d(C(a))U$ is a semigroup right ideal which centralizes a, and if $d(C(a))U \neq \{0\}$. Lemma 2.1(iii) yields $a \in Z$. Assume now that $d(C(a))U = \{0\}$, in which case $d(C(a)) = \{0\}$ and hence $d(F(U)) = \{0\}$. It follows that for all $x \in N$ and $v \in V$, $d(F(xF(v))) = 0 = d(F(x)F(v) + g(x)d(F(v))) = d(F(x)F(v)) = d(F(x))g(F(v)) + F(x)d(F(v)) = d(F(x))F(v)$, so that $d(F(N))F(V) = \{0\}$ and by Lemma 2.13, $d(F(N)) = \{0\}$. By Lemma 2.19

$$d^2(x)d(y) + d(x)d^2(y) = 0 \text{ for all } x, y \in N \text{ and } F(d(N)) = \{0\}. \tag{2.9}$$

As in the proof of Theorem 4.1 of [3], we calculate $F(d(x)d(y))$ in two ways, obtaining $F(d(x)d(y)) = F(d(x))d(y) + g(d(x))d^2(y) = d(g(x))d^2(y) = d(x)d^2(y)$ and $F(d(x)d(y)) = d^2(x)g(d(y)) + d(x)F(d(y)) = d^2(x)d(g(y)) = d^2(x)d(y)$. Comparing the two results, we get $d(x)d^2(y) = d^2(x)d(y)$ for all $x, y \in N$, which together with (2.9) gives $d^2(x)d(y) = 0$ for all $x, y \in N$ and hence $d^2 = 0$. But by Lemma 2.8, this contradicts our assumption that $d \neq 0$; thus $d(C(a))U \neq \{0\}$ and our proof is complete.

3 Some Results Involving Two Generalized Semiderivations

The theorems that we prove in this section extend the results proved in [4].

Theorem 3.1 *Let N be a 3-prime near ring and U be a nonzero semigroup ideal of N. Suppose that N admits a nonzero generalized semiderivation F with associated semiderivation d and a map g associated with d such that $g(U) = U$ and $g(uv) = g(u)g(v)$ for all $u, v \in U$. If $F(U) \subseteq Z$, then $(N, +)$ is abelian. Moreover, if N is 2-torsion free, then N is a commutative ring.*

Proof We begin by showing that $(N, +)$ is abelian, which by Lemma 2.1(iv) is accomplished by producing $z \in Z\backslash\{0\}$ such that $z + z \in Z$. Let a be an element of U such that $F(a) \neq 0$. Then for all $x \in N$, $ax \in U$ and $ax + ax = a(x + x) \in U$, so that $F(ax) \in Z$ and $F(ax) + F(ax) \in Z$; hence we need only to show that there exists $x \in N$ such that $F(ax) \neq 0$. Suppose that this is not the case, so that $F((ax)a) = 0 = F(ax)a + g(ax)d(a) = g(a)g(x)d(a) = axd(a)$ for all $x \in N$. By Lemma 2.2(ii) either $a = 0$ or $d(a) = 0$.

If $d(a) = 0$, then $F(xa) = F(x)a + g(x)d(a)$; that is, $F(xa) = F(x)a \in Z$, for all $x \in N$. Thus, $[F(u)a, y] = 0$ for all $y \in N$ and $u \in U$. This implies that $F(u)[a, y] = 0$ for all $u \in U$ and $y \in N$ and Lemma 2.1(i) gives $a \in Z$. Thus, $0 = F(ax) = F(xa) = F(x)a$ for all $x \in N$. Replacing x by $u \in U$, we have $F(U)a = 0$, and by Lemmas 2.1(i) and 2.11, we get $a = 0$. Thus we have a contradiction.

To complete the proof, we show that if N is 2-torsion free, then N is commutative.

Consider first case $d = 0$. This implies that $F(ux) = F(u)x \in Z$ for all $u \in U$ and $x \in N$. By Lemma 2.11, we have $u \in U$ such that $F(u) \in Z\backslash\{0\}$, so N is commutative by Lemma 2.1(ii).

Now consider the case $d \neq 0$. Let $c \in Z\backslash\{0\}$. This implies that $x \in U$, $F(xc) = F(x)c + g(x)d(c) = F(x)c + xd(c) \in Z$. Thus $(F(x)c + xd(c))y = y(F(x)c + xd(c))$ for all $x, y \in U$ and $c \in Z$. Therefore, by Lemma 2.12(i), $F(x)cy + xd(c)y = yF(x)c + yxd(c)$ for all $x, y \in U$ and $c \in Z$. Since $d(c) \in Z$ and $F(x) \in Z$, we obtain $d(c)[x, y] = 0$ for all $x, y \in U$ and $c \in Z$. Let $d(Z) \neq \{0\}$. Choosing c such that $d(c) \neq 0$ and noting that $d(c)$ is not a zero divisor, we have $[x, y] = 0$ for all $x, y \in U$. By Lemma 2.1(iii), $U \subseteq Z$; hence N is commutative by Lemma 2.3.

The remaining case is $d \neq 0$ and $d(Z) = \{0\}$. Suppose we can show that $U \cap Z \neq \{0\}$. Taking $z \in (U \cap Z)\backslash\{0\}$ and $x \in N$, we have $F(xz) = F(x)z \in Z$; therefore $F(N) \subseteq Z$ by Lemma 2.1(ii). Let $F(x) \in Z$ for all $x \in N$.

Since $d(Z) = 0$, for all $x, y \in N$. We have

$$0 = d(F(xy)).$$

$$0 = d(F(x)y + g(x)d(y)).$$

$$0 = F(x)d(y) + g(x)d^2(y) + d(g(x))g(d(y)) \text{ for all } x, y \in Z.$$

Hence $F(xd(y)) = -d(g(x))g(d(y)) \in Z$ for all $x, y \in N$. By hypothesis, we have $d(x)d(y) \in Z$ for all $x, y \in N$. This implies that

$$d(x)(d(x)d(y) - d(y)d(x)) = 0 \text{ for all } x, y \in N.$$

Left multiplying by $d(y)$, we arrive at

$$d(y)d(x)N(d(x)d(y) - d(y)d(x)) = \{0\} \text{ for all } x, y \in N.$$

Since N is a 3-prime near ring, we get

$$[d(x), d(y)] = 0 \text{ for all } x, y \in N.$$

Using Lemma 2.10, N is a commutative ring.

Assume that $U \cap Z = \{0\}$. For each $u \in U$, $F(u^2) = F(u)u + g(u)d(u) = F(u)u + ud(u) = u(F(u) + d(u)) \in U \cap Z$. So $F(u^2) = 0$, thus for all $u \in U$ and $x \in N$, $F(u^2x) = F(u^2)x + g(u^2)d(x) = u^2d(x) \in U \cap Z$. So $u^2d(x) = 0$ and Lemma 2.7, $u^2 = 0$. Since $F(xu) = F(x)u + g(x)d(u) = F(x)u + xd(u) \in Z$ for all $u \in U$ and $x \in N$. We have $(F(x)u + xd(u))u = u(F(x)u + xd(u))$ and right multipling by u gives $uxd(u)u = 0$. Consequently, $d(u)uNd(u)u = \{0\}$. So that $d(u)u = 0$ for all $u \in U$,so $F(u)u = 0$ for all $u \in U$. But by Lemma 2.11, there exist $u_0 \in U$ for which $F(u_0) \neq 0$; and $F(u_0) \in Z$, we get $u_0 = 0$, contradiction. Therefore, $U \cap Z \neq \{0\}$ as required.

Theorem 3.2 *Let N be a 2-torsion free 3-prime near ring and U be a nonzero semigroup ideal of N. Let F_1 and F_2 be generalized semiderivations on N with associated semiderivations d_1 and d_2 respectively with at least one of d_1, d_2 not zero and a map g associated with d_1 and d_2 such that $g(uv) = g(u)g(v)$ for all $u, v \in U$ and $g(U) = U$. If $F_1(x)d_2(y) + F_2(x)d_1(y) = 0$ for all $x, y \in U$, then $F_1 = 0$ or $F_2 = 0$.*

Proof By hypothesis

$$F_1(x)d_2(y) + F_2(x)d_1(y) = 0 \text{ for all } x, y \in U. \tag{3.1}$$

Replacing x by uv in (3.1), we get

$$(d_1(u)g(v) + uF_1(v))d_2(y) + (d_2(u)g(v) + uF_2(v))d_1(y) = 0 \text{ for all } u, v, y \in U.$$

Using Lemmas 2.12(ii) and 2.14(iv), we conclude that

$$(d_1(u)g(v) + uF_1(v))d_2(y) + (uF_2(v) + d_2(u)g(v))d_1(y) = 0.$$

$$d_1(u)g(v)d_2(y) + uF_1(v)d_2(y) + uF_2(v)d_1(y) + d_2(u)g(v)d_1(y) = 0.$$

$$d_1(u)vd_2(y) + u(F_1(v)d_2(y) + F_2(v)d_1(y)) + d_2(u)vd_1(y) = 0 \text{ for all } u, v, y \in U.$$

Since middle summand is 0 by (3.1), we conclude that

$$d_1(u)vd_2(y) + d_2(u)vd_1(y) = 0 \text{ for all } u, v, y \in U. \tag{3.2}$$

Substituting yt for y in (3.2), we get

$$d_1(u)vd_2(yt) + d_2(u)vd_1(yt) = 0 \text{ for all } u, v, y, t \in U.$$

$$d_1(u)v(d_2(y)g(t) + yd_2(t)) + d_2(u)v(d_1(y)g(t) + yd_1(t)) = 0.$$

Using Lemma 2.14(ii), we have

$$d_1(u)v(d_2(y)g(t) + yd_2(t)) + d_2(u)v(yd_1(t) + d_1(y)g(t)) = 0.$$

This implies that

$$d_1(u)vd_2(y)t + (d_1(u)vyd_2(t) + d_2(u)vyd_1(t)) + d_2(u)vd_1(y)t = 0.$$

Again the middle summand is 0, so

$$d_1(u)vd_2(y)t + d_2(u)vd_1(y)t = 0 \text{ for all } u, v, y, t \in U. \tag{3.3}$$

Replacing t by $td_1(w)$ in (3.3), where $w \in U$, we have

$$d_1(u)v(d_2(y)td_1(w)) + d_2(u)(vd_1(y)t)d_1(w) = 0 \text{ for all } u, v, y, t, w \in U.$$

Using (3.2), we get

$$d_1(u)v(-d_1(y)td_2(w)) - d_1(u)vd_1(y)td_2(w) = 0.$$

This implies that

$$2d_1(u)vd_1(y)td_2(w) = 0 \text{ for all } u, v, y, t, w \in U.$$

Since N is 2-torsion free, we get

$$d_1(u)vd_1(y)td_2(w) = 0 \text{ for all } u, v, y, t, w \in U.$$

Thus $d_1(U)Ud_1(U)Ud_2(U) = \{0\}$; and by Lemmas 2.2(ii) and 2.5, one of d_1, d_2 must be 0. Assuming without loss that $d_1 = 0$, in which case $d_2 \neq 0$, we get $F_1(U)d_2(U) = \{0\}$, so by Lemmas 2.7 and 2.11, we have $F_1 = 0$.

Theorem 3.3 *Let N be a 2-torsion free 3-prime near ring and U be a nonzero semigroup ideal of N. Let F_1 and F_2 be generalized semiderivations on N with associated semiderivations d_1 and d_2 respectively and a map g associated with d_1 and d_2 such that $g(uv) = g(u)g(v)$ for all $u, v \in U$ and $g(U) = U$. If d_1 and d_2 are not both zero and F_1F_2 acts on U as a generalized semiderivation with associated semiderivation d_1d_2 and a map g associated with d_1d_2, then $F_1 = 0$ or $F_2 = 0$.*

Proof By the hypothesis, we have

$$F_1F_2(xy) = F_1F_2(x)y + g(x)d_1d_2(y) \text{ for all } x, y \in U.$$

$$F_1F_2(xy) = F_1F_2(x)y + xd_1d_2(y) \text{ for all } x, y \in U. \tag{3.4}$$

We also have

$$\begin{aligned} F_1F_2(xy) = F_1(F_2(xy)) &= F_1(F_2(x)y + g(x)d_2(y)) \\ &= F_1(F_2(x)y) + F_1(g(x)d_2(y)) \\ &= F_1(F_2(x)y) + F_1(xd_2(y)). \end{aligned}$$

i.e.

$$\begin{aligned} F_1F_2(xy) &= F_1F_2(x)y + g(F_2(x))d_1(y) + F_1(x)d_2(y) + g(x)d_1d_2(y) \\ &= F_1F_2(x)y + F_2(g(x))d_1(y) + F_1(x)d_2(y) + g(x)d_1d_2(y) \\ &= F_1F_2(x)y + F_2(x)d_1(y) + F_1(x)d_2(y) + xd_1d_2(y) \text{ for all } x, y \in U. \end{aligned} \tag{3.5}$$

Comparing (3.4) and (3.5), we get

$$F_2(x)d_1(y) + F_1(x)d_2(y) = 0 \text{ for all } x, y \in U.$$

Hence application of Theorem 3.2 yields that $F_1 = 0$ or $F_2 = 0$.

Theorem 3.4 *Let N be a 2-torsion free 3-prime near ring and U be a nonzero semigroup ideal of N. Let F_1 and F_2 be generalized semiderivations on N with associated semiderivations d_1 and d_2 respectively and a map g associated with d_1 and*

d_2 such that $g(uv) = g(u)g(v)$ for all $u, v \in U$ and $g(U) = U$. If $F_1F_2(U) = \{0\}$, then $F_1 = 0$ or $F_2 = 0$.

Proof By the hypothesis

$$F_1F_2(U) = \{0\}.$$

$$F_1F_2(xy) = F_1(F_2(xy)) = 0 = F_1(F_2(x)y + g(x)d_2(y))$$

$$= F_1(F_2(x)y) + F_1(xd_2(y))$$

$$= F_1F_2(x)y + g(F_2(x))d_1(y) + F_1(x)d_2(y) + g(x)d_1d_2(y)$$

$$= F_2(g(x))d_1(y) + F_1(x)d_2(y) + xd_1d_2(y) \text{ for all } x,\, y \in U.$$

This implies that

$$F_2(x)d_1(y) + xd_1d_2(y) + F_1(x)d_2(y) = 0 \text{ for all } x,\, y \in U. \tag{3.6}$$

Replacing x by zx in (3.6), we have

$$F_2(zx)d_1(y) + zxd_1d_2(y) + F_1(zx)d_2(y) = 0 \text{ for all } x,\, y, z \in U.$$

$$(d_2(z)g(x) + zF_2(x))d_1(y) + zxd_1d_2(y) + (d_1(z)g(x) + zF_1(x))d_2(y) = 0.$$

$$(d_2(z)g(x) + zF_2(x))d_1(y) + zxd_1d_2(y) + (zF_1(x) + d_1(z)g(x))d_2(y) = 0.$$

$$d_2(z)g(x)d_1(y) + zF_2(x)d_1(y) + zxd_1d_2(y) + zF_1(x)d_2(y) + d_1(z)g(x)d_2(y) = 0.$$

$$d_2(z)xd_1(y) + z(F_2(x)d_1(y) + xd_1d_2(y) + F_1(x)d_2(y)) + d_1(z)xd_2(y) = 0.$$

Since the middle summand is 0 by (3.6), we have

$$d_2(z)xd_1(y) + d_1(z)xd_2(y) = 0 \text{ for all } x,\, y, z \in U.$$

But this is just (3.2) of Theorem 3.2, so we argue as in the proof of Theorem 3.2 that $d_1 = 0$ or $d_2 = 0$. It now follows from (3.6) that

$$F_2(x)d_1(y) + F_1(x)d_2(y) = 0 \text{ for all } x,\, y \in U.$$

If one of d_1, d_2 is nonzero, then F_1 or F_2 is 0 by Theorem 3.2, so we assume that $d_1 = d_2 = 0$. Then $F_1F_2(xy) = 0 = F_1(F_2(x)y) = F_2(x)F_1(y)$ for all $x, y \in U$, so that $F_2(U)F_1(U) = \{0\}$. Applying Lemma 2.13, we conclude that $F_1 = 0$ or $F_2 = 0$.

We now consider a somewhat different condition that elements of $F_1(U)$ and $F_2(U)$ anti-commute.

Theorem 3.5 *Let N be a 2-torsion free 3-prime near ring with nonzero semigroup ideal U; and let F_1 and F_2 be generalized semiderivations on N with associated semiderivations d_1 and d_2 respectively such that $F_1(U^2) \subseteq U$ and $F_2(U^2) \subseteq U$ and a map g associated with d_1 and d_2 such that $g(uv) = g(u)g(v)$ for all $u, v \in U$ and $g(U) = U$. If*

$$F_1(x)F_2(y) + F_2(y)F_1(x) = 0 \text{ for all } x, y \in U, \tag{3.7}$$

then $F_1 = 0$ or $F_2 = 0$.

Proof Assume that $F_1 \neq 0$ and $F_2 \neq 0$. Note that if $w \in F_2(U^2)$, $-w \in F_2(U)$; and apply Lemma 2.4 to get $(uv)w = w(uv)$ for all $u, v \in F_1(U)$ and $w \in F_2(U^2)$. It follows by Lemma 2.20 that $F_1(U)F_1(U) \subseteq Z$, and it is easy to see that

$$F_1(x)F_1(y)(F_1(x)F_1(y) - F_1(y)F_1(x)) = 0 \text{ for all } x, y \in U.$$

This implies that

$$F_1(y)F_1(x)(F_1(x)F_1(y) - F_1(y)F_1(x)) = 0 \text{ for all } x, y \in U.$$

Since $F_1(x)F_1(y)$ and $F_1(y)F_1(x)$ are central, Lemma 2.1(i) shows that either both are zero or one can be cancelled to yield

$$F_1(x)F_1(y) = F_1(y)F_1(x).$$

Thus $[F_1(U), F_1(U)] = \{0\}$ and by Lemma 2.20, $F_1(U) \subseteq Z$, hence N is a commutative ring by Theorem 3.1. This fact together with (3.7) gives $F_1(U)F_2(U) = \{0\}$. Contradicting our assumption that $F_1 \neq 0 \neq F_2$. Therefore $F_1 = 0$ or $F_2 = 0$ as required.

If U is closed under addition, then $F(U^2) \subseteq U$ for any generalized semiderivation F; hence we have

Corollary 3.6 *Let N be a 2-torsion free 3-prime near ring and U be a nonzero semigroup ideal of N which is closed under addition. If F_1 and F_2 are generalized semiderivations on N with associated semiderivations d_1 and d_2 respectively and a map g associated with d_1 and d_2 such that $g(uv) = g(u)g(v)$ for all $u, v \in U$ and $g(U) = U$. if*

$$F_1(x)F_2(y) + F_2(y)F_1(x) = 0 \text{ for all } x, y \in U,$$

then $F_1 = 0$ or $F_2 = 0$.

We now replace the hypothesis that $F_1(U) \subseteq U$ and $F_2(U) \subseteq U$ in Theorem 3.5 by some commutativity hypothesis.

Theorem 3.7 *Let N be a 2-torsion free 3-prime near ring with nonzero semigroup ideal U; and let F_1 and F_2 be generalized semiderivations on N with associated semiderivations d_1 and d_2 respectively and a map g associated with d_1 and d_2 such that $g(U) = U$ and $g(uv) = g(u)g(v)$ for all $u, v \in U$. If*

$$F_1(x)F_2(y) + F_2(y)F_1(x) = 0 \, \textit{for all}\ x, y \in U,$$

then $F_1 = 0$ *or* $F_2 = 0$ *and one of the following is satisfied: (a)* $d_1(Z) \neq \{0\}$ *and* $d_2(Z) \neq \{0\}$*; (b)* $U \cap Z \neq \{0\}$.

Proof (a) Let $z_1 \in Z$ such that $d_1(z_1) \neq 0$. Then for all $x, y \in U$, we have

$$F_1(z_1x)F_2(y) + F_2(y)F_1(z_1x) = 0.$$

$$(d_1(z_1)g(x) + z_1F_1(x))F_2(y) + F_2(y)(F_1(x)z_1 + g(x)d_1(z_1)) = 0.$$

$$d_1(z_1)g(x)F_2(y) + z_1F_1(x)F_2(y) + F_2(y)F_1(x)z_1 + F_2(y)g(x)d_1(z_1) = 0.$$

$$d_1(z_1)xF_2(y) + z_1(F_1(x)F_2(y) + F_2(y)F_1(x)) + F_2(y)xd_1(z_1) = 0.$$

It follows that

$$d_1(z_1)xF_2(y) + F_2(y)xd_1(z_1) = 0 \text{ for all } x, y \in U.$$

Choosing $z_2 \in Z$ such that $d_2(z_2) \neq 0$ and using a similar argument, we now get

$$xy + yx = 0 \text{ for all } x, y \in U;$$

and applying Lemma 2.4 with $S = U$ and $T = U^2$ shows that U^2 centralizes U^2, so that $U^2 \subseteq Z$ by Lemma 2.1(iii) and hence N is commutative ring by Lemma 2.3. It now follows that $F_1(x)F_2(y) = F_2(y)F_1(x) = -F_2(y)F_1(x)$ for all $x, y \in U$. Hence $F_1(U)F_2(U) = \{0\}$. Therefore $F_1 = 0$ or $F_2 = 0$.
(b) We assume that $F_1 \neq 0$ and $F_2 \neq 0$. Let $z_0 \in (U \cap Z)\backslash\{0\}$. By Lemma 2.17, $F_1(z_0) \in Z$; hence if $F_1(z_0) \neq 0$ the condition

$$F_1(z_0)F_2(x) + F_2(x)F_1(z_0) = 0 \text{ for all } x \in U$$

gives $2F_2(x) = 0 = F_2(x)$ for all $x \in U$, so that $F_1 = 0$ by Lemma 2.11. Therefore, $F_1(z_0) = 0$ and similarly $F_2(z_0) = 0$. Now $z_0^2 \in (U \cap Z)\backslash\{0\}$ also, so $F_1(z_0^2) = 0 = F_2(z_0^2)$; and since $F_1(z_0^2) = F_1(z_0)z_0 + g(z_0)d_1(z_0) = z_0d_1(z_0)$ and $F_2(z_0^2) = F_2(z_0)z_0 + g(z_0)d_2(z_0) = z_0d_2(z_0)$. we have $d_1(z_0) = d_2(z_0) = 0$. Observing that $F_1(z_0x) = F_1(z_0)x + g(z_0)d_1(x) = F_1(z_0)x + z_0d_1(x)$ and $F_1(xz_0) = F_1(x)z_0 + g(x)d_1(z_0) = F_1(x)z_0 + xd_1(z_0)$ for all $x \in N$, we see that $F_1(x) = d_1(x)$ for all $x \in N$, So that F_1 is a semiderivation; and similarly F_2 is a semiderivation. We can now derive a contradiction as in the proof of Theorem 3.5, with Lemmas 2.8 and 2.18 used instead of Lemma 2.20.

4 Some Commutativity Conditions

The skew-commutativity hypothesis of Theorems 3.4 and 3.5 suggests investigating conditions of the form $F_1(x)F_2(y) + F_2(y)F_1(x) \in Z$ or $xF(y) + F(y)x \in Z$.

Theorem 4.1 *Let N be a 2-torsion free 3-prime near ring and U be a nonzero semigroup ideal of N which is closed under addition.*
(i) Suppose N has nonzero generalized semiderivations F_1, F_2 with associated semiderivations d_1 and d_2 respectively and a map g associated with d_1 and d_2 such that $g(U) = U$ and $g(uv) = g(u)g(v)$ for all $u, v \in U$. If $F_1(x)F_2(y) + F_2(y)F_1(x) \in Z$, for all $x, y \in U$ and at least one of $F_1(U) \cap Z$ and $F_2(U) \cap Z$ is nonzero, then N is a commutative ring.
(ii) If N admits a nonzero generalized semiderivation F with associated semiderivation d and a map g associated with d such that $g(U) = U$ and $g(uv) = g(u)g(v)$ for all $u, v \in U$ and $U \cap Z \neq \{0\}$ and $xF(y) + F(y)x \in Z$, for all $x, y \in U$, then N is commutative ring.

Proof (i) Assume that $F_1(U) \cap Z \neq \{0\}$. Let $x \in U$ such that $F_1(x) \in Z\backslash\{0\}$. Then $F_1(x)F_2(y) + F_2(y)F_1(x) = 2F_1(x)F_2(y) = F_1(x)F_2(2y) \in Z$ for all $y \in U$. Since $F_1(x) \in Z\backslash\{0\}$, Lemma 2.1(ii) gives $F_2(2y) \in Z$ for all $y \in U$ -i.e. $F_2(2U) \subseteq Z$. Since $0 \in Z$, we get $F_2(2U) = \{0\}$ -i.e. $2F_2(U) = \{0\}$. But N is 2-torsion free, we get $F_2(U) = \{0\}$ would contradict our hypothesis that $F_2 \neq 0$; hence $F_2(2U) \neq \{0\}$ and we may choose $y \in U$ such that $F_2(2y) \in Z\backslash\{0\}$. Since $2U \subseteq U$, this shows that $F_2(2y)$ and $2F_2(2y) = F_2(4y)$ are in $F_2(U) \cap Z\backslash\{0\}$, so that for all $x \in U$, $F_1(x)(2F_2(2y)) \in Z$ and hence $F_1(x) \in Z$. Thus, $F_1(U) \subseteq Z$ and by Theorem 3.1, N is a commutative ring.
(ii) Essentially the same argument yields $U \subseteq Z$, and the result follows by Lemma 2.3.

Theorem 4.2 *Let N be a 2-torsion free 3-prime near ring and U be a nonzero semigroup ideal of N which is closed under addition. Suppose N admits nonzero generalized semiderivations F_1 and F_2 with associated semiderivations d_1 and d_2 respectively and a map g associated with d_1 and d_2 such that $g(U) = U$ and $g(uv) = g(u)g(v)$ for all $u, v \in U$. Suppose that $F_1(x)F_2(y) + F_2(y)F_1(x) \in Z$, for all $x, y \in U$ and $F_1(U) \subseteq U$; $F_2(U) \subseteq U$. If $F_1(N) \cap Z \neq \{0\}$ or $F_2(N) \cap Z \neq \{0\}$, then N is a commutative ring.*

Proof By Corollary 3.6, we cannot have $F_1(x)F_2(y) + F_2(y)F_1(x) = 0$ for all $x, y \in U$, hence there exist $x_0, y_0 \in U$ such that $u_0 = F_1(x_0)F_2(y_0) + F_2(y_0)F_1(x_0) \in (Z\backslash\{0\}) \cap U$. Since $F_1(Z)$ and $F_2(Z)$ are central by Lemma 2.17, if $F_1(u_0) \neq 0$ or $F_2(u_0) \neq 0$ we have $F_1(U) \cap Z \neq \{0\}$ or $F_2(U) \cap Z \neq \{0\}$ and our conclusion follows by Theorem 4.1(i).

Assume, therefore, that $F_1(u_0) = F_2(u_0) = 0$. For all $x, y \in U$, $F_1(u_0x)F_2(u_0y) + F_2(u_0y)F_1(u_0x) = u_0^2(d_1(x)d_2(y) + d_2(y)d_1(x)) \in Z$, hence $d_1(x)d_2(y) + d_2(y)d_1(x) \in Z$; and if $d_1(u_0) \neq 0$ or $d_2(u_0) \neq 0$ our desired conclusion follows by Lemma 2.15. Therefore we may assume $d_1(u_0) = d_2(u_0) = 0$. For all $x, y \in N$, $F_1(xu_0)F_2(yu_0) + F_2(yu_0)F_1(xu_0) \in Z$, so $u_0^2(F_1(x)F_2(y) + F_2(y)F_1(x)) \in Z$

and $F_1(x)F_2(y) + F_2(y)F_1(x) \in Z$. Since $F_1(N) \cap Z \neq \{0\}$ or $F_2(N) \cap Z \neq \{0\}$, our result follows by Theorem 4.1(i).

Acknowledgments The authors would like to thank the referee for his valuable suggestions.

References

1. Ali, A., Bell, H.E., Miyan, P.: Generalized derivations on prime near rings. Int. J. Math. Math. Sci. Article ID 170749, 5 p. (2013)
2. Bell, H.E.: On Derivations in Near-Rings II, pp. 191–197. Kluwer Academic Publishers, Netherlands (1997)
3. Bell, H.E.: On prime near-rings with generalized derivation. Int. J. Math. Math. Sci. Article ID 490316, 5 p. (2008)
4. Bell, H.E., Argac, N.: Derivations, products of derivations and commutativity in near rings. Algebr. Colloq. **8**(4), 399–407 (2001)
5. Bergen, J.: Derivations in prime rings. Can. Math. Bull. **26**, 267–270 (1983)

n-Strongly Gorenstein Projective and Injective Complexes

C. Selvaraj and R. Saravanan

Abstract In this paper, we introduce and study the notions of n-strongly Gorenstein projective and injective complexes, which are generalizations of n-strongly Gorenstein projective and injective modules, respectively. Further, we characterize the so-called notions and prove that the Gorenstein projective (resp., injective) complexes are direct summands of n-strongly Gorenstein projective (resp., injective) complexes. Also, we discuss the relationships between n-strongly Gorenstein injective and n-strongly Gorenstein flat complexes, and for any two positive integers n and m, we exhibit the relationships between n-strongly Gorenstein projective (resp., injective) and m-strongly Gorenstein projective (resp., injective) complexes.

Keywords n-SG-projective complex · n-SG-injective complex · n-SG-flat complex

AMS Subject classification (2010) 16E05 · 16E30

1 Introduction

Throughout this paper, let R be an associative ring with identity and $\mathscr{C}$ be the abelian category of complexes of R-modules. Unless stated otherwise, a complex and an R-module will be understood to be a complex of left R-modules and a left R-module respectively.

Bennis and Mahdou [2] introduced the notions of strongly Gorenstein projective, injective and flat modules which are further studied and characterized by Liu [8]. Later, Bennis and Mahdou [3] generalized the notion of strongly Gorenstein projective modules to n-strongly Gorenstein projective modules and [11] Zhao studied the homological behaviors of n-strongly Gorenstein projective, injective and flat modules. Zhang et al. [10] studied the notions of strongly Gorenstein projective

C. Selvaraj (✉) · R. Saravanan
Department of Mathematics, Periyar University, Salem 636 011, TN, India
e-mail: selvavlr@yahoo.com

S.T. Rizvi et al. (eds.), *Algebra and its Applications*, Springer Proceedings in Mathematics & Statistics 174, DOI 10.1007/978-981-10-1651-6_17

and injective complexes. Motivated by the above works in this article, we introduce and study the notions of n-strongly Gorenstein projective and injective complexes, which are generalizations of n-strongly Gorenstein projective and injective modules, respectively. In [2, Theorem 2.7], it is proved that a module is Gorenstein projective if and only if it is a direct summand of a strongly Gorenstein projective module. Using [9, Theorem 2.3], we prove the following.

Theorem *Let G be a complex. Then the following holds:*

(1) G is Gorenstein projective if and only if it is a direct summand of an n-SG-projective complex.
(2) G is Gorenstein injective if and only if it is a direct summand of an n-SG-injective complex.

In [9, Theorem 3.1], the relationship between Gorenstein flat and Gorenstein injective complexes is given. In connection to [9, Theorems 3.1 and 3.3] and [7, Proposition 4.7], we have the following result.

Theorem *Let R be a left artinian ring and let the injective envelope of every simple left R-module be finitely generated. Then the following hold:*

(1) If a complex G of left R-modules is n-SG-injective, then G^+ is an n-SG-flat complex of right R-modules.
(2) If a complex G of right R-modules is n-SG-flat, then G^+ is an n-SG-injective complex of left R-modules.

In Sect. 2, we recall some known definitions and terminologies which will be needed in the sequel.

In Sect. 3, we introduce and study the notions of n-strongly Gorenstein projective and injective complexes. We show that a complex is Gorenstein projective (resp., injective) if and only if it is a direct summand of an n-SG-projective (resp., injective) complex and prove that the modules in an n-SG-projective (resp., injective) complex are precisely the n-SG-projective (resp., injective) modules. Further, over a left artinian ring R, we discuss the relationships between n-SG-injective and n-SG-flat complexes.

In the last section, we study the relationships between n-SG-projective (resp., injective) and m-SG-projective (resp., injective) complexes for any two positive integers n and m.

2 Preliminaries

In this section, we first recall some known definitions and terminologies which we need in the sequel.

In this paper, a complex

$$\cdots \to C^{-1} \xrightarrow{\delta^{-1}} C^0 \xrightarrow{\delta^0} C^1 \xrightarrow{\delta^1} \cdots$$

will be denoted by C or (C, δ). We will use subscripts to distinguish complexes. So if $\{C_i\}_{i\in I}$ is a family of complexes, C_i will be

$$\cdots \to C_i^{-1} \xrightarrow{\delta_i^{-1}} C_i^0 \xrightarrow{\delta_i^0} C_i^1 \xrightarrow{\delta_i^1} \cdots.$$

Given an R-module M, we will denote by $\overline{M}$ the complex

$$\cdots 0 \to 0 \to M \xrightarrow{id} M \to 0 \to 0 \cdots$$

with M in the 1st and 0th degrees. Similarly, we denote by $\underline{M}$ the complex with M in the 0th degree and 0 in the other places. Note that an R-module M is injective (resp., projective) if and only if the complex $\overline{M}$ is injective (resp., projective).

Given a complex C and an integer m, $C[m]$ denotes the complex such that $C[m]^n = C^{m+n}$ and whose boundary operators are $(-1)^m\delta^{m+n}$. The nth cycle of a complex C is defined as $\mathrm{Ker}\delta^n$ and is denoted by Z^nC. The nth boundary of C is defined as $\mathrm{Im}\delta^{n-1}$ and is denoted by B^nC.

Let C be a complex of left R-modules (resp., of right R-modules) and let D be a complex of left R-modules. We denote by $\mathrm{Hom}(C, D)$ (respectively, $C \otimes D$) the usual homomorphism complex (resp., tensor product) of the complexes C and D. The nth degree term of the complex $\mathrm{Hom}(C, D)$ is given by

$$\mathrm{Hom}(C, D)^n = \prod_{t\in\mathbb{Z}} \mathrm{Hom}(C^t, D^{n+t})$$

and whose boundary operators are

$$(\delta^n f)^m = \delta_D^{n+m} f^m - (-1)^n f^{m+1}\delta_C^m.$$

The nth degree term of $C \otimes D$ is given by

$$(C \otimes D)^n = \bigoplus_{t\in\mathbb{Z}} (C^t \otimes_R D^{n-t})$$

and

$$\delta(x \otimes y) = \delta_C^t(x) \otimes y + (-1)^t x \otimes \delta_D^{n-t}(y),$$

for $x \in C^t$ and $y \in D^{n-t}$.

For a complex C of left R-modules, we have a functor $-\otimes \mathscr{C} : \mathscr{C}_R \to \mathscr{C}_{\mathbb{Z}}$, where $\mathscr{C}_R$ denotes the category of right R-modules. The functor $-\otimes \mathscr{C} : \mathscr{C}_R \to \mathscr{C}_{\mathbb{Z}}$ being right exact, we can construct the left derived functors which we denote by

$Tor_i(-, C)$. Given two complexes C and D of $\mathscr{C}$, we use $Ext^i(C, D)$ for $i \geq 0$ to denote the groups we obtain from the right derived functors of Hom and we use C^+ to denote the complex $\underline{Hom}(C, \overline{\mathbb{Q}/\mathbb{Z}})$.

Recall that a complex C is projective (respectively, injective) if C is exact and $Z^n C$ is a projective (respectively, an injective) R-module for each $i \in \mathbb{Z}$. A complex C is flat if C is exact and $Z^n C$ is flat R-module for each $i \in \mathbb{Z}$. Equivalently, a complex C is projective (respectively, injective) if and only if $\mathrm{Hom}(C, -)$ (respectively, Hom $(-, C)$) is exact. Also a complex C is flat if and only if $- \otimes C$ is exact. For unexplained terminologies and notations we refer to [1, 4–6].

Definition 2.1 ([10]) A complex G is called strongly Gorenstein projective (for short SG-projective) if there exists an exact sequence of complexes

$$\mathbb{P} : \cdots \to P \xrightarrow{\delta} P \xrightarrow{\delta} P \xrightarrow{\delta} \cdots$$

such that (i) P is a projective complex;
(ii) $\mathrm{Ker}\, \delta_0 \cong G$;
(iii) $\mathrm{Hom}(\mathbb{P}, Q)$ is exact for any projective complex Q.

Similarly, the SG-injective complexes are defined.

Definition 2.2 ([7]) A complex G of right R-modules is called strongly Gorenstein flat (for short SG-flat) if there exists an exact sequence of complexes of right R-modules

$$\mathbb{F} : \cdots \to F \xrightarrow{\delta} F \xrightarrow{\delta} F \xrightarrow{\delta} \cdots$$

such that (i) F is flat;
(ii) $\mathrm{Ker}\, \delta_0 \cong G$;
(iii) $\mathbb{F} \otimes I$ is exact for any injective complex I.

Definition 2.3 ([7]) Let n be a positive integer. A complex G of right R-modules is said to be an n-SG-flat if there exists an exact sequence of complexes

$$0 \to G \xrightarrow{\delta_{n+1}} F_n \xrightarrow{\delta_n} F_{n-1} \to \cdots \to F_1 \xrightarrow{\delta_1} G \to 0$$

with F_i projective for any $1 \leq i \leq n$, such that $- \otimes I$ leaves the sequence exact whenever I is an injective complex.

Next, we present the characterizations of n-SG-flat complexes in order to use it further.

Proposition 2.4 ([7]) *Let R be a right coherent ring and G be any complex of right R-modules. Then the following are equivalent;*

(1) G is n-SG-flat;

(2) There exists an exact sequence of complexes of right R-modules

$$0 \to G \stackrel{\delta_{n+1}}{\to} F_n \stackrel{\delta_n}{\to} F_{n-1} \to \cdots \to F_1 \stackrel{\delta_1}{\to} G \to 0$$

with F_i flat for any $1 \leq i \leq n$, such that $\bigoplus_{i=2}^{n+1} Im\ \delta_i$ is SG-flat;

(3) There exists an exact sequence of complexes of right R-modules

$$0 \to G \stackrel{\delta_{n+1}}{\to} F_n \stackrel{\delta_n}{\to} F_{n-1} \to \cdots \to F_1 \stackrel{\delta_1}{\to} G \to 0$$

with F_i flat for any $1 \leq i \leq n$, such that $\bigoplus_{i=2}^{n+1} Im\ \delta_i$ is Gorenstein flat.

3 n-Strongly Gorenstein Projective and Injective Complexes

In this section, we introduce and study the n-SG-projective and injective complexes which are generalizations of SG-projective and injective modules, respectively. Also we extend the results in [3, 11] on n-strongly Gorenstein projective and injective modules to that of complexes.

Definition 3.1 Let n be a positive integer. A complex G is said to be an n-strongly Gorenstein projective (for short n-SG-projective) if there exists an exact sequence of complexes

$$0 \to G \stackrel{\delta_{n+1}}{\to} P_n \stackrel{\delta_n}{\to} P_{n-1} \to \cdots \to P_1 \stackrel{\delta_1}{\to} G \to 0$$

with P_i projective for any $1 \leq i \leq n$, such that $\mathrm{Hom}(-, Q)$ leaves the sequence exact whenever Q is a projective complex.

Definition 3.2 Let n be a positive integer. A complex G is said to be an n-strongly Gorenstein injective (for short n-SG-injective) if there exists an exact sequence of complexes

$$0 \to G \stackrel{\alpha_{n+1}}{\to} I_n \stackrel{\alpha_n}{\to} I_{n-1} \to \cdots \to I_1 \stackrel{\alpha_1}{\to} G \to 0$$

with I_i injective for any $1 \leq i \leq n$, such that $\mathrm{Hom}(E, -)$ leaves the sequence exact whenever E is an injective complex.

Note that 1-SG-projective (resp., injective) complexes are just SG-projective (resp., injective) complexes. It is also clear that for any i with $2 \leq i \leq n+1$, the complex Im δ_i (resp., Im α_i) in the above exact sequence is n-SG-projective (resp., injective). The following proposition shows that the class of all n-SG-projective (resp., injective) complexes is between the class of all SG-projective (resp., injective) complexes and the class of all Gorenstein projective (resp., injective) complexes.

Proposition 3.3 *Let n be a positive integer. Then:*

(1) Every SG-projective (resp., injective) complex is an n-SG-projective (resp., injective) complex.
(2) Every n-SG-projective (resp., injective) complex is a Gorenstein projective (resp., injective) complex.

Proof Since the SG-injective complex is the dual notion of SG-projective, we prove the results for SG-projective case.

(1) Let G be an SG-projective complex. There exists an exact sequence of complexes

$$0 \to G \xrightarrow{f} P \xrightarrow{g} G \to 0,$$

where P is a projective complex, such that $\mathrm{Hom}(-, Q)$ leaves the sequence exact whenever Q is a projective complex. Then we get an exact sequence of complexes of the form

$$X : 0 \to G \xrightarrow{f} P \xrightarrow{fg} P \xrightarrow{fg} \cdots \to P \xrightarrow{g} G \to 0$$

such that $\mathrm{Hom}(X, Q)$ is exact for any projective complex Q. Therefore G is an n-SG-projective complex.

(2) Let G be an n-SG-projective complex. There exists an exact sequence of complexes

$$Y : 0 \to G \xrightarrow{\delta_{n+1}} P_n \xrightarrow{\delta_n} P_{n-1} \to \cdots \to P_1 \xrightarrow{\delta_1} G \to 0$$

with P_i projective for any $1 \leq i \leq n$, such that $\mathrm{Hom}(-, Q)$ leaves the sequence exact whenever Q is a projective complex. Thus, we get the following exact sequence of complexes

$$Y' : \cdots \to P_1 \xrightarrow{\delta_{n+1}\delta_1} P_n \xrightarrow{\delta_n} P_{n-1} \to \cdots \to P_1 \xrightarrow{\delta_{n+1}\delta_1} P_n \xrightarrow{\delta_n} \cdots .$$

such that $Im(\delta_{n+1}\delta_1) \cong G$. Let Q be any projective complex. Then the exactness of $\mathrm{Hom}(Y', Q)$ follows from the exactness of $\mathrm{Hom}(Y, Q)$ and hence G is a Gorenstein projective complex. □

Proposition 3.4 *Let $\{G_i\}_I$ be any family of complexes. Then*

(1) If G_i is n-SG-projective for every $i \in I$, then $\bigoplus_I G_i$ is an n-SG-projective complex.
(2) If G_i is n-SG-injective for every $i \in I$, then $\prod_I G_i$ is an n-SG-injective complex.

Proof (1) For each i in I there exists an exact sequence of complexes

$$\mathbb{X}_i : 0 \to G_i \to P_{in} \to P_{in-1} \to \cdots \to P_{i1} \to G_i \to 0$$

with P_{ij} projective for $1 \leq j \leq n$, such that $\mathrm{Hom}(\mathbb{X}_i, Q)$ is exact for any projective complex Q. Since the direct sum of projective complexes is projective, we obtain the following exact sequence of complexes

$$\bigoplus_{i\in I} \mathbb{X}_i : 0 \to \bigoplus_{i\in I} G_i \to \bigoplus_{i\in I} P_{in} \to \cdots \to \bigoplus_{i\in I} P_{i1} \to \bigoplus_{i\in I} G_i \to 0$$

with $\bigoplus_{i\in I} P_{ij}$ projective for $1 \leq j \leq n$. Let Q be any projective complex. Then $\mathrm{Hom}(\bigoplus \mathbb{X}_i, Q) \cong \prod \mathrm{Hom}(\mathbb{X}_i, Q)$ is exact, and hence $\bigoplus G_i$ is an n-SG-projective complex.

(2) The proof is similar to (1). □

In [2, Theorem 2.7], it is proved that a module is Gorenstein projective if and only if it is a direct summand of a strongly Gorenstein projective module. Using [9, Theorem 2.3] and Proposition 3.3, we have the following.

Theorem 3.5 *Let G be a complex. Then the following hold:*

(1) G is Gorenstein projective if and only if it is a direct summand of an n-SG-projective complex.
(2) G is Gorenstein injective if and only if it is a direct summand of an n-SG-injective complex.

Proof (1) Let G be a Gorenstein projective complex. Then it is a direct summand of an SG-projective complex by [10, Theorem 1]. Hence G is a direct summand of an n-SG-projective complex by Proposition 3.3. Conversely, let G be a direct summand of an n-SG-projective complex C. Then C is Gorenstein projective by Proposition 3.3 (2). Since the class of all Gorenstein projective complexes is closed under direct summands by [9, Theorem 2.3], it follows that G is Gorenstein projective.

(2) The proof is similar to (1). □

In [11, Theorem 3.9], Zhao and Huang have given some characterizations of n-SG-projective modules. Now, we have the similar characterization for n-SG-projective complexes in the following.

Proposition 3.6 *Let G be any complex. Then the following are equivalent;*

(1) G is n-SG-projective;
(2) There exists an exact sequence of complexes

$$0 \to G \overset{\delta_{n+1}}{\to} P_n \overset{\delta_n}{\to} P_{n-1} \to \cdots \to P_1 \overset{\delta_1}{\to} G \to 0$$

with P_i projective for any $1 \leq i \leq n$, such that $\bigoplus_{i=2}^{n+1} \mathrm{Im}\, \delta_i$ is SG-projective;

(3) There exists an exact sequence of complexes

$$0 \to G \overset{\delta_{n+1}}{\to} P_n \overset{\delta_n}{\to} P_{n-1} \to \cdots \to P_1 \overset{\delta_1}{\to} G \to 0$$

with F_i projective for any $1 \le i \le n$, such that $\bigoplus_{i=2}^{n+1}$ Im δ_i is Gorenstein projective.

Proof (1) $\Rightarrow$ (2). Let G be an SG-projective complex. Then there exists an exact sequence of complexes

$$0 \to G \overset{\delta_{n+1}}{\to} P_n \overset{\delta_n}{\to} P_{n-1} \to \cdots \to P_1 \overset{\delta_1}{\to} G \to 0$$

with P_i projective for any $1 \le i \le n$, such that Hom$(-, Q)$ leaves the sequence exact whenever Q is a projective complex. Now for each i with $2 \le i \le n+1$, we have an exact sequence of complexes

$$0 \to \operatorname{Im} \delta_i \overset{\alpha_i}{\to} P_{i-1} \overset{\delta_{i-1}}{\to} \cdots \to P_1 \overset{\delta_{n+1}\delta_1}{\longrightarrow} P_n \overset{\delta_n}{\to} \cdots \to P_i \overset{\delta_i}{\to} \operatorname{Im} \delta_i \to 0.$$

By adding these exact sequences, we obtain the following exact sequence

$$0 \to \bigoplus_{i=2}^{n+1} \operatorname{Im} \delta_i \overset{\alpha}{\to} \bigoplus_{i=1}^{n} P_i \overset{\delta}{\to} \cdots \to P_n \oplus P_0 \oplus \cdots \oplus P_{n-1} \to \cdots$$

where $\alpha = \operatorname{diag}\{\alpha_1, \alpha_2, ..., \alpha_n\}$ and $\delta = \operatorname{diag}\,\{\delta_{n+1}\delta_1, \delta_2, ..., \delta_n\}$. Hence it is clear that $\operatorname{Im} \delta \cong \bigoplus_{i=2}^{n+1} \delta_i$ and $Ext_1(\bigoplus_{i=2}^{n+1} \operatorname{Im} \delta_i, Q) \cong \prod_{i=2}^{n+1} Ext_1(\operatorname{Im} \delta_i, Q) = 0$ for any projective complex Q. Therefore $\bigoplus_{i=2}^{n+1} \operatorname{Im} \delta_i$ is SG-flat.

(2) $\Rightarrow$ (3) It follows from the Proposition 3.3.

(3) $\Rightarrow$ (1) It is obvious. □

Similarly, we can characterize the n-SG-injective complexes.

In [9, Theorem 3.1], the relationship between Gorenstein flat and Gorenstein injective complexes is given. In connection to [9, Theorems 3.1 and 3.3] and Proposition 2.4, we have the following.

Theorem 3.7 *Let R be a left artinian ring and let the injective envelope of every simple left R-module be finitely generated. Then the following hold:*

(1) If a complex G of left R-modules is n-SG-injective, then G^+ is an n-SG-flat complex of right R-modules.

(2) If a complex G of right R-modules is n-SG-flat, then G^+ is an n-SG-injective complex of left R-modules.

Proof (1) Let G be an n-SG-injective complex. Then using the characterization of n-SG-injective complexes similar to Proposition 3.6, we get an exact sequence of complexes

$$\mathbf{I}: 0 \to G \overset{\delta_{n+1}}{\to} I_n \overset{\delta_n}{\to} I_{n-1} \to \cdots \to I_1 \overset{\delta_1}{\to} G \to 0$$

where I_j is an injective complex for $1 \leq j \leq n$ and $\bigoplus_{j=2}^{n+1} \operatorname{Im} \delta_j$ is Gorenstein injective. Thus we have the following exact sequence of right R-modules

$$\mathbf{I}^+: 0 \to G^+ \overset{\delta_1^+}{\longrightarrow} I_1^+ \overset{\delta_2^+}{\longrightarrow} I_2^+ \to \cdots \to I_n^+ \overset{\delta_{n+1}^+}{\longrightarrow} G^+ \to 0$$

where I_j^+ is a flat complex for $1 \leq j \leq n$. Since G is Gorenstein injective by Proposition 3.3, we have that $\operatorname{Im} \delta_1^+ \cong G^+$ is Gorenstein flat by [9, Theorem 3.5]. Since $\bigoplus_{j=2}^{n+1} \operatorname{Im} \delta_j$ is Gorenstein injective, we get that $\operatorname{Im} \delta_j$ is Gorenstein injective for $2 \leq j \leq n+1$ by [9, Theorem 2.10]. Thus for every j with $1 \leq j \leq n$, $\operatorname{Im} \delta_j^+$ is Gorenstein flat by [9, Theorem 3.5]. Hence $\bigoplus_{j=1}^{n} \operatorname{Im} \delta_j^+$ is Gorenstein flat since Gorenstein flat complexes are closed under direct sums. Therefore G^+ is n-SG-flat by Proposition 2.4.

(2) Let G be an n-SG-flat complex. Then by Proposition 2.4, we get an exact sequence of complexes of right R-modules

$$\mathbf{F}: 0 \to G \overset{\delta_{n+1}}{\to} F_n \overset{\delta_n}{\to} F_{n-1} \to \cdots \to F_1 \overset{\delta_1}{\to} G \to 0$$

where F_j is a flat complex for $1 \leq j \leq n$ and $\bigoplus_{j=2}^{n+1} \operatorname{Im} \delta_j$ is Gorenstein flat. Thus we have the following exact sequence of complexes of R-modules

$$\mathbf{F}^+: 0 \to G^+ \overset{\delta_1^+}{\longrightarrow} F_1^+ \overset{\delta_2^+}{\longrightarrow} F_2^+ \to \cdots \to F_n^+ \overset{\delta_{n+1}^+}{\longrightarrow} G^+ \to 0$$

where F_j^+ is an injective complex for $1 \leq j \leq n$. Since G is Gorenstein flat by [7, Proposition 4.2], we have that $\operatorname{Im} \delta_1^+ \cong G^+$ is Gorenstein injective by [9, Theorem 3.1]. Since $\bigoplus_{j=2}^{n+1} \operatorname{Im} \delta_j$ is Gorenstein flat, we get that $\operatorname{Im} \delta_j$ is Gorenstein flat for $2 \leq j \leq n+1$ by [9, Theorem 3.3]. Thus for every j with $1 \leq j \leq n$, $\operatorname{Im} \delta_j^+$ is Gorenstein injective by [9, Theorem 3.1]. Hence $\bigoplus_{j=1}^{n} \operatorname{Im} \delta_j^+$ is Gorenstein injective since Gorenstein injective complexes are closed under finite direct sums. Therefore G^+ is n-SG-injective by Proposition 3.3. □

Corollary 3.8 *Let R be a left artinian ring and let the injective envelope of every simple left R-module be finitely generated. Then the following hold:*

(1) If a complex G of R-modules is n-SG-injective, then G^{++} is an n-SG-injective complex.
(2) If a complex G of right R-modules is n-SG-flat, then G^{++} is an n-SG-flat complex.

Proof The proof follows from Theorem 3.7. □

The following result shows the relationship between n-SG-projective complexes and n-SG-projective modules.

Proposition 3.9 *Let G be a complex. If G is n-SG-projective, then G^i is an n-SG-projective R-module for all $i \in \mathbb{Z}$.*

Proof Suppose G is an n-SG-projective complex. By Proposition 3.6, there exists an exact sequence of complexes

$$0 \to G \overset{\delta_{n+1}}{\to} P_n \overset{\delta_n}{\to} P_{n-1} \to \cdots \to P_1 \overset{\delta_1}{\to} G \to 0$$

where P_j is a projective complex for $1 \le j \le n$ and $\bigoplus_{j=2}^{n+1} \operatorname{Im} \delta_j$ is Gorenstein projective. Then for each $i \in \mathbb{Z}$, we get an exact sequence of modules

$$0 \to G^i \xrightarrow{\delta_{n+1}^i} P_n^i \xrightarrow{\delta_n^i} P_{n-1}^i \to \cdots \to P_1^i \xrightarrow{\delta_1^i} G^i \to 0$$

such that P_j^i is a projective R-module for $1 \le j \le n$. Since $\bigoplus_{j=2}^{n+1} \operatorname{Im} \delta_j$ is a Gorenstein projective complex if and only if $\operatorname{Im} \delta_j$ is a Gorenstein projective complex for $2 \le j \le n+1$ by [9, Theorem 2.3]. Then by [9, Theorem 2.2], we have $\operatorname{Im} \delta_j$ is a Gorenstein projective complex if and only if $\operatorname{Im} \delta_j^i$ is a Gorenstein projective R-module for every $i \in \mathbb{Z}$ and $2 \le j \le n+1$. Thus we get that $\bigoplus_{j=2}^{n+1} \operatorname{Im} \delta_j^i$ is a Gorenstein projective R-module since the class of all Gorenstein projective modules is closed under direct sums. Therefore the result follows from [11, Theorem 3.9]. □

Corollary 3.10 *Let M be an R-module. Then M is n-SG-projective if and only if the complex $\overline{M}$ is n-SG-projective.*

Proof Suppose M is an n-SG-projective module. Then there exists an exact sequence of R-modules

$$X : 0 \to M \to P_n \to P_{n-1} \to \cdots \to P_1 \to M \to 0,$$

where P_i is a projective R-module for $1 \le i \le n$, such that $\operatorname{Hom}_R(-, Q)$ leaves the sequence exact for any projective module Q. Thus, we get an exact sequence of complexes

$$\overline{X} : 0 \to \overline{M} \to \overline{P}_n \to \overline{P}_{n-1} \to \cdots \to \overline{P}_1 \to \overline{M} \to 0$$

with $\overline{P}_i$ a projective complex for $1 \leq i \leq n$. Now let Q' be any projective complex. Then it is a direct product of complexes of the form $\overline{P}[n]$ for some projective module P and $n \in \mathbb{Z}$. Then

$$\begin{aligned} Hom(\overline{X}, Q') &\cong Hom(\overline{X}, \prod_{n \in \mathbb{Z}} \overline{P}[n]) \\ &\cong \prod_{n \in \mathbb{Z}} Hom(\overline{X}, \overline{P}[n]) \end{aligned}$$

is exact for all $n \in \mathbb{Z}$ and hence $\overline{M}$ is n-SG-projective. The converse follows from Proposition 3.9.

The following example describes that there are 2-SG-projective complexes which are not necessarily 1-SG-projective.

Example 3.11

(1) Let R be a local ring and consider the ring $S = R[X, Y]/(XY)$. Let $[X]$ and $[Y]$ be the residue classes in S of X and Y respectively. Then by [3, Example 2.6], we observe that the R-modules $[X]$ and $[Y]$ are 2-SG-projective but are not 1-SG-projective. Then by Corollary 3.10, the complexes $\overline{[X]}$ and $\overline{[Y]}$ are 2-SG-projective but are not SG-projective.
(2) In general, n-SG-projective complexes need not be m-SG-projective whenever $n \nmid m$. Based on the assumptions in [11, Example 3.2], we observe that the modules S_i $(1 \leq i \leq n)$ are n-strongly Gorenstein projective but are not m-strongly Gorenstein projective. Then by the Corollary 3.10, we see that the complexes $\overline{S_i}$ are n-SG-projective but are not m-SG-projective whenever $n \nmid m$.

4 n-SG-Projective and m-SG-Projective Complexes

In this section, we study the relationships between n-SG-projective (resp., injective) and m-SG-projective (resp., injective) complexes for any two positive integers n and m.

Lemma 4.1 *Let m, n and r be any positive integers such that $m = rn$. Then the class of all m-SG-projective (resp., injective) complexes contains the class of all n-SG-projective (resp., injective) complexes.*

Proof Let G be an n SG-projective complex. Then there exists an exact sequence of complexes

$$\mathbf{X} : 0 \to G \overset{\alpha_{n+1}}{\to} I_n \overset{\alpha_n}{\to} I_{n-1} \to \cdots \to I_1 \overset{\alpha_1}{\to} G \to 0$$

with I_j injective for any $1 \leq j \leq n$, such that $\bigoplus_{j=2}^{n+1} \text{Im}\, \alpha_j$ is a Gorenstein projective complex. So $\text{Im}\, \delta_j$ is Gorenstein projective for every $1 \leq j \leq n$ by [9, Theorem 2.3]. Using the exact sequence $\mathbf{X}$ for r times, we have the following exact sequence

$$\mathbf{Y}: 0 \to G \overset{\alpha_{n+1}}{\to} I_n \overset{\alpha_n}{\to} I_{n-1} \to \cdots \to I_1 \overset{\delta}{\to} I_n \to \cdots I_1 \overset{\alpha_1}{\to} G \to 0$$

with I_j injective for any $1 \leq j \leq n$ and $\delta = \alpha_{n+1}\alpha_1$. Then $\bigoplus_{j=2}^{n+1} \text{Im}\, \alpha_j$ is Gorenstein projective since $\text{Im}\, \alpha_j$ and G are Gorenstein projective. □

For any positive integer n, we use n-SG-Proj($\mathscr{C}$) (resp., n-SG-Inj($\mathscr{C}$)) to denote the subcategory of ${}_R\mathscr{C}$ consisting of n-SG-projective (resp., injective) complexes of left R-modules. The following results extend [11, Proposition 3.4 (2) and Theorem 3.5] to that of complexes.

Proposition 4.2 *Let n and m be positive integers. Then the following hold:*

(1) If $n|m$, then n-SG-Proj($\mathscr{C}$) $\bigcap$ m-SG-Proj($\mathscr{C}$) = n-SG-Proj($\mathscr{C}$).
(2) If $n \nmid m$ and $m = kn + j$, where k is a positive integer and $0 < j < n$, then n-SG-Proj($\mathscr{C}$) $\bigcap$ m-SG-Proj($\mathscr{C}$) $\subseteq$ j-SG-Proj($\mathscr{C}$).

Proof (1) It follows from Lemma 4.1.

(2) By Lemma 4.1, we have that m-SG-Proj($\mathscr{C}$) $\bigcap$ n-SG-Proj($\mathscr{C}$) $\subseteq$ m-SG-Proj($\mathscr{C}$) $\bigcap$ kn-SG-Proj($\mathscr{C}$). Suppose that a complex G is in m-SG-Proj($\mathscr{C}$) $\bigcap$ kn-SG-Proj($\mathscr{C}$). Then there exists an exact sequence of complexes

$$\mathbb{P}: 0 \to G \to P_m \to \cdots \to P_2 \to P_1 \to 0$$

with P_i projective for any $1 \leq i \leq m$. Put $L_i = \text{Ker}(P_i \to P_{i-1})$ for any $2 \leq i \leq m$. Since G is kn-SG-projective, we see that G and L_{kn} are projectively equivalent, i.e., there exist projective complexes P and Q in $\mathscr{C}$ such that $G \oplus P \cong Q \oplus L_{kn}$.

Now consider the following pullback diagram:

$$\begin{array}{ccccccccc}
 & & & & 0 & & 0 & & \\
 & & & & \downarrow & & \downarrow & & \\
 & & & & Q & = & Q & & \\
 & & & & \downarrow & & \downarrow & & \\
0 & \to & L_{kn+1} & \to & X & \to & G \oplus P & \to & 0 \\
 & & \| & & \downarrow & & \downarrow & & \\
0 & \to & L_{kn+1} & \to & P_{kn} & \to & L_{kn} & \to & 0 \\
 & & & & \downarrow & & \downarrow & & \\
 & & & & 0 & & 0. & &
\end{array}$$

Then X is a projective complex. Next, consider the following pullback diagram

$$\begin{array}{ccccccccc}
 & & & & 0 & & 0 & & \\
 & & & & \downarrow & & \downarrow & & \\
0 & \longrightarrow & L_{kn+1} & \longrightarrow & Y & \longrightarrow & G & \longrightarrow & 0 \\
 & & \| & & \downarrow & & \downarrow & & \\
0 & \longrightarrow & L_{kn+1} & \longrightarrow & X & \longrightarrow & G \oplus P & \longrightarrow & 0 \\
 & & & & \downarrow & & \downarrow & & \\
 & & & & P & = & P & & \\
 & & & & \downarrow & & \downarrow & & \\
 & & & & 0 & & 0. & &
\end{array}$$

Hence Y is also projective. Combining the exact sequence $\mathbb{P}$ and the first row in the above diagram, we get the following exact sequence of complexes

$$0 \to G \to P_m \to \cdots \to P_{kn+1} \to Y \to G \to 0$$

such that Hom$(-, Q')$ leaves the sequence exact for any projective complex Q'. Thus G is j-SG-projective and hence n-SG-Proj$(\mathscr{C}) \bigcap m$-SG-Proj$(\mathscr{C}) \subseteq j$-$SG$-Proj$(\mathscr{C})$. □

Dually, we have the following result for n-SG-injective complexes.

Proposition 4.3 *Let n and m be positive integers. Then the following hold:*

(1) If $n|m$, then n-SG-Inj$(\mathscr{C}) \bigcap m$-SG-Inj$(\mathscr{C}) = n$-SG-Inj$(\mathscr{C})$.
(2) If $n \nmid m$ and $m = kn + j$, where k is a positive integer and $0 < j < n$, then n-SG-Inj$(\mathscr{C}) \bigcap m$-SG-Inj$(\mathscr{C}) \subseteq j$-SG-Inj$(\mathscr{C})$.

For any two positive integers m and n, we use (m, n) (resp., $[m, n]$) to denote the greatest common divisor (resp., least common multiple) of m and n.

Proposition 4.4 *For any two positive integers m and n, we have the following:*

(1) m-SG-Proj$(\mathscr{C}) \bigcap n$-SG-Proj$(\mathscr{C}) = (m, n)$-SG-Proj$(\mathscr{C})$.
(2) m-SG-Proj$(\mathscr{C}) \bigcap (m + 1)$-SG-Proj$(\mathscr{C}) = 1$-SG-Proj$(\mathscr{C})$.

Proof (1) If $n|m$, then the result follows from Proposition 4.3 (1). Now suppose $n \nmid m$ and $m = k_0 n + j_0$, where k_0 is a positive integer and $0 < j_0 < n$. By Proposition 4.3 (2), we have that m-SG-Proj$(\mathscr{C}) \bigcap n$-SG-Proj$(\mathscr{C}) \subseteq j_0$-$SG$-Proj$(\mathscr{C})$. If $j_0 \nmid n$ and $n = k_1 j_0 + j_1$, with $0 < j_1 < j_0$, then by Proposition 4.3 (2) again, we have that m-SG-Proj$(\mathscr{C}) \bigcap n$-SG-Proj$(\mathscr{C}) \subseteq n$-$SG$-Proj$(\mathscr{C}) \bigcap j_0$-$SG$-Proj$(\mathscr{C})$ $\subseteq j_1$-SG-Proj$(\mathscr{C})$. Continuing the process, after finite steps, there exists a positive integer t such that $j_t = k_{t+2} j_{t+1}$ and $j_{t+1} = (m, n)$. Thus m-SG-Proj$(\mathscr{C}) \bigcap$

n-SG-Proj($\mathscr{C}$) $\subseteq j_t$-SG-Proj($\mathscr{C}$) $\bigcap j_{t+1}$-SG-Proj($\mathscr{C}$) $= j_{t+1}$-SG-Proj($\mathscr{C}$)=(m,n)-SG-Proj($\mathscr{C}$). Then the result follows from the fact that (m,n)-SG-Proj($\mathscr{C}$) $\subseteq m$-SG-Proj($\mathscr{C}$) $\bigcap$ n-SG-Proj($\mathscr{C}$).

(2) It follows from (1). □

Corollary 4.5 *For any two positive integers m and n, we have the following: m-SG-Proj*($\mathscr{C}$) $\bigcup$ *n-SG-Proj*($\mathscr{C}$) $\subseteq [m,n]$*-SG-Proj*($\mathscr{C}$).

Proof It is clear from the fact that every n-SG-projective complex is m-SG-projective whenever $n|m$. □

For the case of n-SG-injective complexes, we have the following.

Proposition 4.6 *For any two positive integers m and n, we have the following:*

(1) m-SG-Inj($\mathscr{C}$) $\bigcap$ *n-SG-Inj*($\mathscr{C}$) $= (m,n)$*-SG-Inj*($\mathscr{C}$).
(2) m-SG-Inj($\mathscr{C}$) $\bigcap$ $(m+1)$*-SG-Inj*($\mathscr{C}$) = *1-SG-Inj*($\mathscr{C}$).

Proof The proof is similar to Proposition 4.4. □

Corollary 4.7 *For any two positive integers m and n, we have the following: m-SG-Inj*($\mathscr{C}$) $\bigcup$ *n-SG-Inj*($\mathscr{C}$) $\subseteq [m,n]$*-SG-Inj*($\mathscr{C}$).

Proof It is clear from the fact that every n-SG-injective complex is m-SG-injective whenever $n|m$. □

References

1. Anderson, F.W., Fuller, K.R.: Rings and Categories of Modules. Graduate Texts in Mathematics, vol. 13, 2nd edn. Springer, New York (1992)
2. Bennis, D., Mahdou, N.: Strongly Gorenstein projective, injective and flat modules. J. Pure Appl. Algebra **210**, 437–445 (2007)
3. Bennis, D., Mahdou, N.: A generalization of strongly Gorenstein projective modules. J. Algebra Appl. **8**, 219–227 (2009)
4. Enochs, E., García Rozas, J.R.: Tensor products of complexes. Math. J. Okayama Univ. **39**, 17–39 (1997)
5. Enochs, E., Jenda, O.M.G.: Relative Homological Algebra. de Gruyter Expositions in Mathematics, vol. 30. Walter de Gruyter, Berlin (2000)
6. García Rozas, J.R.: Covers and Envelopes in the Category of Complexes of Modules. CRC Press, Boca Raton (1999)
7. Selvaraj, C., Saravanan, R.: SG-flat complexes and their generalizations, communicated to Vietnam J. Math
8. Yang, X., Liu, Z.: Strongly Gorenstein projective, injective and flat modules. J. Algebra **320**, 2659–2674 (2008)
9. Yang, X., Liu, Z.: Gorenstein projective, injective and flat complexes. J. Commut. Algebra **39**, 1705–1721 (2011)
10. Zhang, D., Ouyang, B.: Strongly Gorenstein projective and injective complexes. J. Nat. Sci. Hunan Norm. Univ. **35**, 1–5 (2012)
11. Zhao, G., Huang, Z.: n-strongly Gorenstein projective, injective and flat modules. Commun. Algebra **39**, 3044–3062 (2011)

Generalized Derivations with Nilpotent Values on Multilinear Polynomials in Prime Rings

Basudeb Dhara

Abstract Let R be a prime ring with Utumi quotient ring U and extended centroid C, F a nonzero generalized derivation of R, I a nonzero right ideal of R, $f(r_1, \ldots, r_n)$ a multilinear polynomial over C and $s \geq 1, t \geq 1$ be fixed integers. If $(F(f(r_1, \ldots, r_n))^s - f(r_1, \ldots, r_n)^s)^t = 0$ for all $r_1, \ldots, r_n \in I$, then one of the following holds:

(1) $IC = eRC$ for some idempotent $e \in soc(RC)$ and $f(x_1, \ldots, x_n)$ is central-valued on $eRCe$;
(2) there exist $a, b \in U$ such that $F(x) = ax + xb$ for all $x \in R$ and $(a - \alpha)I = (0)$, $(b - \beta)I = (0)$ for some $\alpha, \beta \in C$ with $(\alpha + \beta)^s = 1$.

Keywords Prime ring · Derivation · Generalized derivation · Extended centroid · Utumi quotient ring

Mathematics Subject Classification 2010 16W25 · 16N60

1 Introduction

Let R be an associative prime ring with center $Z(R)$. Throughout this paper, U will denote the Utumi quotient ring of R and $C = Z(U)$, the center of U, which is called extended centroid of R. For $x, y \in R$, the symbol $[x, y]$ stands for the commutator $xy - yx$ and $x \circ y$ stands for anti-commutator $xy + yx$.

An additive mapping $d : R \to R$ is called a derivation if $d(xy) = d(x)y + xd(y)$ holds for all $x, y \in R$. The concept of derivation is extended to generalized derivation. The generalized derivation means an additive mapping $F : R \to R$ such that

This work is supported by a grant from University Grant Commission, INDIA. Grant No. is PSW-131/14-15 (ERO), dated 03.02.2015.

B. Dhara (✉)
Department of Mathematics, Belda College, Belda Paschim Medinipur
721424, West Bengal, India
e-mail: basu_dhara@yahoo.com

S.T. Rizvi et al. (eds.), *Algebra and its Applications*, Springer Proceedings in Mathematics & Statistics 174, DOI 10.1007/978-981-10-1651-6_18

$F(xy) = F(x)y + xd(y)$ for all $x, , y \in R$, where d is a derivation of R. For some fixed $a, b \in R$, the maps $F(x) = ax + xb$ for all $x \in R$, is an example of generalized derivation. This kind of generalized derivations are called inner generalized derivations.

Daif and Bell [7] proved that if R is a semiprime ring with a nonzero ideal I such that $d([x, y]) = \pm[x, y]$ for all $x, y \in I$, then I is central ideal. In particular, if $I = R$, then R is commutative.

Recently, Quadri et al. [24] generalized this result replacing derivation d with a generalized derivation in a prime ring R. More precisely, they proved the following:

Let R be a prime ring and I a nonzero ideal of R. If R admits a generalized derivation F associated with a nonzero derivation d such that any one of the following holds: (i) $F([x, y]) = [x, y]$ for all $x, y \in I$; (ii) $F([x, y]) = -[x, y]$ for all $x, y \in I$; (iii) $F(x \circ y) = (x \circ y)$ for all $x, y \in I$; (iv) $F(x \circ y) = -(x \circ y)$ for all $x, y \in I$; then R is commutative.

In [9], I studied all these cases of [24] in semiprime ring.

On the other hand, Ashraf et al. [3] proved that the prime ring R must be commutative, if $F(xy) \pm xy \in Z$ for all $x, y \in I$, where F is a generalized derivation of R associated with a nonzero derivation d and I is a nonzero two-sided ideal of R. Recently, in [11], these results were generalized for multiplicative (generalized)-derivations in semiprime rings.

In [2], Argac and Inceboz studied the situation $d(x \circ y)^n = x \circ y$ for all x, y in some nonzero ideal of prime ring R. In [8], De Filippis and Huang studied the situation $(F([x, y]))^n = [x, y]$ for all $x, y \in I$, where I is a nonzero ideal in a prime ring R, F a generalized derivation of R and $n \geq 1$ fixed integer. Then in [1], Ali et al. investigated the situation when a prime ring R satisfies $(F(x \circ y))^m = (x \circ y)^n$ for all x, y in some suitable subsets of R, where F is a generalized derivation of R associated with a derivation d. More precisely, they proved the following:

Let R be a prime ring, I a nonzero right ideal of R and F a generalized derivation of R. If $(F(x \circ y))^n = (x \circ y)^n$ for all $x, y \in I$, where $n \geq 1$ is fixed integer, then one of the following holds: (1) $[I, I]I = (0)$; (2) there exists $a \in U$ and $\alpha \in C$ such that $F(x) = ax$ for all $x \in R$, with $(a - \alpha)I = (0)$ and $\alpha^n = 1$. (see [1, Theorem 2])

It is natural to investigate above situation for multilinear polynomials in prime rings. Recently in [10] Dhara et al. have proved the following:

Let R be a prime ring, I be a nonzero right ideal of R and $f(r_1, \ldots, r_n)$ a nonzero multilinear polynomial over C. Suppose that d is a derivation of R such that $(d(f(x_1, \ldots, x_n))^m - f(x_1, \ldots, x_n))^p = 0$ for all $x_1, \ldots, x_n \in I$ and $m \geq 1, n \geq 1$ are fixed integers, then $IC = eRC$ for some idempotent element $e \in Soc(RC)$ and $f(x_1, \ldots, x_n)$ is a polynomial identity for $eRCe$.

It is natural to consider the more general situation when $(F(f(x_1, \ldots, x_n))^m - f(x_1, \ldots, x_n)^m)^p = 0$ for all $x_1, \ldots, x_n \in I$, where F is a generalized derivation of R, I is a nonzero right ideal of R and $f(x_1, \ldots, x_n)$ is a multilinear polynomial on R over C. In the present paper our main objective is to investigate this situation.

Let R be a prime ring and U be the Utumi quotient ring of R and $C = Z(U)$, the center of U. Note that U is also a prime ring with C a field. Let $f(x_1, \ldots, x_n)$ be a multilinear polynomial over C. We can write it as

$$f(x_1, \ldots, x_n) = x_1 x_2 \ldots x_n + \sum_{I \neq \sigma \in S_n} \alpha_\sigma x_{\sigma(1)} \ldots x_{\sigma(n)},$$

where S_n is the permutation group over n elements and any $\alpha_\sigma \in C$. We denote by $f^d(x_1, \ldots, x_n)$ the polynomial obtained from $f(x_1, \ldots, x_n)$ by replacing each coefficient α_σ with $d(\alpha_\sigma.1)$. In this way we have

$$d(f(x_1, \ldots, x_n)) = f^d(x_1, \ldots, x_n) + \sum_i f(x_1, \ldots, d(x_i), \ldots, x_n).$$

Now we include some facts which will be used to prove our theorems.

Fact 1. It is well known that any derivation of R can be uniquely extended to a derivation of U (see [18, Lemma 2]).

Fact 2. Let ρ be a nonzero right ideal of R. Then ρ, ρC, ρU satisfy the same generalized polynomial identities with coefficients in U (see [5]).

Fact 3. Let ρ be a nonzero right ideal of R. Then ρ, ρR and ρU satisfy the same differential identities with coefficients in U (see [18, Theorem 2]).

Fact 4. Let ρ be a nonzero right ideal of R. If ρ satisfies a nontrivial polynomial identity, then RC is a primitive ring with $soc(RC) \neq 0$ and $\rho C = eRC$ for some idempotent $e = e^2 \in soc(RC)$ (see [19, Proposition])

Fact 5. Let R be a dense ring of linear transformations of a vector space V over a division ring D and $a \in R$. If for any $v \in V$, av and v are linearly D-dependent, then there exists a $\beta \in D$ such that $av = v\beta$ for all $v \in V$.

Proof For any $v \in V$, $av = v\alpha_v$ for some $\alpha_v \in D$. Now we prove that α_v is independent of the choice of $v \in V$. Let u be a fixed vector of V. Then $au = u\alpha$. Let v be any vector of V. Then $av = v\alpha_v$, where $\alpha_v \in D$. If u and v are linearly D-dependent, then $u = v\beta$, for $\beta \in D$. In this case, we see that $u\alpha = au = av\beta = (v\alpha_v)\beta = (v\beta)\alpha_v = u\alpha_v$, implying $\alpha = \alpha_v$.

Now if u and v are linearly D-independent, then we have $(u + v)\alpha_{u+v} = a(u + v) = au + av = u\alpha + v\alpha_v$, which implies $u(\alpha_{u+v} - \alpha) + v(\alpha_{u+v} - \alpha_v) = 0$. Since u and v are linearly D-independent, we have $\alpha_{u+v} - \alpha = 0 = \alpha_{u+v} - \alpha_v$ and so $\alpha = \alpha_v$. Thus $av = v\alpha$ for all $v \in V$, where $\alpha \in D$ independent of the choice of $v \in V$.

Fact 6. Let I be a nonzero right ideal of R and $a \in U$. If for every $x \in I$, ax and x are linearly C-dependent, then there exists $\alpha \in C$ such that $(a - \alpha)I = (0)$.

The proof of Fact 6 is similar to that of Fact 5, so we omit it here.

Remark 1 Now we mention a result of Lee in [20] which will be used to prove our main theorem. In [20], Lee extended the definition of generalized derivation as follows: generalized derivation means an additive mapping $g : \rho \to U$ such that $g(xy) = g(x)y + x\delta(y)$ for all $x, y \in \rho$, where ρ is a dense right ideal of R and δ ia a derivation from ρ into U. The author proved that every generalized derivation of R can be uniquely extended to generalized derivation of U and has the form $g(x) = ax + \delta(x)$ for all $x \in U$, where $a \in U$ and δ is a derivation of U [20, Theorem 3]. For more details about generalized derivations we refer to [14, 20, 21].

2 Main Results

First we study the case when F is inner generalized derivation of R, that is, for some $a, b \in U$, $F(x) = ax + xb$ for all $x \in R$.

Lemma 2.1 *Let $R = M_k(C)$, $k \geq 2$, be the set of all $k \times k$ matrices over a field C and $f(x_1, \ldots, x_n)$ be a noncentral multilinear polynomial over C. If for some $a, b \in R$, $((af(x_1, \ldots, x_n) + f(x_1, \ldots, x_n)b)^s - f(x_1, \ldots, x_n)^s)^t = 0$ for all $x_1, \ldots, x_n \in R$, then $a, b \in C.I_k$ with $(a + b)^s = I_k$.*

Proof Let $a = (a_{ij})_{k\times k}$, $b = (b_{ij})_{k\times k}$. Since $f(x_1, \ldots, x_n)$ is not central valued on R, by [22, Lemma 2, Proof of Lemma 3] there exists a sequence of matrices $r = (r_1, \ldots, r_n)$ in R such that $f(r_1, \ldots, r_n) = \gamma e_{ij}$ with $0 \neq \gamma \in C$ and $i \neq j$. Since the set $f(R) = \{f(x_1, \ldots, x_n), x_i \in R\}$ is invariant under the action of all inner automorphisms of R, for all $i \neq j$ there exists a sequence of matrices $r = (r_1, \ldots, r_n)$ such that $f(r) = \gamma e_{ij}$. Thus

$$((af(x_1, \ldots, x_n) + f(x_1, \ldots, x_n)b)^s - f(x_1, \ldots, x_n)^s)^t = 0$$

gives $0 = ((a\gamma e_{ij} + \gamma e_{ij}b)^s - (\gamma e_{ij})^s)^t$ i.e., $0 = ((ae_{ij} + e_{ij}b)^s - (e_{ij})^s)^t$. Left multiplying by e_{ij} yields $a_{ji}^{st} = 0$ and right multiplying by e_{ij} yields $b_{ji}^{st} = 0$. Thus, we have $a_{ji} = 0$ and $b_{ji} = 0$ for any $i \neq j$, that is, a and b are diagonal matrices.

Now for any C-automorphism θ of R, we have

$$((a^\theta f(x_1, \ldots, x_n) + f(x_1, \ldots, x_n)b^\theta)^s - f(x_1, \ldots, x_n)^s)^t = 0$$

for all $x_1, \ldots, x_n \in R$. Then by above argument a^θ and b^θ must be diagonal. Write, $a = \sum_{i=0}^{k} a_{ii}e_{ii}$ and $b = \sum_{i=0}^{k} b_{ii}e_{ii}$; then for $p \neq q$, we have

$$(1 + e_{qp})a(1 - e_{qp}) = \sum_{i=0}^{m} a_{ii}e_{ii} + (a_{pp} - a_{qq})e_{qp}$$

diagonal and

$$(1 + e_{qp})b(1 - e_{qp}) = \sum_{i=0}^{m} b_{ii}e_{ii} + (b_{pp} - b_{qq})e_{qp}$$

diagonal, implying $a_{pp} = a_{qq}$, $b_{pp} = b_{qq}$ and so $a, b \in C.I_k$.

Then our assumption

$$((af(x_1,\dots,x_n)+f(x_1,\dots,x_n)b)^s - f(x_1,\dots,x_n)^s)^t = 0$$

for all $x_1,\dots,x_n \in R$, reduces to $((a+b)^s - I_k)^t f(x_1,\dots,x_n)^{st} = 0$. This implies either $((a+b)^s - I_k)^t = 0$ or $f(x_1,\dots,x_n)^{st} = 0$ for all $x_1,\dots,x_n \in R$. But by [22, Corollary 5], $f(x_1,\dots,x_n)^{st} = 0$ for all $x_1,\dots,x_n \in R$, implies that $f(x_1,\dots,x_n) = 0$ for all $x_1,\dots,x_n \in R$, a contradiction. Hence $((a+b)^s - I_k)^t = 0$. Since $(a+b)^s - I_k \in C.I_k$, we conclude that $(a+b)^s - I_k = 0$

Hence, the lemma is proved. □

Proposition 2.2 *Let R be a prime ring with Utumi quotient ring U and extended centroid C, and $f(r_1,\dots,r_n)$ be a multilinear polynomial over C which is not central valued on R. If for some $a,b \in U$, $((af(r)+f(r)b)^s - f(r)^s)^t = 0$ for all $r = (r_1,\dots,r_n) \in R^n$, where $s \geq 1$, $t \geq 1$ are fixed integers, then $a,b \in C$ with $(a+b)^s - 1 = 0$.*

Proof Since R and U satisfy same generalized polynomial identity (see [5]), U satisfies

$$h(x_1,\dots,x_n) = ((af(x_1,\dots,x_n)+f(x_1,\dots,x_n)b)^s - f(x_1,\dots,x_n)^s)^t = 0.$$

Suppose that $h(x_1,\dots,x_n)$ is a trivial GPI for U. Let $T = U *_C C\{x_1,\dots,x_n\}$, the free product of U and $C\{x_1,\dots,x_n\}$, the free C-algebra in noncommuting indeterminates $x_1,\dots,x_n$. Then,

$$((af(x_1,\dots,x_n)+f(x_1,\dots,x_n)b)^s - f(x_1,\dots,x_n)^s)^t$$

is zero element in T. If $a \notin C$, then a and 1 are linearly independent over C. Then expanding the above identity, it will imply

$$(af(x_1,\dots,x_n))^s((af(x_1,\dots,x_n)+f(x_1,\dots,x_n)b)^s - f(x_1,\dots,x_n)^s)^{t-1} = 0$$

in T. Again, since a and 1 are linearly independent over C, this implies that

$$(af(x_1,\dots,x_n))^{2s}((af(x_1,\dots,x_n)+f(x_1,\dots,x_n)b)^s - f(x_1,\dots,x_n)^s)^{t-2} = 0$$

and so $(af(x_1,\dots,x_n))^{ts} = 0$, implying $a = 0$, a contradiction. Hence, $a \in C$. Then our generalized polynomial identity (GPI) reduces to $((f(x_1,\dots,x_n)(a+b))^s - f(x_1,\dots,x_n)^s)^t = 0$ in T. If $a+b \notin C$, then $a+b$ and 1 are linearly independent over C. Then by same argument as above, $(f(x_1,\dots,x_n)(a+b))^{st} = 0$, which is a nontrivial generalized polynomial identity for R, a contradiction. Thus, $a+b \in C$ and hence $b \in C$. Then our GPI becomes $\{(a+b)^s - 1\}^t f(x_1,\dots,x_n)^{st} = 0$, which is a trivial GPI for R, implying $(a+b)^s - 1 = 0$.

Next suppose that $h(x_1, \ldots, x_n)$ is a nontrivial GPI for R and so for U. In case C is infinite, we have $h(x_1, \ldots, x_n) = 0$ for all $x_1, \ldots, x_n \in U \otimes_C \overline{C}$, where $\overline{C}$ is the algebraic closure of C. Since both U and $U \otimes_C \overline{C}$ are prime and centrally closed [12, Theorems 2.5 and 3.5], we may replace R by U or $U \otimes_C \overline{C}$ according to C finite or infinite. Then R is centrally closed over C and $h(x_1, \ldots, x_n) = 0$ for all $x_1, \ldots, x_n \in R$. By Martindale's theorem [23], R is then a primitive ring with nonzero socle $soc(R)$ and with C as its associated division ring. Then, by Jacobson's theorem [15, p. 75], R is isomorphic to a dense ring of linear transformations of a vector space V over C. Assume first that V is finite dimensional over C, that is, $\dim_C V = m$. By density of R, we have $R \cong M_m(C)$. Since $f(r_1, \ldots, r_n)$ is not central valued on R, R must be noncommutative and so $m \geq 2$. In this case, by Lemma 2.1, we obtain our required conclusion.

Now, if V is infinite dimensional over C, then as in lemma 2 in [25], the set $f(R)$ is dense on R and so from

$$((af(r_1, \ldots, r_n) + f(r_1, \ldots, r_n)b)^s - f(r_1, \ldots, r_n)^s)^t = 0$$

for all $r_1, \ldots, r_n \in R$, we have

$$((ar + rb)^s - r^s)^t = 0$$

for all $r \in R$. Let v and bv be linearly C-independent for some $v \in V$. Then by density there exists $r \in R$ such that $rv = 0, rbv = v$. Therefore, we have $0 = ((ar + rb)^s - r^s)^t v = v$, a contradiction. Hence, v and bv are linearly C-dependent for all $v \in V$. By Fact 5, we can write $bv = v\alpha$ for all $v \in V$ and $\alpha \in C$ fixed.

Now let $r \in R$, $v \in V$. Since $bv = v\alpha$,

$$[b, r]v = (br)v - (rb)v = b(rv) - r(bv) = (rv)\alpha - r(v\alpha) = 0.$$

Thus $[b, r]v = 0$ for all $v \in V$ i.e., $[b, r]V = 0$. Since $[b, r]$ acts faithfully as a linear transformation on the vector space V, $[b, r] = 0$ for all $r \in R$. Therefore, $b \in C$. Then we obtain

$$(((a + b)r)^s - r^s)^t = 0$$

for all $r \in R$. Let v and $(a + b)v$ be linearly C-independent for some $v \in V$. By density, we may choose $r \in R$ such that $rv = v, r(a + b)v = 0$. Then we have $(((a + b)r)^s - r^s)^t v = 0$. But we see that for $s = 1$, $((a + b)r - r)^t v = (-1)^t v$ if $t \geq 2$ and $(a + b)v - v$ if $t = 1$. On the other hand for $s \geq 2$, $(((a + b)r)^s - r^s)^t v = (-1)^t v$. In any case we have a contradiction. Hence, v and $(a + b)v$ are linearly C-dependent for all $v \in V$, which implies as before that $a + b \in C$ and so $a \in C$. Therefore, $\{(a + b)^s - 1\}^t r^{st} = 0$ for all $r \in R$. Since V is infinite dimensional over C, $(a + b)^s - 1 = 0$. □

Proposition 2.3 *Let R be a prime ring with Utumi quotient ring U and extended centroid C, I a nonzero right ideal of R and $f(r_1, \ldots, r_n)$ a multilinear polynomial over C. If for some $a, b \in U$, $((af(r) + f(r)b)^s - f(r)^s)^t = 0$ for all $r = (r_1, \ldots, r_n) \in I^n$, then one of the following holds:*

(1) $IC = eRC$ for some idempotent $e \in soc(RC)$ and $f(x_1, \ldots, x_n)$ is central-valued on $eRCe$;

(2) there exist $\alpha, \beta \in C$ such that $(a - \alpha)I = (0)$ and $(b - \beta)I = (0)$ with $(\alpha + \beta)^s = 1$.

Proof Let $u \in I$. Then R satisfies the GPI

$$((af(ux_1, \ldots, ux_n) + f(ux_1, \ldots, ux_n)b)^s - f(ux_1, \ldots, ux_n)^s)^t = 0. \quad (1)$$

Now we consider following two cases:

Case-I: R does not satisfy any nontrivial GPI

Then (1) is a trivial GPI for R, that is,

$$((af(ux_1, \ldots, ux_n) + f(ux_1, \ldots, ux_n)b)^s - f(ux_1, \ldots, ux_n)^s)^t \quad (2)$$

is zero element in $R *_C C\{x_1, \ldots, x_n\}$. Suppose first that there exists $u \in I$ such that $\{bu, u\}$ is linearly C-independent. Then $b \notin C$, and hence above GPI implies that

$$((af(ux_1, \ldots, ux_n) + f(ux_1, \ldots, ux_n)b)^s - f(ux_1, \ldots, ux_n)^s)^{t-1}(f(ux_1, \ldots, ux_n)b)^s = 0.$$

Now since $\{bu, u\}$ is linearly C-independent, we see expanding the above expression that $(f(ux_1, \ldots, ux_n)b)^{st}$ appears nontrivially, a contradiction. Hence bu and u are linearly C-dependent for all $u \in I$. Then by Fact 6, there exists $\beta \in C$ such that $(b - \beta)I = (0)$. Next suppose that there exists $u \in I$ such that $\{au, u\}$ is linearly C-independent. Then from (2), we obtain that

$$(af(ux_1, \ldots, ux_n))^s((af(ux_1, \ldots, ux_n) + f(ux_1, \ldots, ux_n)b)^s - f(ux_1, \ldots, ux_n)^s)^{t-1} = 0.$$

Expanding the above expression we find that the term $\{af(ux_1, \ldots, ux_n)\}^{st}$ appears nontrivially, a contradiction. Hence we conclude that au and u are linearly C-dependent for all $u \in I$. By Fact 6, there exists $\alpha \in C$ such that $(a - \alpha)I = (0)$.

Then (1) reduces to

$$((f(ux_1, \ldots, ux_n)(\alpha + b))^s - f(ux_1, \ldots, ux_n)^s)^t = 0. \quad (3)$$

Right multiplying by $f(ux_1, \ldots, ux_n)$ and then using $(b - \beta)I = (0)$, it follows that

$$((\alpha + \beta)^s - 1)^t f(ux_1, \ldots, ux_n)^{st+1)} = 0. \quad (4)$$

Since this is trivial GPI for R, $(\alpha + \beta)^s - 1 = 0$.

Case-II: R satisfies a nontrivial GPI

Now assume first that $[f(I), I]I = (0)$, that is $[f(x_1, \ldots, x_n), x_{n+1}]x_{n+2} = 0$ for all $x_1, x_2, \ldots, x_{n+2} \in I$. Then by Fact 4, $IC = eRC$ for some idempotent $e \in soc(RC)$. Since $[f(I), I]I = (0)$, we have $[f(IR), IR]IR = (0)$ and hence $[f(IU), IU]IU = (0)$ by [5, Theorem 2]. In particular, $[f(IC), IC]IC = (0)$, or equivalently, $[f(eRC), eRC]eRC = (0)$. Then $[f(eRCe), eRCe] = (0)$, that is, $f(x_1, \ldots, x_n)$ is central-valued on $eRCe$ and hence conclusion (1) is obtained.

So, we assume that $[f(I), I]I \neq (0)$, that is, $[f(x_1, \ldots, x_n), x_{n+1}]x_{n+2}$ is not an identity for I. In this case R is a prime GPI-ring and so is U (see [5]). Since U is centrally closed over C, it follows from [23] that U is a primitive ring with $H = Soc(U) \neq (0)$. Then $[f(IH), IH]IH \neq (0)$. For otherwise, $[f(IU), IU]IU = (0)$ by [5], a contradiction. Choose $u_1, \ldots, u_{n+2} \in IH$ such that $[f(u_1, \ldots, u_n), u_{n+1}]u_{n+2} \neq 0$. Let $u \in IH$. Since H is a regular ring, there exists $e^2 = e \in H$ such that $eH = uH + u_1H + \cdots + u_{n+2}H$. Then $e \in IH$ and $u = eu$, $u_i = eu_i$ for $i = 1, \ldots, n+2$. Thus, we have $(0) \neq [f(eH), eH]eH = [f(eHe), eHe]H$ i.e., $f(r_1, \ldots, r_n)$ is not central-valued in eHe.

By our assumption and by [5], we may also assume that

$$((af(x_1, \ldots, x_n) + f(x_1, \ldots, x_n)b)^s - f(x_1, \ldots, x_n)^s)^t = 0$$

is an identity for IU. In particular,

$$((af(x_1, \ldots, x_n) + f(x_1, \ldots, x_n)b)^s - f(x_1, \ldots, x_n)^s)^t = 0$$

is an identity for IH and so for eH. It follows that, for all $r_1, \ldots, r_n \in H$,

$$((af(er_1, \ldots, er_n) + f(er_1, \ldots, er_n)b)^s - f(er_1, \ldots, er_n)^s)^t = 0. \quad (5)$$

We may write

$$f(x_1, \ldots, x_n) = \sum_i t_i(x_1, \ldots, x_{i-1}, x_{i+1}, \ldots, x_n)x_i,$$

where t_i is a suitable multilinear polynomial in $n-1$ variables and x_i never appears in any monomials of t_i. Since $f(eHe) \neq (0)$, there exists some t_i which does not vanish in eHe. Without loss of generality, we assume that $t_n(eHe) \neq (0)$. Let $r \in H$. Then replacing r_n with $r(1-e)$ in (5), we have

$$\begin{aligned}0 = &((at_n(er_1, \ldots, er_{n-1})er(1-e) + t_n(er_1, \ldots, er_{n-1})er(1-e)b)^s \\ &- (t_n(er_1, \ldots, er_{n-1})er(1-e))^s)^t. \end{aligned} \quad (6)$$

Left multiplying by $(1-e)$, we obtain $(1-e)(at_n(er_1, \ldots, er_{n-1})er(1-e))^{st} = 0$, that is, $\{(1-e)at_n(er_1, \ldots, er_{n-1})er\}^{st+1} = 0$ for all $r \in H$. By [13], $(1-e)at_n(er_1e, \ldots, er_{n-1}e) = 0$ for all $r_1, \ldots, r_{n-1} \in H$. Since eHe is a simple Artinian ring and $t_n(eHe) \neq (0)$ is invariant under the action of all inner automorphisms of eHe, by [6, Lemma 2], $(1-e)ae = 0$. Now again right multiplying by e in (6), we obtain $(t_n(er_1, \ldots, er_{n-1})er(1-e)b)^{st}e = 0$ that is, $\{(1-e)bt_n(er_1, \ldots, er_{n-1})er)^{st+1} = 0$ for all $r \in H$, implying $(1-e)bt_n(er_1e, \ldots, er_{n-1}e) = 0$ for all $r_1, \ldots, r_{n-1} \in H$. By above argument we conclude that $(1-e)be = 0$.

In particular, from (5), we can write that H satisfies

$$e\{(af(er_1e, \ldots, er_ne) + f(er_1e, \ldots, er_ne)b)^s \\ -f(er_1e, \ldots, er_ne)^s\}^t e = 0 \tag{7}$$

and so using the facts $(1-e)ae = 0$ and $(1-e)be = 0$, we have, prime ring eHe satisfies

$$((eaef(r_1, \ldots, r_n) + f(r_1, \ldots, r_n)ebe)^s - f(r_1, \ldots, r_n)^s)^t = 0. \tag{8}$$

By Proposition 2.2, since $f(r_1, \ldots, r_n)$ is not central-valued in eHe, we conclude $eae, ebe \in Ce$ with $(eae + ebe)^s - e = 0$. Therefore, $ae = eae \in Ce$ and $be = ebe \in Ce$. Thus $au = aeu = eaeu \in Cu$ and hence au, u are linearly C-dependent for each $u \in I$. So $(a-\alpha)I = (0)$ for some $\alpha \in C$. Similarly, $(b-\beta)I = (0)$ for some $\beta \in C$.

Thus our hypothesis $((af(x_1, \ldots, x_n) + f(x_1, \ldots, x_n)b)^s - f(x_1, \ldots, x_n)^s)^t = 0$ for all $x_1, \ldots, x_n \in I$, implies by right multiplying $f(x_1, \ldots, x_n)$ that $\{(\alpha+\beta)^s - 1\}^t f(x_1, \ldots, x_n)^{st} = 0$ for all $x_1, \ldots, x_n \in I$. By Lemma 2 in [4], either $f(I)I = (0)$ or $(\alpha+\beta)^s - 1 = 0$. If $f(I)I = (0)$, then by Fact 4, conclusion (1) is obtained. If $(\alpha+\beta)^s - 1 = 0$, then conclusions (2) is obtained. □

We are now ready to prove our main theorem.

Theorem 2.4 *Let R be a prime ring with Utumi quotient ring U and extended centroid C, F a nonzero generalized derivation of R, I a nonzero right ideal of R, $f(r_1, \ldots, r_n)$ a multilinear polynomial over C and $s \geq 1, t \geq 1$ be fixed integers. If $(F(f(r_1, \ldots, r_n))^s - f(r_1, \ldots, r_n)^s)^t = 0$ for all $r_1, \ldots, r_n \in I$, then one of the following holds:*

(1) $IC = eRC$ for some idempotent $e \in soc(RC)$ and $f(x_1, \ldots, x_n)$ is central-valued on $eRCe$;

(2) there exist $a, b \in U$ such that $F(x) = ax + xb$ for all $x \in R$ and $(a-\alpha)I = (0)$, $(b-\beta)I = (0)$ for some $\alpha, \beta \in C$ with $(\alpha+\beta)^s = 1$.

Proof If F is a inner generalized derivation of R, then result follows by Proposition 2.3. Assume that F is not U-inner. Then by Remark 1, we may assume that for all $x \in U$, $F(x) = ax + d(x)$, where $a \in U$ and d is a derivation of U. By our assumption, I satisfies $(F(f(x_1, \ldots, x_n))^s - f(x_1, \ldots, x_n)^s)^t = 0$. Since I and IU satisfy the same generalized polynomial identities (see [5]) as well as the same differential identities (see [18]), we may assume for $u_1, \ldots, u_n \in I$ that U satisfies

$$\{(af(u_1x_1,\ldots,u_nx_n)+d(f(u_1x_1,\ldots,u_nx_n)))^s-f(u_1x_1,\ldots,u_nx_n)^s\}^t=0.$$

Since F is not inner, d can not be inner derivation of U. Then from above we have

$$\Bigg\{\Bigg\{af(u_1x_1,\ldots,u_nx_n)+f^d(u_1x_1,\ldots,u_nx_n) \\ +\sum_j f(u_1x_1,\ldots,d(u_j)x_j+u_jd(x_j),\ldots,u_nx_n)\Bigg\}^s \\ -f(u_1x_1,\ldots,u_nx_n)^s\Bigg\}^t=0. \tag{9}$$

By Kharchenko's theorem [16], we have that U satisfies

$$\Bigg\{\Bigg\{af(u_1x_1,\ldots,u_nx_n)+f^d(u_1x_1,\ldots,u_nx_n) \\ +\sum_j f(u_1x_1,\ldots,d(u_j)x_j+u_jy_j,\ldots,u_nx_n)\Bigg\}^s \\ -f(u_1x_1,\ldots,u_nx_n)^s\Bigg\}^t=0. \tag{10}$$

In particular, putting $x_1=0$, U satisfies

$$0=f(u_1y_1,\ldots,u_nx_n)^{st}. \tag{11}$$

Since I and IU satisfy the same polynomial identities, we have that I satisfies $f(x_1,\ldots,x_n)^{st}=0$. By [6, Main Theorem], $f(I)I=(0)$ and hence conclusion (1) is obtained by using Fact 4. Hence the theorem is proved. □

It is well known that if R is a prime ring and L is a non-central Lie ideal of R, then there exists a nonzero two-sided ideal I of R such that $0\neq[I,R]\subseteq L$, unless char $(R)=2$ and R satisfies the standard identity s_4. Thus from above theorem following corollary is straightforward.

Corollary 2.5 *Let R be a prime ring with Utumi quotient ring U and extended centroid C, F a nonzero generalized derivation of R, L a noncentral Lie ideal of R and $n\geq1, s\geq1$ be fixed integers. If $(F(u)^s-u^s)^n=0$ for all $u\in L$, then one of the following holds:*

(1) char $(R)=2$ and R satisfies s_4, standard identity of four variables.
(2) there exists $\lambda\in C$ such that $F(x)=\lambda x$ for all $x\in R$ with $\lambda^s=1$.

Now we prove our next corollary.

Corollary 2.6 *Let R be a prime ring with Utumi quotient ring U and extended centroid C, F a nonzero generalized derivation of R, I a nonzero right ideal of*

R and $f(r_1,\dots,r_n)$ be a multilinear polynomial over C. If $((F(f(r_1,\dots,r_n))^2 - f(r_1,\dots,r_n)^2)^t = 0$ for all $r_1,\dots,r_n \in I$, for some $t \geq 1$, then one of the following holds:

(1) $IC = eRC$ for some idempotent $e \in soc(RC)$ and $f(x_1,\dots,x_n)$ is central-valued on $eRCe$;
(2) there exists $a \in U$ such that $F(x) = xa$ for all $x \in I$ with $(a \mp 1)I = (0)$.

Proof By Theorem 2.4, we have only to consider the case when $F(x) = ax + xb$ for all $x \in R$ and $(a-\alpha)I = (0)$, $(b-\beta)I = (0)$ for some $\alpha, \beta \in C$ with $(\alpha+\beta)^2 = 1$ that is $\alpha + \beta = \pm 1$. Then $F(x) = ax + xb = \alpha x + xb = x(\alpha + b)$ for all $x \in I$ with $(0) = (b-\beta)I = (b + \alpha \mp 1)I$. Hence we obtain our conclusion (2). □

Corollary 2.7 *Let R be a prime ring with extended centroid C, F a nonzero generalized derivation of R, $f(r_1,\dots,r_n)$ a noncentral multilinear polynomial over C and $t \geq 1$ fixed integer. If $((F(f(r_1,\dots,r_n))^2 - f(r_1,\dots,r_n)^2)^t = 0$ for all $r_1,\dots,r_n \in R$, then $F(x) = \pm x$ for all $x \in R$.*

Corollary 2.8 *Let R be a prime ring with extended centroid C, d a derivation of R, $f(r_1,\dots,r_n)$ a noncentral multilinear polynomial over C and $t \geq 1$ fixed integer. If $((d(f(r_1,\dots,r_n))^2 - f(r_1,\dots,r_n)^2)^t = 0$ for all $r_1,\dots,r_n \in R$, then $d = 0$.*

Corollary 2.9 *Let R be a prime ring with extended centroid C, I a nonzero ideal of R, F a generalized derivation of R and $m \geq 1, n \geq 1$. If $(F(xy)^n - (xy)^n)^m = 0$ for all $x, y \in I$, then one of the following holds:*

(1) R is commutative;
(2) there exists $\lambda \in C$ such that $F(x) = \lambda x$ for all $x \in R$ with $\lambda^n = 1$.

Proof If $F = 0$, then $(xy)^{nm} = 0$ for all $x, y \in I$. Since I and R satisfies the same polynomial identities, R satisfies $(xy)^{nm} = 0$. Then by [17, Lemma 1], $R \subseteq M_k(E)$, matrix ring of all $k \times k$ matrices over a field E, $k \geq 1$ and $M_k(E)$ satisfies $(xy)^{nm} = 0$. But by choosing $x = y = e_{11}$, we have that $0 = (xy)^{nm} = e_{11}$, a contradiction for $k \geq 2$. Thus $k = 1$ that is, R is commutative.

If $F \neq 0$, then by Theorem 2.4, we have either (i) $xy \in C$ for all $x, y \in R$ or (ii) there exists $\lambda \in C$ such that $F(x) = \lambda x$ for all $x \in R$ with $\lambda^n = 1$. If $xy \in C$ for all $x, y \in R$, then $[xy, z] = 0$ for all $x, y, z \in R$. Replacing y with yu, we have $0 = [xyu, z] = xy[u, z] + [xy, z]u = xy[u, z]$ for all $x, y, z, u \in R$. This implies $[u, z] = 0$ for all $u, z \in R$, that is R is commutative. Thus the conclusions are obtained. □

Corollary 2.10 *Let R be a prime ring with extended centroid C, I a nonzero ideal of R, F a generalized derivation of R and $m \geq 1, n \geq 1$. If $(F([x, y])^n - ([x, y])^n)^m = 0$ for all $x, y \in I$, then one of the following holds:*

(1) R is commutative;
(2) there exists $\lambda \in C$ such that $F(x) = \lambda x$ for all $x \in R$ with $\lambda^n = 1$.

Proof If $F = 0$, then $[x, y]^{nm} = 0$ for all $x, y \in I$. Since I and R satisfies the same polynomial identities, R satisfies $[x, y]^{nm} = 0$. Then by [17, Lemma 1], $R \subseteq M_k(E)$, matrix ring of all $k \times k$ matrices over a field E, $k \geq 1$ and $M_k(E)$ satisfies $[x, y]^{nm} = 0$. But by choosing $x = e_{12}$, $y = e_{21}$, we have that $[x, y]^{nm} = e_{11} + (-1)^{nm} e_{22} \neq 0$. Thus for $k \geq 2$, we have a contradiction. Hence $k = 1$ and then R is commutative.

If $F \neq 0$, then by Theorem 2.4, we have either (i) $[x, y] \in C$ for all $x, y \in R$ or (ii) there exists $\lambda \in C$ such that $F(x) = \lambda x$ for all $x \in R$ with $\lambda^n = 1$. If $[x, y] \in C$ for all $x, y \in R$, then $[[x, y], z] = 0$ for all $x, y, z \in R$. Then again by same argument as before, $R \subseteq M_k(E)$ and $M_k(E)$ satisfies $[[x, y], z] = 0$, where E is a field. But for $k \geq 2$, $0 = [[x, y], z] = [[e_{11}, e_{12}], e_{22}] = e_{12}$, a contradiction. Hence $k = 1$, that is, R is commutative. Thus the conclusions are obtained. □

Similarly, we can prove the following:

Corollary 2.11 *Let R be a prime ring with extended centroid C, I a nonzero ideal of R, F a generalized derivation of R and $m \geq 1, n \geq 1$. If $(F(x \circ y)^n - (x \circ y)^n)^m = 0$ for all $x, y \in I$, then one of the following holds:*

(1) R is commutative;
(2) there exists $\lambda \in C$ such that $F(x) = \lambda x$ for all $x \in R$ with $\lambda^n = 1$.

Corollary 2.12 *Let R be a prime ring with extended centroid C, I a nonzero ideal of R, d a derivation of R and $m \geq 1, n \geq 1$. If $(d(x \circ y)^n + (x \circ y)^n)^m = 0$ for all $x, y \in I$ or $(d([x, y])^n - [x, y]^n)^m = 0$ for all $x, y \in I$, then R is commutative.*

References

1. Ali, A., Dhara, B., Khan, S.: Generalized derivations in prime rings and Banach algebras. Discuss. Math.-Gen. Algebra Appl. **34**, 125–138 (2014)
2. Argac, N., Inceboz, H.G.: Derivations of prime and semiprime rings. J. Korean Math. Soc. **46**(5), 997–1005 (2009)
3. Ashraf, M., Ali, A., Ali, S.: Some commutativity theorems for rings with generalized derivations. Southeast Asian Bull. Math. **31**, 415–421 (2007)
4. Chang, C.M.: Power central values of derivations on multilinear polynomials. Taiwan. J. Math. **7**(2), 329–338 (2003)
5. Chuang, C.L.: GPIs having coefficients in Utumi quotient rings. Proc. Am. Math. Soc. **103**(3), 723–728 (1988)
6. Chuang, C.L., Lee, T.K.: Rings with annihilator conditions on multilinear polynomials. Chin. J. Math. **24**(2), 177–185 (1996)
7. Daif, M.N., Bell, H.E.: Remarks on derivations on semiprime rings. Int. J. Math. Math. Sci. **15**(1), 205–206 (1992)
8. De Filippis, V., Huang, S.: Generalized derivations on semiprime rings. Bull. Korean Math. Soc. **48**, 1253–1259 (2011)
9. Dhara, B.: Remarks on generalized derivations in prime and semiprime rings. Int. J. Math. Math. Sci. **2010**, 6 (2010). Article ID 646587
10. Dhara, B., Ali, S.: On multiplicative (generalized)-derivations in prime and semiprime rings. Aequ. Math. **86**(1–2), 65–79 (2013)

11. Dhara, B., Kar, S., Mondal, S.: Power values of derivations on multilinear polynomials in primerings. Acta Math. Vietnamica **41**, 121–131 (2016)
12. Erickson, T.S., Martindale III, W.S., Osborn, J.M.: Prime nonassociative algebras. Pac. J. Math. **60**, 49–63 (1975)
13. Felzenszwalb, B.: On a result of Levitzki. Canad. Math. Bull. **21**, 241–242 (1978)
14. Hvala, B.: Generalized derivations in rings. Commun. Algebra **26**(4), 1147–1166 (1998)
15. Jacobson, N.: Structure of Rings, vol. 37. American Mathematical Society Colloquium Publications, Providence (1964)
16. Kharchenko, V.K.: Differential identity of prime rings. Algebra Logic. **17**, 155–168 (1978)
17. Lanski, C.: An Engel condition with derivation. Proc. Am. Math. Soc. **118**(3), 731–734 (1993)
18. Lee, T.K.: Semiprime rings with differential identities. Bull. Inst. Math. Acad. Sin. **20**(1), 27–38 (1992)
19. Lee, T.K.: Power reduction property for generalized identities of one sided ideals. Algebra Colloq. **3**, 19–24 (1996)
20. Lee, T.K.: Generalized derivations of left faithful rings. Commun. Algebra **27**(8), 4057–4073 (1999)
21. Lee, T.K., Shiue, W.K.: Identities with generalized derivations. Commun. Algebra **29**(10), 4437–4450 (2001)
22. Leron, U.: Nil and power central valued polynomials in rings. Trans. Am. Math. Soc. **202**, 97–103 (1975)
23. Martindale III, W.S.: Prime rings satisfying a generalized polynomial identity. J. Algebra **12**, 576–584 (1969)
24. Quadri, M.A., Khan, M.S., Rehman, N.: Generalized derivations and commutativity of prime rings. Indian J. Pure Appl. Math. **34**(9), 1393–1396 (2003)
25. Wong, T.L.: Derivations with power central values on multilinear polynomials. Algebra Colloq. **3**(4), 369–378 (1996)

Properties of Semi-Projective Modules and their Endomorphism Rings

Manoj Kumar Patel

Abstract In this paper, we have studied the properties of semi-projective module and its endomorphism rings related with Hopfian, co-Hopfian, and directly finite modules. We have provide an example of module which are semi-projective but not quasi-projective. We also prove that for semi-projective module M with $dimM < \infty$ or $CodimM < \infty$, M^n is Hopfian for every integer $n \geq 1$. Apart from this we have studied the properties of pseudo-semi-injective module and observed that for pseudo-semi-injective module, co-Hopficity weakly co-Hopficity and directly finiteness are equivalent. Finally proved that for pseudo-semi-injective module M, N be fully invariant M-cyclic submodule of M with N is essential in M, then N is weakly co-Hopfian if and only if M is weakly co-Hopfian.

Keywords Semi-projective · Pseudo-semi-injective · Hopfian · Co-Hopfian

2000 Mathematics Subject Classification 16D10 · 16D40 · 16D60 · 16D70

1 Introduction

The notion of quasi-principally projective module was introduced by Wisbauer [14] under the terminology of semi-projective modules. Tansee and Wongwai [11] introduced the idea of M-principally projective module and defined a module M quasi-principally projective if it is M-principally projective. They also established several properties of the endomorphism ring of such modules and proved that quasi-principally projective modules are equivalent to semi-projective module. In this paper, we have established some properties of endomorphism ring of quasi-principally projective module in terms of Hopfian modules and proved that a quasi-principally projective module M is Hopfian if and only if M/N is Hopfian, where N is fully invariant small submodule of M.

M.K. Patel (✉)
Department of Mathematics, NIT Nagaland, Dimapur 797103, India
e-mail: mkpitb@gmail.com

S.T. Rizvi et al. (eds.), *Algebra and its Applications*, Springer Proceedings in Mathematics & Statistics 174, DOI 10.1007/978-981-10-1651-6_19

2 Preliminaries

Throughout this paper, by a ring R we always mean an associative ring with identity and every R-module M is an unitary right R-module. Let M be an R-module; a module N is called M-generated, if there is an epimorphism $M^{(I)} \longrightarrow N$ for some index set I. If I is finite then N is called finitely M-generated. In particular, a submodule N of M is called an M-cyclic submodule of M if $N = s(M)$ for some $s \in EndM_R$ or if there exist an epimorphism from M to N, equivalently it is isomorphic to M/L for some submodule L of M. A submodule K of an R-module M is said to be small in M, written $K \ll M$, if for every submodule $L \subseteq M$ with $K + L = M$ implies $L = M$. A nonzero R-module M is called hollow if every proper submodule of it is small in M. A submodule N of M is called fully invariant submodule of M, if $f(N) \subseteq N$ for any $f \in S = EndM_R$. A module M is called indecomposable, if $M \neq 0$ and cannot be written as a direct sum of nonzero submodules.

Consider the following conditions for an R-module M:
(D_1): For every submodule A of M there is a decomposition $M = M_1 \bigoplus M_2$ such that $M_1 \subseteq A$ and $A \cap M_2 \ll M$.
(D_2): If $A \subseteq M$ such that M/A is isomorphic to a summand of M, then A is a summand of M.
(D_3): If M_1 and M_2 are summands of M with $M_1 + M_2 = M$, then $M_1 \cap M_2$ is a summand of M.

An R-module M is called a lifting module if M satisfies (D_1), M is called discrete module if it satisfies (D_1) and (D_2) and quasi-discrete if it satisfies (D_1) and (D_3).

We will freely make use of the standard notations, terminologies, and results of [1, 3, 14].

3 M-Principally Projective Module

Let M be a right R-module. A right R-module N is called M-principally projective

$$\begin{array}{ccccc} & & N & & \\ & g\swarrow & \downarrow f & & \\ M & \xrightarrow[s]{} & s(M) & \longrightarrow & 0 \end{array}$$

if every R-homomorphism f from N to an M-cyclic submodule $s(M)$ of M can be lifted to an R-homomorphism g from N to M, such that the above diagram is commutative, i.e., $s \cdot g = f$. A right R-module M is called quasi-principally projective (or semi-projective) if it is M-principally projective. Some examples of semi-projective modules are $\mathbb{Z}_4$, $\mathbb{Z}_6$ over $\mathbb{Z}$ (set of integers). Clearly, every projective module and quasi-projective module are semi-projective. But converse need not be true:

1. The $\mathbb{Z}$-module $\mathbb{Q}$ is semi-projective but not quasi-projective.
2. Let R be any integral domain with quotient field $F \neq R$. Then $M = F \oplus R$ is semi-projective (but in general not quasi-projective).
3. For any prime p in $\mathbb{Z}$, the Prufer p-group $\mathbb{Z}(p\infty)$ is not semi-projective.

Now, we provide an example of semi-projective module which is not M-principally projective module.

Example 3.1 Let $M_1 = \mathbb{Z}/p\mathbb{Z}$ and $M_2 = \mathbb{Z}/p^2\mathbb{Z}$ for any prime $p \in \mathbb{Z}$ be modules over $\mathbb{Z}$. Then we can easily check that both M_1 and M_2 are semi-projective modules. However M_1 is not M_2-principally projective.

Proposition 3.2 *If M is quasi-projective module and K is fully invariant submodule of M then M/K is semi-projective module.*

Proof The Proof is straightforward and hence we omit it.

An R-module M is called Hopfian (resp. co-Hopfian), if every surjective (resp. injective) R-homomorphism $f : M \longrightarrow M$ is an automorphism. For example, every Noetherian R-modules are Hopfian and every Artinian R-modules are co-Hopfian. A module M is called directly finite, if M is not isomorphic to a proper summand of itself.

Lemma 3.3 *(Proposition 3.25, Mohamed and Muller* (1990)[6]*) An R-module M is directly finite if and only if $f \cdot g = 1$ implies $g \cdot f = 1$ for any $f, g \in End M_R$.*

In the following propositions, we relate semi-projective module with Hopfian, co-Hopfian and directly finite modules.

Proposition 3.4 *Let M be semi-projective co-Hopfian, then it is Hopfian.*

Proof Let f be surjective endomorphism on M and $I_M : M \longrightarrow M$ be an identity map on M. By semi-projectivity of M there exists an R-homomorphism $g : M \longrightarrow M$ such that $f \cdot g = I_M$, implies that g is monomorphism. Since M is co-Hopfian, then it follows that $f = g^{-1}$ is an automorphism on M. Therefore M is Hopfian.

Proposition 3.5 *For the semi-projective modules M, the following statements are equivalent:*
(i) M is Hopfian;
(ii) M is co-Hopfian;
(iii) M is directly finite.

Proof Proof is trivial.

Proposition 3.6 *Let M be semi-projective and N is fully invariant small submodule of M. Then M is Hopfian if and only if M/N is Hopfian.*

Proof Assume that M/N is Hopfian. Let $f : M \longrightarrow M$ be any epimorphism, then semi-projectivity of M implies that there exist an homomorpshim $g : M \longrightarrow M$ such that $f \cdot g = I_M$. Hence $M \cong M \oplus (kerf)$ hence $K = (kerf)$ is direct summand of M. Since N is fully invariant implies $f(N) \subseteq N$, now we have induced a map $f' : M/N \longrightarrow M/N$ which is clearly an epimorphism, the Hopficity of M/N implies that $f' : M/N \longrightarrow M/N$ is an isomorphism. Now by $(f'.\pi)(K) = (\pi \cdot f)(K) = 0$, where $\pi : M \longrightarrow M/N$ be natural epimorphism, we see that $\pi(K) = 0$, it means $K \subseteq N$, but $K \subseteq N \ll M$ implies that $K \ll M$. Since M is semi-projective there exist a splitting for f, i.e., $K = kerf$ is direct summand of M. Therefore $K = kerf = 0$, implies that M is Hopfian.

Conversely, assume that M is Hopfian and $N \ll M$ if $f : M/N \longrightarrow M/N$ is an epimorphism. We have $f \cdot \pi : M \longrightarrow M/N$, where π is natural epimorphism from $M \longrightarrow M/N$. Then by semi-projectivity of M, there exists $g \in EndM_R$, such that $\pi \cdot g = f \cdot \pi$ implies that g is an epimorphism by 19.2, Wisbauer (1991) [14] as π is a small epimorphism. Since M is Hopfian then g is an isomorphism.

Assume $kerf \neq 0$, then there exists $x \in M$ such that $f(x + N) = N$ implies $f.\pi(x) = \pi.g(x) = g(x) + N = N$ gives that $g(x) \in N \Rightarrow x \in g^{-1}(N) \subseteq N$. It follows that $kerf = N$, therefore M/N is Hopfian.

Corollary 3.7 *Let M be finitely generated semi-projective module. Then M is Hopfian if and only if $M/J(M)$ is Hopfian.*

Proof We know that $J(M)$ is fully invariant submodule of M. If M is finitely generated then we have $J(M) \ll M$. Thus by the above proposition proof is obvious.

Corollary 3.8 *Let M be semi-projective, N and L are submodules of M such that $N + L = M$ and $N \cap L \ll M$. Then M/N and M/L are Hopfian.*

Proof We have $M/(N \cap L) = N/(N \cap L) \oplus L/(N \cap L)$, by above Proposition 3.6, $M/(N \cap L)$ is Hopfian, hence so its direct summand, as $N/(N \cap L) \cong (N + L)/L = M/L$, similarly $L/(N \cap L) \cong (N + L)/N = M/N$ is Hopfian.

The next proposition is the generalization of Pandeya et.al. (Proposition 3.8) [7], whose proof is straightforward and hence we omit it.

Proposition 3.9 *Let M be finitely generated semi-projective hollow module then M is directly finite if and only if each homomorphic image is directly finite.*

For any module M, we denote the Goldie dimension of M by $dimM$ and the dual Goldie dimension of M by $CodimM$.

Proposition 3.10 *Let M be semi-projective modules with $dimM < \infty$ or $CodimM < \infty$. Then M^n is Hopfian for every integer $n \geq 1$.*

Proof We can easily seen that M^n satisfies the hypothesis of the statement, since $dimM^n = n(dimM)$, $CodimM^n = n(CodimM)$, and M is semi-projective module implies that M^n is semi-projective. Hence it remains to prove that M is Hopfian. Let $f : M \longrightarrow M$ be any epimorphism, then semi-projectivity of M implies that there exist an homomorpshim $g : M \to M$ such that $f \cdot g = I_M$. Hence $M \cong M \oplus (kerf)$. This yields $dimM = dimM + dim(kerf)$ and $CodimM = CodimM + Codim(kerf)$. If $dimM < \infty$ then first of these equations will imply that $dim(kerf) = 0$, hence $kerf = 0$ that is f is an automorphism. If $CodimM < \infty$, then second of these equations will imply that $Codim(kerf) = 0$, hence $kerf = 0$ that is f is an automorphism. Thus in both cases, we get our assumed surjective endomorphism is an automorphism that is M is Hopfian implies that M^n is Hopfian.

Corollary 3.11 *Let M be semi-projective modules with $CodimM < \infty$. Then for any fully invariant submodule K of M and any integer $n \geq 1$, the module $(M/K)^n$ is Hopfian.*

Proof Immediate consequence of Propositions 3.2 and 3.10.

Corollary 3.12 *Let R be a ring with $dimR_R < \infty$. Then $M_n(R)$ is directly finite for every integer $n \geq 1$.*

Proof Since R_R is projective, assume that $dimR_R < \infty$ then by Proposition 3.9, we see that R^n is Hopfian for all integer $n \geq 1$. Then it is proved by the observation that M is Hopfian then $EndM_R$ is directly finite.

Lemma 3.13 *Let N be a submodule of a semi-projective module M. Then N is a summand if M/N is isomorphic to a summand of M.*

Proof The Proof is straightforward and hence we omit it.

Therefore, we say that a semi-projective module satisfies (D_2) condition. In general, we have the following implication:

Projective $\Rightarrow$ Quasi-projective $\Rightarrow$ semi-projective $\nRightarrow$ Discrete.

Corollary 3.14 *Let M be semi-projective module, then the following statements are equivalent:*
(1)*M is discrete;*
(2)*M is quasi-discrete;*
(3)*M is lifting.*

Proof (1) $\Rightarrow$ (2) $\Rightarrow$ (3) are clear from definitions and (3) $\Rightarrow$ (1) immediate from Lemma 3.13.

Corollary 3.15 *An indecomposable semi-projective module M is discrete if and only if M is hollow.*

Proof The Proof is straightforward and hence we omit it.

4 Pseudo-Semi-Injective Modules

Let M be a right R-module. M is called semi-injective if for any M-cyclic submodule N of M, monomorphism $g : N \longrightarrow M$ and corresponding to any homomorphism $f : N \longrightarrow M$ there exists a map $h \in EndM_R$, such that $h \cdot g = f$, i.e., diagram is commutative.

We wish to consider the situation where the map h in this definition is required to be a monomorphism. For this to happen, a map f must be a monomorphism. This leads to the following definition.

A right R-module M is called pseudo-M-principally injective (or pseudo-semi-injective) if for any M-cyclic submodule N of M and R-monomorphism $f, g : N \longrightarrow M$ there exists a monomorphism $h \in EndM_R$, such that $h \cdot g = f$.

It is easy to show that if M is pseudo-semi-injective module, then every monomorphism in $EndM_R$ is an automorphism, that is every pseudo-semi-injective module is co-Hopfian.

It is clear that every semi-injective module is pseudo-semi-injective, however, converse need not be true. In the following Proposition, we impose the uniformness on pseudo-semi-injective module that is desirable to make it semi-injective modules.

Proposition 4.1 *Every uniform pseudo-semi-injective module is semi-injective.*

Proof Let M be uniform pseudo-semi-injective module and N be M-cyclic submodule of M, let $f : N \longrightarrow M$ be any homomorphism implies that $kerf \subseteq N$. If $kerf = N$ case is trivial. If $kerf = 0$, then f is a monomorphism which extend to a homomorphism h from M to M. If $kerf \neq 0$, since N is uniform then it can be easily checked that $g = I_N - f : N \longrightarrow M$ is injective map that is $kerg = 0$, where $I_N : N \longrightarrow M$ be the inclusion map. By definition of pseudo-semi-injectivity of M, there exists an extension h of g from M to M such that $g = I_N - f = h \cdot i$ implies that $f = (1 - h) \cdot i$, which gives that $(1 - h)$ is an extension of f to M. Thus, we conclude that M is semi-injective module.

Corollary 4.2 *Every semi-simple pseudo-semi-injective module is semi-injective.*

Proposition 4.3 *Let M be a pseudo-semi-injective module and $f : M \longrightarrow M$ be a monomorphism. Then $f(M)$ is a direct summand of M.*

Proof The proof is straightforward and hence we omit it.

Proposition 4.4 *Let N be indecomposable pseudo M-principally injective modules, then every element $f \in EndN_R$ is invertible if and only if $kerf = 0$.*

Proof The invertible in $EndN_R$ is just the R-isomorphism from N to N. Thus it is clear that, if f is an invertible elements of $EndN_R$ then $kerf = 0$. Conversely suppose that $kerf = 0$ then f is a monomorphism and $f(N)$ is injective and so pseudo M-principally injective module. Then $f(N)$ is a direct summand of every extension of itself, thus $f(N)$ is a direct summand of N, and $f(N) \neq 0$ so $f(N) = N$, since N is indecomposable. Therefore f is a surjective homomorphism and so f is an invertible element of $EndN_R$.

A R-module M is called weakly co-Hopfian if any injective endomorphism f of M is essential, i.e., $f(M) \subseteq^e M$. The set of Integer $\mathbb{Z}$ is weakly co-Hopfian but not co-Hopfian.

Proposition 4.5 *Let M be pseudo-semi-injective module, then the following statements are equivalent:*
$(i)M$ is co-Hopfian;
$(ii)M$ is weakly co-Hopfian;
$(iii)M$ is directly finite.

Proof (1) $\Rightarrow$ (2) $\Rightarrow$ (3) are trivial. For (3) $\Rightarrow$ (1) Assume that $f : M \longrightarrow M$ be an injective endomorphism, then $f(M) \cong M$ and so $f(M)$ is pseudo-M-principally injective. Thus, $f(M)$ is direct summand of M that is there exist a submodule K of M such that $f(M) \oplus K = M$. Hence, $M \oplus K \cong M \Rightarrow K = 0$ since M is directly finite. Therefore, $f(M) = M$ implies that f is surjective and hence M is co-Hopfian.

Corollary 4.6 *If M is indecomposable pseudo-semi-injective module, then it is co-Hopfian.*

Proposition 4.7 *Let M be pseudo-semi-injective and nonsingular module. Then M Hopfian if and only if M co-Hopfian.*

Proof Let M is co-Hopfian and $f : M \longrightarrow M$ be surjective endomorphism of M. Then $M/kerf$ is nonsingular, and so $kerf$ is essentially closed in M. since M is pseudo-semi-injective modules, then $kerf$ is also pseudo-semi-injective. Thus, $M \cong M \oplus kerf$. As M is co-Hopfian, it is directly finite module by Proposition 4.5, so the above isomorphism implies that $kerf = 0$, i.e., f is an automorphism. Thus M is Hopfian. Conversely, It is well known that every Hopfian and co-Hopfian modules is directly finite so prove is done in the light of Proposition 4.5.

Proposition 4.8 *Let M be pseudo-semi-injective module and N be fully invariant M-cyclic submodule of M with N is essential in M. Then N is weakly co-Hopfian if and only if M is weakly co-Hopfian.*

Proof A sume that N is weakly co-Hopfian. Let $f : M \longrightarrow M$ be an injective endomorphism then by Proposition 2.3, $f(M)$ is direct summand of M. Since N is fully invariant $f(N) \subseteq N$. Thus $f|_N : N \longrightarrow N$ is an injective homomophism, the weakly co-Hopficity of N implies that $f(N) \subseteq^e N$, since $N \subseteq^e M$ we deduce that $f(N) \subseteq^e M$ and we have $f(N) \subseteq f(M) \subseteq M$, thus $f(M) \subseteq^e M$ therefore M is weakly co-Hopfian.

Conversely, let $f : N \longrightarrow N$ be an injective endomorphism and $i : N \longrightarrow M$ be an inclusion map. Since M is pseudo-semi-injective module, there exists a monomorphism $h : M \longrightarrow M$ such that $i \cdot f = h \cdot i$. Since M is weakly co-Hopfian by Proposition 4.5, M is co-Hopfian, so h is an isomorphism. N is fully invariant M-cyclic submodule of M so it is pseudo-semi-injective and $h(N) \subseteq N \Rightarrow h^{-1}(N) \subseteq N$ so $h(N) = N$. But $f = h|_N$ hence $f : N \longrightarrow N$ is surjective, so N is co-Hopfian then by Proposition 4.5, proof is complete.

References

1. Anderson, F.W., Fuller, K.R.: Rings and Categories of Modules. Springer, New York (1974)
2. Birkenmeier, G.F., Muller, B.J., Rizvi, S.T.: Modules in which every fully invariant submodule is essential in a direct summand. Commun. Algebr. **30**, 1395–1415 (2002)
3. Clark, J., Lomp, C., Vanaja, N., Wisbauer, R.: Lifting Modules Supplements and Projectivity in Module Theory. Birkhauser, Boston (2006)
4. Ghorbani, A., Haghany, A.: Generalized hopfian modules. J. Algebr. **255**, 324–341 (2002)
5. Haghany, A., Vedadi, M.R.: Study of semi-projective retractable modules. Algebr. Coll. **14**(3), 489–496 (2007)
6. Mohamed, S.H., Muller, B.J.: Continuous and Discrete module. Cambridge University Press, Cambridge (1990)
7. Pandeya, B.M., Pandey, A.K.: Almost perfect ring and directly finite module. Int. J. Math. Sci. **1**(1–2), 111–115 (2002)
8. Patel, M.K., Pandeya, B.M., Gupta, A.J., Kumar, V.: Generalization of semi-projective modules. Int. J. Comput. Appl. **83**(8), 1–6 (2013)
9. Patel, M.K., Pandeya, B.M., Gupta, A.J., Kumar, V.: Quasi-principally injective modules. Int. J. Algebr. **4**, 1255–1259 (2010)
10. Rangaswami, K.M., Vanaja, N.: Quasi Projective in abelian categories. Pac. J. Math. **43**, 221–238 (1972)
11. Tansee, H., Wongwai, S.: A note on semi projective modules. Kyungpook Math. J. **42**, 369–380 (2002)
12. Varadarajan, K.: Properties of endomorphism rings. Acta. Math. Hungar. **74**(1–2), 83–92 (1997)
13. Kumar, V., Gupta, A.J., Pandeya, B.M., Patel, M.K.: M-SP-Injective module. Asian Eur. J. Math. **5**, 1–11 (2011)
14. Wisbauer, R.: Foundation of Module and Ring Theory. Gordon and Breach, London (1991)

Labeling of Sets Under the Actions of $\overrightarrow{S_n}$ and $\overrightarrow{A_n}$

Ram Parkash Sharma, Rajni Parmar and V.S. Kapil

Abstract We prove that distinguishing number $D_{\overrightarrow{S_n}}(X)$ can be at most $n + 1 + [\frac{n}{6}]$ for $n < 36$ and find the complete sets of distinguishing numbers $D_{\overrightarrow{S_2}}(X)$ and $D_{\overrightarrow{A_2}}(X)$. The distinguishing numbers of the actions of $\overrightarrow{S_3}$ and $\overrightarrow{A_3}$ are also discussed.

Keywords Distinguishing number · Distinguishing group actions · Labeling of sets and graphs

1 Introduction

The wreath product $\overrightarrow{S_n}$ is defined as

Definition 1.1 Let $\mathbb{Z}_2^n = \{f \mid f : \{1, 2, \ldots, n\} \longrightarrow \mathbb{Z}_2\}$. Define

$$\overrightarrow{S_n} = \mathbb{Z}_2 \wr S_n = \{(f, \pi) \mid f : \{1, 2, \ldots, n\} \longrightarrow \mathbb{Z}_2, \pi \in S_n\},$$

where S_n is the symmetric group on n symbols. $\overrightarrow{S_n}$ is a group under the composition defined by

$$(f, \pi)(f', \pi') = (f' f_{\pi'^{-1}}, \pi\pi'),$$

where $(ff')(i) = f(i) + f'(i)$, $i \in \{1, 2, \ldots, n\}$ and $f_{\pi'^{-1}} = f \circ \pi'$, for $\pi' \in S_n$ and $f \in \mathbb{Z}_2^n$. This group of type B_n is called the wreath product of $\mathbb{Z}_2$ by S_n.

The group $\overrightarrow{S_n}$ has a presentation with generators $S = \{s_1, s_2, \ldots, s_n\}$, satisfying

(1) $s_i^2 = 1$, for every $i \geq 1$
(2) $\left(s_i s_j\right)^2 = 1$ if $|i - j| \neq 1$
(3) $(s_i s_{i+1})^3 = 1$ for every $i \geq 2$, and
(4) $(s_1 s_2)^4 = 1$.

This group $\overrightarrow{S_n}$ is isomorphic to the group of signed Brauer diagrams having no horizontal edges. For more detail about Brauer and signed Brauer algebras one can refer to [4, 5, 7–9, 11].

R.P. Sharma (✉) · R. Parmar · V.S. Kapil
Department of Mathematics, Himachal Pradesh University,
Summer Hill, Shimla 171005, India
e-mail: rp_math_hpu@yahoo.com

S.T. Rizvi et al. (eds.), *Algebra and its Applications*, Springer Proceedings in Mathematics & Statistics 174, DOI 10.1007/978-981-10-1651-6_20

Since the subgroup of $\overrightarrow{S_n}$ that is generated by s_i, $i = 2, 3, \ldots, n$ is isomorphic to S_n, this subgroup of $\overrightarrow{S_n}$ is identified with S_n by taking each s_i, $i \geq 2$ to the basic transposition $(i-1, i)$.

The group $\overrightarrow{S_n}$ becomes a subgroup of S_{2n} as observed in [3]

Definition 1.2 For any integer $n \geq 2$, the group $\overrightarrow{S_n}$ can be identified to be the subgroup of S_{2n} as follows:

$$\overrightarrow{S_n} = \{\theta \in S_{2n} | \theta(i) + \theta(-i) = 0, \text{ for all } i, 1 \leq i \leq n\}.$$

Here, the set $\{1, 2, 3, \ldots, n, n+1, \ldots, 2n\}$ is identified by $\{1, 2, \ldots, n, -1, -2, \ldots -n\}$. In this paper, we use the former notation. In [6], the authors found the elements of $\overrightarrow{S_n}$ which correspond to even permutations in S_{2n}. The set of such elements form a normal subgroup of $\overrightarrow{S_n}$ of order $2^{n-1}n!$ denoted by $\overrightarrow{A_n}$. We are interested to find the distinguishing numbers of the actions of these groups on various sets.

A labeling of the vertices of a graph $\hat{G}$, $\phi : V(\hat{G}) \longrightarrow \{1, 2, 3, \ldots, r\}$ is said to be r-distinguishing provided no automorphism of the graph preserves all of the vertex labels. That is, for every nontrivial $\sigma \in Aut(\hat{G})$ there exists x in $V = V(\hat{G})$ such that $\phi(x) \neq \phi(\sigma(x))$. The distinguishing number of a graph $\hat{G}$, denoted by $D(\hat{G})$, is the minimum r such that $\hat{G}$ has an r-distinguishing labeling. That is,

$D(\hat{G}) = \min\{r | \hat{G}$ has a labeling that is r-distinguishing (See [1, 2]).

The main algebraic difference between distinguishing groups and distinguishing graphs is: many groups do not act faithfully (i.e., stabilizer is nontrivial) while the automorphism group of a graph always has trivial stabilizer. Distinguishing labeling of graphs can naturally be extended in the same way to a group action of a group G on a set X, if G acts faithfully on X ($St_G(X) = e$); that is, the labeling ϕ of X is said to be r-distinguishing with respect to a faithful action of G if for every $g \neq e \in G$, there is an element $x \in X$ such that $\phi(x) \neq \phi(g(x))$. In case, $St_G(X) \neq e$, then $\phi(x) \neq \phi(g(x))$ for every g that does not belongs to $St_G(X)$.

The distinguishing number of a group G on a set X is defined by

$$D_G(X) = \min\{r : \text{ there exists an } r\text{-distinguishing labeling of X}\}.$$

It is proved in [10] that $D_{S_n}(X)$, for any set on which S_n acts, is at most n. We started this paper with the aim of finding an upper bound for the distinguishing numbers $D_{\overrightarrow{S_n}}(X)$ for $\overrightarrow{S_n}$ actions similar to that of S_n having $D_{S_n}(X)$ at most n. We could answer this question for $n \leq 36$. We establish that $D_{\overrightarrow{S_n}}(X)$ can be at most $n + 1 + [\frac{n}{6}]$ for $n \leq 36$. Once we know the utmost value of $D_{\overrightarrow{S_n}}(X)$ for $n \leq 36$, it is a natural question of finding the complete set of the numbers $D_{\overrightarrow{S_n}}(X)$ for $n = 2$ or 3. We find a complete set of distinguishing numbers $D_{\overrightarrow{S_2}}(X)$, that is, $D_{\overrightarrow{S_2}}(X)$ is either 1, 2, or 3 for any set X with $\overrightarrow{S_2}$ action on it. Regarding $D_{\overrightarrow{S_3}}(X)$, we searched for different sets with action of $\overrightarrow{S_3}$ on them, but $D_{\overrightarrow{S_3}}(X)$ turns out to be 3. Still it remains unsolved.

Whether $D_{\overrightarrow{S_3}}(X) = 4$ for some X with action of $\overrightarrow{S_3}$ on it?

Similarly for $\overrightarrow{A_3}$ actions, We get $D_{\overrightarrow{A_3}}(X) = 1, 2$ and 3.

2 Alternating Subgroup $\overrightarrow{A_n}$ of $\overrightarrow{S_n}$

In order to find out the elements of $\overrightarrow{A_n}$, the authors of [6] fixed a notation as

$$i^* = \begin{cases} i + n, o < i \leq n \\ i - n, n < i \leq 2n \end{cases}$$

Using this notation, Definition 1.2 can be rewritten as follows:

Definition 2.1 We have $\overrightarrow{S_n} = \{\theta \in S_{2n} |$ if $\theta(i) = k$, then $\theta(i^*) = k^*\}$.

Example 2.2 The eight elements of $\overrightarrow{S_2}$ identified in S_4 according to the above definition are as follows:

$$\left\{ \begin{pmatrix} 1&2&3&4 \\ 1&2&3&4 \end{pmatrix}, \begin{pmatrix} 1&2&3&4 \\ 1&4&3&2 \end{pmatrix}, \begin{pmatrix} 1&2&3&4 \\ 3&2&1&4 \end{pmatrix}, \begin{pmatrix} 1&2&3&4 \\ 3&4&1&2 \end{pmatrix}, \atop \begin{pmatrix} 1&2&3&4 \\ 2&1&4&3 \end{pmatrix}, \begin{pmatrix} 1&2&3&4 \\ 2&3&4&1 \end{pmatrix}, \begin{pmatrix} 1&2&3&4 \\ 4&1&2&3 \end{pmatrix}, \begin{pmatrix} 1&2&3&4 \\ 4&3&2&1 \end{pmatrix} \right\}.$$

After rearranging, these elements can be written as

$$\overrightarrow{S_2} = \{e, (13), (24), (12)(34), (13)(24), (14)(23), (1234), (1432)\}.$$

The alternating subgroup $\overrightarrow{A_2}$ consists of the even permutations of $\overrightarrow{S_2}$ in S_4, and hence

$$\overrightarrow{A_2} = \{e, (12)(34), (13)(24), (14)(23)\}.$$

We also need the elements of $\overrightarrow{S_3}$ and $\overrightarrow{A_3}$ to be used in the subsequent results.

Example 2.3 The elements of $\overrightarrow{S_3}$ and $\overrightarrow{A_3}$ are as follows:

$\overrightarrow{S_3} = \{e$, (14), (25), (36), (14)(25), (14)(36), (25)(36), (12)(45), (15)(24), (26)(35), (23)(56), (13)(46), (16)(34), (14)(25)(36), (15)(24)(36), (14)(26)(35), (14)(23)(56), (25)(13)(46), (25)(16)(34), (36)(12)(54), (126)(345), (153)(264), (132)(465), (135)(246), (162)(354), (165)(243), (123)(564), (234)(156), (14)(2356), (14)(2653), (25)(1346), (25)(1643), (36)(1245), (36)(1542), (1245), (1542), (2356), (2653), (1346), (1643), (156423), (132465), (123456), (165432), (126453), (135462), (153426), (162435)$\}$, and

$\overrightarrow{A_3} = \{e, (14)(25), (14)(36), (25)(36), (12)(45), (15)(24), (26)(35), (23)(56), (13)(46), (16)(34), (126)(345), (153)(264), (132)(465), (135)(246), (162)(354), (165)(243), (123)(564), (234)(156), (14)(2356), (14)(2653), (25)(1346), (25)(1643), (36)(1245), (36)(1542)\}.$

3 Labeling of Sets Under the Actions of $\overrightarrow{S_n}$ and $\overrightarrow{A_n}$

This section contains the main results of this paper, that is, an upper bound for distinguishing numbers $D_{\overrightarrow{S_n}}(X)$ is found for $n \leq 36$ and the complete sets of distinguishing numbers $D_{\overrightarrow{S_2}}(X)$ and $D_{\overrightarrow{A_2}}(X)$ are also found.

We need

Proposition 3.1 $2^n n! < (n + 2 + [\frac{n}{6}])!$ *for* $n \geq 0$. *Further, for* $1 < n < 36$, *the integer* $(n + 2 + [\frac{n}{6}])$ *is the smallest positive integer satisfying* $2^n n! < (n + 2 + [\frac{n}{6}])!$.

Proof We prove the inequality

$$2^n n! < \left(n + 2 + \left[\frac{n}{6}\right]\right)! \tag{1}$$

for $n = 6m + r$, $m \geq 4$, $0 \leq r \leq 5$ as this can be easily verified for $m = 0, 1, 2, 3$. We have

$$2^{6m+r}\,(6m + r)! < (7m + r + 2)!$$

iff

$$2^{6m+r}\,(6m + r)! < (6m + r + 1)\,(6m + r + 2)\ldots(7m + r + 2)\,(6m + r)!$$

iff

$$2^{6m+r} < (6m + r + 1)\,(6m + r + 2)\ldots(7m + r + 2)\,. \tag{2}$$

We have

$$(6m + r + 1)\,(6m + r + 2)\ldots(7m + r + 2) > (6m + r + 1)^{m+2}\,. \tag{3}$$

Note that the number of terms in RHS of (3) is $m + 2$. So, to prove (2) it suffices to show that

$$(6m + r + 1)^{m+2} > 2^{6m+r}. \tag{4}$$

Since $\log x$ is continuous and increases for $x > 0$, therefore $\log x > \log y$ iff $x > y$. So (4) holds iff

$$(m+2)\log(6m+r+1) > (6m+r)\log 2. \tag{5}$$

To prove (5) we show that the function

$$f(m) = (m+2)\log(6m+r+1) - (6m+r)\log 2 \tag{6}$$

is positive for $m \geq 4$ and $r = 0, 1, \ldots, 5$.

We show that $f(m)$ increases for $m \geq 4$. We have from (6)

$$f'(m) = 6\frac{m+2}{6m+r+1} + \log(6m+r+1) - 6\log 2. \tag{7}$$

At $m = 2$ and $r = 0, 1, 2, 3, 4, 5$, we have $f'(m) = \frac{24}{17} + \log 17 - 6\log 2 = 8.6095 \times 10^{-2} > 0$. So, to show that $f'(m)$ is positive we have to show that $f'(m)$ increases for $m \geq 2$. From (7), we have

$$f''(m) = 6\frac{(6m+2r+-10)}{(6m+r+1)^2} \geq 0 \text{ for } m \geq 2.$$

Hence the proof of (1) is complete.

In order to prove that $(n+2+[\frac{n}{6}])$ is the smallest positive integer satisfying the inequality (1) for $1 < n < 36$, it suffices to show that $2^n n! \not< (n+1+[\frac{n}{6}])!$, for these values. Clearly $g(n) = (n+1+[\frac{n}{6}])!$ is a continuous and strictly increasing function for $n > 1$ and $g(2) = 3! = 6 < 2^2 2! = 8$. Note that

$$g(35) = 41! = 3.3453 \times 10^{49} < 2^{35}35! = 3.5504 \times 10^{50},$$

but

$$g(36) = 43! = 6.0415 \times 10^{52} \not< 2^{36}36! = 2.5563 \times 10^{52}.$$

Hence the result.

Theorem 3.2 **(i)** *For the group* $\overrightarrow{S_n}$, $1 < n < 36$, *the distinguishing number* $D_{\overrightarrow{S_n}}(X)$ *is at most* $n+1+[\frac{n}{6}]$.

(ii) *The distinguishing number* $D_{\overrightarrow{A_3}}(X)$ *is at most* 4.

Proof **(i)** By [10, Corollary 2.1], for any finite group G, the distinguishing number $D_G(X)$ is at most m, where m is the largest positive integer such that $|G| \geq m!$. Since for $1 < n < 36$, the number $(n+1+[\frac{n}{6}])$ is the largest positive integer such that $|\overrightarrow{S_n}| = 2^n n! \geq (n+1+[\frac{n}{6}])!$, therefore, the distinguishing number $D_{\overrightarrow{S_n}}(X)$ is at most $n+1+[\frac{n}{6}]$ for $1 < n < 36$, where X is any set on which $\overrightarrow{S_n}$ acts.

(ii) The largest value of $D_{\overrightarrow{A_3}}(X)$ is 4, because $|\overrightarrow{A_3}| = 4!$.

By the above theorem, if $\overrightarrow{S_2}$ acts on a set X then $D_{\overrightarrow{S_2}}(X)$ is at most 3. In the following theorem, we find the complete set of distinguishing numbers for $\overrightarrow{S_2}$ actions. For the element of $\overrightarrow{S_2}$, one can see Example 2.2.

Theorem 3.3 *If $\overrightarrow{S_2}$ acts on X, then the distinguishing number $D_{\overrightarrow{S_2}}(X)$ is either* 1, 2 *or* 3.

Proof The trivial $\overrightarrow{S_2}$ action on a one element set has distinguishing number 1. If $\overrightarrow{S_2}$ acts on itself by translation, its distinguishing number is 2 by [10, Proposition 2.2]. If $\overrightarrow{S_2}$ act on $X = \{1, 2, 3, 4\}$, then we show that the distinguishing number $D_{\overrightarrow{S_2}}(X)$ $= 3$. For, define a labeling $\phi : X \longrightarrow \{1, 2, 3\}$ by $\phi(1) = 1$, $\phi(2) = 2$. If we take $\phi(3) = 1$, then $(13) \in \overrightarrow{S_2}$ preserves the labeling. So we take $\phi(3) = 2$. If we take $\phi(4) = 1$, then $\sigma = (14)(23) \in \overrightarrow{S_2}$ preserves the labeling. But $\phi(4) \neq 2$ as in that case $(24) \in \overrightarrow{S_2}$ preserves the labeling. Hence, there is only one choice left, that is, $\phi(4) = 3$. Therefore, for $\overrightarrow{S_2}$ action on the set $\{1, 2, 3, 4\}$, we need minimum 3 labels to distinguish its action. Hence $D_{\overrightarrow{S_2}}(X) = 3$.

Theorem 3.4 *Let $\overrightarrow{A_2}$ act on a set $X = \{1, 2, 3, 4\}$. Then $D_{\overrightarrow{A_2}}(X) = 2$.*

Proof Obviously $D_{\overrightarrow{A_2}}(X) > 1$. Let $\phi(1) = 1$. Since $(13)(24) \in \overrightarrow{A_2}$, $\phi(3) \neq 1$ or $\phi(2) \neq \phi(4)$. So we label 3 as $\phi(3) = 2$ and $\phi(4) = 2 = \phi(2)$. Clearly, any $\sigma \epsilon \overrightarrow{A_2}$ does not preserve the labeling. Hence for $n = 2$, we need minimum 2 labels to distinguish the set $\{1, 2, 3, 4\}$ by $\overrightarrow{A_2}$, so $D_{\overrightarrow{A_2}}(X) = 2$.

Finally, in this section, we find the distinguishing number of $\overrightarrow{A_3}$ action on the set $X = \{1, 2, 3, \ldots, 6\}$.

Theorem 3.5 *Let $X = \{1, 2, 3, \ldots, 6\}$ and $\overrightarrow{A_3}$ act on X by permuting the numbers. Then $D_{\overrightarrow{A_3}}(X) = 3$.*

Proof (i) The element $(14)(25) \in \overrightarrow{A_3}$ forces $\phi(1) \neq \phi(4)$, but here we can have $\phi(2) = \phi(5)$,

(ii) $(25)(36) \in \overrightarrow{A_3}$ forces $\phi(2) = \phi(5)$, because if $\phi(2) \neq \phi(5)$, then it contradicts (i). So we can take $\phi(3) \neq \phi(6)$.

(iii) $(14)(36) \in \overrightarrow{A_3}$ forces $\phi(3) \neq \phi(6)$ and $\phi(1) \neq \phi(4)$, because if $\phi(3) = \phi(6)$,then it contradicts (ii) and if $\phi(1) = \phi(4)$, then it contradicts (i).

(iv) $(12)(45) \in \overrightarrow{A_3}$ forces $\phi(1) \neq \phi(2)$, but here we can have

(v) $\phi(4) = \phi(5)$,

The element $(15)(24) \in \overrightarrow{A_3}$ forces $\phi(1) \neq \phi(5)$, suppose we take $\phi(1) = \phi(5)$, then using (i) we have $\phi(2) = \phi(1)$, a contradiction to (iv). We can take $\phi(2) = \phi(4)$.

(vi) $(26)(53) \in \overrightarrow{A_3}$ forces $\phi(3) \neq \phi(5)$ and we can have $\phi(2) = \phi(6)$.

(vii) $(23)(56) \in \overrightarrow{A}_3$ forces $\phi(2) \neq \phi(3)$. Suppose we take $\phi(2) = \phi(3)$, then using (i) we have $\phi(3) = \phi(5)$, a contradiction (vi). Therefore, we also have

$(viii)$ $\phi(5) = \phi(6)$.

(ix) $(13)(46) \in \overrightarrow{A_3}$ forces $\phi(4) = \phi(6)$ because if $\phi(4) \neq \phi(6)$, then using $(viii)$ we have $\phi(4) \neq \phi(5)$ which contradicts (v). So we can take

(x) $\phi(1) \neq \phi(3)$.

(xi) $(16)(34) \in \overrightarrow{A_3}$ forces $\phi(1) \neq \phi(6)$ because if $\phi(1) = \phi(6)$, then using (vi) we have $\phi(2) = \phi(1)$ which is a contradiction to (iv). If we take $\phi(3) = \phi(4)$, then by (ix) we get $\phi(3) = \phi(6)$ which is a contradiction to (ii). Therefore,

(xii) $\phi(3) \neq \phi(4)$.

The labeling $\phi : X \to \{1, 2, 3\}$ defined by $\phi(1) = 1, \phi(3) = 2$. $\phi(2) = \phi(4) = \phi(5) = \phi(6) = 3$ satisfies all the conditions given above, and hence distinguishes the action of $\overrightarrow{A_3}$ on X. Moreover, the conditions (i),(x) and (xii) give that $\phi(1) \neq \phi(4) \neq \phi(3)$. So the action of $\overrightarrow{A_3}$ on X is not 2-distinguishable, hence $D_{\overrightarrow{A_3}}(X) = 3$.

4 The Sets with $D_{\overrightarrow{S_3}}(X) = 3$

In this section, we examine the distinguishing numbers of $\overrightarrow{S_3}$ actions on various sets. But in all the cases $D_{\overrightarrow{S_3}}(X) = 3$. By [10, Proposition 2.1], $D_{\overrightarrow{S_3}}(X) = 1$ when $\overrightarrow{S_3}$ acts trivially on one element set and By [10, Proposition 2.1], $D_{\overrightarrow{S_3}}(X) = 2$ when $\overrightarrow{S_3}$ acts on itself by translation. By Theorem 3.2.(i), the distinguishing number $D_{\overrightarrow{S_3}}(X)$ is at most $n + 1 + [\frac{n}{6}] = 4$. Hence the question finding a set X with group action of $\overrightarrow{S_3}$ on it such that $D_{\overrightarrow{S_3}}(X) = 4$ still remains unsolved.

First, we examine the distinguishing numbers of conjugacy action of $\overrightarrow{S_3}$ on the conjugacy classes of the various elements of $\overrightarrow{S_3}$ when it acts by conjugation on itself.

Theorem 4.1 *Let* $\overrightarrow{S_3}$ *act on* $X = C(14)(2356) = \{(14)(2356), (14)(2653), (36)(1245), (36)(1542), (25)(1346), (25)(1643)\}$. *Then* $D_{\overrightarrow{S_3}}(X) = 3$.

Proof It is easy to see that $St_{\overrightarrow{S_3}}(X) = \{e, (14)(25)(36)\}$. Therefore, any labeling ϕ of X distinguishes the action of $\overrightarrow{S_3}$ on X, if there exist $\sigma \in X$ for every τ that does not belongs to $St_{\overrightarrow{S_3}}(X)$ such that $\phi(\sigma) \neq \phi(\tau(\sigma))$. We calculated the stabilizers of all the elements belonging to X and found that the order of stabilizer of every element belonging to X is 8. So we take

$$St_{(14)(2356)} = \{e, (14), (14)(25)(36), (25)(36), (2356), (2653), (14)(2356), (14)(2653)\} = St_{(14)(2653)}.$$

If we label $\phi(14)(2356) = 1$, then there are 40 elements of $\overrightarrow{S_3}$ which do not preserve the labeling provided the remaining elements of X are not labeled by

1. Take $X_1 = X \backslash \{(14)(2356), (14)(2653)\}$ and $G_1 = St_{(14)(2356)}$, then X_1 is G_1-invariant. For, let $\tau \in X_1$ and $\sigma \in G_1$. If $\sigma\tau\sigma^{-1} = (14)(2356)$ or $(14)(2653)$, then $\sigma\tau\sigma^{-1} = \sigma(14)(2356)\sigma^{-1}$ or $\sigma(14)(2653)\sigma^{-1}$, which gives that $\tau = (14)(2356)$ or $(14)(2653)$, which is a contradiction. Hence X_1 is G_1-invariant and G_1 acts on X_1. Now the problem of labeling is reduced to the labeling of the set X_1 by G_1 with the numbers > 1. When G_1 acts on X_1, we have

$$St_{(36)(1245)} = St_{(36)(1542)} = St_{(25)(1346)} = St_{(25)(1643)} = \{e, (14)(25)(36)\}.$$

Therefore, if we label any element of X_1 by 2 and the remaining elements of X_1 by 3, then this labeling is 3-distinguishable by $\overrightarrow{S_3}$ action. Obviously, this is the minimum number to have the action of $\overrightarrow{S_3}$ on X distinguishable. Thus, we define $\phi : X \longrightarrow \{1, 2, 3\}$ by $\phi((14)(2356)) = 1$, $\phi((36)(1245)) = 2$ and $\phi(\sigma) = 3$ for the remaining elements of X.

Theorem 4.2 *Let* $\overrightarrow{S_3}$ *act on*

$$X = C(1245) = \{(1245), (1542), (2356), (2653), (1346), (1643)\}.$$

Then $D_{\overrightarrow{S_3}}(X) = 3$.

Proof Here also the order of stabilizer of each $\sigma \in X$ is 8. So we take
$St_{(1245)} = \{e, (36), (14)(25)(36), (36)(1245), (36)(1542), (1245), (1542), (14)(25)\}$
and $X_1 = X \backslash \{(1245), (1542)\}$. As in the above theorem, X_1 is G_1-invariant. When G_1 acts on X_1, we have

$$St_{(2356)} = St_{(2653)} = St_{(1346)} = St_{(1643)} = \{e, (14)(25)(36)\}.$$

Therefore as in the above theorem, $D_{\overrightarrow{S_3}}(X) = 3$.

Theorem 4.3 *Let* $\overrightarrow{S_3}$ *act on* $X = C(14) \cup C(14)(25) = \{(14), (25), (36), (14)(25), (14)(36), (14)(25)\}$. *Then* $D_{\overrightarrow{S_3}}(X) = 3$.

Proof It is easy to see that

$$St_{\overrightarrow{S_3}}(X) = \{e, (14), (25), (36), (14)(25)(36), (14)(25), (14)(36), (25)(36)\}.$$

Therefore any labeling ϕ of X distinguishes the action of $\overrightarrow{S_3}$ on X, if there exist $\sigma \in X$ for every τ that does not belongs to $St_{\overrightarrow{S_3}}(X)$ such that $\phi(\sigma) \neq \phi(\tau(\sigma))$. We calculated the stabilizers of all the elements belonging to X and found that the order of stabilizer of every element belonging to X is 16 because X is the union of two conjugacy classes each having three elements. So we take

$St_{(14)(36)} = \{e, (14), (25), (36), (14)(25), (25)(36), (14)(36), (13)(46), (16)(34), (14)(25)(36), (1346), (1643), (25)(1346), (25)(1643), (25)(16)(34), (25)(13)(46)\}.$

If we label $\phi((14)(36)) = 1$, then there are 32 elements of $\overrightarrow{S_3}$ which do not preserve the labeling provided the remaining elements of X are not labeled by 1. Take $X_1 = X \backslash \{(14)(36)\}$ and $G_1 = St_{(14)(36)}$, then X_1 is G_1-invariant and G_1 acts on X_1. Now the problem of labeling is reduced to the labeling of the set X_1 by G_1 with the numbers > 1. When G_1 acts on X_1, we have

$$St_{(14)(25)} = St_{(25)(36)} = St_{(14)} = St_{(25)} = St_{(36)} = \{e, (14), (25), (36), (14)(25)(36), (14)(25), (14)(36), (25)(36)\}.$$

The elements of the above stabilizers preserve any labeling as they are in $St_{\overrightarrow{S_3}}(X)$.

Therefore, if we label any element of X_1 by 2 and the remaining elements of X_1 by 3, then this labeling is 3-distinguishable by $\overrightarrow{S_3}$ action. Obviously, this is the minimum number to have the action of $\overrightarrow{S_3}$ on X distinguishable. Thus, we define $\phi : X \longrightarrow \{1, 2, 3\}$ by $\phi((14)(36)) = 1$, $\phi((25)) = 2$ and $\phi(\sigma) = 3$ for the remaining elements of X.

Theorem 4.4 *Let X be the set of conjugacy class* {(123456), (165432), (135462), (126453), (132465), (156423), (162435),
(153426)} *of the permutation* (123456) *and* $\overrightarrow{S_3}$ *act on X by conjugation. Then* $D_{\overrightarrow{S_3}}(X) = 3$.

Proof Here, we have 4 pairs of elements, each pair consisting of σ and its inverse σ^{-1}:
{(123456),(165432)},
{(135462),(126453)},
{(132465),(156423)},
{(162435),(153426)}.
Further, $St_{\overrightarrow{S_3}}(X) = \{e, (14)(25)(36)\}$. First, we show that X is not 2-distinguishable. Suppose ϕ gives both elements of two different pairs the same label, that is $\phi(156423) = \phi(132465)$, $\phi(162435) = \phi(153426)$, $\phi(123456) = \phi(135462)$ and $\phi(165432) = \phi(126453)$.Then the action of (12)(45) on X preserves each component, exchanging each of these pairs while fixing the first two components, so ϕ does not distinguish X. That is, (12)(45) preserve the labeling as

$$(12)(45)(123456)(12)(45) = (135462)$$
$$(12)(45)(165432)(12)(45) = (126453)$$
$$(12)(45)(135462)(12)(45) = (123456)$$
$$(12)(45)(126453)(12)(45) = (165432)$$
$$(12)(45)(156423)(12)(45) = (132465)$$
$$(12)(45)(162453)(12)(45) = (153426)$$
$$(12)(45)(132465)(12)(45) = (156423)$$
$$(12)(45)(153426)(12)(45) = (162435).$$

Suppose ϕ gives both elements of two pairs the same label, say without loss of generality that $\phi(132465) = \phi(156423)$ and $\phi(162435) = \phi(153426)$. Then the action of (15)(24)(36) exchanges the vertices of (132465), (156423) and (162435), (153426) among themselves that is preserving the two components and exchanging the other two components. That is,

$$\begin{aligned}
(15)(24)(36)(123456)(15)(24)(36) &= (135426)\\
(15)(24)(36)(165432)(15)(24)(36) &= (126453)\\
(15)(24)(36)(135462)(15)(24)(36) &= (123456)\\
(15)(24)(36)(126453)(15)(24)(36) &= (165432)\\
(15)(24)(36)(156423)(15)(24)(36) &= (132465)\\
(15)(24)(36)(132465)(15)(24)(36) &= (156423)\\
(15)(24)(36)(162435)(15)(24)(36) &= (153426)\\
(15)(24)(36)(153426)(15)(24)(36) &= (162435).
\end{aligned}$$

Thus (15)(24)(36) preserves the labeling.

So, ϕ doesnot distinguish X. Suppose, that ϕ gives different label to the two elements of three components, that is $\phi(135462) = \phi(132465) = \phi(162435)$ and $\phi(126453) = \phi(156423) = \phi(153426)$. Then the action of (153)(264) cyclically permutes the three components and preserving the first component, i.e.,

$$\begin{aligned}
(153)(264)(123456)(135)(246) &= (123456)\\
(153)(264)(165432)(135)(246) &= (165432)\\
(153)(264)(135462)(135)(246) &= (162435)\\
(153)(264)(126453)(135)(246) &= (153426)\\
(153)(264)(156423)(135)(246) &= (126453)\\
(153)(264)(132465)(135)(246) &= (135462)\\
(153)(264)(162435)(135)(246) &= (132465)\\
(153)(264)(153426)(135)(246) &= (156423).
\end{aligned}$$

That is, (153)(264) preserves the labeling, so ϕ does not distinguish X.

Suppose that $\phi(123456) = \phi(126453), \phi(135462) = \phi(165432)$ and $\phi(132465) = \phi(162435), \phi(156423) = \phi(153426)$. Then the action of (14)(25) preserves the labeling as

$$\begin{aligned}
(14)(25)(123456)(14)(25) &= (126453)\\
(14)(25)(165432)(14)(25) &= (135462)\\
(14)(25)(132465)(14)(25) &= (162435)\\
(14)(25)(156423)(14)(25) &= (153426)\\
(14)(25)(135462)(14)(25) &= (165432)\\
(14)(25)(126453)(14)(25) &= (123456)\\
(14)(25)(162435)(14)(25) &= (132465)\\
(14)(25)(153426)(14)(25) &= (156423).
\end{aligned}$$

Hence, X is not 2-distinguishable under $\overrightarrow{S_3}$ action. Define labeling $\phi : X \longrightarrow \{1, 2, 3\}$ by $\phi(165432) = \phi(132465) = \phi(153426) = 1$, $\phi(123456) = \phi(126453) = \phi(162435) = 2$ and $\phi(135462) = \phi(156423) = 3$. Then this labeling is 3-distinguishable by $\overrightarrow{S_3}$ action on X.

Theorem 4.5 *If $\overrightarrow{S_3}$ acts on $X = \{1, 2, 3, 4, 5, 6\}$, then the distinguishing numbers $D_{\overrightarrow{S_3}}(X)$ is* 3.

Proof Define a labeling ϕ on X by $\phi(1) = 1$, then $\phi(4) \neq 1$ because if we take $\phi(4) = 1$, then $(14) \in \overrightarrow{S_3}$ preserves the labeling. We also observe that

(i) $(25) \in \overrightarrow{S_3}$ forces $\phi(2) \neq \phi(5)$,
(ii) $(36) \in \overrightarrow{S_3}$ forces $\phi(3) \neq \phi(6)$,
(iii) $(12)(45) \in \overrightarrow{S_3}$ forces $\phi(1) \neq \phi(2)$, but here we can have
(iv) $\phi(4) = \phi(5)$,

The element $(15)(24) \in \overrightarrow{S_3}$ forces $\phi(1) \neq \phi(5)$ or $\phi(2) \neq \phi(4)$.

Suppose we take $\phi(1) = \phi(5)$, then from (iii) we have $\phi(4) = \phi(1)$, a contradiction as observed above. Suppose we take $\phi(2) = \phi(4)$, then from (iii) we have $\phi(4) = \phi(5)$ which give $\phi(2) = \phi(5)$, a contradiction to (i). Therefore, we have

(v) $\phi(1) \neq \phi(5)$ and $\phi(2) \neq \phi(4)$.
(vi) $(26)(53) \in \overrightarrow{S_3}$ forces $\phi(3) \neq \phi(5)$ and we can have $\phi(2) = \phi(6)$.
(vii) $(23)(56) \in \overrightarrow{S_3}$ forces $\phi(2) \neq \phi(3)$. Suppose we take $\phi(5) = \phi(6)$, then from (vi) we have $\phi(2) = \phi(6)$ which give $\phi(2) = \phi(5)$, a contradiction (i). Therefore, we also have
$(viii)$ $\phi(5) \neq \phi(6)$.
(ix) $(13)(46) \in \overrightarrow{S_3}$ forces $\phi(4) \neq \phi(6)$ because if $\phi(4) = \phi(6)$, then using (vi) we have $\phi(4) = \phi(2)$ which contradicts (v). So we can take
(x) $\phi(1) = \phi(3)$.
(xi) $(16)(34) \in \overrightarrow{S_3}$ forces $\phi(1) \neq \phi(6)$ because if $\phi(1) = \phi(6)$, then using (x) we have $\phi(3) = \phi(6)$ which is a contradiction to (ii). If we take $\phi(3) = \phi(4)$, then by (x) we get $\phi(1) = \phi(4)$, which is a contradiction as observed in the beginning. Therefore,
(xii) $\phi(3) \neq \phi(4)$.

The labeling $\phi : X \to \{1, 2, 3\}$ defined by $\phi(1) = \phi(3) = 1$. $\phi(2) = \phi(6) = 2$ and $\phi(4) = \phi(5) = 3$ satisfies all the conditions given above and hence distinguish the action of $\overrightarrow{S_3}$ on X. As the action of $\overrightarrow{S_3}$ on X is not 2-distinguishable, $D_{\overrightarrow{S_3}}(X) = 3$.

Theorem 4.6 *Let $\overrightarrow{S_3}$ act on set $X = \{\{1, 2, 3, 4, 5\}, \{1, 2, 3, 4, 6\}, \{1, 2, 3, 5, 6\}, \{1, 2, 4, 5, 6\}, \{1, 3, 4, 5, 6\}, \{2, 3, 4, 5, 6\}\}$, then the distinguishing number $D_{\overrightarrow{S_3}}(X)$ is* 3.

Proof The element $(14) \in \overrightarrow{S_3}$ forces

(i) $\phi(\{2,3,4,5,6\}) \neq \phi(\{1,2,3,5,6\})$

(ii) $(25) \in \overrightarrow{S_3}$ forces $\phi(\{1,2,3,4,6\}) \neq \phi(\{1,3,4,5,6\})$

(iii) $(36) \in \overrightarrow{S_3}$ forces $\phi(\{1,2,4,5,6\}) \neq \phi(\{1,2,3,4,5\})$

(iv) $(12)(45) \in \overrightarrow{S_3}$ forces $\phi(\{1,3,4,5,6\}) \neq \phi(\{2,3,4,5,6\})$, but here we can have

(v) $\phi(\{1,2,3,4,6\}) = \phi(\{1,2,3,5,6\})$.

The element $(15)(24) \in \overrightarrow{S_3}$ forces $\phi(\{1,3,4,5,6\}) \neq \phi(\{1,2,3,5,6\})$ or $\phi(\{1,2,3,4,6\}) \neq \phi(\{2,3,4,5,6\})$.

Suppose we take $\phi(\{1,3,4,5,6\}) = \phi(\{1,2,3,5,6\})$, then from (v) we have $\phi(\{1,2,3,4,6\}) = \phi(\{1,3,4,5,6\})$, a contradiction to (ii). Suppose we take $\phi(\{1,2,3,4,6\}) = \phi(\{2,3,4,5,6\})$, then from (v) we have $\phi(\{2,3,4,5,6\}) = \phi(\{1,2,3,5,6\})$ which is a contradiction to (i). Therefore, we have

(vi) $\phi(\{1,3,4,5,6\}) \neq \phi(\{1,2,3,5,6\})$ and $\phi(\{1,2,3,4,6\}) \neq \phi(\{2,3,4,5,6\})$.

(vii) $(26)(35) \in \overrightarrow{S_3}$ forces $\phi(\{1,2,4,5,6\}) \neq \phi(\{1,2,3,4,6\})$ and we can have $\phi(\{1,3,4,5,6\}) = \phi(\{1,2,3,4,5\})$.

$(viii)$ $(23)(56) \in \overrightarrow{S_3}$ forces $\phi(\{1,3,4,5,6\}) \neq \phi(\{1,2,4,5,6\})$. Suppose we take $\phi(\{1,2,3,4,6\}) = \phi(\{1,2,3,4,5\})$, then from (vii) we have $\phi(\{1,2,3,4,6\})$ $= \phi(\{1,3,4,5,6\})$ which is a contradiction (ii). Therefore, we also have

(ix) $\phi(\{1,2,3,4,6\}) \neq \phi(\{1,2,3,4,5\})$.

(ix) $(13)(46) \in \overrightarrow{S_3}$ forces $\phi(\{1,2,3,5,6\}) \neq \phi(\{1,2,3,4,5\})$ because if $\phi(\{1,2,3,5,6\}) = \phi(\{1,2,3,4,5\})$, then using (vi) we have $\phi(\{1,2,3,5,6\}) = \phi(\{1,3,4,5,6\})$ which contradicts (vi). So we can take

(x) $\phi(\{2,3,4,5,6\}) = \phi(\{1,2,4,5,6\})$.

(xi) $(16)(34) \in \overrightarrow{S_3}$ forces $\phi(\{2,3,4,5,6\}) \neq \phi(\{1,2,3,4,5\})$ because if $\phi(\{2,3,4,5,6\}) = \phi(\{1,2,3,4,5\}$, then using (x) we have $\phi(\{1,2,4,5,6\}) = \phi(\{1,2,3,4,5\})$ which is a contradiction to (iii). If we take $\phi(\{1,2,4,5,6\}) = \phi(\{1,2,3,5,6\})$, then by (x) we get $\phi(\{2,3,4,5,6\}) = \phi(\{1,2,3,5,6\})$ which is a contradiction to (i). Therefore,

(xii) $\phi(\{1,2,4,5,6\}) \neq \phi(\{1,2,3,5,6\})$.

The labeling ϕ defined by $\phi(\{1,2,3,4,6\}) = \phi(\{1,2,3,5,6\}) = 1$, $\phi(\{1,3,4,5,6\}) = \phi(\{1,2,3,4,5\}) = 2$ and $\phi(\{2,3,4,5,6\}) = \phi(\{1,2,4,5,6\}) = 3$ distinguishes the action of $\overrightarrow{S_3}$ on X. As the action of $\overrightarrow{S_3}$ on X is not 2-distinguishable, $D_{\overrightarrow{S_3}}(X) = 3$.

5 Distinguishing $\overrightarrow{A_3}$ Actions on Graphs

In this section, we find the distinguishing number $D(\hat{G})$ of a graph on which an action of $\overrightarrow{A_3}$ consists of graph automorphisms. The conjugacy class $C(14)(2356)$ will be used to compute the distinguishing number of $\overrightarrow{A_3}$ actions on a graph $\hat{G}$ with vertices belonging to $C(14)(2356)$.

Theorem 5.1 *Let $\hat{G}$ be a graph whose vertex set is the conjugacy class of the permutation* $(14)(2356)$, *that is,*

$$V = \{(14)(2356), (14)(2653), (25)(1346), (25)(1643), (36)(1245), (36)(1542)\}$$

and whose edge set consists of (v, v') such that permutation v is the inverse of v'. Let $\overrightarrow{A_3}$ act on V by conjugation. Then $D(\hat{G}) = 3$.

Proof The graph X given below shows a labeling φ indicated by the numbers 1, 2 and 3 with the vertices.

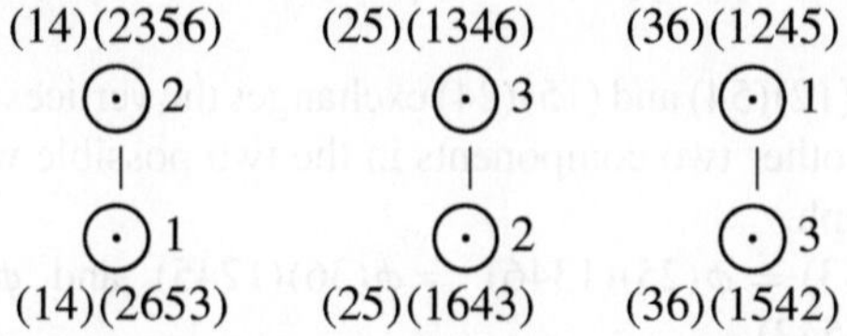

Let $\overrightarrow{A_3}$ act on V by conjugation. This faithful action of $\overrightarrow{A_3}$ on V consists of graph automorphisms since conjugation preserves inverses. The figure given above shows a 3-distinguishing labeling of $\hat{G}$ under the conjugation action of $\overrightarrow{A_3}$.

It remains to show that no 2-labeling distinguishes $\hat{G}$. Suppose ϕ gives both vertices of a component the same label for two different components, i.e., say $\phi(25)(1346) = \phi(25)(1643)$ and $\phi(14)(2356) = \phi(14)(2653)$. Then we check whether the action of $(14)(25)$ on $\hat{G}$ distinguishes the labeling? The action of $(14)(25)$ on V is given as

$$\begin{aligned}
(14)(25)(36)(1245)(14)(25) &= (36)(1245),\\
(14)(25)(36)(1542)(14)(25) &= (36)(1542),\\
(14)(25)(14)(2356)(14)(25) &= (14)(2653),\\
(14)(25)(14)(2653)(14)(25) &= (14)(2356),\\
(14)(25)(25)(1643)(14)(25) &= (25)(1346),\\
(14)(25)(25)(1346)(14)(25) &= (25)(1643).
\end{aligned}$$

Thus, the action of $(14)(25)$ on $\hat{G}$ preserves each component, exchanging each of these pairs while fixing the first component, so ϕ does not distinguish $\hat{G}$.

Suppose both vertices of one component share the same label, say $\phi(36)(1245) = \phi(36)(1542)$.

Then we check whether the actions of $(12)(45)$ and $(15)(24)$ on $\hat{G}$ distinguish the labeling? The action of $(12)(45)$ on V is given as

$$\begin{aligned}
(12)(45)(36)(1245)(12)(45) &= (36)(1542),\\
(12)(45)(36)(1542)(12)(45) &= (36)(1245),\\
(12)(45)(14)(2356)(12)(45) &= (25)(1346),\\
(12)(45)(14)(2653)(12)(45) &= (25)(1643),\\
(12)(45)(25)(1346)(12)(45) &= (14)(2356),\\
(12)(45)(25)(1643)(12)(45) &= (14)(2653).
\end{aligned}$$

And the action of $(15)(24)$ on X is given as

$$\begin{aligned}
(15)(24)(36)(1245)(15)(24) &= (36)(1542),\\
(15)(24)(36)(1542)(15)(24) &= (36)(1245),\\
(15)(24)(14)(2356)(15)(24) &= (25)(1643),\\
(15)(24)(14)(2653)(15)(24) &= (25)(1346),\\
(15)(24)(25)(1346)(15)(24) &= (14)(2653),\\
(15)(24)(25)(1643)(15)(24) &= (14)(2356).
\end{aligned}$$

Thus, action of $(12)(54)$ and $(15)(24)$ exchanges the vertices $(36)(1245)$, $(36)(1542)$ and switching the other two components in the two possible ways. Thus, ϕ can not distinguish the graph.

Let $\phi(14)(2653) = \phi(25)(1346) = \phi(36)(1245)$ and $\phi(14)(2356) = \phi(25)(1643) = \phi(36)(1542)$.

Then $(126)(345)$ cyclically permutes the three components

$$\begin{aligned}
\text{i.e., } (126)(345)(14)(2356)(162)(354) &= (36)(1542),\\
(126)(345)(36)(1542)(162)(354) &= (25)(1643),\\
(126)(345)(25)(1643)(162)(354) &= (14)(2356).
\end{aligned}$$

And

$$\begin{aligned}
(126)(345)(14)(2653)(162)(354) &= (36)(1245),\\
(126)(345)(36)(1245)(162)(354) &= (25)(1346),\\
(126)(345)(25)(1346)(162)(354) &= (14)(2653).
\end{aligned}$$

So, ϕ preserves the labeling. Therefore X is not 2-distinguishable and 3-distinguishing.

Acknowledgments The second author gratefully acknowledges financial assistance from UGC under BSR fellowship.

References

1. Albertson, M.O., Collin, K.: An introduction to symmetry breaking in graphs. Graph Theory Notes N. Y. **30**, 6–7 (1996)
2. Albertson, M.O., Collin, K.: Symmetry breaking in graphs. Electron. J. Comb. **3**(# R18), 17 p (1996)
3. Musli, C.: Representations of Finite Groups. Hindustan book agency, New Delhi (1993)
4. Parvathi, M., Kamaraj, M.: Signed Brauer's algebras. Commun. Algebra **26**(3), 839–855 (1998)
5. Parvathi, M., Selvaraj, C.: Signed Brauer's algebras as centralizer algebras. Commun. Algebra **27**(12), 5985–5998 (1999)
6. Sharma, R.P., Parmar, R.: A new Class of Subgroups of Symmetric Groups using Wreath Products. arXiv:1509.03728v1 [math.RA]
7. Sharma, R.P., Kapil, V.S.: Irreducible $\overrightarrow{S_n}$-modules and a cellular structure of the signed brauer algebras. Southeast Asian Bull. Math. **35**, 497–522 (2011)
8. Sharma, R.P., Kapil, V.S.: Generic irreducibles of the Brauer algebras. Contemp. Math. (AMS) **456**, 189–204 (2008)
9. Sharma, R.P., Kapil, V.S.: Prime ideals of a ring R with Brauer diagrams acting on it. Recent Advances in Continuum Mechanics and Algebra, Chapter 15. Regal Publications, New Delhi (2011)
10. Tymoczko, J.: Distinguishing numbers for graphs and groups. Electron. J. Comb. **11**(# R63), 13 p (2004)
11. Wenzl, H.: On the structure of Brauer's centralizer algebras. Ann. Math. **128**, 173–193 (1988)

Zero-Divisor Graphs of Laurent Polynomials and Laurent Power Series

Anil Khairnar and B.N. Waphare

Abstract In this paper, we examine the preservation of diameter and girth of the zero-divisor graph under extension to Laurent polynomial and Laurent power series rings.

Keywords Zero-divisor graphs · Laurent polynomials · Laurent power series

2010 Mathematics Subject Classication Primary 05C99 · Secondary 13B99

1 Introduction

The concept of the zero-divisor graph of a commutative ring was first introduced by Beck [3], and later redefined in [1] by Anderson and Livingston. We adopt the approach used by D.F. Anderson and Livingston [1], which consider only nonzero zero-divisors as vertices of the graph.

It is an interesting question to consider the preservation of graph theoretic properties under various ring theoretic extensions. Work on polynomial and power series extensions is done by M. Axtell, J. Coykendall, J. Stickles [2]. In this paper, we examine the preservation of the diameter of the zero-divisor graph under extensions to Laurent polynomial and Laurent power series rings. Also, we consider the effects of the same extensions on the girth of the graph.

We use $Z(R)$ to denote the set of zero-divisors of R; we use $Z^*(R)$ to denote the set of nonzero zero-divisors of R. By the zero-divisor graph of R, denoted $\Gamma(R)$, we mean the graph whose vertices are the nonzero zero-divisors of R, and for distinct $r, s \in Z^*(R)$, there is an edge connecting r and s if and only if $rs = 0$. For two

A. Khairnar (✉)
Department of Mathematics, Abasaheb Garware College, Pune 411004, India
e-mail: anil_maths2004@yahoo.com

B.N. Waphare
Center For Advanced Studies in Mathematics, Department of Mathematics, Savitribai Phule Pune University, Pune 411007, India
e-mail: bnwaph@math.unipune.ac.in; waphare@yahoo.com

S.T. Rizvi et al. (eds.), *Algebra and its Applications*, Springer Proceedings in Mathematics & Statistics 174, DOI 10.1007/978-981-10-1651-6_21

distinct vertices a and b in a graph Γ, the distance between a and b, denoted $d(a, b)$, is the length of the shortest path connecting a and b, if such a path exists; otherwise, $d(a, b) = \infty$. The diameter of a graph Γ is $diam(\Gamma) = sup\{d(a, b) \mid a$ and b are distinct vertices of $\Gamma\}$. We will use the notation $diam(\Gamma(R))$ to denote the diameter of the graph $\Gamma(R)$. The girth of a graph Γ, denoted $g(\Gamma)$, is the length of the shortest cycle in Γ, provided Γ contains a cycle; otherwise, $g(\Gamma) = \infty$. A graph is said to be *connected* if there exists a path between any two distinct vertices, and a graph is *complete* if it is connected with diameter one. A graph G is said to be *k-connected* (or k-vertex connected) if there does not exist a set of $k - 1$ vertices of G whose removal disconnects the graph. Therefore a connected graph is 1-connected. A ring is said to be reduced if it does not have any nonzero nilpotent elements.

Throughout this paper, R denotes a commutative ring with identity, $R[x]$ denotes a polynomial ring, $R[x, x^{-1}]$ denotes a Laurent polynomial ring, $R[[x]]$ denotes a power series ring, $R[[x, x^{-1}]]$ denotes a Laurent power series ring, $\mathbb{N}$ is the set of natural numbers and $\mathbb{W}$ is the set of nonnegative integers.

2 Diameter of $\Gamma(R)$, $\Gamma(R[x, x^{-1}])$ and $\Gamma(R[[x, x^{-1}]])$

Note that $\Gamma(\mathbb{Z}_6)$ is connected but not 2-connected and $\Gamma(\mathbb{Z}_3 \times \mathbb{Z}_3)$ is 2-connected but not 3-connected. In this section we prove that $\Gamma(R[x])$ is n-connected for any $n \in \mathbb{N}$.

First, we recall the following result now known as McCoy's Theorem; see [4].

Theorem 2.1 ([4, Theorem 2]) *Let R be a commutative ring. If $g(x)$ is a zero-divisor in $R[x]$, then there exists a nonzero element $c \in R$ such that $g(x)c = 0$.*

Theorem 2.2 *Let R be a commutative ring. Then $\Gamma(R[x])$ is n-connected for any $n \in \mathbb{N}$.*

Proof Let $g_1, g_2, \ldots, g_n \in Z^*(R[x])$ for $n \in \mathbb{N}$ and $f_1, f_2 \in Z^*(R[x]) \backslash \{g_1, g_2, \ldots, g_n\}$. By Theorem 2.1, there exist nonzero $r_1, r_2 \in R$ such that $f_1 r_1 = f_2 r_2 = 0$. Let $m = max\{deg(g_1), deg(g_2), \ldots, deg(g_n)\}$. If $r_1 r_2 = 0$ then $f_1 - r_1 x^{m+1} - r_2 x^{m+1} - f_2$ is a path in $\Gamma(R[x])$ not containing any vertex from $\{g_1, g_2, \ldots, g_n\}$. If $r_1 r_2 \neq 0$ then $f_1 - r_1 r_2 x^{m+1} - f_2$ is a path in $\Gamma(R[x])$ not containing any vertex from $\{g_1, g_2, \ldots, g_n\}$. Hence $\Gamma(R[x]$ is n-connected for any $n \in \mathbb{N}$. □

Recall the following result due to D.F. Anderson and Livingston [1].

Theorem 2.3 ([1, Theorem 2.3]) *Let R be a commutative ring, not necessarily with identity, with $Z^*(R) \neq \emptyset$. Then $\Gamma(R)$ is connected and $diam(\Gamma(R)) \leq 3$.*

Now, we prove the result which shows that; $diam(\Gamma(R[x, x^{-1}])) = diam(\Gamma(R[x]))$.

Theorem 2.4 *Let R be a commutative ring. Then $diam(\Gamma(R[x, x^{-1}])) = diam(\Gamma(R[x]))$ and $diam(\Gamma(R[[x, x^{-1}]])) = diam(\Gamma(R[[x]]))$.*

Proof Let $f_1(x), f_2(x) \in Z^*(R[x, x^{-1}])$. Let $g_1(x), g_2(x) \in Z^*(R[x])$ be such that $f_1(x) = x^{-t_1}g_1(x)$ and $f_2(x) = x^{-t_2}g_2(x)$ for some $t_1, t_2 \in \mathbb{W}$. If $diam(\Gamma(R[x])) = 1$, then $g_1(x)g_2(x) = 0$. Therefore $f_1(x)f_2(x) = 0$. This yields $diam(\Gamma(R[x, x^{-1}])) = 1$.

Suppose $diam(\Gamma(R[x])) = 2$. If $f_1(x)f_2(x) = 0$ then we are through. Suppose $f_1(x)f_2(x) \neq 0$. Then $g_1(x)g_2(x) \neq 0$. Since $diam(\Gamma(R[x])) = 2$, there exists $g_3(x) \in Z^*(R[x])$ such that $g_1(x) - g_2(x) - g_3(x)$ is a path in $\Gamma(R[x])$. So $g_3(x) \in Z^*(R[x, x^{-1}])$ be such that $f_1(x) - g_2(x) - f_2(x)$ is a path in $\Gamma(R[x, x^{-1}])$. Therefore $diam(\Gamma(R[x, x^{-1}])) = 2$.

Similarly we can prove that if $diam(\Gamma(R[x])) = 3$ then $diam(\Gamma(R[x, x^{-1}])) = 3$. Thus by using Theorem 2.3, we get $diam(\Gamma(R[x, x^{-1}])) = diam(\Gamma(R[x]))$. Similarly we can prove $diam(\Gamma(R[[x, x^{-1}]])) = diam(\Gamma(R[[x]]))$. □

Observe that $\Gamma(R)$ is a subgraph of $\Gamma(R[x, x^{-1}])$, which is a subgraph of $\Gamma(R[[x, x^{-1}]])$. Since $Z^*(R) \subseteq Z^*(R[x]) \subseteq Z^*([x, x^{-1}]) \subseteq Z^*([[x, x^{-1}]])$, we have $diam(\Gamma(R)) \leq diam(\Gamma(R[x])) \leq diam(\Gamma(R[x, x^{-1}])) \leq diam(\Gamma(R[[x, x^{-1}]]))$.

Example 2.5 There exists a ring R such that $\Gamma(R)$ is complete but $\Gamma(R[x, x^{-1}])$ is not complete and hence $\Gamma(R[[x, x^{-1}]])$ is not complete. Let $R = \mathbb{Z}_2 \times \mathbb{Z}_2$ then $\Gamma(R)$ is complete. However, $\Gamma(R[x, x^{-1}])$ is not complete. Since for $a = (1, 0) + (1, 0)x$, $b = (1, 0) + (1, 0)x^2 \in Z^*(R[x, x^{-1}])$, $ab \neq 0$. Therefore $\Gamma(R[x, x^{-1}])$ is not complete.

The following theorem follows from the results of Axtell et al. [2] and Theorem 2.4.

Theorem 2.6 *Let R be a commutative ring.*

(1) If $R \neq \mathbb{Z}_2 \times \mathbb{Z}_2$ then $\Gamma(R[[x, x^{-1}]])$ is complete if and only if $\Gamma(R[x, x^{-1}])$ is complete if and only if $\Gamma(R)$ is complete.
(2) If $diam(\Gamma(R)) = 2$ and $Z(R) = P_1 \cup P_2$ is the union of precisely two maximal primes in $Z(R)$, then $diam(\Gamma(R[x, x^{-1}])) = 2$.
(3) If R is a Noetherian ring, $Z(R) = P$ is a prime ideal and $diam(\Gamma(R)) = 2$, then $diam(\Gamma(R[x, x^{-1}])) = 2$.
(4) If R is a Noetherian ring and $diam(\Gamma(R)) = 2$, then $diam(\Gamma(R[[x, x^{-1}]])) = 2$.
(5) If for some $n \in \mathbb{N}$ with $n > 2$, $(Z(R))^n = 0$, then $diam(\Gamma(R[[x, x^{-1}]])) = diam(\Gamma(R[x, x^{-1}])) = diam(\Gamma(R)) = 2$.

Does there exists a ring R with $diam(\Gamma(R)) = 2$ but either $diam(\Gamma(R[x])) = 3$ or $diam(\Gamma(R[[x]])) = 3$? In [2], Axtell et al. have given an example of a commutative ring R with identity, such that $diam(\Gamma(R)) = 2$ and $diam(\Gamma(R[[x]])) = 3$. The following is an example of a commutative ring R without identity, such that $diam(\Gamma(R)) = 2$ and $diam(\Gamma(R[[x]])) = 3$. Our example is simple than example given by Axtell et al. [2]. In [6] Rege and Chhawchharia introduced Armendariz rings. First we recall the definition introduced by Zhang [5].

Definition 2.7 A ring R is said to be Strong Armendariz (also called an Armendariz ring of power series type), if

$$f(x) = \sum a_i x^i, \quad g(x) = \sum b_j x^j \in R[[x]]$$

satisfy $f(x)g(x) = 0$, then $a_i b_j = 0, \ \forall\, i, j$.

Example 2.8 Let $R = \{X \subseteq \mathbb{N} \mid |X| < \infty\}$. Consider a ring $(R, +, .)$ where, $A + B = (A - B) \cup (B - A)$ and $A.B = A \cap B$. Observe that $diam(\Gamma(R)) = 2$.

Let $f(x) = A_0 + A_1 x + A_2 x^2 + \cdots, g(x) = B_0 + B_1 x + B_2 x^2 + \cdots$ with $A_0 = \{1\}, A_n = \{2n\}, B_0 = \{1\}, B_n = \{2n + 1\}$, for all $n = 1, 2, 3 \cdots$. Since $f(x).\{3\} = \emptyset$ and $g(x).\{2\} = \emptyset$, therefore $f(x), g(x) \in Z^*(R[[x]])$. Note that $f(x).g(x) = \{1\} \neq \emptyset$. Assume that there exists $h(x) \in Z^*(R[[x]])$ such that $f(x).h(x) = \phi = g(x).h(x)$. As R is a reduced ring, it is a strong Armendariz ring. This yields $A_i.C_j = \emptyset = B_i.C_j$ for all i and j. Consequently $C_j = \emptyset$ for all j. That is $h(x) = \emptyset$, a contradiction. Thus $diam(\Gamma(R[[x, x]])) \geq 3$. Thus by Theorem 2.3, we get $diam(\Gamma(R[[x]])) = 3$. Inview of of Theorem 2.4, it is clear that $diam(\Gamma(R[[x, x^{-1}]])) = 3$.

3 Girth of $\Gamma(R[x, x^{-1}])$ and $\Gamma(R[[x, x^{-1}]])$

For a non-constant polynomial $f \in R[x, x^{-1}]$ does there exists a cycle containing vertex f and at least one vertex of $\Gamma(R)$? We answer this question in sequel. First, we recall the following theorem due to Zhang [5].

Theorem 3.1 ([5, Corollary 3.3]) *Let R be a commutative ring. If $g(x)$ is a zero-divisor in $R[x, x^{-1}]$, then there exists a nonzero element $c \in R$ such that $g(x)c = 0$.*

Theorem 3.2 *Let R be a commutative ring not necessarily with identity. If $f \in Z^*(R[x, x^{-1}])$ is a non-constant polynomial, then there exists a cycle of length 3 or 4 in $\Gamma(R[x, x^{-1}])$ with f as one vertex and some $a \in Z^*(R)$ as another.*

Proof Let $f \in Z^*(R[x, x^{-1}])$ be a non-constant polynomial. By Theorem 3.1, there exists $a \in Z^*(R)$ such that $af = 0$. Hence there exists a non-constant polynomial $g \in Z^*(R[x, x^{-1}])$ different than f such that $fg = 0$ (we can take $g(x) = ax^{1+deg(f)}$). By Theorem 3.1, there exists $b \in Z^*(R)$ such that $bg = 0$.
Case(1): $ab = 0$ and $bf \neq 0$, then $a - f - g - b - a$ is a required cycle.
Case(2): $ab = 0$ and $bf = 0$, $a - f - b - a$ is a required cycle.
Case(3): $ab \neq 0$ and $bf = 0$, then $b - f - g - b$ is a required cycle.
Case(4): $ab \neq 0$ and $bf \neq 0$. Let $f = c_{-m}x^{-m} + \cdots + c_0 + \cdots + c_n x^n$. There exists some j such that $bc_j \neq 0$. Thus, $a - f - g - bc_j - a$ is a required cycle.
Thus in any case there exists a cycle of length 3 or 4 in $\Gamma(R[x, x^{-1}])$ with f as one vertex and some $a \in Z^*(R)$ as another.

In [2] Axtell et al. have given an example of a ring R such that $g(\Gamma(R)) \neq g(\Gamma(R[x]))$ and they also proved the following result.

Theorem 3.3 ([2, Theorem 7]) *Let R be a commutative ring, not necessarily with identity. Then $g(\Gamma(R)) \geq g(\Gamma(R[x])) = g(\Gamma(R[[x]]))$. In addition, if R is reduced and $\Gamma(R)$ contains a cycle, then $g(\Gamma(R)) = g(\Gamma(R[x])) = g(\Gamma(R[[x]]))$.*

Theorem 3.4 *Let R be a commutative ring, not necessarily with identity. Then $g(\Gamma(R[x])) = g(\Gamma(R[x, x^{-1}])) = g(\Gamma(R[[x]])) = g(\Gamma(R[[x, x^{-1}]]))$.*

Proof Since $Z^*(R) \subseteq Z^*(R[x]) \subseteq Z^*([x, x^{-1}])$, we have $g(\Gamma(R)) \geq g(\Gamma(R[x])) \geq g(\Gamma([x, x^{-1}]))$. Hence it is sufficient to prove, $g(\Gamma(R[x])) \leq g(\Gamma([x, x^{-1}]))$, when $g(\Gamma([x, x^{-1}]))$ is finite, say k. Let $f_1 - f_2 - \cdots - f_k - f_1$ be a k-cycle in $\Gamma(R[x, x^{-1}])$. Let t be the least power of x occurs in $f_1, f_2, \ldots, f_k$. This gives a k-cycle, $f_1x^t - f_2x^t - \cdots - f_kx^t - f_1x^t$ in $\Gamma(R[x])$. Therefore $g(\Gamma(R[x])) \leq k = g(\Gamma(R[x, x^{-1}]))$. Thus $g(\Gamma(R[x])) = g(\Gamma(R[x, x^{-1}]))$.

On the similar lines, we can prove that $g(\Gamma(R[[x]])) = g(\Gamma(R[[x, x^{-1}]]))$. Thus by using Theorem 3.3, we get $g(\Gamma(R[x])) = g(\Gamma(R[x, x^{-1}])) = g(\Gamma(R[[x]])) = g(\Gamma(R[[x, x^{-1}]]))$. □

Theorem 3.5 *Let R be a commutative ring, not necessarily with identity. Then $g(\Gamma(R)) \geq g(\Gamma(R[x, x^{-1}])) = g(\Gamma(R[[x, x^{-1}]]))$. In addition, if R is reduced and $\Gamma(R)$ contains a cycle, then $g(\Gamma(R)) = g(\Gamma(R[x, x^{-1}])) = g(\Gamma(R[[x, x^{-1}]]))$.*

Proof Follows from Theorems 3.3 and 3.4. □

References

1. Anderson, D.F., Livingston, P.S.: The zero-divisor graph of a commutative ring. J. Algebra **217**, 434–447 (1999)
2. Axtell, M., Coykendall, J., Stickles, J.: Zero-divisor graphs of polynomials and power series over commutative rings. Commun. Algebra **33**(6), 2043–2050 (2005)
3. Beck, I.: Coloring of commutative rings. J. Algebra **116**, 208–226 (1988)
4. McCoy, N.H.: Remarks on divisors of zero. Am. Math. Mon. **49**(5), 286–295 (1942)
5. Mian-mian, Z.: Some Properties on Baer, PP and PS Rings. J. Math. Res. Expos. **26**, 685–693 (2006)
6. Rege, M.B., Robson, S.C.: Armendariz rings. Proc. Japan Acad. Ser. Math. Sci. **73**, 14–17 (1997)

Pair of Generalized Derivations and Lie Ideals in Prime Rings

Basudeb Dhara, Asma Ali and Shahoor Khan

Abstract Let R be a prime ring and $F, G : R \to R$ be two generalized derivations of R such that $F^2 + G$ is n-commuting or n-skew-commuting on a nonzero square closed Lie ideal U of R. In the present paper we prove under certain conditions that $U \subseteq Z(R)$.

Keywords Prime rings · Lie ideals · Generalized derivations

2010 Mathematics Subject Classification 16W25 · 16R50 · 16N60

1 Introduction

Throughout the paper R will denote an associative prime ring with center $Z(R)$ and extended centroid C, RC the central closure of R. A ring R is said to be prime if $aRb = \{0\}$ implies either $a = 0$ or $b = 0$. A ring R is said to be n-torsion free if $nx = 0$ for $x \in R$ implies that $x = 0$. We shall write for any pair of elements $x, y \in R$ the commutator $[x, y] = xy - yx$, and skew-commutator $x \circ y = xy + yx$. We will frequently use the basic commutator and skew-commutator identities: (i) $[xy, z] = x[y, z] + [x, z]y$ and $[x, yz] = y[x, z] + [x, y]z$ for all $x, y, z \in R$

First author is supported by a grant from National Board for Higher Mathematics (NBHM), India. Grant No. is NBHM/R.P. 26/ 2012/Fresh/1745 dated 15.11.12.
Second author is supported by a grant from Science and Engineering Research Board (SERB), DST, New Delhi, India. Grant No. SR/S4/MS:852/13.

B. Dhara
Department of Mathematics, Belda College, Belda, Paschim Medinipur 721424, India
e-mail: basu_dhara@yahoo.com

A. Ali (✉) · S. Khan
Department of Mathematics, Aligarh Muslim University, Aligarh 202002, India
e-mail: asma_ali2@rediffmail.com

S. Khan
e-mail: shahoor.khan@rediffmail.com

S.T. Rizvi et al. (eds.), *Algebra and its Applications*, Springer Proceedings in Mathematics & Statistics 174, DOI 10.1007/978-981-10-1651-6_22

and (ii) $x \circ yz = (x \circ y)z - y[x, z] = y(x \circ z) + [x, y]z$ and $xy \circ z = x(y \circ z) - [x, z]y = (x \circ z)y + x[y, z]$, for all $x, y, z \in R$. An additive mapping $d : R \to R$ is said to be a derivation if $d(xy) = d(x)y + xd(y)$ holds for all $x, y \in R$. An additive mapping $F : R \to R$ is said to be a right generalized derivation with associated derivation d on R if $F(xy) = F(x)y + xd(y)$ holds for all $x, y \in R$ and F is said to be a left generalized derivation with associated derivation d on R if $F(xy) = d(x)y + xF(y)$ holds for all $x, y \in R$. F is said to be a generalized derivation with associated derivation d on R if F is both a right and a left generalized derivation with associated derivation d on R. (Note that this definition differs from the one given by Hvala in [5], his generalized derivations are our right generalized derivations.) Every derivation is a generalized derivation. An additive subgroup U of R is said to be a Lie ideal of R, if $[u, r] \in U$ for all $u \in U$ and $r \in R$. The Lie ideal U of R is said to be square closed if $u^2 \in U$ for all $u \in U$. Deng and Bell [2] defined n-centralizing and n-commuting mappings, the concept more general than centralizing and commuting mappings. Let $n \geq 1$ be a fixed integer and S be a nonempty subset of R. A mapping $f : R \to R$ is said to be n-centralizing (resp. n-commuting) on S, if $[f(x), x^n] \in Z(R)$ (resp. $[f(x), x^n] = 0$) for all $x \in S$. Analogously a mapping $f : R \to R$ is said to be n-skew-centralizing (resp. n-skew-commuting) on S, if $f(x)x^n + x^n f(x) \in Z(R)$ (resp. $f(x)x^n + x^n f(x) = 0$) for all $x \in S$.

In [7], Park et al. proved that if d, g are two derivations in a Banach algebra A such that $\alpha d^2 + g$ is n-commuting on A, then both d and g map A into $rad(A)$, the Jacobson radical of A. More recently in [4], Fosner and Vukman proved the following: If R is a 2-torsion free semiprime ring and $f : R \to R$ is an additive mapping which is 2-commuting, then f is commuting. As an application of this result, authors showed that if d, g are two derivations such that $d^2 + g$ is 2-commuting then d and g map R into its center. In the present paper we study the situations when $F^2 + G$ is n-commuting or n-skew-commuting on a nonzero square closed Lie ideal U of R, where F, G are two generalized derivations of R with associated nonzero derivations d and g respectively and then we show under certain conditions that $U \subseteq Z(R)$.

2 Preliminaries

Let U be a Lie ideal of R such that $x^2 \in U$ for all $x \in U$. Then for all $x, y \in U$, we have $xy + yx = (x + y)^2 - x^2 - y^2 \in U$. Again by the definition of Lie ideal, we have $xy - yx \in U$. Combining these two we get $2xy \in U$ for all $x, y \in U$.

We begin with several Lemmas, most of which have been proved elsewhere.

Lemma 2.1 ([1, Lemma 2]) *Let R be a 2-torsion free prime ring. If $U \nsubseteq Z(R)$ is a Lie ideal of R, then $C_R(U) = Z(R)$.*

Lemma 2.2 ([1, Lemma 3]) *Let R be a 2-torsion free prime ring. If U is a Lie ideal of R, then $C_R([U, U]) = C_R(U)$.*

Lemma 2.3 ([1, Lemma 4]) *Let R be a 2-torsion free prime ring. If $U \nsubseteq Z(R)$ is a Lie ideal of R and $aUb = (0)$, then $a = 0$ or $b = 0$.*

Lemma 2.4 ([1, Lemma 5]) *Let R be a 2-torsion free prime ring and U be a nonzero Lie ideal of R. If d is a nonzero derivation of R such that $d(U) = (0)$, then $U \subseteq Z(R)$.*

Lemma 2.5 ([6, Theorem 5]) *Let R be a 2-torsion free prime ring and U a nonzero Lie ideal of R. If d is a nonzero derivation of R such that $[u, d(u)] \in Z(R)$ for all $u \in U$, then $U \subseteq Z(R)$.*

Lemma 2.6 ([2, Lemma 1]) *Let n be a fixed positive integer, R be $n!$-torsion free ring and ϕ be an additive map on R. For $i = 1, 2, \ldots, n$, let $P_i(X, Y)$ be a generalized polynomial homogeneous of degree i in the noncommuting indeterminates X and Y. Suppose that $a \in R$, and (a) is the additive subgroup generated by a. If*

$$P_n(x, \phi(x)) + P_{n-1}(x, \phi(x)) + \cdots + P_1(x, \phi(x)) \in Z(R) \text{ for all } x \in (a),$$

then $P_i(a, \phi(a)) \in Z(R)$ for $i = 1, 2, \ldots, n$.

Lemma 2.7 ([8, Theorem 6]) *Let R be a prime ring and U be a nonzero Lie ideal of R such that $U \nsubseteq Z(R)$. Let d be a nonzero derivation of R and $a \in R$ such that $a[d(x^n), x^n]_k = 0$ for all $x \in U$, where k and n are fixed positive integers. Then $a = 0$ except when $dim_C RC = 4$, C the extended centroid of R.*

Lemma 2.8 ([3, Lemma 2.5]) *Let R be a 2-torsion free prime ring and U be a nonzero Lie ideal of R. If d is a nonzero derivation of R and $V = \{u \in U \mid d(u) \in U\}$, then V is also a nonzero Lie ideal of R. Moreover, if $U \nsubseteq Z(R)$, then $V \nsubseteq Z(R)$.*

Lemma 2.9 *Let R be a 2-torsion free prime ring and U be a nonzero square closed Lie ideal of R. If d is a nonzero derivation of R such that $d(x)x + xd(x) = 0$ for all $x \in U$, then $U \subseteq Z(R)$.*

Proof On contrary, we assume that $U \nsubseteq Z(R)$. By hypothesis

$$d(x)x + xd(x) = 0 \text{ for all } x \in U. \tag{1}$$

Linearization of (1) yields that

$$d(x)y + d(y)x + xd(y) + yd(x) = 0 \text{ for all } x, y \in U. \tag{2}$$

Substituting $2yx$ for y in (2) and using 2-torsion freeness of R, we get

$$d(x)yx + d(y)x^2 + yd(x)x + xd(y)x + xyd(x) + yxd(x) = 0 \text{ for all } x, y \in U. \tag{3}$$

Using (1), we get

$$d(x)yx + d(y)x^2 + xd(y)x + xyd(x) = 0 \text{ for all } x, y \in U. \tag{4}$$

Right multiplying (2) by x and then subtracting from (4), we get

$$yd(x)x - xyd(x) = 0 \text{ for all } x, y \in U. \tag{5}$$

Replacing y by $2zy$ in (5) and then using 2-torsion freeness of R, we find that

$$zyd(x)x - xzyd(x) = 0 \text{ for all } x, y, z \in U. \tag{6}$$

Left multiplying (5) by z and then subtracting from (6), we obtain

$$[z, x]yd(x) = 0 \text{ for all } x, y, z \in U. \tag{7}$$

By Lemma 2.3, we obtain either $[z, x] = 0$ or $d(x) = 0$. Now $\{x \in U \mid [U, x] = 0\}$ and $\{x \in U \mid d(x) = 0\}$ form additive subgroups of U such that their union is U. Since a group can not be union of its two proper subgroups, we conclude that either $[U, U] = (0)$ or $d(U) = (0)$. By Lemma 2.1, $[U, U] = (0)$ gives $U \subseteq Z(R)$, a contradiction. By Lemma 2.4, $d(U) = (0)$, leads $U \subseteq Z(R)$, a contradiction.

3 Main Results

Theorem 3.1 *Let $n \geq 1$ be a fixed integer. Let R be a prime ring, U a nonzero square closed Lie ideal of R and $F, G : R \to R$ two generalized derivations of R with associated nonzero derivations d and g of R respectively. (1) If R is 2-torsion free and $F^2 + G$ is commuting on U, then $U \subseteq Z(R)$. (2) If $n \geq 2$, R is $n!$-torsion free and $F^2 + G$ is n-commuting on U, then $U \subseteq Z(R)$, except when $dim_C RC = 4$.*

Proof Assume that $U \nsubseteq Z(R)$. By hypothesis, we have

$$[\Delta(x), x^n] = 0 \text{ for all } x \in U, \tag{8}$$

where $\Delta = F^2 + G$. Consider an integer k with $1 \leq k \leq n$. Replacing x by $x + ky$ in (8), we obtain

$$kP_1(x, y) + k^2 P_2(x, y) + k^3 P_3(x, y) + \cdots + k^n P_n(x, y) = 0 \text{ for all } x, y \in U, \tag{9}$$

where $P_i(x, y)$ denotes the sum of those terms in which y appears as a term in the product i times. By Lemma 2.6, we have

$$P_1(x, y) = [\Delta(y), x^n] + [\Delta(x), x^{n-1}y] + [\Delta(x), x^{n-2}yx] + \cdots + [\Delta(x), yx^{n-1}] = 0 \tag{10}$$

for all $x, y \in U$. Replacing y by $2xy$ in (10), we get

$$\begin{aligned}2[x\Delta(y)+2d(x)F(y)+H(x)y,x^n]+2x[\Delta(x),x^{n-1}y]+2[\Delta(x),x]x^{n-1}y\\ +2x[\Delta(x),x^{n-2}yx]+2[\Delta(x),x]x^{n-2}yx+\cdots+2x[\Delta(x),yx^{n-1}]\\ +2[\Delta(x),x]yx^{n-1}=0 \text{ for all } x,y\in U,\end{aligned} \tag{11}$$

where $H(x)=(d^2+g)(x)$. Using 2-torsion freeness of R, it gives

$$\begin{aligned}x[\Delta(y),x^n]+[2d(x)F(y),x^n]+H(x)[y,x^n]+[H(x),x^n]y+x[\Delta(x),x^{n-1}y]\\ +[\Delta(x),x]x^{n-1}y+x[\Delta(x),x^{n-2}yx]+[\Delta(x),x]x^{n-2}yx+\cdots+x[\Delta(x),yx^{n-1}]\\ +[\Delta(x),x]yx^{n-1}=0 \text{ for all } x,y\in U.\end{aligned} \tag{12}$$

Left multiplying (10) by x, we get

$$x[\Delta(y),x^n]+x[\Delta(x),x^{n-1}y]+x[\Delta(x),x^{n-2}yx]+\cdots+x[\Delta(x),yx^{n-1}]=0 \tag{13}$$

for all $x, y\in U$.

Subtracting (12) from (13), we obtain

$$\begin{aligned}[2d(x)F(y),x^n]+H(x)[y,x^n]+[H(x),x^n]y+[\Delta(x),x]x^{n-1}y\\ +[\Delta(x),x]x^{n-2}yx+\cdots+[\Delta(x),x]yx^{n-1}=0 \text{ for all } x,y\in U.\end{aligned} \tag{14}$$

Substituting $2yx$ for y in (14), we have

$$\begin{aligned}2[2d(x)F(y),x^n]x+2[2d(x)yd(x),x^n]+2H(x)[y,x^n]x+2[H(x),x^n]yx\\ +2[\Delta(x),x]x^{n-1}yx+2[\Delta(x),x]x^{n-2}yx^2+\cdots+2[\Delta(x),x]yx^{n-1}x\\ =0 \text{ for all } x,y\in U.\end{aligned} \tag{15}$$

Since R is 2-torsion free, it gives

$$\begin{aligned}[2d(x)F(y),x^n]x+[2d(x)yd(x),x^n]+H(x)[y,x^n]x+[H(x),x^n]yx\\ +[\Delta(x),x]x^{n-1}yx+[\Delta(x),x]x^{n-2}yx^2+\cdots+[\Delta(x),x]yx^{n-1}x\\ =0 \text{ for all } x,y\in U.\end{aligned} \tag{16}$$

Right multiplying (14) by x, we get

$$\begin{aligned}[2d(x)F(y),x^n]x+H(x)[y,x^n]x+[H(x),x^n]yx+[\Delta(x),x]x^{n-1}yx\\ +[\Delta(x),x]x^{n-2}yx^2+\cdots+[\Delta(x),x]yx^{n-1}x=0 \text{ for all } x,y\in U.\end{aligned} \tag{17}$$

Subtracting (16) from (17), we obtain

$$2[d(x)yd(x),x^n]=0 \text{ for all } x,y\in U. \tag{18}$$

Since R is 2-torsion free, it gives

$$d(x)yd(x)x^n - x^n d(x)yd(x) = 0 \text{ for all } x, y \in U. \tag{19}$$

Set $V = \{x \in U \mid d(x) \in U\}$. Then by Lemma 2.8, V is a noncentral nonzero Lie ideal of R. Since $V \subseteq U$, it follows from (19) that

$$d(x)yd(x)x^n - x^n d(x)yd(x) = 0 \text{ for all } y \in U, \text{ for all } x \in V. \tag{20}$$

Replacing y by $2yz$ in (20) and using 2-torsion freeness of R, we get

$$d(x)yzd(x)x^n - x^n d(x)yzd(x) = 0 \text{ for all } y, z \in U, \text{ for all } x \in V. \tag{21}$$

For $x \in V$, replace z by $2d(x)z$ in (21), and then get

$$2d(x)yd(x)zd(x)x^n - 2x^n d(x)yd(x)zd(x) = 0.$$

Since R is 2-torsion free, using (20), we get

$$d(x)yx^n d(x)zd(x) - d(x)yd(x)x^n zd(x) = 0 \text{ for all } y, z \in U, \text{ for all } x \in V, \tag{22}$$

which gives

$$d(x)y[d(x), x^n]zd(x) = 0 \text{ for all } y, z \in U, \text{ for all } x \in V. \tag{23}$$

By Lemma 2.3, for each $x \in V$, either $d(x) = 0$ or $[d(x), x^n] = 0$. Both of these two conditions together implies that

$$[d(x), x^n] = 0 \text{ for all } x \in V. \tag{24}$$

Taking $n = 1$ in (24) we have $[d(x), x] = 0$ for all $x \in V$. In this case, if R is 2-torsion free, Lemma 2.5 yields that $V \subseteq Z(R)$, a contradiction.

Let $n \geq 2$. In this case $[d(x), x^n] = 0$ for all $x \in V$, implies that $[d(x^n), x^n] = 0$ for all $x \in V$. Then by Lemma 2.7, since V is noncentral Lie ideal of R, we have that $dim_C RC = 4$. Thus the theorem is proved.

Theorem 3.2 *Let $n \geq 1$ be a fixed integer. Let R be a prime ring, U a nonzero square closed Lie ideal of R and $F, G : R \to R$ two generalized derivations of R with associated nonzero derivations d and g of R respectively. (1) If R is 2-torsion free and $F^2 + G$ is skew-commuting on U, then $U \subseteq Z(R)$. (2) If $n \geq 2$, R is $n!$-torsion free and $F^2 + G$ is n-skew-commuting on U, then $U \subseteq Z(R)$, except when $dim_C RC = 4$.*

Proof Assume that $U \nsubseteq Z(R)$. By hypothesis, we have

$$\Delta(x) \circ x^n = 0 \text{ for all } x \in U, \tag{25}$$

where $\Delta = F^2 + G$. Consider an integer k with $1 \leq k \leq n$. Replacing x by $x + ky$ in (25), we obtain

$$kP_1(x, y) + k^2 P_2(x, y) + k^3 P_3(x, y) + \cdots + k^n P_n(x, y) = 0 \text{ for all } x, y \in U, \tag{26}$$

where $P_i(x, y)$ denotes the sum of those terms in which y appears as a term in the product i times. By Lemma 2.6, we have

$$\begin{aligned} P_1(x, y) = (\Delta(y) \circ x^n) + (\Delta(x) \circ x^{n-1}y) + (\Delta(x) \circ x^{n-2}yx) + \cdots \\ +(\Delta(x) \circ yx^{n-1}) = 0 \text{ for all } x, y \in U. \end{aligned} \tag{27}$$

Replacing y by $2xy$ in (27), we get

$$\begin{aligned} 2((x\Delta(y) + 2d(x)F(y) + H(x)y) \circ x^n) + 2x(\Delta(x) \circ x^{n-1}y) + 2[\Delta(x), x]x^{n-1}y \\ +2x(\Delta(x) \circ x^{n-2}yx) + 2[\Delta(x), x]x^{n-2}yx + \cdots + 2x(\Delta(x) \circ yx^{n-1}) \\ +2[\Delta(x), x]yx^{n-1} = 0 \text{ for all } x, y \in U, \end{aligned} \tag{28}$$

where $H(x) = (d^2 + g)(x)$. Since R is 2-torsion free, it gives

$$\begin{aligned} x(\Delta(y) \circ x^n) + (2d(x)F(y) \circ x^n) + (H(x) \circ x^n)y + H(x)[y, x^n] \\ +x(\Delta(x) \circ x^{n-1}y) + [\Delta(x), x]x^{n-1}y + x(\Delta(x) \circ x^{n-2}yx) + [\Delta(x), x]x^{n-2}yx \\ + \cdots + x(\Delta(x) \circ yx^{n-1}) + [\Delta(x), x]yx^{n-1} = 0 \text{ for all } x, y \in U. \end{aligned} \tag{29}$$

Left multiplying (27) by x, we get

$$\begin{aligned} x(\Delta(y) \circ x^n) + x(\Delta(x) \circ x^{n-1}y) + x(\Delta(x) \circ x^{n-2}yx) \\ + \cdots + x(\Delta(x) \circ yx^{n-1}) = 0 \text{ for all } x, y \in U. \end{aligned} \tag{30}$$

Subtracting (30) from (29), we obtain

$$\begin{aligned} (2d(x)F(y) \circ x^n) + (H(x) \circ x^n)y + H(x)[y, x^n] + [\Delta(x), x]x^{n-1}y \\ +[\Delta(x), x]x^{n-2}yx + \cdots + [\Delta(x), x]yx^{n-1} = 0 \text{ for all } x, y \in U. \end{aligned} \tag{31}$$

Substituting $2yx$ in place of y in (31), we find that

$$\begin{aligned} 2((2d(x)F(y))x \circ x^n) + 2(2d(x)yd(x) \circ x^n) + 2(H(x) \circ x^n)yx \\ +2H(x)[y, x^n]x + 2[\Delta(x), x]x^{n-1}yx + 2[\Delta(x), x]x^{n-2}yx^2 \\ + \cdots + 2[\Delta(x), x]yx^{n-1}x = 0 \text{ for all } x, y \in U. \end{aligned} \tag{32}$$

Using 2-torsion freeness of R, we have

$$\begin{aligned} (2d(x)F(y)) \circ x^n)x + (2d(x)yd(x) \circ x^n) + (H(x) \circ x^n)yx + H(x)[y, x^n]x \\ +[\Delta(x), x]x^{n-1}yx + [\Delta(x), x]x^{n-2}yx^2 + \cdots + [\Delta(x), x]yx^{n-1}x \\ = 0 \text{ for all } x, y \in U. \end{aligned} \tag{33}$$

Right multiplying (31) by x, we obtain

$$(2d(x)F(y) \circ x^n)x + (H(x) \circ x^n)yx + H(x)[y, x^n]x + [\Delta(x), x]x^{n-1}yx \\ +[\Delta(x), x]x^{n-2}yx^2 + \cdots + [\Delta(x), x]yx^{n-1}x = 0 \text{ for all } x, y \in U. \quad (34)$$

Subtracting (34) from (33), we get

$$(2d(x)yd(x) \circ x^n) = 0 \text{ for all } x, y \in U. \quad (35)$$

Since R is 2-torsion free, it gives

$$d(x)yd(x)x^n + x^n d(x)yd(x) = 0 \text{ for all } x, y \in U. \quad (36)$$

Set $V = \{x \in U \mid d(x) \in U\}$. Then by Lemma 2.8, V is a noncentral nonzero Lie ideal of R. Since $V \subseteq U$, it follows from (36) that

$$d(x)yd(x)x^n + x^n d(x)yd(x) = 0 \text{ for all } y \in U, \text{ for all } x \in V. \quad (37)$$

Replacing y by $2yz$ in (37) and using 2-torsion freeness of R, we get

$$d(x)yzd(x)x^n + x^n d(x)yzd(x) = 0 \text{ for all } y, z \in U, \text{ for all } x \in V. \quad (38)$$

Replacing z by $2d(x)z$ in (38) and using 2-torsion freeness of R, we have

$$d(x)yd(x)zd(x)x^n + x^n d(x)yd(x)zd(x) = 0 \text{ for all } y, z \in U, \text{ for all } x \in V. \quad (39)$$

Using (37), we get

$$-d(x)yx^n d(x)zd(x) - d(x)yd(x)x^n zd(x) = 0 \quad (40)$$

which is

$$d(x)y(d(x) \circ x^n)zd(x) = 0 \text{ for all } y, z \in U, \text{ for all } x \in V. \quad (41)$$

By Lemma 2.3, we have for each $x \in V$, either $d(x) = 0$ or $(d(x) \circ x^n) = 0$. Since $d(x) = 0$ yields $(d(x) \circ x^n) = 0$, we conclude that in any case

$$(d(x) \circ x^n) = 0 \text{ for all } x \in V. \quad (42)$$

If $n = 1$ in (42), and R is 2-torsion free, then we have $(d(x) \circ x) = 0$ for all $x \in V$ which implies $V \subseteq Z(R)$ by Lemma 2.9, a contradiction.

If $n \geq 2$, then $(d(x) \circ x^n) = 0$ yields $[d(x), x^{2n}] = 0$ for all $x \in V$ and hence $[d(x^{2n}), x^{2n}] = 0$ for all $x \in V$. Then by Lemma 2.7, since V is noncentral Lie ideal of R, we conclude that $dim_C RC = 4$, as desired.

Theorem 3.3 *Let n be a fixed positive integer. Let R be a $(n+1)!$-torsion free prime ring, U a nonzero square closed Lie ideal of R and $F, G : R \to R$ two generalized derivations of R with associated nonzero derivations d and g of R respectively. If $F^2 + G$ is n-centralizing on U, then $U \subseteq Z(R)$, except when $dim_C RC = 4$.*

Proof Let $x \in U$ and take $t = [\Delta(x), x^n]$, where $\Delta = F^2 + G$. Then $t \in Z(R)$. By our hypothesis, we have

$$[\Delta(x), x^n] \in Z(R) \text{ for all } x \in U. \tag{43}$$

Consider an integer k with $1 \le k \le n$. Replacing x by $x + ky$ in (43), we obtain

$$kP_1(x, y) + k^2 P_2(x, y) + k^3 P_3(x, y) + \cdots + k^n P_n(x, y) \in Z(R) \text{ for all } x, y \in U, \tag{44}$$

where $P_i(x, y)$ denotes the sum of those terms in which y appears as a term in the product i times. By Lemma 2.6, we have

$$P_1(x, y) = [\Delta(y), x^n] + [\Delta(x), x^{n-1}y + x^{n-2}yx + \cdots + yx^{n-1}] \in Z(R) \tag{45}$$

for all $x, y \in U$. Since $x, x^2 \in U$, we have $2x.x^2 = 2x^3 \in U, 2(2x^3).x = 2^2x^4 \in U$ and so $2^{n-1}x^{n+1} \in U$. Substituting $2^{n-1}x^{n+1}$ for y in (45), we obtain

$$2^{n-1}[\Delta(x^{n+1}), x^n] + 2^{n-1}[\Delta(x), x^{2n} + x^{2n} + \cdots + x^{2n}] \in Z(R). \tag{46}$$

Since R is 2-torsion free, it can be written as

$$[\Delta(x)x^n + 2F(x)d(x^n) + xH(x^n), x^n] + n[\Delta(x), x^{2n}] \in Z(R), \tag{47}$$

where $H = d^2 + g$. Since $[\Delta(x)x^n, x^n] = [\Delta(x), x^n]x^n = tx^n$ and $[\Delta(x), x^{2n}] = x^n[\Delta(x), x^n] + [\Delta(x), x^n]x^n = 2tx^n$, we have from above relation that

$$(2n+1)tx^n + [2F(x)d(x^n), x^n] + x[H(x^n), x^n] \in Z(R)$$

for all $x \in U$. Now, we suppose

$$z = (2n+1)tx^n + [2F(x)d(x^n), x^n] + x[H(x^n), x^n] \in Z(R). \tag{48}$$

This implies that

$$\sum_{i=0}^{n} x^{ni}zx^{n(n-i)} = \sum_{i=0}^{n} x^{ni}((2n+1)tx^n)x^{n(n-i)} + \sum_{i=0}^{n} x^{ni}[2F(x)d(x^n), x^n]x^{n(n-i)} + \sum_{i=0}^{n} x^{ni}(x[H(x^n), x^n])x^{n(n-i)}. \tag{49}$$

Since $z, t \in Z(R)$, it reduces to

$$(n+1)zx^{n^2} = (n+1)(2n+1)tx^{n(n+1)} + [2F(x)d(x^n), x^{n(n+1)}] \\ +x[H(x^n), x^{n(n+1)}]. \tag{50}$$

Since $2^{n-1}x^{n+1} \in U$, replacing x by $2^{n-1}x^{n+1}$ in our assumption, we get 2^{n^2-1} $[\Delta(x^{n+1}), x^{n(n+1)}] \in Z(R)$. Since R is 2-torsion free, this implies

$$\begin{aligned}[\Delta(x^{n+1}), x^{n(n+1)}] &= [\Delta(x)x^n + 2F(x)d(x^n) + xH(x^n), x^{n^2+n}] \\ &= [\Delta(x), x^{n^2+n}]x^n + [2F(x)d(x^n), x^{n^2+n}] \\ &\quad +x[H(x^n), x^{n^2+n}] \in Z(R). \end{aligned} \tag{51}$$

We notice that

$$\begin{aligned}[\Delta(x), x^{n^2+n}] &= \sum_{i=0}^{n} x^{ni}[\Delta(x), x^n]x^{n(n-i)} \\ &= \sum_{i=0}^{n} x^{ni}tx^{n(n-i)} \\ &= (n+1)x^{n^2}t. \end{aligned} \tag{52}$$

Applying (50)–(52) yields that

$$2n(n+1)x^{n^2+n}t - (n+1)zx^{n^2} \in Z(R). \tag{53}$$

Since R is $(n+1)$-torsion free, we have

$$2nx^{n^2+n}t - zx^{n^2} \in Z(R). \tag{54}$$

Now commuting x^{kn} with $\Delta(x)$ successively, we get

$$[\Delta(x), x^{kn}] = \sum_{i=0}^{k-1} x^{ni}[\Delta(x), x^n]x^{n(k-1-i)} = \sum_{i=0}^{k-1} x^{ni}tx^{n(k-1-i)} = ktx^{(k-1)n}$$

and

$$[\Delta(x), [\Delta(x), x^{kn}]] = kt[\Delta(x), x^{(k-1)n}] = k(k-1)t^2x^{(k-2)n} = \frac{k!}{(k-2)!}t^2x^{(k-2)n}.$$

Thus commuting x^{kn} with $\Delta(x)$ successively m-times, we find that

$$[\Delta(x), \ldots, [\Delta(x), x^{kn}]] = \frac{k!}{(k-m)!}t^m x^{(k-m)n}.$$

Using this fact and commuting both sides of (54) successively n-times with $\Delta(x)$ we have

$$2n\frac{(n+1)!}{1!}t^{n+1}x^n - n!zt^n = 0. \tag{55}$$

Again commuting with $\Delta(x)$, we obtain

$$2n(n+1)!t^{n+2} = 0. \tag{56}$$

Since R is $(n+1)!$-torsion free, this expression yields that $t^{n+2} = 0$. Since center of a semiprime ring contains no nonzero nilpotent elements, we have $t = 0$ that is, $[\Delta(x), x^n] = 0$ for all $x \in U$ and Theorem 3.1 completes the proof.

Theorem 3.4 *Let n be a fixed positive integer. Let R be a $(2n)!$-torsion free prime ring, U a nonzero square closed Lie ideal of R and $F, G : R \to R$ two generalized derivations of R with associated nonzero derivations d and g of R respectively. If $F^2 + G$ is n-skew-centralizing on U, then $U \subseteq Z(R)$, except when $dim_C RC = 4$.*

Proof Let $\Delta = F^2 + G$. By hypothesis, we have

$$\Delta(x) \circ x^n \in Z(R) \tag{57}$$

for all $x \in U$. This implies

$$\begin{aligned} 0 &= [\Delta(x) \circ x^n, x^n] \\ &= [\Delta(x), x^{2n}] \end{aligned} \tag{58}$$

for all $x \in U$. Since R is $(2n)!$-torsion free, we obtain our conclusion by Theorem 3.1.

The following example illustrates that the above Theorems do not hold for arbitrary rings and torsion condition in the hypothesis is not superfluous.

Example Let $R = \left\{ \begin{pmatrix} a & b \\ 0 & c \end{pmatrix} \mid a, b, c \in \mathbb{Z}_2 \right\}$ and $U = \left\{ \begin{pmatrix} a & b \\ 0 & a \end{pmatrix} \mid a, b \in \mathbb{Z}_2 \right\}$. Define $F : R \longrightarrow R$ by $F\begin{pmatrix} a & b \\ 0 & c \end{pmatrix} = \begin{pmatrix} a & 0 \\ 0 & 0 \end{pmatrix}$. Then F is a generalized derivation with associated nonzero derivation d given by $d\begin{pmatrix} a & b \\ 0 & c \end{pmatrix} = \begin{pmatrix} 0 & b \\ 0 & 0 \end{pmatrix}$ satisfying the hypothesis of Theorems 3.1–3.4 for $F = G$. But U is not central.

References

1. Bergen, J., Herstein, I.N., Kerr, J.W.: Lie ideals and derivations of prime rings. J. Algebra **71**, 259–267 (1981)
2. Deng, Q., Bell, H.E.: On derivations and commutativity in semiprime rings. Commun. Algebra **23**(10), 3705–3713 (1995)
3. Dhara, B., Kar, S., Mondal, S.: A result on generalized derivations on Lie ideals in prime rings. Beitr. Algebra Geom. doi:10.1007/s13366-012-0128-0 (Published Online)
4. Fosner, A., Vukman, J.: Some results concerning additive mappings and derivations on semiprime rings. Publ. Math. Debr. **78**, 3–4 (2011)
5. Hvala, B.: Generalized derivations in rings. Commun. Algebra **26**, 1147–1166 (1998)
6. Lee, P.H., Lee, T.K.: Lie ideals of prime rings with derivations. Bull. Inst. Math. Acad. Sin. **11**, 75–79 (1983)
7. Park, K.H., Jung, Y.S., Bae, J.H.: Derivations in Banach algebras. Int. J. Math. Math. Sci. **29**(10), 579–583 (2002)
8. Shiue, W.K.: Annihilators of derivations with Engel conditions on one-sided ideals. Publ. Math. Debr. **62**(1–2), 237–243 (2003)

On Domination in Graphs from Commutative Rings: A Survey

T. Tamizh Chelvam, T. Asir and K. Selvakumar

Abstract Zero-divisor graphs and total graphs are most popular graph constructions from commutative rings. Through these constructions, the interplay between algebraic structures and graphs are studied. Indeed, it is worthwhile to relate algebraic properties of commutative rings to the combinatorial properties of assigned graphs. The concept of dominating sets and domination parameters is very important in graph theory due to varied applications. Several authors extensively studied about domination parameters for zero-divisor graphs and total graphs from commutative rings. In this survey article, we present results obtained with regard to domination for zero-divisor graphs and total graphs from commutative rings.

Keywords Zero-divisor graph · Total graph · Domination number · Dominating sets

Mathematics Subject Classification Primary: 05C75 · 05C25 · Secondary: 13A15 · 13M05

1 Introduction

The study of algebraic structures using the properties of graphs has become an exciting research topic in the last twenty years. The benefit of studying these graphs is that one may find some results about the algebraic structures and vice versa. There

T. Tamizh Chelvam (✉) · K. Selvakumar
Department of Mathematics, Manonmaniam Sundaranar University,
Tirunelveli 627012, Tamil Nadu, India
e-mail: tamche59@gmail.com

K. Selvakumar
e-mail: selva_158@yahoo.co.in

T. Asir
Department of Mathematics (DDE),
Madurai Kamaraj University, Madurai 625021, Tamil Nadu, India
e-mail: asirjacob75@gmail.com

S.T. Rizvi et al. (eds.), *Algebra and its Applications*, Springer Proceedings in Mathematics & Statistics 174, DOI 10.1007/978-981-10-1651-6_23

are three major problems in this area such as characterization of the resulting graphs, characterization of the algebraic structures with isomorphic graphs, and realization of the connections between the structures and the corresponding graphs. There are so many ways to construct graphs from commutative rings. Some of them to mention are zero-divisor graph of a commutative ring [3], Cayley graph of a commutative ring, total graph of a commutative ring [5], unit graph of a commutative ring, intersection graph of ideals of rings and comaximal graph of a commutative ring. The domination properties in zero-divisor graphs are studied in [1, 2, 5, 12, 13, 16, 17, 20, 22, 27], whereas the domination in total graphs from commutative rings are studied in [3, 4, 6, 7, 21, 23–26, 28]. The goal of this survey article is to enclose many of the main results on the domination in zero-divisor graph and total graph of commutative rings.

Throughout this paper, R denotes a commutative ring with nonzero identity 1. Then $Z(R)$ denotes its set of zero-divisors, $Nil(R)$ denotes its ideal of nilpotent elements, $Reg(R)$ denotes its set of nonzero-divisors (i.e., $Reg(R) = R \setminus Z(R)$), and $U(R)$ denotes its group of units. For $A \subseteq R$, let $A^* = A \setminus \{0\}$. We say that R is reduced if $Nil(R) = \{0\}$. Note that an element $x \in R$ is said to be *nilpotent* if $x^m = 0$ for some $m \in \mathbb{Z}^+$. An element $x \in R$ is said to be a *zero-divisor* if there exists $0 \neq y \in R$ such that $xy = 0$ where 0 is the additive identity. According to Kaplansky [15], $Z(R) = \cup P_i$, where each P_i is a prime ideal of R. If $a \in R$, then $\{x \in R : ax = 0\}$ is called the *annihilator of a* and it is denoted by $ann(a)$. It is well known that $Z(R) = \bigcup_{0 \neq x \in R} ann(x)$. A commutative ring R is said to be *Noetherian* if every ascending sequence of ideals in R is finite. General references for ring theory are [14, 15].

The definition along with name for zero-divisor graph was first introduced by Anderson and Livingston [3] in 1999, after modifying the definition of Beck [9]. In their attempt, Anderson and Livingston investigated certain basic features of $\Gamma(R)$. It may be noted that in the original definition, Beck took all elements of the ring as vertices of the graph $\Gamma(R)$. The modified definition of the zero-divisor graph is given below:

Definition 1.1 ([3]) Let R be a commutative ring. The *zero-divisor graph* of R, denoted by $\Gamma(R)$, is the undirected graph with vertex set $Z^*(R)$ and two distinct vertices x and y are adjacent if $xy = 0$.

In last twenty years, there are many research articles that have been published on zero graphs of commutative rings. Moreover, zero-divisor graphs were defined and studied for noncommutative rings, near rings, semigroups, modules, lattices, and posets. In variation to the concept of zero-divisor graphs, Anderson and Badawi [5] introduced and studied the total graph of a commutative ring. The definition of the total graph of R is given below.

Definition 1.2 ([5]) The *total graph* of a commutative ring R, denoted by $T_\Gamma(R)$, is the undirected graph with all elements of R as vertices and for distinct $x, y \in R$, the vertices x and y are adjacent if $x + y \in Z(R)$. The three (induced) subgraphs $Nil_\Gamma(R)$, $Z_\Gamma(R)$, and $Reg_\Gamma(R)$ of $T_\Gamma(R)$ are the induced subgraphs with vertex sets $Nil(R)$, $Z(R)$ and $Reg(R)$, respectively.

For a prime p and an integer $m \geq 2$, let $R = \mathbb{Z}_{p^m}$ and $\lambda = |Z(R)| = p^{m-1}$. Then

$$T_\Gamma(R) = \begin{cases} 2K_\lambda & \text{if } p = 2; \\ K_\lambda \cup \underbrace{K_{\lambda,\lambda} \cup K_{\lambda,\lambda} \cup \ldots \cup K_{\lambda,\lambda}}_{\frac{p-1}{2} \text{ copies}} & \text{otherwise.} \end{cases}$$

The study of the total graph of a commutative ring and related graph problems is one of the interesting concepts in both algebra and graph theory. In recent years, many research articles have been published on total graphs from rings, for which one can refer the survey by Badawi [8].

For the sake of completeness, we state some definitions and notations used throughout to keep this article as self contained as possible. Let $G = (V, E)$ be a simple graph. The *open neighborhood of a vertex* v in G is the set of vertices of G which are adjacent with v and it is denoted by $N(v)$. The *closed neighborhood* of v is defined by $N[v] = N(v) \cup \{v\}$. For a subset $S \subseteq V$, the open neighborhood of S is defined by $N(S) = \underset{v \in S}{\cup} N(v)$ and the closed neighborhood $N[S] = N(S) \cup S$. For basic definitions in graph theory, one may refer [10]. The *complement* $\overline{G}$ of G is the graph whose vertex set is $V(G)$ and such that for a pair u, v of vertices of G, uv is an edge of $\overline{G}$ if and only if uv is not an edge of G. The *Cartesian product* of two graphs G and H, denoted by $G \Box H$, is a graph with vertex set $V(G \Box H) = V(G) \times V(H)$ and edge set $E(G \Box H) = \{((u_1, v_1), (u_2, v_2)) : (u_1, u_2) \in E(G)$ with $v_1 = v_2$ (or) $(v_1, v_2) \in E(H)$ with $u_1 = u_2\}$.

A nonempty subset S of V is called a *dominating set* if every vertex in $V - S$ is adjacent to at least one vertex in S. A subset S of V is called a *total dominating set* if every vertex in V is adjacent to some vertex in S. A dominating set S is called a *connected* (or *clique*) *dominating set* if the subgraph induced by S is connected (or complete). A dominating set S is called an *independent dominating set* if no two vertices of S are adjacent. A dominating set S is called a *perfect dominating set* if every vertex in $V - S$ is adjacent to exactly one vertex in S. A dominating set S is called an *efficient dominating set* if S is both independent and perfect. A dominating set S is called a *strong* (or *weak*) *dominating set* if for every vertex $u \in V - S$, there is a vertex $v \in S$ with $deg(v) \geq deg(u)$ (or $deg(v) \leq deg(u)$) and u is adjacent to v.

The *domination number* γ of G is defined to be the minimum cardinality of a dominating set in G and the corresponding dominating set is called as a γ-set of G. In a similar way, we define the *total dominating number* γ_t, *connected dominating number* γ_c, *clique dominating number* γ_{cl}, *independent dominating number* γ_i, *perfect dominating number* γ_p, *efficient dominating number* γ_{eff}, *strong dominating number* γ_s, and the *weak dominating number* γ_w.

A graph G is called *excellent* if for every vertex $v \in V(G)$, there is a γ-set S containing v. A *domatic partition* of G is a partition of $V(G)$ into dominating sets of G. The maximum number of sets in a domatic partition of G is called the *domatic number* of G and the same is denoted by $d(G)$. In a similar way, we define the *perfect domatic number* $d_p(G)$, *independent domatic number* $d_i(G)$ and the *total domatic number* $d_t(G)$. A graph G is called *domatically full* if $d(G) = \delta(G) + 1$, which is maximum possible order of a domatic partition of V. The *bondage number* $b(G)$ is the minimum number of edges whose removal increases the domination number. A set of vertices $S \subseteq V$ is said to be *independent* if no two vertices in S are adjacent in G. The *independent number* $\beta_0(G)$, is the maximum cardinality of an independent set in G. A graph G is called *well-covered* if $\beta_0(G) = i(G)$. The *disjoint domination number* $\gamma\gamma(G)$ defined by $\gamma\gamma(G) = \min\{|S_1| + |S_2| : S_1, S_2$ are disjoint dominating sets of $G\}$. Similarly, we can define *disjoint independent domination number* $ii(G)$ and $\gamma i(G)$. For results concerning domination parameters, one can refer to Haynes et al. [11].

2 Domination in Zero-Divisor Graphs of Commutative Rings

In this section, let us review results concerning domination parameters in the zero-divisor graph of a commutative ring. The study of dominating set in zero-divisor graph of a commutative ring has been initiated by Redmond [20]. The following is the main result in this connection:

Theorem 2.1 ([20, Theorem 5.1]) *Let R be a commutative Artinian ring with identity that is not a domain. If the radius of $\Gamma(R)$ is at most 1, then the domination number of $\Gamma(R)$ is 1. If the radius is 2, then the domination number is equal to the number of factors in the Artinian decomposition of R. (In particular, the domination number is finite and at least two.)*

One immediate consequence of the above theorem is that if the radius of $\Gamma(R)$ is 2, then the domination number is equal to the number of distinct maximal ideals of R. However, this need not be true if the radius of $\Gamma(R)$ is 1. For any field F, $\Gamma(\mathbb{Z}_2 \times F)$ is a star graph, which has radius 1, but $\mathbb{Z}_2 \times F$ has two distinct maximal ideals. The following corollary shows that a more precise relationship between the domination number and the number of maximal ideals occurs in the finite case.

Corollary 2.2 ([20, Corollary 5.2]) *Let R be a finite commutative ring with identity that is not a domain. If $\Gamma(R)$ is not a star graph, then R has $\gamma(\Gamma(R))$ distinct maximal ideals. If $\Gamma(R)$ is a star graph, then either R has 2 distinct maximal ideals or R is isomorphic to one of the five following local rings: $\mathbb{Z}_9$, $\frac{\mathbb{Z}_3[x]}{(x^2)}$, $\mathbb{Z}_8$, $\frac{\mathbb{Z}_2[x]}{(x^3)}$, or $\frac{\mathbb{Z}_4[x]}{(2x,x^2-2)}$. (In other words, if $\Gamma(R)$ is a star graph, then R has $\gamma(\Gamma(R))$ distinct maximal ideals if R is local and $\gamma(\Gamma(R)) + 1$ distinct maximal ideals if R is reduced.)*

Corollary 2.3 ([20, Corollary 5.3]) *Let R be a finite commutative ring with identity that is not a domain. If $R \not\cong \mathbb{Z}_2 \times F$ for any finite field F, then the domination number of $\Gamma(R)$ equals the number of distinct maximal ideals of R. If $R \cong \mathbb{Z}_2 \times F$ for some finite field F, then the domination number is one less than the number of maximal ideals of R.*

It is of interest to note that these dominating sets were connected, showing that the connected domination number of $\Gamma(R)$ equals the domination number of $\Gamma(R)$. The dominating set in zero-divisor graph has implicitly been studied by Jafari Rad, Jafari and Mojdeh in [13]. They first determined the domination number for the zero-divisor graph of the product of two commutative rings with identity.

Proposition 2.4 ([13, Proposition 2.2]) *If R is an integral domain, then $\gamma(\Gamma(\mathbb{Z}_2 \times R)) = 1$.*

A *semi-total dominating set* in $\Gamma(R)$ is a subset $S \subseteq Z(R)$ such that S is a dominating set for $\Gamma(R)$ and for any $x \in S$ there is a vertex $y \in S$ (not necessarily distinct) such that $xy = 0$. The *semi-total domination number* $\gamma_{st}(\Gamma(R))$ of $\Gamma(R)$ is the minimum cardinality of a semi-total dominating set in $\Gamma(R)$. (Note that for all rings R, $\gamma(\Gamma(R)) \leq \gamma_{st}(\Gamma(R)) \leq 2\gamma(\Gamma(R))$). For a commutative ring R with 1, let

$$a(R) = \begin{cases} 1 & \text{if } Z(R) = 0; \\ \gamma_{st}(\Gamma(R)) & \text{otherwise.} \end{cases}$$

Theorem 2.5 ([13, Theorem 2.6]) *If R_1, R_2 are commutative rings with 1 and $\mathbb{Z}_2 \notin \{R_1, R_2\}$, then $\gamma(\Gamma(R_1 \times R_2)) = a(R_1) + a(R_2)$.*

Corollary 2.6 ([13, Corollaries 2.7, 2.8]) *If R_1, R_2 are commutative rings with 1 and $\mathbb{Z}_2 \notin \{R_1, R_2\}$, then $\gamma(\Gamma(R_1 \times R_2)) = \gamma_t(\Gamma(R_1 \times R_2)) = \gamma_c(\Gamma(R_1 \times R_2)) = \gamma_{st}(\Gamma(R_1 \times R_2))$.*

Corollary 2.7 ([13, Corollary 2.11]) *Let $R_1, \ldots, R_n$ be local commutative Artinian rings with identity. If $R = R_1 \times \ldots \times R_n$, where $R \ncong F$ or $\mathbb{Z}_2 \times F$ for a field F, then $\gamma(\Gamma(R)) = n$.*

The study on properties of dominating sets of zero-divisor graphs was continued by Mojdeh and Rahimi [17].

Theorem 2.8 ([17, Proposition 8]) *Suppose for a fixed integer $n \geq 2$, that $R = R_1 \times \ldots \times R_n$, where R_i is an integral domain for each $i = 1, \ldots, n$. Then*

(a) $\gamma(\Gamma(R)) = n$ *if* $n \geq 3$;
(b) $\gamma(\Gamma(R)) = 2$ *if* $n = 2$ *and* $min\{|R_1|, |R_2|\} \geq 3$;
(c) $\gamma(\Gamma(R)) = 1$ *if* $n = 2$ *and* $min\{|R_1|, |R_2|\} = 2$.

Corollary 2.9 ([17, Corollary 9]) *For any given positive integer k, there exists a commutative ring R whose zero-divisor graph has domination number is equal to k.*

Theorem 2.10 ([17, Theorem 11]) *Let R be a commutative Artinian ring (in particular, R could be a finite commutative ring). Suppose that $R = R_1 \times \ldots \times R_n$, where R_i is a local ring for each $i = 1, \ldots, n$. Then:*

(a) $\gamma(\Gamma(R)) = n$ *if* $n \geq 3$;
(b) $\gamma(\Gamma(R)) = 2$ *if* $n = 2$ *and* $\min\{|R_1|, |R_2|\} \geq 3$ *or* $|R_1| = 2$ *and the maximal ideal of* R_2 *is nonzero;*
(c) $\gamma(\Gamma(R)) = 1$ *if* $n = 1$ *and the maximal ideal of* R_2 *is nonzero;*
(d) $\gamma(\Gamma(R)) = 1$ *if* $n = 2$, $|R_1| = \mathbb{Z}_2$ *and* R_2 *is a field.*

The next corollary states certain necessary and sufficient conditions for $\gamma(\Gamma(\mathbb{Z}_n)) = k$, where $n = p_1^{t_1} \ldots p_k^{t_k}$ for distinct primes $p_1, \ldots, p_k$.

Corollary 2.11 ([17, Corollary 13]) *For any fixed integer* $k \geq 1$, *let* $n = p_1^{t_1} \ldots p_k^{t_k}$ *for distinct primes* $p_1, \ldots, p_k$ *and positive integers* $t_1, \ldots, t_k$. *Then*

(a) $\gamma(\Gamma(\mathbb{Z}_n)) = k \geq 3$ *if and only if* $n = p_1^{t_1} \ldots p_k^{t_k}$ *for distinct primes* $p_1, \ldots, p_k$ *and positive integers* $t_1, \ldots, t_k$;
(b) $\gamma(\Gamma(\mathbb{Z}_n)) = 2$ *if and only if* $n = p_1^{t_1} p_2^{t_2}$, *where either* $t_1 \geq 2$ *or* $t_2 \geq 2$; *or* $t_1 = t_2 = 1$ *and* $p_1, p_2 \geq 3$;
(c) $\gamma(\Gamma(\mathbb{Z}_n)) = 1$ *if and only if* $n = p_1^{t_1}$ *where* $t_1 \geq 2$; *or* $n = p_1 p_2$, *where either* $p_1 = 2$, $p_2 \geq 3$ *or* $p_1 \geq 3$, $p_2 = 2$.

Theorem 2.12 ([17, Theorem 15]) *Let R be a finite reduced commutative ring which is not a field. If* $\gamma(\Gamma(R)) \neq 1$, *then* $\gamma(\Gamma(R))$ *is equal to the number of minimal prime ideals of R. In addition, if R has* $k \geq 3$ *minimal prime ideals, then* $\gamma(\Gamma(R)) = k$.

Further Kiani, Maimani, Nikandish [16], investigated the domination, total domination, and semi-total domination numbers of a zero-divisor graph of commutative Noetherian rings.

3 Domination in Zero-Divisor Graph of Generalized Structures

The graph of zero-divisors for commutative rings has been generalized to the ideal-based zero-divisor graph and annihilating-ideal graph of commutative rings. Also the zero-divisor graph has been generalized to commutative semirings and modules over commutative rings. A generalization of the zero-divisor graph called *the ideal-based zero-divisor graph* for commutative rings. In 2001, Redmond [19] introduced the following definition as a generalization of zero-divisor graphs.

Definition 3.1 ([19]) Let R be a commutative ring with nonzero identity, and let I be an ideal of R. The *ideal-based zero-divisor graph* of R, denoted by $\Gamma_I(R)$, is the graph whose vertices are the set $\{x \in R \setminus I : xy \in I$ for some $y \in R \setminus I\}$ and two distinct vertices x and y are adjacent if and only if $xy \in I$.

In the case $I = \{0\}$, $\Gamma_0(R)$ is nothing but the zero-divisor graph $\Gamma(R)$. Also, $\Gamma_I(R)$ is empty if and only if I is prime. Note that $\Gamma_I(R) = \emptyset$ if and only if $\frac{R}{I}$ is an integral domain. In this connection, Mojdeh and Rahimi [17] studied the domination number of the zero-divisor graph with respect to an ideal. Actually, they explored a relationship between domination numbers of $\Gamma_I(R)$ and $\Gamma(\frac{R}{I})$.

Theorem 3.2 ([17, Theorem 19]) *Let R be a commutative ring. Let S be a nonempty subset of R/I. If S is a dominating set of $\Gamma_I(R)$, then $S + I = \{s + I : s \in S\}$ is a dominating set of $\Gamma(\frac{R}{I})$.*

The converse of Theorem 3.2 is not necessarily true. For example, let $R = \mathbb{Z}_6 \times \mathbb{Z}_3$ and $I = \{0\} \times \mathbb{Z}_3$. Then it is clear that $S = \{(3, 0) + I\}$ is a dominating set of $\Gamma(\frac{R}{I})$, but $\{(3, 1), (4, 1)\}$ is a dominating set of $\Gamma_I(R)$ and $\Gamma_I(R)$ cannot be dominated by any set of one vertex. Readers are recommended to refer [16], for some more relations between the domination numbers of $\Gamma_I(R)$ and $\Gamma(\frac{R}{I})$.

As mentioned earlier, zero-divisor graphs were defined and studied for noncommutative rings, near rings, semigroups, modules, lattices and posets. Actually Redmond [18] extended zero-divisor graph concept to noncommutative rings. The corresponding definition is given below:

Definition 3.3 [18] Let R be a noncommutative ring. The *zero-divisor graph of a noncommutative ring* is a directed graph with vertex set $Z(R)^*$, where for distinct vertices x and y of $Z(R)^*$ there is a directed edge from x to y if and only if $xy = 0$ in R.

Let us review some definitions and notation from domination parameters of directed graph. Let $D = (V, A)$ be a digraph(directed graph) with vertex set V and arc set A. The indegree and outdegree of a vertex v are, respectively, denoted by $id(v)$ and $od(v)$. For a subset S of vertices of D, the out-neighborhood $N^+(S)$ of S consists of all those vertices w in $D - S$ such that (v, w) is an arc of D for some $v \in S$. The in-neighborhood $N^-(S)$ consists of all those vertices $u \in D - S$ such that (u, v) is an arc of D for some $v \in S$. For a digraph $D = (V, A)$, a subset S of V is called an *out-dominating set* of D if for every $v \in V - S$, there exists $u \in S$ such that $(u, v) \in A$. The out-dominating set of a digraph D is commonly called as *dominating set* of D. A subset S of V is called an *in-dominating set* of D if for every $v \in V - S$, there exists $u \in S$ such that $(v, u) \in A$. A subset S of V is called a *twin dominating set* if S is both an out-dominating and an in-dominating set. A dominating set S of V is called an *independent* if the sub digraph induced by S has no arcs. A dominating set S of V is called a *total dominating set* if the induced sub digraph $< S >$ has no isolated vertices. A dominating set S of V is called an *open dominating set* of D if for every $v \in V$, there exists $u \in S$ such that $(u, v) \in A$.

The *out-domination number* (resp. *upper out-domination number*) of a digraph D, denoted by $\gamma^+(D)$ (resp. $\Gamma^+(D)$), is the minimum (resp. maximum) cardinality of a out-dominating set of D. In a similar way, one can define the *in-domination number* γ^-, the *twin domination number* γ^*, the *independent domination number* γ_i, the *open domination number* γ_o, the *total domination number* γ_t, and the *weakly connected domination number* γ_{wc}. An out-dominating set S in a digraph D with cardinality γ^+ is called γ^+- set of D. The *irredundance number* $ir(D)$ and the *upper irredundance number* $IR(D)$ are, respectively, the minimum and maximum cardinalities of a maximal irredundant set. An irredundant set S in a digraph D with cardinality ir is called ir- set of D. The *out-domatic (resp. in-domatic) number* $d^+(D)$ (resp. $d^-(D)$) of a digraph D to be the maximum number of elements in a partition of $V(D)$ into out-dominating (resp. in-dominating) sets. The *total domatic number* $d_t(D)$ of a digraph D to be the maximum number of elements in a partition of $V(D)$ into total dominating sets. A digraph D is *domatically full* if $d^+(D) = 1 + \delta(D)$. The *reinforcement number* $r(D)$ of a digraph D is the minimum number of extra arcs whose addition to D results in a graph D' with $\gamma^+(D') < \gamma^+(D)$.

Tamizh Chelvam and Selvakumar [22], obtained the values of certain domination parameters for the directed zero-divisor graph D on $M_2(\mathbb{Z}_p)$, the ring of all 2×2 matrices over $\mathbb{Z}_p$, where p is a prime number. Some of the results on this work are given below:

Theorem 3.4 ([22, Theorem 2.1]) *Let p be a prime number and D be the directed zero-divisor graph of $M_2(\mathbb{Z}_p)$. Then $ir(D) = IR(D) = p + 1$.*

Proposition 3.5 ([22, Proposition 2.1]) *Let D be the directed zero-divisor graph of $M_2(\mathbb{Z}_p)$. Then $\gamma_t(D) = \gamma_{wc}(D) = \gamma_o(D) = p + 1$.*

Theorem 3.6 ([22, Theorem 2.2]) *Let D be the directed zero-divisor graph of $M_2(\mathbb{Z}_p)$. If Ω is a minimal dominating set of D, then Ω is independent if and only if $A^2 = 0$ for all $A \in \Omega$.*

Theorem 3.7 ([22, Theorem 2.3]) *Let D be the directed zero-divisor graph on $M_2(\mathbb{Z}_p)$. Then $d^+(D) = d^-(D) = d_t(D) = p^2 - 1$.*

Proposition 3.8 ([22, Proposition 3.1]) *Let D be the directed zero-divisor graph of $R = M_2(\mathbb{Z}_p)$. Then $\gamma^*(D) = \gamma_p(D) = \gamma_e(D) = p + 1$.*

Corollary 3.9 ([22, Lemma 3.1]) *Let D be a directed zero-divisor graph of $R = M_2(\mathbb{Z}_p)$. The number of efficient dominating sets in D is $p - 1$.*

Theorem 3.10 ([22, Theorem 3.1]) *Let D be the directed zero-divisor graph of $R = M_2(\mathbb{Z}_p)$. If $Z(R)^* = \cup_{i=0}^{p} O_l(M_i)$, then Ω is a η-set of D if and only if*

(a) *each element of Ω belongs to different orbit of $Z(R)^*$;*
(b) *$S^2 \neq 0$ for all $S \in \Omega$;*
(c) *Ω is independent.*

Recently Tamizh Chelvam and Selvakumar [27], studied the domination parameters on the directed zero-divisor graph of $M_2(\mathbb{F})$, where $\mathbb{F}$ is a finite field.

Theorem 3.11 ([27, Theorem 2.2]) *Let $\mathbb{F}$ be a finite field with $|\mathbb{F}| = p^m$ where p is a prime and $m \geq 1$ and $D = \Gamma(M_2(\mathbb{F}))$. Then $ir(D) = IR(D) = p^m + 1$.*

Proposition 3.12 ([27, Proposition 2.8]) *Let $\mathbb{F}$ be a finite field with $|\mathbb{F}| = p^m$ where p is a prime and $m \geq 1$. Let $D = \Gamma(M_2(\mathbb{F}))$. Then $\gamma_t(D) = \gamma_{wc}(D) = \gamma_o(D) = p^m + 1$.*

Proposition 3.13 ([27, Proposition 2.9]) *Let $\mathbb{F}$ be a finite field with $|\mathbb{F}| = p^m$ where p is a prime and $m \geq 1$. Let $D = \Gamma(M_2(\mathbb{F}))$. If Ω is an open dominating set of D, then Ω contains no nilpotent elements.*

Theorem 3.14 ([27, Theorem 3.1]) *Let $\mathbb{F}$ be a finite field with $|\mathbb{F}| = p^m$ where p is a prime and $m \geq 1$. Let Ω is a minimal dominating set of $\Gamma(M_2(\mathbb{F}))$. Then Ω is independent if and only if $a^2 = 0$ for every $a \in \Omega$.*

Theorem 3.15 ([27, Theorem 3.2]) *Let $\mathbb{F}$ be a finite field with $|\mathbb{F}| = p^m$ where p is a prime number and $m \geq 1$. Let $D = \Gamma(M_2(\mathbb{F}))$. Then $d^+(D) = d^-(D) = d_t(D) = p^{2m} - 1$ and hence D is domatically full.*

Theorem 3.16 ([27, Theorem 3.3]) *Let $\mathbb{F}$ be a finite field with $|\mathbb{F}| = p^m$ where p is a prime number and $m \geq 1$. Let $D = \Gamma(M_2(\mathbb{F}))$. Then $\gamma^*(D) = \gamma_p(D) = \gamma_e(D) = p^m + 1$.*

Corollary 3.17 ([27, Corollary 3.4]) *Let $\mathbb{F}$ be a finite field with $|\mathbb{F}| = p^m$ where p is a prime number and $m \geq 1$. If Ω is a minimal dominating set of $\Gamma(M_2(\mathbb{F}))$, then Ω is perfect (and so efficient) if and only if $a^2 = 0$ for all $a \in \Omega$.*

The domination parameters of zero-divisor graphs of matrix rings over a commutative ring with identity has been discussed in a recent article by Heidar Jafari and Jafari Rad [12]. Remaining part of the section lists the results from [12].

Theorem 3.18 ([12, Lemma 2.5]) *For any commutative ring R, $\gamma_o(\Gamma(M_n(R))) = \gamma_i(\Gamma(M_n(R)))$.*

Theorem 3.19 ([12, Lemma 2.6]) *If A is an out-dominating set for $\Gamma(M_n(R))$, then there exists an out-dominating set B for $\Gamma(M_n(R))$ such that $|B| \leq |A|$ and any element of B is of rank 1.*

Theorem 3.20 ([12, Corollary 2.8]) *Let $R = M_n(\mathbb{F})$, where $\mathbb{F}$ is a finite field. Then*

$$S = \{A = (a_{ij})_{n\times n} : a_{ij} = \delta_{1j}(\lambda_j), \text{ where } \lambda_1, \ldots, \lambda_n \in F \text{ and } \lambda_j = 1 \text{ for some } j\}$$

is a $\gamma_o(\Gamma(R))$-set.

Theorem 3.21 ([12, Corollary 2.9]) *Let $\mathbb{F}$ be a finite field with $|\mathbb{F}| = q$. For any n, $\gamma_o(M_n(\mathbb{F})) = \gamma_i(M_n(\mathbb{F})) = \frac{q^n-1}{q-1}$.*

Theorem 3.22 ([12, Theorem 3.6]) *Let $(R, \mathfrak{m})$ be a local commutative ring with identity and let $R/\mathfrak{m}$ be finite. Then $\gamma_o(\Gamma(M_n(R))) \leq \gamma_o(\Gamma(M_n(\frac{R}{\mathfrak{m}})))$.*

Theorem 3.23 ([12, Theorem 3.7]) *Let $(R, \mathfrak{m})$ be a finite local commutative ring with identity and $\mathfrak{m}$ be cyclic as an R-module. Then $\gamma_o(\Gamma(M_n(R))) = \gamma_o(\Gamma(M_n(\frac{R}{\mathfrak{m}})))$.*

Theorem 3.24 ([12, Theorem 3.10]) *Let $R = R_1 \times \ldots \times R_t$, where R_i is a commutative ring with identity such that the unique maximal ideal of R_i is principal. Then*

$$\gamma_o(\Gamma(M_n(R))) = \gamma_o(\Gamma(M_n(R_1))) + \ldots + \gamma_o(\Gamma(M_n(R_t))).$$

4 Domination in the Total Graph of $\mathbb{Z}_n$

As mentioned earlier, Anderson and Badawi [5] introduced the total graph of a commutative ring. The *total graph* of a commutative ring R, denoted by $T_\Gamma(R)$, is the undirected graph with all elements of R as vertices and for distinct $x, y \in R$, the vertices x and y are adjacent if $x + y \in Z(R)$. In this section, we shall discuss about domination concepts in the total graph of $\mathbb{Z}_n$.

First let us see domination parameters of the total graph of a ring of integer modulo $\mathbb{Z}_n$. In this regard, Tamizh Chelvam and Asir [21] have initiated the study on domination parameters in $T_\Gamma(\mathbb{Z}_n)$. Let $p > 2$ be prime. Then $Z(\mathbb{Z}_p) = \{0\}$ so that 0 is an isolated vertex and the neighborhood $N(x) = \{-x\}$ for all $0 \neq x \in \mathbb{Z}_p$. Thus $\gamma(T_\Gamma(\mathbb{Z}_p)) = \frac{p+1}{2}$. Also note that $\gamma(T_\Gamma(\mathbb{Z}_2)) = 2$. Due to these facts, hereafter assume that n is a composite integer and $n = p_1^{\alpha_1} p_2^{\alpha_2} \ldots p_m^{\alpha_m}$ where p_i are distinct prime numbers for $1 \leq i \leq m$ with $p_1 < p_2 < \ldots < p_m$. First let us the see the domination number of $T_\Gamma(\mathbb{Z}_n)$.

Theorem 4.1 ([21, Theorem 2.3]) *Let n be a composite number and p_1 be the smallest prime divisor of n. Then $\gamma(T_\Gamma(\mathbb{Z}_n)) = p_1$.*

Corollary 4.2 ([21, Corollary 2.5]) *Let n be a composite integer. Then $T_\Gamma(\mathbb{Z}_n)$ is domatically full if and only if $n = p^k$ for some prime p and $1 < k \in \mathbb{Z}^+$.*

Corollary 4.3 ([21, Corollary 2.6]) *For an composite integer $n > 1$,*

$$i(T_\Gamma(\mathbb{Z}_n)) = \begin{cases} 2 & if\ n\ is\ even \\ \frac{n-p^{k-1}}{2} + 1 & if\ n = p^k\ where\ p > 2\ is\ prime\ and\ k > 1. \end{cases}$$

Corollary 4.4 ([21, Corollary 2.7]) *For any composite integer n, $d_i(T_\Gamma(\mathbb{Z}_n)) = \frac{n}{2}$ if n is even and $d_i(Reg(T_\Gamma(\mathbb{Z}_n))) = 2$ if $n = p^k$ where $p > 2$ is prime.*

Corollary 4.5 ([21, Corollary 2.8]) *For any composite integer $n > 2$, $\gamma_s(T_\Gamma(\mathbb{Z}_n)) = \gamma_w(T_\Gamma(\mathbb{Z}_n)) = p_1$, where p_1 is the smallest prime divisor of n.*

The following theorem characterizes all γ-sets in $T_\Gamma(\mathbb{Z}_n)$.

Theorem 4.6 ([21, Theorem 2.9]) *Let n be a composite integer and p_1 be the smallest prime divisor of n. A set $S = \{x_1, x_2, \ldots, x_{p_1}\} \subset V(T_\Gamma(\mathbb{Z}_n))$ is a γ-set of $T_\Gamma(\mathbb{Z}_n)$ if and only if $x_i + lp_1 \notin S$ for all $i = 1, ..., p_1$ and $l \in \mathbb{Z}^+$.*

From the above theorem, one can observe that for any integer n, $T_\Gamma(\mathbb{Z}_n)$ is an excellent graph ([21, Corollary 2.10]). The following theorem obtains the total domination number of $T_\Gamma(\mathbb{Z}_n)$.

Theorem 4.7 ([21, Theorem 2.11]) *Let $n = p_1^{\alpha_1} p_2^{\alpha_2} ... p_m^{\alpha_m}$ be a composite number. Then*

$$\gamma_t(T_\Gamma(\mathbb{Z}_n)) = \begin{cases} 4 & if\ n = 2^k\ for\ some\ k\ with\ 1 < k \in \mathbb{Z}^+; \\ p+1 & if\ n = p^k\ for\ some\ prime\ p > 2\ and\ k\ with\ 1 < k \in \mathbb{Z}^+; \\ p_1 & otherwise. \end{cases}$$

Next result gives the value of prefect domination number of $T_\Gamma(\mathbb{Z}_n)$.

Theorem 4.8 ([21, Theorem 2.13]) *Let $n \neq 2$ be an integer. Then*

$$\gamma_p(T_\Gamma(\mathbb{Z}_n)) = \begin{cases} \frac{p+1}{2} & if\ n = p\ for\ some\ prime\ p; \\ p & if\ n = p^k\ for\ some\ prime\ p\ and\ an\ integer\ k > 1; \\ 2 & if\ n = 2p\ for\ some\ prime\ p > 2; \\ n & otherwise. \end{cases}$$

Note that, if $n = 2p$ for some prime $p > 2$, then the sets $\{i, p - i\}$ for all $i \in \mathbb{Z}_n$ are the only γ_p-sets of $T_\Gamma(\mathbb{Z}_n)$ and i is adjacent to $p - i$ for all $i \in \mathbb{Z}_n$. Therefore, we have the following result:

Corollary 4.9 ([21, Corollary 2.14]) *Let n be a composite integer. Then*

$$\gamma_{eff}(T_\Gamma(\mathbb{Z}_n)) = \begin{cases} \frac{n-p^{k-1}}{2} + 1 & if\ n = p^k\ for\ some\ prime\ p\ and\ k \geq 1; \\ does\ not\ exists & otherwise. \end{cases}$$

Corollary 4.10 ([21, Corollary 2.15]) *Let n be a composite integer. Then*

$$d_p(T_\Gamma(\mathbb{Z}_n)) = \begin{cases} p^{k-1} & if\ n = p^k\ for\ some\ prime\ p\ and\ integer\ k > 1; \\ p & if\ n = 2p\ for\ some\ prime\ p > 2; \\ does\ not\ exists & if\ n = p\ for\ some\ prime\ p > 2; \\ 1 & otherwise. \end{cases}$$

Next, we see that the connected and clique domination number of $T_\Gamma(\mathbb{Z}_n)$ is equal to the domination number of $T_\Gamma(\mathbb{Z}_n)$.

Theorem 4.11 ([21, Theorem 2.16]) *Let $n \geq 2$ be any integer and not a prime power and $T_\Gamma(\mathbb{Z}_n)$. Then $\gamma_c(T_\Gamma(\mathbb{Z}_n) = \gamma_{cl}(T_\Gamma(\mathbb{Z}_n) = p_1$ where p_1 is a smallest prime divisor of n.*

Next, we list certain results concerning domination parameters of the complement of the total graph on $\mathbb{Z}_n$ i.e., $\overline{T_\Gamma(\mathbb{Z}_n)}$. If n is prime, then clearly $\gamma(\overline{T_\Gamma(\mathbb{Z}_n)}) = 1$.

Theorem 4.12 ([21, Theorem 3.1]) *Let n be any composite integer. Then*

$$\gamma(\overline{T_\Gamma(\mathbb{Z}_n)}) = \begin{cases} 2 & if\ n = p^k\ for\ some\ prime\ p\ and\ k > 1; \\ 2 & if\ n = 2p\ for\ some\ prime\ p > 3. \end{cases}$$

Corollary 4.13 ([21, Corollary 3.2]) *For any positive integer n,*

$$i(\overline{T_\Gamma(\mathbb{Z}_n)}) = \begin{cases} p^{k-1} & if\ n = p^k\ for\ some\ prime\ p\ and\ k > 1; \\ 2 & if\ n = 2p\ for\ some\ prime\ p > 3. \end{cases}$$

Corollary 4.14 ([21, Corollary 3.3]) *Let n be a composite integer. Then*

$$\gamma_c(\overline{T_\Gamma(\mathbb{Z}_n)}) = \begin{cases} 2 & if\ n = p^k\ for\ some\ prime\ p\ and\ k > 1; \\ 4 & if\ n = 2p\ for\ some\ prime\ p > 3. \end{cases}$$

Next theorem shows that, for every composite integer n, there exists a positive integer ℓ such that $\gamma(\overline{T_\Gamma(\mathbb{Z}_n)}) \leq \ell + 1$.

Theorem 4.15 ([21, Theorem 3.4]) *Let n be a composite integer. If $Z(\mathbb{Z}_n)$ contains at most ℓ consecutive integers of $\mathbb{Z}_n$ for some $\ell \in \mathbb{Z}^+$, then $\gamma(\overline{T_\Gamma(\mathbb{Z}_n)}) \leq \ell + 1$.*

Note that in the above theorem if no consecutive integer exists in $Z(\mathbb{Z}_n)$, then we take $\ell = 1$. For any integer n, $1 \leq \gamma(\overline{T_\Gamma(\mathbb{Z}_n)}) \leq \ell + 1$. The lower bound is attained in the case of $n = p$, where p is a prime number. On the other hand, if $n = p^k$ for some $k > 1$, then $\ell = 1$ and $\gamma(\overline{T_\Gamma(\mathbb{Z}_n)}) = 2$. Also if $R = \mathbb{Z}_{12}$, then $\ell = 3$ and $\gamma(\overline{T_\Gamma(R)}) = 4$. Hence the bounds are sharp ([21, Remark 3.5]).

5 Domination in Total Graph of a Commutative Ring

In this section, we enumerate results on the domination parameters of the total graph of a commutative ring. Actually these lists exhibit that the domination number of the total graph of an Artin ring equals the upper bound. In this connection, a conjecture was posed and the same is given here.

Let I be a maximum annihilator ideal of R. That means, I is a maximal annihilator ideal of R such that $|\frac{R}{I}| = \min\{|\frac{R}{A}| : A$ is a maximal annihilator ideal of $R\}$. We begin with the following theorem, which exhibits a relation between the product of rings and the product of corresponding total graphs. More specifically, the relation is concerning the domination number of the total graph of the direct product of two rings and the domination number of Cartesian product of the total graphs of rings.

Theorem 5.1 ([25, Theorem 2.1]) *Let R_1 and R_2 be two commutative rings with identity. Then $\gamma(T_\Gamma(R_1 \times R_2)) \leq \gamma(T_\Gamma(R_1) \square T_\Gamma(R_2))$.*

For any integral domain R, the maximum degree $\Delta(T_\Gamma(R)) \leq 1$. If R is a finite integral domain, then $\gamma(T_\Gamma(R)) = \frac{|R|-k}{2} + k$, where $k = |\{a \in R : a = -a\}|$. If R is infinite, then there exists no positive integer k such that $\gamma(T_\Gamma(R)) = k$. So hereafter, we assume throughout this section that all rings are commutative which is not an integral domain. In the following theorem, we obtain lower and upper bounds for the domination number of the total graph of a commutative ring.

Lemma 5.2 ([25, Lemma 2.2]) *Let R be a commutative ring (not necessarily finite) with identity, I be a maximum annihilator ideal of R and $|R/I| = \mu$(finite). Then $2 \leq \gamma(T_\Gamma(R)) \leq \mu$.*

It is also shown that the lower and upper bounds are sharp ([25, Example 2.3]).

If $Z(R)$ is an ideal of R, then the maximal annihilator ideal in R is $Z(R)$ and so the following are proved:

Lemma 5.3 ([25, Lemma 2.4]) *If R is a commutative rings with identity, $Z(R)$ is an ideal of R and $|\frac{R}{Z(R)}| = \mu$, then $\gamma(T_\Gamma(R)) = \mu$.*

Theorem 5.4 ([25, Theorem 2.5]) *Let R be an Artin ring, I be a maximum annihilator ideal of R and $|\frac{R}{I}| = \mu$. Then $\gamma(T_\Gamma(R)) = \mu$.*

Since every finite ring is an Artin ring, the following is obtained.

Corollary 5.5 ([25, Corollary 2.6]) *Let R is a finite commutative ring, I be a maximum annihilator ideal of R and $|\frac{R}{I}| = \mu$. Then $\gamma(T_\Gamma(R)) = \mu$.*

Using [25, Theorem 2.5], the domination number for the total graph of certain classes of commutative rings is determined and proved that there are families of infinite graphs whose domination number is finite.

Corollary 5.6 ([25, Corollary 2.7])

(a) *If n is a composite integer, then $\gamma(T_\Gamma(\mathbb{Z}_n)) = p$ where p is the smallest prime divisor of n;*
(b) *For any $n, k \in \mathbb{Z}^+$, $\gamma(T_\Gamma(\frac{\mathbb{Z}_n[x]}{<x^k>})) = \gamma(T_\Gamma(\frac{\mathbb{Z}_n[x,y]}{<x^k,xy,y^k>}) = p$ where p is the smallest prime divisor of n;*
(c) *If R_i's are finite integral domains, then $\gamma(T_\Gamma(R_1 \times R_2 \times \ldots \times R_k)) = \min\{|R_1|, |R_2|, \ldots, |R_k|\}$;*
(d) *If n is a composite positive integer, then $\gamma(T_\Gamma(\mathbb{Z}_n \times \mathbb{Z} \times \ldots \times \mathbb{Z})) = p$ where p is the smallest prime divisor of n;*
(e) *Let n be a composite positive integer, $k \in \mathbb{Z}^+$ and $\mathbb{F}$ be a field. If $R = \mathbb{Z}_n \times \mathbb{F} \times \ldots \times \mathbb{F}$ or $R = \frac{\mathbb{Z}_n[x]}{<x^k>} \times \mathbb{Z} \times \ldots \times \mathbb{Z}$, then $\gamma(T_\Gamma(R)) = p$ where p is the smallest prime divisor of n.*

Having obtained the domination number of the total graph of some classes of rings, a conjecture was proposed by Tamizh Chelvam and Asir and the same is given below:

Conjecture 5.7 ([25, Conjecture 2.8]) Let R be a commutative ring with identity which is not an Artin ring, $Z(R)$ be not an ideal of R and I_i's are maximal annihilator ideals of R. If $|\frac{R}{I_i}| = finite$ for some i, then $\gamma(T_\Gamma(R)) = \min\{|\frac{R}{I_i}| : I_i$ is a maximal annihilator ideal of $R\}$, where the minimum is taken over all I_i for which $|\frac{R}{I_i}|$ is finite.

Next, we list certain properties on the domination parameters of $T_\Gamma(R)$ under the assumption that $\gamma(T_\Gamma(R)) = \mu$. As mentioned earlier, I is a maximum annihilator ideal in R, $|I| = \lambda$ and $|\frac{R}{I}| = \mu$.

Lemma 5.8 ([25, Lemma 3.1]) *Let R be a commutative ring. If $\gamma(T_\Gamma(R)) = \mu$, then the set $S = \{x_1, x_2, \ldots, x_\mu\} \subset V(T_\Gamma(R))$ is a γ-set of G where $x_j \notin x_i + I$ for all $i, j = 1, \ldots, \beta$ and $i \neq j$.*

Corollary 5.9 ([25, Corollary 3.2]) *Let R be a commutative ring. If $\gamma(T_\Gamma(R)) = \mu$, then*

(a) $\gamma'(T_\Gamma(R)) = \mu$, *where $\gamma'(G)$ is the inverse domination number of G;*
(b) $T_\Gamma(R)$ *is excellent;*
(c) *the domatic number $d(T_\Gamma(R)) = \lambda$.*

Theorem 5.10 ([25, Theorem 3.3]) *For a commutative ring R, if $Z(R)$ is not an ideal of R, $R =< Z(R) >$(i.e., R is generated by $Z(R)$) and $\gamma(T_\Gamma(R)) = \mu$, then $\gamma_t(T_\Gamma(R)) = \gamma_c(T_\Gamma(R)) = \mu$.*

Let R be a commutative ring and $G = T_\Gamma(R)$. If R is not an integral domain, then G satisfies $\gamma(G - v) = \gamma(G)$ for all $v \in V(G)$. Using this, the bondage number of the total graph was obtained.

Theorem 5.11 ([25, Theorem 3.3]) *For a finite commutative ring R, if $\gamma(T_\Gamma(R)) = \mu$, then bondage number $b(T_\Gamma(R)) = |Z(R)| - 1$.*

Now, we see the results regarding domination parameters of $T_\Gamma(R)$ and $\overline{T_\Gamma(R)}$ when $Z(R)$ is an ideal of R. Let $Z(R)$ be an ideal of R and so $I = Z(R)$, $\lambda = \alpha$, $\mu = \beta$ and $\gamma(T_\Gamma(R)) = \beta$.

Lemma 5.12 ([25, Lemma 4.2]) *Let R be a finite commutative ring such that $Z(R)$ is an ideal of R. Then $\gamma(\overline{T_\Gamma(R)}) = 2$.*

Let R be a finite commutative ring and $G =< Reg(R) >\subseteq \overline{T_\Gamma(R)}$. If $2 \in Z(R)$ and $\beta = |\frac{R}{Z(R)}| = 2$, then $G = \overline{K_\alpha}$ and so $\gamma(G) = \alpha$. All remaining cases of R, we have $\gamma(G) = 2$. Therefore

$$\gamma(< Reg(R) >) = \begin{cases} \alpha & \text{if } 2 \in Z(R) \text{ and } \beta = 2; \\ 2 & \text{otherwise.} \end{cases}$$

The following corollary determines the inverse domination number:

Corollary 5.13 ([25, Corollary 4.4]) *(i) Let R be a commutative ring except the one with $2 \in Z(R)$, $\alpha > 2$, $\beta = 2$ and $G =< Reg(R) >$ in $\overline{T_\Gamma(R)}$. Then $\gamma'(G) = 2$.*
(ii) For any commutative ring R, $\gamma'(\overline{T_\Gamma(R)}) = 2$.

Theorem 5.14 ([25, Theorem 4.5]) *Let R be a commutative ring such that $Z(R)$ is an ideal of R and $G = T_\Gamma(R)$. A set $S = \{x_1, x_2, \ldots, x_\beta\} \subset V(G)$ is a γ-set of G if and only if $x_j \notin x_i + Z(R)$ for all $1 \leq i, j \leq \beta$ and $i \neq j$.*

Corollary 5.15 ([25, Corollary 4.7]) *Let R be a finite commutative ring with $Z(R)$ is an ideal of R. Then*

(a) $T_\Gamma(R)$ *and* $\overline{T_\Gamma(R)}$ *are excellent;*
(b) $d(T_\Gamma(R)) = \alpha$ *and* $d(\overline{T_\Gamma(R)}) = \left\lfloor \frac{|R|}{2} \right\rfloor$;
(c) *If* $G_1 = \overline{Reg_\Gamma(R)}$, *then*

$$d(G_1) = \begin{cases} 1 & if\ 2 \in Z(R)\ and\ \beta = 2 \\ \left\lfloor \frac{|Reg(R)|}{2} \right\rfloor & otherwise; \end{cases}$$

(d) $T_\Gamma(R)$ *is domatically full.*

Theorem 5.16 ([25, Theorem 4.8]) *Let* R *be a finite commutative ring with* $Z(R)$ *is an ideal of* R *and* $G = T_\Gamma(R)$. *Then* G *and* $\overline{G}$ *are well covered.*

Corollary 5.17 ([25, Corollary 4.9]) *If* R *is a finite commutative ring such that* $Z(R)$ *is an ideal of* R *and* $|Z(R)| = \alpha$, *then* $\omega(T_\Gamma(R)) = \alpha$.

Theorem 5.18 ([25, Theorem 4.10]) *Let* R *be a finite commutative ring such that* $Z(R)$ *is an ideal of* R, $|Z(R)| = \alpha$, $|\frac{R}{Z(R)}| = \beta$ *and* $G = T_\Gamma(R)$. *Then*

(a) $\gamma_t(G) = \begin{cases} 2\beta & if\ 2 \in Z(R) \\ \beta + 1 & otherwise; \end{cases}$
(b) $\gamma_t(\overline{G}) = 2$;
(c) $\gamma_c(G) = 2$;
(d) $\gamma_s(G) = \gamma_w(G) = \beta$ *and* $\gamma_s(\overline{G}) = \gamma_w(\overline{G}) = 2$;
(e) $\gamma_p(G) = \beta$;
(f) $\gamma_p(\overline{G}) = 2$ *if* $\beta = 2$;
(g) *If* $G_1 =< Reg(R) >$ *in* $\overline{T_\Gamma(R)}$, $\beta = 2$ *and* $2 \notin Z(R)$, *then* $\gamma_p(G_1) = 2$.

Finally, we see the double domination parameters of $T_\Gamma(R)$.

Theorem 5.19 ([25, Theorem 4.11]) *Let* R *be a finite commutative ring with* $Z(R)$ *is an ideal of* R, $|Z(R)| = \alpha$, $|R/Z(R)| = \beta$ *and* $G = T_\Gamma(R)$. *Then*

(a) $\gamma\gamma(G) = 2\beta$;
(b) $\gamma i(G) = \begin{cases} 2\beta & if\ 2 \in Z(R) \\ \beta + (\frac{\beta-1}{2})\alpha + 1 & otherwise; \end{cases}$
(c) $ii(G) = \begin{cases} 2\beta & if\ 2 \in Z(R) \\ 2(\frac{\beta-1}{2})\alpha + 2 & otherwise; \end{cases}$
(d) $tt(G) = \begin{cases} 4\beta & if\ 2 \in Z(R)\ and\ \alpha \geq 4 \\ 2(\beta + 1) & if\ 2 \notin Z(R) \\ does\ not\ exists & otherwise. \end{cases}$

Acknowledgments This work is supported by the UGC-BSR One-time grant(F.19-109/2013 (BSR)) of University Grants Commission, Government of India for the first author. Second author's research was in part supported by the Start-Up grant(F.30-118/2015(BSR)) from University Grants Commission, Government of India. Also the work reported here is supported by the UGC Major Research Project(F.42-8/2013(SR)) awarded to the third author by the University Grants Commission, Government of India.

References

1. Abbasi, A., Habibi, S.: The total graph of a commutative ring with respect to proper ideals. J. Korean Math. Soc. **49**(1), 85–98 (2012)
2. Akbari, S., Kiani, D., Mohammadi, F., Moradi, S.: The total graph and regular graph of a commutative ring. J. Pure Appl. Algebra **213**, 2224–2228 (2009)
3. Anderson, D.F., Livingston, P.S.: The zero-divisor graph of a commutative ring. J. Algebra **217**, 434–447 (1999)
4. Anderson, D.F., Frazier, A., Lauve, A., Livingston, P.S.: The zero-divisor graph of a commutative ring-II. Lecture Notes Pure Appl. Math. **202**, 61–72 (2001)
5. Anderson, D.F., Badawi, A.: The total graph of a commutative ring. J. Algebra **320**, 2706–2719 (2008)
6. Asir, T.: Tamizh Chelvam, T.: On the total graph and its complement of a commutative ring. Comm. Algebra **41**(10), 3820–3835 (2013)
7. Asir, T., Tamizh Chelvam, T.: On the intersection graph of gamma sets in the total graph II. J. Algebra Appl. **12**(4), Article No. 1250199, 14 p. (2013). doi:10.1142/S021949881250199X
8. Badawi, A.: On the total graph of a ring and its related graphs: A survey. In: Fontana, M., et al. (eds.) Commutative Algebra: Recent Advances in Commutative Rings, Integer-Valued Polynomials, and Polynomial Functions. Springer Science+Business Media, New York (2014). doi:10.1007/978-1-4939-0925-4_3
9. Beck, I.: Coloring of a commutative ring. J. Algebra **116**, 208–226 (1988)
10. Chartrand, G., Zhang, P.: Introduction to Graph Theory. Tata McGraw-Hill Edition, New Delhi (2006)
11. Haynes, T.W., Hedetniemi, S.T., Slatar, P.J.: Fundamental of domination in graphs. Marcel Dekker. Inc., New York (1998)
12. Heidar Jafari, S., Jafari Rad, N.: On domination of zero-divisor graphs of matrix rings. Canad. Math. Bull. **58**(2), 271–275 (2015)
13. Jafari Rad, N., Jafari, S.H., Mojdeh, D.A.: On domination in zero-divisor graphs: Canad. Math. Bull. **56**, 407–411 (2013)
14. Huckaba, J.A.: Commutative Rings with Zero Divisors. Marcel Dekker, New York (1988)
15. Kaplansky, I.: Commutative Rings. Polygonal Publishing House, Washington (1974)
16. Kiani, S., Maimani, H.R., Nikandish, R.: Some results on the domination number of a zero-divisor graph. Canad. Math. Bull. **57**, 573–578 (2014)
17. Mojdeh, D.A., Rahimi, A.M.: Dominating sets of some graphs associated to commutative rings. Comm. Algebra **40**(9), 3389–3396 (2012)
18. Redmond, S.P.: The zero-divisor graph of a non-commutative ring. Int. J. Commutative Rings **1**(4), 203–211 (2002)
19. Redmond, S.P.: An ideal-based zero-divisor graph of a commutative ring. Comm. Algebra **31**, 4425–4443 (2003)
20. Redmond, S.P.: Central sets and radii of the zero-divisor graphs of commutative rings. Comm. Algebra **34**, 2389–2401 (2006)
21. Tamizh Chelvam, T., Asir, T.: Domination in total graph on Z_n. Discrete Math. Algorithms Appl. **3**(4), 1–9 (2011)
22. Tamizh Chelvam, T., Selvakumar, K.: Domination in zero-divisor graph of matrices. J. Adv. Res. Pure Math. **3**(2), 31–39 (2011). doi:10.5373/jarpm.541.081710

23. Tamizh Chelvam, T., Asir, T.: Intersection graph of gamma sets in the total graph. Discuss. Math. Graph Theory **32**, 341–356 (2012)
24. Tamizh Chelvam, T., Asir, T.: On the genus of the total graph of a commutative ring. Comm. Algebra **41**(1), 142–153 (2013)
25. Tamizh Chelvam, T., Asir, T.: Domination in the total graph of a commutative ring. J. Combin. Math. Combin. Comput. **87**, 147–158 (2013)
26. Tamizh Chelvam, T., Asir, T.: On the intersection graph of gamma sets in the total graph I. J. Algebra Appl. **12**(4), Article No. 1250198, 18 p. (2013). doi:10.1142/S0219498812501988
27. Tamizh Chelvam, T., Selvakumar, K.J.: Domination in directed zero-divisor graph of matrices. J. Combin. Math. Combin. Comput. **91**, 155–163 (2014)
28. Tamizh Chelvam, T., Selvakumar, K.: On the intersection graph of gamma sets in the zero-divisor graph. Discret. Math. Algorithms Appl. **7**(1), 14 p. (2015). doi:10.1142/S1793830914500670

On Iso-Retractable Modules and Rings

A.K. Chaturvedi

Abstract Beaumont studied groups with isomorphic proper subgroups (see Beaumont: Bull Amer Math Soc 51, 381–387 1945 [1]). In Beaumont et al.: Trans Amer Math Soc 91(2), 209–219 1959 [2], Beaumont and Pierce consider the problem of determining all R-modules M over a principal ideal domain R which have proper isomorphic submodules. Such modules are called I-modules. In Chaturvedi: Iso-retractable Modules and Rings (to appear) [3], we investigate iso-retractable modules that is the modules which are isomorphic to their nonzero submodules. Also, a ring R is said to be iso-retractable if R_R is an iso-retractable module. The class of iso-retractable modules lies in between simple modules and the uniform modules. In the present paper, our main objective is to investigate general properties of iso-retractable modules and rings. Finally, we show that being iso-retractable is a Morita invariant property.

Keywords Iso-retractable modules · Simple modules · Semi-simple rings · Hereditary rings · von-Numann regular rings

Mathematics Subject Classification 16D50 · 16D70 · 16D80

1 On Iso-Retractable Modules and Rings

All rings are associative with unit element and all modules are unitary right modules. We refer to [5, 9] for all undefined notions used in the text.

Beaumont studied groups with isomorphic proper subgroups (see [1]). In [2] the problem of determining all R-modules M over a principal ideal domain R which have proper isomorphic submodules is considered. Such modules are called I-modules. In [3], modules which are isomorphic to their nonzero submodules are said to be iso-retractable modules. In [8], authors call such modules elastic. Equivalently, an R-module M_R is *iso-retractable* if for every nonzero submodule N of M_R there exists

A.K. Chaturvedi (✉)
Department of Mathematics, University of Allahabad, Allahabad 211002, India
e-mail: akc99@rediffmail.com; akchaturvedi.math@gmail.com

S.T. Rizvi et al. (eds.), *Algebra and its Applications*, Springer Proceedings in Mathematics & Statistics 174, DOI 10.1007/978-981-10-1651-6_24

an isomorphism $\varphi : M \to N$. Also, a ring R is said to be iso-retractable if R_R is an iso-retractable module.

Iso-retractable modules are concerned mainly with infinite modules which do not satisfy the descending chain condition. We provide some examples of iso-retractable modules.

Example 1.1 1. Infinite cyclic modules are iso-retractable because infinite cyclic modules have isomorphic proper submodules.
2. Division rings and fields are natural examples of iso-retractable rings.

Recall, an R-module M is epi-retractable if every submodule of M_R is a homomorphic image of M. Some application of epi-retractable modules studied in [6]. By [4, 6.9.3], an R-module M is called *compressible* if for every nonzero submodule N of M there exists a monomorphism from M to N. The concept of epi-retractable modules is dual to the concept of compressible modules. There exist some epi-retractable modules which are not compressible.

Remark 1.2 The class of iso-retractable modules lies in between compressible and epi-retractable modules. Therefore we have the following implications:

slightly compressible $\Leftarrow$ compressible $\Leftarrow$ iso-retractable $\Rightarrow$ epi-retractable.

But the reverse implications are not true in all cases. A nonzero semi-simple module is epi-retractable. But it is not compressible therefore it is not iso-retractable. We note that every iso-retractable module is a *slightly compressible module* (see [7]) but the converse need not be true.

Remark 1.3 The class of iso-retractable modules lies in between simple modules and the uniform modules. Therefore the following implications hold:

simple $\Rightarrow$ iso-retractable $\Rightarrow$ uniform $\Rightarrow$ indecomposable.

Trivially simple modules are iso-retractable but the converse need not be true. For example, Z_Z is iso-retractable but not simple. In the following we observe following sufficient conditions:

Proposition 1.4 *If an iso-retractable module is finite then it must be simple.*

Proof It is obvious. □

Proposition 1.5 *If M_R is a simple module then the following are equivalent:*

1. *M_R is epi-retractable.*
2. *M_R is compressible.*
3. *M_R is iso-retractable.*

Proof It is clear. □

In [3], we investigate iso-retractable modules and provide sufficient conditions for iso-retractable modules to be simple. Recall that a module M is said to be semi-coHopfian if any injective endomorphism of M has a direct summand image. M is regular if every cyclic submodule of M is a direct summand of M. In the following we state one result of [3] that characterizes simple modules in terms of iso-retractable modules.

Theorem 1.6 *Let R be a ring. The following are equivalent for an R-module M_R.*

1. *M_R is simple,*
2. *M_R is regular and iso-retractable,*
3. *M_R is semi-coHopfian and iso-retractable,*
4. *M_R is continuous and iso-retractable.*

Now we discuss some general properties of iso-retractable modules.

Proposition 1.7 *The submodule of an iso-retractable module is iso-retractable.*

Proof Let N be a submodule of an iso-retractable module M. Then N is isomorphic to M. Let K be any submodule of N then obviously K is isomorphic to N. Hence N is also an iso-retractable module. □

Remark 1.8 In general, quotient of an iso-retractable module need not be iso-retractable. For example, Z_Z is iso-retractable but $Z/4Z \cong Z_4$ is not iso-retractable.

In the following we show that when quotients are iso-retractable:

Proposition 1.9 *Let M_R be an iso-retractable module. Then for any fully invariant submodule N of M_R, the factor module $(M/N)_R$ is iso-retractable.*

Proof Let K/N be any submodule of $(M/N)_R$. There is an isomorphism $\phi : M \to K$. Now $\phi(N) \subseteq N$ by our assumption, and so $\alpha : M/N \to K/N$ with $\alpha(m + N) = \alpha(m) + N$ is an isomorphism. □

Recall that the right R-module M is a duo module provided every submodule of M is fully invariant. As a consequence of the above result, we have the following:

Corollary 1.10 *Let M_R be an iso-retractable and duo module. Then the factor module $(M/N)_R$ is iso-retractable for any submodule N of M_R.*

Remark 1.11 Direct sum of two iso-retractable modules need not be iso-retractable. For example, consider $Z_6 = \{0, 2, 4\} \oplus \{0, 3\}$. Z_6 as a Z-module is not an iso-retractable module. But every direct summand of Z_6 is simple, therefore they are iso-retractable.

In the following, we observe that every iso-retractable module is cyclic, Noetherian and uniform.

Theorem 1.12 *Let R be a ring and M_R be an iso-retractable module. Then*

1. *M_R is a cyclic module,*
2. *M_R is Noetherian,*
3. *M_R is a uniform module.*

Proof 1. We observe that for all nonzero $x \in M$, $M \cong xR$. Since homomorphic image of a cyclic module is cyclic therefore M is cyclic.
2. Let N be a submodule of M_R. Then $N \cong M$ and for all nonzero $x \in M$, $M \cong xR$. Therefore $M \cong xR \cong N$ and every submodule N is cyclic. Hence M is Noetherian.
3. If M_R is iso-retractable then it is Noetherian by (2). Hence M is with $u.dim < \infty$. Then M has a uniform submodule U (see [5, Proposition 6.4]). But $M \cong U$, therefore M is uniform.

□

Remark 1.13 Every iso-retractable module is cyclic but the converse need not be true. For example, if m is not prime then Z_m is a cyclic Z-module but not iso-retractable. Also, an iso-retractable module is uniform but the converse need not be true. For example, Z_4 as a Z-module is uniform but not iso-retractable. Iso-retractable modules are concerned mainly with infinite modules which do not satisfy the descending chain condition. But they satisfy ascending chain condition.

Finally, we discuss Morita invariant property.

Theorem 1.14 *Being iso-retractable is a Morita invariant property.*

Proof It is well known fact that any category equivalence preserves mono-morphisms and epi-morphisms. Therefore, it is clear by the observation that a module M_R is iso-retractable if and only if for any $N \in$Mod-R there is an isomorphism from M_R to N_R. □

References

1. Beaumont, R.A.: Groups with isomorphic proper subgroups. Bull. Amer. Math. Soc. **51**, 381–387 (1945)
2. Beaumont, R.A., Pierce, R.S.: Partly transitive modules and modules with proper isomorphic submodules. Trans. Amer. Math. Soc. **91**(2), 209–219 (1959)
3. Chaturvedi, A.K.: Iso-retractable Modules and Rings (to appear)
4. Mc Connell, J.C., Robson, J.C.: Non-commutative Noetherian Rings, Wiley-Interscience, New York (1987)
5. Lam, T.Y.: Lectures on Modules and Rings, Graduate Texts in Mathematics, vol. 139. Springer, Berlin (1998)
6. Pandeya, B.M., Chaturvedi, A.K., Gupta, A.J.: Applications of epi-retractable modules. Bull. Iranian Math. Soc. **38**(2), 469–477 (2012)
7. Smith, P.F.: Modules with many homomorphisms. J. Pure Appl. Algebra **197**, 305–321 (2005)

8. Tiwary, A.K., Pandeya, B.M.: Modules whose non zero endomorphisms are monomorphisms. In: Proceedings of the International Symposium Algebra and its Applications, New Delhi, 1981, Lecturer Notes Pure Applications Mathematics, vol. 91, pp. 199-203. Marcel Dekker, New York (1984)
9. Wisbauer, R.: Foundations of Module and Ring Theory. Gordon and Breach, London (1991)

Normal Categories from Completely Simple Semigroups

P.A. Azeef Muhammed

Abstract In this paper, we characterize the normal categories associated with a completely simple semigroup $S = \mathscr{M}[G; I, \Lambda; P]$ and show that the semigroup of normal cones $T\mathcal{L}(S)$ is isomorphic to the semi-direct product $G^\Lambda \ltimes \Lambda$. We characterize the principal cones in this category and the Green's relations in $T\mathcal{L}(S)$.

Keywords Normal category · Completely simple semigroup · Normal cones · Cross-connections

AMS 2010 Mathematics Subject Classification 20M17 · 20M10 · 20M50

1 Introduction

A semigroup S is said to be (von-neumann) *regular* if for every $a \in S$, there exists b such that $aba = a$. In the study of the structure theory of regular semigroups, there are mainly two approaches. The first approach inspired by the work of WD Munn (cf. [9]) uses the set of idempotents E of the semigroup to construct the semigroup as a full-subsemigroup of the semigroup of the principal-ideal isomorphisms of E. The biggest contribution of Kerala to the world of semigroup theory lies at the heart of this construction wherein KSS Nambooripad (cf. [10]) abstractly characterized the set of idempotents of a (regular) semigroup as a *(regular) biordered set*. It was later proved by D. Easdown (cf. [3]) that infact the idempotents of any arbitrary semigroup form a biordered set.

The second approach initiated by Hall (cf. [6]) uses the ideal structure of the regular semigroup to analyze its structure. PA Grillet (cf. [4]) refined Hall's theory to abstractly characterize the ideals as *regular partially ordered sets* and constructing the

The author wishes to acknowledge the financial support (via JRF and SRF) of the Council for Scientific and Industrial Research, New Delhi in the preparation of this paper.

P.A. Azeef Muhammed (✉)
Department of Mathematics, University of Kerala, Trivandrum, Kerala 695581, India
e-mail: azeefp@gmail.com

S.T. Rizvi et al. (eds.), *Algebra and its Applications*, Springer Proceedings in Mathematics & Statistics 174, DOI 10.1007/978-981-10-1651-6_25

fundamental image of the regular semigroup as a cross-connection semigroup. Again Nambooripad (cf. [11]) generalized the idea to any arbitrary regular semigroups by characterizing the ideals as *normal categories*.

A cross-connection between two normal categories $\mathcal{C}$ and $\mathcal{D}$ is a *local isomorphism* $\Gamma : \mathcal{D} \to N^*\mathcal{C}$ where $N^*\mathcal{C}$ is the normal dual of the category $\mathcal{C}$. A cross-conection Γ determines a cross-connection semigroup $\tilde{S}\Gamma$ and conversely every regular semigroup is isomorphic to a cross-connection semigroup for a suitable cross-connection.

A completely simple semigroup is a semigroup without zero which has no proper ideals and contains a primitive idempotent. It is known that S is a regular semigroup and any completely simple semigroup is isomorphic to the Rees matrix semigroup $\mathscr{M}[G; I, \Lambda; P]$ where G is a group I and Λ are sets and $P = (p_{\lambda i})$ is a $\Lambda \times I$ matrix with entries in G. (cf. [13]). Then $S = G \times I \times \Lambda$ with the binary operation

$$(a, i, \lambda)(b, j, \mu) = (ap_{\lambda j}b, i, \mu).$$

In this paper, we characterize the normal categories involved in the construction of a completely simple semigroup as a cross-connection semigroup. We show that the category of principal left ideals of S - $\mathcal{L}(S)$ has Λ as its set of objects and G as the set of morphisms between any two objects. We observe that it forms a normal category and we characterize the semigroup of normal cones arising from this normal category. We show that this semigroup is equal to the semi-direct product of $G^\Lambda \ltimes \Lambda$. We characterize the principal cones in this category and show that the principal cones form a regular subsemigroup of $G^\Lambda \ltimes \Lambda$. We also show for $\gamma_1 = (\bar{\gamma_1}, \lambda_k), \gamma_2 = (\bar{\gamma_2}, \lambda_l) \in T\mathcal{L}(S)$, $\gamma_1 \mathscr{L} \gamma_2$ if and only if $\lambda_k = \lambda_l$ and $\gamma_1 \mathscr{R} \gamma_2$ if and only if $\bar{\gamma_1} G = \bar{\gamma_2} G$.

2 Preliminaries

In the sequel, we assume familiarity with the definitions and elementary concepts of category theory (cf. [8]). The definitions and results on cross-connections are as in [11]. For a category $\mathcal{C}$, we denote by $v\mathcal{C}$ the set of objects of $\mathcal{C}$.

Definition 2.1 Let $\mathcal{C}$ and $\mathcal{D}$ be two categories and $F : \mathcal{C} \to \mathcal{D}$ be a functor. We shall say that a functor F is *v-injective* if vF is injective. F is said to be *v-surjective* if vF is surjective. F is said to be an isomorphism if it is v-injective, v-surjective, full and faithful.

Definition 2.2 A *preorder* $\mathcal{P}$ is a category such that for any $p, p' \in \mathcal{P}$, the hom-set $\mathcal{P}(p, p')$ contains atmost one morphism.

In this case, the relation $\subseteq$ on the class $v\mathcal{P}$ of objects of $\mathcal{P}$ defined by $p \subseteq p' \iff \mathcal{P}(p, p') \neq \emptyset$ is a quasi-order. $\mathcal{P}$ is said to be a strict preorder if $\subseteq$ is a partial order.

Definition 2.3 Let $\mathcal{C}$ be a category and $\mathcal{P}$ be a subcategory of $\mathcal{C}$. Then $(\mathcal{C}, \mathcal{P})$ is called a *category with subobjects* if $\mathcal{P}$ is a strict preorder with $v\mathcal{P} = v\mathcal{C}$ such that every $f \in \mathcal{P}$ is a monomorphism in $\mathcal{C}$ and if $f, g \in \mathcal{P}$ and if $f = hg$ for some $h \in \mathcal{C}$, then $h \in \mathcal{P}$. In a category with subobjects, if $f : c \to d$ is a morphism in $\mathcal{P}$, then f is said to be an *inclusion*. We denote this inclusion by $j(c, d)$.

Definition 2.4 A morphism $e : d \to c$ is called a *retraction* if $c \subseteq d$ and $j(c, d)$ $e = 1_c$.

Definition 2.5 A *normal factorization* of a morphism $f \in \mathcal{C}(c, d)$ is a factorization of the form $f = euj$ where $e : c \to c'$ is a retraction, $u : c' \to d'$ is an isomorphism and $j = j(d', d)$ for some $c', d' \in v\mathcal{C}$ with $c' \subseteq c, d' \subseteq d$.

Definition 2.6 Let $d \in v\mathcal{C}$. A map $\gamma : v\mathcal{C} \to \mathcal{C}$ is called a *cone from the base $v\mathcal{C}$ to the vertex d* if $\gamma(c) \in \mathcal{C}(c, d)$ for all $c \in v\mathcal{C}$ and whenever $c' \subseteq c$ then $j(c', c)\gamma(c) = \gamma(c')$. The cone γ is said to be *normal* if there exists $c \in v\mathcal{C}$ such that $\gamma(c) : c \to c_\gamma$ is an isomorphism.

Given the cone γ we denote by c_γ the the *vertex* of γ and for each $c \in v\mathcal{C}$, the morphism $\gamma(c) : c \to c_\gamma$ is called the *component* of γ at c. We define $M_\gamma = \{c \in \mathcal{C} \mid \gamma(c) \text{ is an isomorphism}\}$.

Definition 2.7 A *normal category* is a pair $(\mathcal{C}, \mathcal{P})$ satisfying the following:

1. $(\mathcal{C}, \mathcal{P})$ is a category with subobjects.
2. Any morphism in $\mathcal{C}$ has a normal factorization.
3. For each $c \in v\mathcal{C}$ there is a normal cone σ with vertex c and $\sigma(c) = 1_c$.

Theorem 1 (cf. [11]) *Let $(\mathcal{C}, \mathcal{P})$ be a normal category and let $T\mathcal{C}$ be the set of all normal cones in $\mathcal{C}$. Then $T\mathcal{C}$ is a regular semigroup with product defined as follows: For $\gamma, \sigma \in T\mathcal{C}$*

$$(\gamma * \sigma)(a) = \gamma(a)(\sigma(c_\gamma))^\circ \tag{1}$$

where $(\sigma(c_\gamma))^\circ$ is the epimorphic part of the $\sigma(c_\gamma)$. Then it can be seen that $\gamma * \sigma$ is a normal cone. $T\mathcal{C}$ is called the semigroup of normal cones *in $\mathcal{C}$.*

Let S be a regular semigroup. The category of principal left ideals of S is described as follows. Since every principal left ideal in S has at least one idempotent generator, we may write objects (vertexes) in $\mathcal{L}(S)$ as Se for $e \in E(S)$. Morphisms $\rho : Se \to Sf$ are right translations $\rho = \rho(e, s, f)$ where $s \in eSf$ and ρ maps $x \mapsto xs$. Thus

$$v\mathcal{L}(S) = \{Se : e \in E(S)\} \quad \text{and} \quad \mathcal{L}(S) = \{\rho(e, s, f) : e, f \in E(S),\ s \in eSf\}. \tag{2}$$

Proposition 1 (cf. [11]) *Let S be a regular semigroup. Then $\mathcal{L}(S)$ is a normal category. $\rho(e, u, f) = \rho(e', v, f')$ if and only if $e \mathscr{L} e'$, $f \mathscr{L} f'$, $u \in eSf$, $v \in e'Sf'$ and $v = e'u$. Let $\rho = \rho(e, u, f)$ be a morphism in $\mathcal{L}(S)$. For any $g \in R_u \cap \omega(e)$ and $h \in E(L_u)$, $\rho = \rho(e, g, g)\rho(g, u, h)\rho(h, h, f)$ is a normal factorization of ρ.*

Proposition 2 (cf. [11]) *Let S be a regular semigroup, $a \in S$ and $f \in E(L_a)$. Then for each $e \in E(S)$, let $\rho^a(Se) = \rho(e, ea, f)$. Then ρ^a is a normal cone in $\mathcal{L}(S)$ with vertex Sa. $M_{\rho^a} = \{Se : e \in E(R_a)\}$. ρ^a is an idempotent in $T\mathcal{L}(S)$ iff $a \in E(S)$. The mapping $a \mapsto \rho^a$ is a homomorphism from S to $T\mathcal{L}(S)$. Further if S is a monoid, then S is isomorphic to $T\mathcal{L}(S)$.*

3 Normal Categories in a Completely Simple Semigroup

Given a completely simple semigroup S, it is known that S is isomorphic to the Rees matrix semigroup $\mathscr{M}[G; I, \Lambda; P]$ where G is a group I and Λ are sets and $P = (p_{\lambda i})$ is a $\Lambda \times I$ matrix with entries in G.(cf. [13]). Then $S = G \times I \times \Lambda$ with the binary operation

$$(a, i, \lambda)(b, j, \mu) = (ap_{\lambda j}b, i, \mu).$$

It is easy to see that $(g_1, i_1, \lambda_1)\mathscr{L}(g_2, i_2, \lambda_2)$ if and only if $\lambda_1 = \lambda_2$ (cf. [7]). Observe that (g, i, λ) is an idempotent in S if and only if $g = p_{\lambda i}^{-1}$. Hence, the set of objects of $v\mathcal{L}(S)$ is equal $S(p_{\lambda i}^{-1}, i, \lambda)$ (see Eq. 2). Now given an idempotent $(p_{\lambda_1 i_1}^{-1}, i_1, \lambda_1)$, for an arbitrary $s = (g_s, i_s, \lambda_s) \in S$, $(g_s, i_s, \lambda_s)(p_{\lambda_1 i_1}^{-1}, i_1, \lambda_1) = (g_s p_{\lambda_s i_1} p_{\lambda_1 i_1}^{-1}, i_s, \lambda_1)$. Now since g_s and i_s in the product are arbitrary we see that principal left ideal generated by the idempotent will have elements of the form (g, i, λ_1) with g and i arbitrary elements of G and I, respectively. The ideal will be of the form $G \times I \times \lambda_1$. Hence any principal left ideal will be of the form $G \times I \times \lambda$ such that $\lambda \in \Lambda$. So

$$v\mathcal{L}(S) = \{G \times I \times \lambda : \lambda \in \Lambda\}.$$

Henceforth, we will denote the left ideal $S(p_{\lambda i}^{-1}, i, \lambda) = G \times I \times \lambda$ by $\bar{\lambda}$ and the set $v\mathcal{L}(S)$ will be denoted by $\bar{\Lambda}$.

Recall that any morphism from $Se = \bar{\lambda_1}$ to $Sf = \bar{\lambda_2}$ will be of the form $\rho(e, u, f)$ where $u \in eSf$ (see Eq. 2). Without loss of generality, let us assume that $e = (p_{\lambda_1 i_1}^{-1}, i_1, \lambda_1)$ and $f = (p_{\lambda_2 i_2}^{-1}, i_2, \lambda_2)$. Then since $u \in eSf$, for some $s = (g_s, i_s, \lambda_s) \in S$, u will be of the form $(p_{\lambda_1 i_1}^{-1}, i_1, \lambda_1)(g_s, i_s, \lambda_s)(p_{\lambda_2 i_2}^{-1}, i_2, \lambda_2) = (p_{\lambda_1 i_1}^{-1} p_{\lambda_1 i_s} g_s p_{\lambda_s i_2} p_{\lambda_2 i_2}^{-1}, i_1, \lambda_2)$. Again since $g_s \in G$ in the product is arbitrary we see that u will be of the form (g_u, i_1, λ_2) for an arbitrary $g_u \in G$. Hence any morphism from $\bar{\lambda_1}$ to $\bar{\lambda_2}$ will be of the form $\rho((p_{\lambda_1 i_1}^{-1}, i_1, \lambda_1), (g_u, i_1, \lambda_2), (p_{\lambda_2 i_2}^{-1}, i_2, \lambda_2))$ where $g_u \in G$. since for any morphism in $\mathcal{L}(S)$, $(g_u, i_1, \lambda_2) \in (p_{\lambda_1 i_1}^{-1}, i_1, \lambda_1)S(p_{\lambda_2 i_2}^{-1}, i_2, \lambda_2)$ and $s = (g_s, i_s, \lambda_s) \in S$ is arbitrary.

For an element $x \in Se$, since a morphism $\rho(e, u, f)$ maps $x \mapsto xu \in Sf$, a morphism $\rho((p_{\lambda_1 i_1}^{-1}, i_1, \lambda_1), (g_u, i_1, \lambda_2), (p_{\lambda_2 i_2}^{-1}, i_2, \lambda_2))$ will map $(g_x, i_x, \lambda_1) \in \bar{\lambda_1}$ to $(g_x, i_x, \lambda_1)(g_u, i_1, \lambda_2) = (g_x p_{\lambda_1 i_1} g_u, i_x, \lambda_2) \in \bar{\lambda_2}$. Since $g_u \in G$ was chosen arbitrarily, $p_{\lambda_1 i_1} g_u$ will be an arbitrary element of the group G. So if $g = p_{\lambda_1 i_1} g_u$, the morphism maps $(g_x, i_x, \lambda_1) \mapsto (g_x g, i_x, \lambda_2)$. Observe that morphism involves only the right translation of the group element g_x to $g_x g$ and the rest of the component is fixed. Hence, the morphism is essentially $g_x \mapsto g_x g$ such that $g \in G$ for some arbitrary $g \in G$. So the set of morphisms from $\bar{\lambda_1}$ to $\bar{\lambda_2}$ is

$$\mathcal{L}(S)(\bar{\lambda_1}, \bar{\lambda_2}) = \{(g_x, i_s, \lambda_1) \mapsto (g_x g, i_s, \lambda_2) : g \in G\}.$$

Hence, the set of morphisms from $\bar{\lambda_1}$ to $\bar{\lambda_2}$ is the set G and each $g \in G$ will map $(g_x, i_x, \lambda_1) \mapsto (g_x g, i_x, \lambda_2)$. We will denote this morphism as ρ_g when there is no ambiguity regarding the domain and range.

Remark 1 Observe that for every morphism ρ_g in $\mathcal{L}(S)$ from $\bar{\lambda_1}$ to $\bar{\lambda_2}$, since $g^{-1} \in G$, there exists a morphism $\rho_{g^{-1}}$ between $\bar{\lambda_2}$ and $\bar{\lambda_1}$. Then for any $(g_x, i_x, \lambda_1) \in \bar{\lambda_1}$, $(g_x, i_x, \lambda_1)\rho_g\rho_{g^{-1}} = (g_x g, i_x, \lambda_2)\rho_{g^{-1}} = (g_x g g^{-1}, i_x, \lambda_1) = (g_x, i_x, \lambda_1)$.

Also for any $(g_y, i_y, \lambda_2) \in \bar{\lambda_2}$, $(g_y, i_y, \lambda_2)\rho_{g^{-1}}\rho_g = (g_y g^{-1}, i_y, \lambda_1)\rho_g = (g_y g^{-1} g, i_y, \lambda_2) = (g_y, i_y, \lambda_2)$.

So $(\bar{\lambda_1})\rho_g\rho_{g^{-1}} = 1_{\bar{\lambda_1}}$ and $(\bar{\lambda_2})\rho_{g^{-1}}\rho_g = 1_{\bar{\lambda_2}}$. Hence ${\rho_g}^{-1} = \rho_{g^{-1}}$ and every morphism ρ_g has an inverse $\rho_{g^{-1}}$. Consequently every morphism will be an isomorphism and hence $\mathcal{L}(S)$ will be a groupoid. So in $\mathcal{L}(S)$ there will be no inclusions and the epimorphic component of every morphism will be the morphism itself.

Now, we characterize the normal cones in $\mathcal{L}(S)$. Recall that a cone in $\mathcal{C}$ with vertex $d \in v\mathcal{C}$ is a map $\gamma : v\mathcal{C} \to \mathcal{C}$ such that $\gamma(c) \in \mathcal{C}(c, d)$ for all $c \in v\mathcal{C}$ and whenever $c' \subseteq c$ then $j(c', c)\gamma(c) = \gamma(c')$. Since there are no inclusions in $\mathcal{L}(S)$, the second condition is redundant. Since $v\mathcal{L}(S)$ is bijective with Λ and morphisms in $\mathcal{L}(S)$ are just right multiplication by elements of G; a cone with vertex $\bar{\lambda}$ is a map $\gamma : \bar{\Lambda} \to G$ such that $\gamma(\bar{\lambda_k}) \in G$ for all $\bar{\lambda_k} \in \bar{\Lambda}$. Hence, a cone γ is a $|\Lambda|$-tuple of elements of G along with the vertex represented by an element $\lambda \in \Lambda$. Hence any normal cone γ with vertex $\bar{\lambda}$ can be represented by $(\bar{\gamma}, \lambda)$ where $\bar{\gamma} \in G^{\Lambda}$, $\lambda \in \Lambda$. Then $\bar{\lambda_k}$ is right multiplied by the group element g_k, the kth coordinate of $\bar{\gamma}$. Also since every morphism is an isomorphism, every cone in $\mathcal{L}(S)$ will be a normal cone.

Now we proceed to look at $T\mathcal{L}(S)$, the semigroup of all normal cones in $\mathcal{L}(S)$. Given $\gamma_1 = (\bar{\gamma_1}, \lambda_k)$, $\gamma_2 = (\bar{\gamma_2}, \lambda_l) \in T\mathcal{L}(S)$, the multiplication is defined as (see Eq. 1):

$$\gamma_1 * \gamma_2 = (\bar{\gamma_1}.\bar{g_k}, \lambda_l) \tag{3}$$

where g_k is the kth coordinate of $\bar{\gamma_2}$ and $\bar{g_k} = (g_k, g_k, ..., g_k)$. Hence $T\mathcal{L}(S)$ is isomorphic to $G^{\Lambda} \times \Lambda$ with multiplication as defined above. We further observe that the semigroup obtained here can be realized as a semi-direct product of semigroups.

Definition 3.1 (cf. [5]) Let S and T be semigroups. A *(left) action* of T on S is a map $S \times T \to S$, $(s, t) \mapsto {}^t s$ satisfying: (i) ${}^{t_1 t_2} s = {}^{t_1}({}^{t_2} s)$ and (ii) ${}^t s_1 s_2 = {}^t(s_1){}^t(s_2)$ for all $t, t_1, t_2 \in T$ and $s, s_1, s_2 \in S$.

Definition 3.2 (cf. [5]) The *semidirect product* $S \ltimes T$ of S and T, with respect to a left action of T on S, has as its underlying set $S \times T$ with multiplication defined by

$$(s_1, t_1)(s_2, t_2) = (s_1 {}^{t_1} s_2, t_1 t_2)$$

It is well known that $S \ltimes T$ is a semigroup. It is trivially verified that the idempotents in $S \ltimes T$ are the pairs (s, t) such that $t \in E(T)$ and $s\, {}^t s = s$.

Now we show that the semigroup of normal cones in $\mathcal{L}(S)$ is the semigroup of a semi-direct product of $G^\Lambda \ltimes \Lambda$.

Firstly, we look at the semigroups G^Λ and Λ. Since G is a group; G^Λ will form a group under component-wise multiplication defined as follows. For $(g_1, g_2...g_{|\Lambda|})$, $(h_1, h_2...h_{|\Lambda|}) \in G^\Lambda$

$$(g_1, g_2...g_{|\Lambda|})(h_1, h_2...h_{|\Lambda|}) = (g_1h_1, g_2h_2, ..., g_{|\Lambda|}h_{|\Lambda|})$$

and hence in particular G^Λ is also a semigroup. The set Λ (coming from the Rees matrix semigroup) admits a right zero semigroup structure and hence has an in-built multiplication given by $\lambda_k\lambda_l = \lambda_l$ for every $\lambda_k, \lambda_l \in \Lambda$. Now we define a left action of Λ on G^Λ, $\phi : G^\Lambda \times \Lambda \to G^\Lambda$ as follows. For $(g_1, g_2, ..., g_{|\Lambda|}) \in G^\Lambda$ and $\lambda_k \in \Lambda$,

$$((g_1, g_2, ..., g_{|\Lambda|}), \lambda_k)\phi = (g_k, g_k, ..., g_k) \tag{4}$$

Lemma 1 *The function $\phi : G^\Lambda \times \Lambda \to G^\Lambda$ as defined in Eq. 4 is a left action of Λ on G^Λ.*

Proof Clearly the function is well-defined. Now for $\bar{g} = (g_1, g_2, ..., g_{|\Lambda|}) \in G^\Lambda$, $(\bar{g}, \lambda_k\lambda_l)\phi = ((g_1, g_2, ..., g_{|\Lambda|}), \lambda_k\lambda_l)\phi = ((g_1, g_2, ..., g_{|\Lambda|}), \lambda_l)\phi = (g_l, g_l, ..., g_l)$. (since $\lambda_k\lambda_l = \lambda_l$).

Also $((\bar{g}, \lambda_l)\phi, \lambda_k)\phi = (((g_1, g_2, ..., g_{|\Lambda|}), \lambda_l)\phi, \lambda_k)\phi = ((g_l, g_l, ..., g_l), \lambda_k)$ $\phi = (g_l, g_l, ..., g_l)$.

Hence $(\bar{g}, \lambda_k\lambda_l)\phi = ((\bar{g}, \lambda_l)\phi, \lambda_k)\phi$ for $\bar{g} \in G^\Lambda$ and $\lambda_k, \lambda_l \in \Lambda$.

Then for $\bar{g} = (g_1, g_2, ..., g_{|\Lambda|})$, $\bar{h} = (h_1, h_2, ..., h_{|\Lambda|}) \in G^\Lambda$, and $\lambda_k \in \Lambda$, $(\bar{g}\bar{h}, \lambda_k)$ $\phi = ((g_1, g_2...g_{|\Lambda|})(h_1, h_2...h_{|\Lambda|}), \lambda_k)\phi = ((g_1h_1, g_2h_2, ..., g_{|\Lambda|}h_{|\Lambda|}), \lambda_k)\phi = (g_k h_k, g_kh_k, ..., g_kh_k)$. $(\bar{g}, \lambda_k)\phi(\bar{h}, \lambda_k)\phi = ((g_1, g_2, ..., g_{|\Lambda|}), \lambda_k)\phi((h_1, h_2, ..., h_{|\Lambda|}), \lambda_k)\phi = (g_k, g_k, ..., g_k)(h_k, h_k, ..., h_k) = (g_kh_k, g_kh_k, ..., g_kh_k)$.

Hence $(\bar{g}\bar{h}, \lambda_k)\phi = (\bar{g}, \lambda_k)\phi(\bar{h}, \lambda_k)\phi$ for $\bar{g}, \bar{h} \in G^\Lambda$, and $\lambda_k \in \Lambda$.

Thus ϕ is a left semigroup action of Λ on G^Λ. □

Proposition 3 *$T\mathcal{L}(S)$ is the semi-direct product $G^\Lambda \ltimes \Lambda$ with respect to the left action ϕ.*

Proof The semi-direct product of $G^\Lambda \ltimes \Lambda$ with respect to ϕ is given by, for $(\bar{\gamma_1}, \lambda_k)$, $(\bar{\gamma_2}, \lambda_l) \in G^\Lambda \times \Lambda$, the multiplication is defined as (see Definition 3.2): $(\bar{\gamma_1}, \lambda_k) * (\bar{\gamma_2}, \lambda_l) = (\bar{\gamma_1}.(\bar{\gamma_2}, \lambda_k)\phi, \lambda_k\lambda_l) = (\bar{\gamma_1}.\bar{g_k}, \lambda_l)$ where g_k is the kth coordinate of $\bar{\gamma_2}$ and $\bar{g_k} = (g_k, g_k, ..., g_k)$. So if we take $\gamma_1 = (\bar{\gamma_1}, \lambda_k)$ and $\gamma_2 = (\bar{\gamma_2}, \lambda_l)$ then, the multiplication defined above is exactly the same multiplication defined in Eq. 3 and hence $T\mathcal{L}(S)$ is the semi-direct product $G^\Lambda \ltimes \Lambda$ with respect ϕ. □

The idempotents in $T\mathcal{L}(S)$ can be characterized by the following lemma.

Lemma 2 *$\gamma = (\bar{\gamma}, \lambda_k) \in T\mathcal{L}(S)$ is an idempotent if and only if $g_k = e$ where g_k is the kth coordinate of $\bar{\gamma}$ and e is the identity of the group G.*

Proof Suppose $\gamma = (\bar{\gamma}, \lambda_k) \in T\mathcal{L}(S)$ be an idempotent. Then, $(\bar{\gamma}, \lambda_k) * (\bar{\gamma}, \lambda_k) = (\bar{\gamma}, \lambda_k)$. .i.e $(\bar{\gamma}.\bar{g_k}, \lambda_k) = (\bar{\gamma}, \lambda_k)$. Hence $\bar{\gamma}.\bar{g_k} = \bar{\gamma}$. Now since $\bar{\gamma} \in G^\Lambda$ was arbitrary; this is possible only if $g.g_k = g$ for every $g \in G$. Hence $g_k = g^{-1}g = e$.

Conversely if $g_k = e$, $\gamma^2 = (\bar{\gamma}, \lambda_k) * (\bar{\gamma}, \lambda_k) = (\bar{\gamma}.\bar{e}, \lambda_k) = (\bar{\gamma}, \lambda_k) = \gamma$. Hence γ is an idempotent. □

Observe that the lemma can also be obtained by appealing to the characterization of idempotents in the semi-direct product of the semigroups. Every morphism in $\mathcal{L}(S)$ being an isomorphism is already a normal factorization. Also since for any $\lambda_k \in \Lambda$, γ such that $\gamma(\bar{\lambda_k}) = e$ gives an idempotent normal cone with vertex $\bar{\lambda_k}$, we see that for any vertex $\lambda_k \in \Lambda$, we have an idempotent normal cone with that vertex. Hence $\mathcal{L}(S)$ explicitly satisfies all the properties to be a normal category. Now we proceed to characterize the principal cones in $T\mathcal{L}(S)$.

Proposition 4 *The principal cones in $T\mathcal{L}(S)$ forms a regular subsemigroup of $G^\Lambda \ltimes \Lambda$.*

Proof Given $a = (g_a, i_a, \lambda_a)$, the principal cone ρ^a (see Proposition 2) will be a normal cone with vertex $\bar{\lambda}_a$ such that each left ideal λ_k is right multiplied by (g_a, i_a, λ_a). But since the morphisms in $\mathcal{L}(S)$ involves right multiplication by the product of a sandwich element of the matrix and the group element; at each $\bar{\lambda}_k \in \Lambda$, $\rho^a(\bar{\lambda}_k) = p_{\lambda_k i_a} g_a$ (where the sandwich matrix $P = (p_{\lambda i})_{\lambda \in \Lambda, i \in I}$).

Hence ρ^a can be represented by $(p_{\lambda_1 i_a} g_a, p_{\lambda_2 i_a} g_a, p_{\lambda_3 i_a} g_a, \dots, p_{\lambda_{|\Lambda|} i_a} g_a; \lambda_a) \in G^\Lambda \ltimes \Lambda$. Observe that $\bar{\rho^a} \in G^\Lambda$ is the right translation of the i_a-th column of the sandwich matrix P with group element g_a.

Now $\rho^a.\rho^b = (p_{\lambda_1 i_a} g_a, p_{\lambda_2 i_a} g_a, p_{\lambda_3 i_a} g_a, \dots, p_{\lambda_{|\Lambda|} i_a} g_a; \lambda_a).(p_{\lambda_1 i_b} g_b, p_{\lambda_2 i_b} g_b, p_{\lambda_3 i_b} g_b, \dots, p_{\lambda_{|\Lambda|} i_b} g_b; \lambda_b)$ $= (p_{\lambda_1 i_a} g_a p_{\lambda_a i_b} g_b, p_{\lambda_2 i_a} g_a p_{\lambda_a i_b} g_b, p_{\lambda_3 i_a} g_a p_{\lambda_a i_b} g_b, \dots, p_{\lambda_{|\Lambda|} i_a} g_a p_{\lambda_a i_b} g_b; \lambda_b)$. Also $a.b = (g_a, i_a, \lambda_a)(g_b, i_b, \lambda_b) = (g_a p_{\lambda_a i_b} g_b, i_a, \lambda_b)$. So $\rho^{ab} = (p_{\lambda_1 i_a} g_a p_{\lambda_a i_b} g_b, p_{\lambda_2 i_a} g_a p_{\lambda_a i_b} g_b, p_{\lambda_3 i_a} g_a p_{\lambda_a i_b} g_b, \dots, p_{\lambda_{|\Lambda|} i_a} g_a p_{\lambda_a i_b} g_b; \lambda_b)$.

Hence $\rho^a.\rho^b = \rho^{ab}$ and consequently the map $a \mapsto \rho^a$ from $S \to T\mathcal{L}(S)$ is a homomorphism. So the set of principal cones in $\mathcal{L}(S)$ forms a subsemigroup of $T\mathcal{L}(S)$. Since S is regular, the semigroup of principal cones forms a regular subsemigroup of $G^\Lambda \ltimes \Lambda$. □

If S is a regular monoid, then it is known that S is isomorphic to $T\mathcal{L}(S)$ (see Proposition 2). But in general, this may not be true. But always there exists a homomorphism from $S \to T\mathcal{L}(S)$ mapping $a \mapsto \rho^a$ which may not be one-one or onto. But in the case of completely simple semigroups, there exists some special cases where this map is one-one and consequently S can be realized as a subsemigroup of $T\mathcal{L}(S)$.

Theorem 2 *$S = \mathscr{M}[G; I, \Lambda; P]$ is isomorphic to the semigroup of principal cones in $T\mathcal{L}(S)$ if and only if for every $g \in G$ and $i_1 \neq i_2 \in I$, there exists $\lambda_k \in \Lambda$ such that $p_{\lambda_k i_1} \neq p_{\lambda_k i_2} g$.*

Proof As seen above, there exists a semigroup homomorphism from $\psi :\to T\mathcal{L}(S)$ mapping $a \mapsto \rho^a$. Now S can be seen as a subsemigroup of $T\mathcal{L}(S)$ if and only if ψ is one-one.

We claim ψ is one-one only if for every $i_1 \neq i_2 \in I$ and every $g \in G$, there exists $\lambda_k \in \Lambda$ such that $p_{\lambda_k i_1} \neq p_{\lambda_k i_2} g$. Suppose not. i.e, there exists $i_1 \neq i_2 \in I$ and a $g \in G$ such that $p_{\lambda_k i_1} = p_{\lambda_k i_2} g$ for every $\lambda_k \in \Lambda$. Then without loss of generality, assume $g = g_2 g_1^{-1}$ for some $g_1, g_2 \in G$. Then $p_{\lambda_k i_1} = p_{\lambda_k i_2} g_2 g_1^{-1}$ for some $g_1, g_2 \in G$. .i.e $p_{\lambda_k i_1} g_1 = p_{\lambda_k i_2} g_2$ for every $\lambda_k \in \Lambda$. Hence, if $a = (g_1, i_1, \lambda)$ and $b = (g_2, i_2, \lambda)$; then since $i_1 \neq i_2$, $a \neq b$. Then $\rho^a = (p_{\lambda_1 i_1} g_1, p_{\lambda_2 i_1} g_1, p_{\lambda_3 i_1} g_1, ..., p_{\lambda_{|\Lambda|} i_1} g_1; \lambda)$ and $\rho^b = .(p_{\lambda_1 i_2} g_2, p_{\lambda_2 i_2} g_2, p_{\lambda_3 i_2} g_2, ..., p_{\lambda_{|\Lambda|} i_2} g_2; \lambda)$. Since $p_{\lambda_k i_1} g_1 = p_{\lambda_k i_2} g_2$ for every $\lambda_k \in \Lambda$, $a \neq b$ but $\rho^a = \rho^b$. Hence ψ is not one-one. Hence S is not isomorphic to the semigroup of principal cones in $T\mathcal{L}(S)$.

Conversely suppose for every $i_1 \neq i_2 \in I$ and every $g \in G$, there exists $\lambda_k \in \Lambda$ such that $p_{\lambda_k i_1} \neq p_{\lambda_k i_2} g$, we need to show ψ is one-one so that S is isomorphic to the semigroup of principal cones. Suppose if $a = (g_a, i_a, \lambda_a)$ and $b = (g_b, i_b, \lambda_b)$ and $\psi(a) = \psi(b)$. So $\rho^a = \rho^b$; .i.e, $(p_{\lambda_1 i_a} g_a, p_{\lambda_2 i_a} g_a, p_{\lambda_3 i_a} g_a, ..., p_{\lambda_{|\Lambda|} i_a} g_a; \lambda_a) = (p_{\lambda_1 i_b} g_b, p_{\lambda_2 i_b} g_b, p_{\lambda_3 i_b} g_b, ..., p_{\lambda_{|\Lambda|} i_b} g_b; \lambda_b)$. So clearly $\lambda_a = \lambda_b$. $p_{\lambda_k i_a} g_a = p_{\lambda_k i_b} g_b$ for every $\lambda_k \in \Lambda$. Now if $g_a \neq g_b$, then $p_{\lambda_k i_a} = p_{\lambda_k i_b} g_b g_a^{-1}$ and so $p_{\lambda_k i_a} = p_{\lambda_k i_b} g$ for every $\lambda_k \in \Lambda$ (taking $g = g_b g_a^{-1}$). But this will contradict our supposition unless $i_a = i_b$. But then $p_{\lambda_k i_a} = p_{\lambda_k i_a} g$ for every $\lambda_k \in \Lambda$. Now this is possible only if $g = e$; which implies $g_b g_a^{-1} = e$ .i.e, $g_b = g_a$; which is again a contradiction. Hence $g_a = g_b$.

Now if $i_a \neq i_b$; then, we have $p_{\lambda_k i_a} g_a = p_{\lambda_k i_b} g_b$ for every $\lambda_k \in \Lambda$. Then taking $g = g_b g_a^{-1}$; we have $p_{\lambda_k i_a} = p_{\lambda_k i_b} g$ for every $\lambda_k \in \Lambda$; which is again a contradiction to our supposition. Hence $i_a = i_b$.

So $(g_a, i_a, \lambda_a) = (g_b, i_b, \lambda_b)$ and so $a = b$. Thus $\rho^a = \rho^b$ implies $a = b$ making ψ one-one. So if for every $i_1 \neq i_2 \in I$ and every $g \in G$, there exists $\lambda_k \in \Lambda$ such that $p_{\lambda_k i_1} \neq p_{\lambda_k i_2} g$, then S is isomorphic to the semigroup of principal cones. Hence the proof. □

Now we proceed to characterize the Green's relations in $T\mathcal{L}(S)$.

Proposition 5 *If* $\gamma_1 = (\bar{\gamma_1}, \lambda_k), \gamma_2 = (\bar{\gamma_2}, \lambda_l) \in T\mathcal{L}(S)$, *then* $\gamma \mathscr{L} \gamma'$ *if and only if* $\lambda_k = \lambda_l$.

Proof Suppose if $\lambda_k \neq \lambda_l$. Then for an arbitrary $\delta = (\bar{\delta}, \lambda_m) \in T\mathcal{L}(S)$; $\delta\gamma_1 = (\bar{\delta}, \lambda_m)$
$(\bar{\gamma_1}, \lambda_k) = (\bar{\delta}.\bar{g_m}, \lambda_k)$ where $g_m = \gamma_1(\bar{\lambda_m})$. Hence $\delta\gamma_1 \in G^\Lambda \times \lambda_k$. Since $\bar{\delta} \in G^\Lambda$ is arbitrary, $T\mathcal{L}(S)\gamma_1 = G^\Lambda \times \lambda_k$. Similarly $T\mathcal{L}(S)\gamma_2 = G^\Lambda \times \lambda_l$. So $T\mathcal{L}(S)\gamma_1 \neq T\mathcal{L}(S)\gamma_2$.

Conversely if $\lambda_k = \lambda_l$, then $G^\Lambda \times \lambda_k = G^\Lambda \times \lambda_l$ and so $T\mathcal{L}(S)\gamma_1 = T\mathcal{L}(S)\gamma_2$. Thus $\gamma_1 \mathscr{L} \gamma_2$. Hence the proof. □

Now we proceed to get a characterization of Green's $\mathscr{R}$ relation in the semigroup $T\mathcal{L}(S)$. For this end, begin by observing that G^Λ is a group with component-wise multiplication defined as follows. For $(g_1, g_2 ... g_{|\Lambda|}), (h_1, h_2 ... h_{|\Lambda|}) \in G^\Lambda$

$$(g_1, g_2 ... g_{|\Lambda|})(h_1, h_2 ... h_{|\Lambda|}) = (g_1 h_1, g_2 h_2, ..., g_{|\Lambda|} h_{|\Lambda|})$$

Then G may be viewed as a subgroup (not necessarily normal) of G^Λ by identifying $g \mapsto (g, g, ..., g) \in G^\Lambda$. Then for some $\bar{\gamma} = (g_1, g_2...g_{|\Lambda|}) \in G^\Lambda$ we look at the left coset $\bar{\gamma}G$ of G in G^Λ with respect to $\bar{\gamma}$. It is defined as

$$\bar{\gamma}G = \{(g_1g, g_2g, ..., g_{|\Lambda|}g) \mid g \in G\}$$

Observe that these cosets form a partition of G^Λ. Now we show that these left cosets of G in G^Λ infact gives the characterization of Green's $\mathscr{R}$ relation in $T\mathcal{L}(S)$.

Let $\delta = (\bar{\delta}, \lambda_m) \in T\mathcal{L}(S)$; then $\gamma_1\delta = (\bar{\gamma_1}, \lambda_k)(\bar{\delta}, \lambda_m) = (\bar{\gamma_1}.\bar{g_k}, \lambda_m)$ where $g_k = \bar{\delta}(\lambda_k)$. Hence $\gamma_1\delta \in (\bar{\gamma_1}G) \times \Lambda$. Since $\delta \in G^\Lambda \times \Lambda$ is arbitrary, both λ_m and g_k can be arbitrarily chosen. Hence $\gamma_1 T\mathcal{L}(S) = (\bar{\gamma_1}G) \times \Lambda$. This gives us the following characterization of the $\mathscr{R}$ relation in $T\mathcal{L}(S)$.

Proposition 6 *For $\gamma_1 = (\bar{\gamma_1}, \lambda_k), \gamma_2 = (\bar{\gamma_2}, \lambda_l) \in T\mathcal{L}(S)$, $\gamma_1\mathscr{R}\gamma_2$ if and only if $\bar{\gamma_1}$ $G = \bar{\gamma_2}G$.*

Proof Suppose $\bar{\gamma_1}G = \bar{\gamma_2}G$; then $(\bar{\gamma_1}G) \times \Lambda = (\bar{\gamma_2}G) \times \Lambda$ and by the above discussion $\gamma_1 T\mathcal{L}(S) = \gamma_2 T\mathcal{L}(S)$ and hence $\gamma_1\mathscr{R}\gamma_2$.

Conversely, suppose $\bar{\gamma_1}G \neq \bar{\gamma_2}G$. i.e., there exists a $\delta = (\bar{\delta}, \lambda_m) \in T\mathcal{L}(S)$ such that $\bar{\delta} \in \bar{\gamma_1}G$ but $\bar{\delta} \notin \bar{\gamma_2}G$. Hence $\delta \in (\bar{\gamma_1}G) \times \Lambda$ but $\delta \notin (\bar{\gamma_1}G) \times \Lambda$. Hence $\gamma_1 T\mathcal{L}$ $(S) \neq \gamma_2 T\mathcal{L}(S)$. So γ_1 and γ_2 are not $\mathscr{R}$ related.

Hence the proof. □

Observe that the $\mathscr{R}$ relation of $\gamma_1 = (\bar{\gamma_1}, \lambda_k) \in T\mathcal{L}(S)$ depends only on $\bar{\gamma_1}$ and not on λ_k and so we will denote the $\mathscr{R}$ classes of $T\mathcal{L}(S)$ by $R_{\bar{\gamma}}$ such that $\bar{\gamma} \in G^\Lambda$. We further show that if G is abelian, the $\mathscr{R}$ classes of $T\mathcal{L}(S)$ infact has an additional structure of that of a group.

Theorem 3 *The $\mathscr{R}$ classes of $T\mathcal{L}(S)$ form a group if and only if G is abelian.*

Proof Recall that G is normal in G^Λ if and only if G is abelian.

Suppose G is abelian, then G is normal in G^Λ. Then define the operation on the $\mathscr{R}$ classes - $R_{\bar{\gamma_1}}$ and $R_{\bar{\gamma_1}}$ as follows.

$$R_{\bar{\gamma_1}} * R_{\bar{\gamma_2}} = R_{\bar{\gamma_1}.\bar{\gamma_2}}$$

Since the binary operation $\bar{\gamma_1}G * \bar{\gamma_2}G = \bar{\gamma_1}\bar{\gamma_2}G$ defines a group structure on G^Λ/G; so does the operation defined above. Hence $\mathscr{R}(T\mathcal{L}(S))$ is isomorphic to G^Λ/G as abelian groups.

Conversely, if G is non-abelian, G is not a normal subgroup of G^Λ, and hence, the coset multiplication is not well-defined; and hence G^Λ/G fails to be a group. □

Remark 2 It must be noted here that $R_{\gamma_1} * R_{\gamma_2} \neq R_{\gamma_1.\gamma_2}$ since the operation $\gamma_1.\gamma_2$ is a semi-direct product operation whereas $\bar{\gamma_1}.\bar{\gamma_2}$ is direct product multiplication.

4 Conclusion

We have characterized the normal categories associated with a completely simple semigroup. Now further these categories can be used to construct the cross-connection semigroups using the appropriate *local isomorphisms* between the categories. Different local isomorphisms will give rise to different completely simple semigroups (cf. [1]). This correspondence can shed more light on the structure of completely simple semigroups; and may be generalized to give satisfactory structure theorems in certain classes of regular semigroups like completely regular semigroups, (inverse) clifford semigroups, etc.

References

1. Azeef Muhammed, P.A., Rajan, A.R.: Cross-connections of completely simple semigroups. Asian Eur. J. Math. (Accepted for publication) (2015)
2. Clifford, A.H., Preston, G.B.: The Algebraic theory of semigroups. Math Surveys, vol. I, No. 7. American Mathematical Society, Providence (1961)
3. Easdown, D.: Bi-ordered sets comes from semigroups. J. Algebra **96**, 581–591 (1985)
4. Grillet, P.A.: Structure of regular semigroups I. A representation II. Cross connections III. The reduced case. Semigroup Forum, 8, pp. 177-183, 254-259, 260-265 (1974)
5. Grillet, P.A.: Semigroups - An Introduction to the Structure Theory. Marcel Dekker, New York (1995)
6. Hall, T.E.: On regular semigroups. J. Algebra **24**, 1–24 (1973)
7. Howie, J.M.: Fundamentals of Semigroup Theory. Clarendon, Oxford (1995)
8. Mac Lane, S.: Categories for the Working Mathematician. Springer, New York (1971)
9. Munn, W.D.: Fundamental inverse semigroups. Quart. J. Math. Oxford **21**, 157–170 (1970)
10. Nambooripad, K.S.S: Structure of regular semigroups I. Mem. Am. Math. Soc. **224** (1979)
11. Nambooripad, K.S.S.: Theory of Cross-connections. Publication No. 28. Centre for Mathematical Sciences, Trivandrum (1994)
12. Rajan, A.R., Azeef Muhammed, P.A.: Normal categories from the transformation semigroup. Bull. Allahabad Math. Soc. **30**(1), 61–74 (2015)
13. Rees, D.: On semigroups. Proc. Camb. Phil. Soc. **36**, 387–400 (1940)

Ordered Semigroups Characterized in Terms of Intuitionistic Fuzzy Ideals

Noor Mohammad Khan and Mohammad Aasim Khan

Abstract In the present paper, the notions of $(\in, \in \vee q_k)$-intuitionistics fuzzy ideal, $(\in, \in \vee q_k)$-intuitionistics fuzzy bi-ideal, and $(\in, \in \vee q_k)$-intuitionistics fuzzy generalized bi-ideal of an ordered semigroup are introduced. Then, we characterize these various intuitionistic fuzzy ideals. Finally, using the properties of these intuitionistic fuzzy ideals, we have characterized different classes of ordered semigroups.

Keywords Ordered semigroup · Regular ordered semigroup · Intra-regular ordered semigroup · Intuitionistic fuzzy ideals

1 Introduction

Zadeh [26] introduced the notion of a fuzzy subset of a set in 1965. This seminal paper has opened up new insights and application in a wide range of scientific fields. Rosenfeld [22] used the notion of a fuzzy subset to put forth cornerstone papers in several areas of mathematics, among other disciplines. Kuroki initiated the theory of fuzzy semigroups in his paper [17, 18]. The monograph by Mordeson et al. [19] deals with the theory of fuzzy semigroups and their use in fuzzy codes, fuzzy finite state machines, and fuzzy languages. In [12], the concept of a fuzzy bi-ideal of an ordered semigroup was introduced by Kehayopulu and Tsingelis and developed a theory of fuzzy generalized sets on ordered semigroups. Murali [20] defined the concept of belongingness of a fuzzy point to a fuzzy subset under a natural equivalence on a fuzzy subset. The idea of quasi-coincidence of a fuzzy point with a fuzzy set played a vital role in generating different types of fuzzy subgroups. Using these ideas, Bhakat and Das [3–5] introduced the concept of (α, β)-fuzzy subgroups by using the 'belong to' $(\in)$ relation and 'quasi-coincident with' (q) relation between a fuzzy point and a fuzzy subgroup and introduced the concept of $(\in, \in \vee q)$-fuzzy subgroup. Davvaz defined $(\in, \in \vee q)$-fuzzy subnearrings and ideals of a near ring in

N.M. Khan (✉) · M.A. Khan
Department of Mathematics, Aligarh Muslim University, Aligarh 202002, India
e-mail: nm_khan123@yahoo.co.in

M.A. Khan
e-mail: khan.aasim09@gmail.com

S.T. Rizvi et al. (eds.), *Algebra and its Applications*, Springer Proceedings in Mathematics & Statistics 174, DOI 10.1007/978-981-10-1651-6_26

[8]. Kazanci and Yamak [11] studied $(\in, \in \vee q)$-fuzzy bi-ideals of a semigroup. In [10] Jun et al characterized the ordered semigroups by fuzzy ideals. Kehayopulu in [12–15] characterized regular, left regular, and right regular ordered semigroups by means of fuzzy left, right, and quasi ideals respectively. As an important generalization of the notion of a fuzzy set, Atanassov [1, 2] introduced the concept of an intuitionistic fuzzy set. Out of several higher order fuzzy sets, intuitionistic fuzzy sets have been found to be highly useful to deal with vagueness. Biswas in [6] studies fuzzy sets and intuitionistic fuzzy sets. Davvaz et al. [7] used this concept to H_v-modules. They introduced the concept of an intuitionistic fuzzy H_v-submodule of an H_v-module. Kim and Jun [16], introducing the concept of an intuitionistic fuzzy ideal of a semigroup, characterized the properties of semigroups. In [9] Hong et al. characterized regular ordered semigroup in terms of intuitionistic fuzzy sets. The theory of intuitionistic fuzzy sets on ordered semigroups has been recently developed (see [19, 23, 25]). In the present paper, we define the notions of $(\in, \in \vee q_k)$-intuitionistics fuzzy ideal, $(\in, \in \vee q_k)$-intuitionistics fuzzy bi-ideal, and $(\in, \in \vee q_k)$-intuitionistics fuzzy generalized bi-ideal of an ordered semigroup. Then, we characterize these various intuitionistic fuzzy ideals. Finally, using the properties of these intuitionistic fuzzy ideals, we characterize regular and intra-regular ordered semigroups.

2 Preliminaries

A partial ordered semigroup (briefly ordered semigroup) is a pair $(S, \cdot)$ comprising a semigroup S and a partial order $\leq$(on S) that is compatible with the binary operation, i.e., for all $a_1, a_2, b_1, b_2 \in S$, $a_1 \leq b_1, a_2 \leq b_2$ implies $a_1 \cdot a_2 \leq b_1 \cdot b_2$.

A nonempty subset A of an ordered semigroup S is called a *subsemigroup* of S if $A^2 \subseteq A$. For any ordered semigroup S and $A \subseteq S$, define

$$(A] = \{a \in S |\, a \leq b \text{ for some } b \in A\}.$$

A nonempty subset A of an ordered semigroup S is called a left (resp. right) ideal of S if (i) $SA \subseteq A$ (resp. $AS \subseteq A$) and (ii) for any $a \in S$ and $b \in A$, if $a \leq b$, then $a \in A$. If A is both a left and right ideal of S, then A is called an ideal of S. A nonempty subset A of an ordered semigroup S is called a *bi-ideal* of S if (i) $\forall\, a, b \in A \Rightarrow ab \in$A; (ii) $(\forall a \in S)(\forall b \in A)\; a \leq b \Rightarrow a \in A$ and (iii) $ASA \subseteq A$. A nonempty subset A of an ordered semigroup S is called a generalized bi-ideal of S if (i) $ASA \subseteq A$ and (ii) If $a \in S$ and $b \in A$, $a \leq b$ implies $a \in A$. An intuitionistic fuzzy set (in short IFS) A in a nonempty set X, is an object having the form $A = (\mu, \nu)$, where $\mu : X \to [0, 1]$ and $\nu : X \to [0, 1]$ are functions denoting the degree of membership(namely $\mu(x)$) and the degree of nonmembership (namely $\nu(x)$) for each element $x \in X$ respectively and $0 \leq \mu(x) + \nu(x) \leq 1$ for all $x \in X$.

An intuitionistic fuzzy subset $A = (\mu, \nu)$ in a set X of the form

$$\mu(y) = \begin{cases} u \in (0, 1] \text{ if } y = x; \\ 0 \qquad\qquad \text{if } y \neq x. \end{cases} \qquad \nu(y) = \begin{cases} v \in [0, 1) \text{ if } y = x; \\ 1 \qquad\qquad \text{if } y \neq x. \end{cases}$$

is said to be an intuitionistic fuzzy point (or in short IFP) with support x and value u, v and is denoted by x_u^v. For a intuitionistic fuzzy point x_u^v and an intuitionistic fuzzy set $A = (\mu, \nu)$ in a set X, we extend the meaning given by Pu and Liu [21] to the symbol $x_u \alpha A$ to the symbol $x_u^v \alpha A$, where $\alpha \in \{\in, q, \in \vee q, \in \wedge q\}$, by saying that an intuitionistic fuzzy point x_u^v belongs to (resp. quasi-coincident with) a intuitionistic fuzzy set A, written as $x_u^v \in A$ (resp. $x_u^v q A$), if $\mu(x) \geq u$, $\nu(x) \leq v$ (resp. $\mu(x) + u > 1$, $\nu(x) + v < 1$). Further we define $x_u^v \in \vee q A$ (resp. $x_u^v \in \wedge q A$) to mean that $x_u^v \in A$ or $x_u^v q A$ (resp. $x_u^v \in A$ and $x_u^v q A$), while $x_u^v \overline{\alpha} A$ is defined to mean that $x_u^v \alpha A$ does not hold.

For any two IFSs $A = (\mu, \nu)$ and $B = (\mu', \nu')$ of an ordered semigroup S, we say that $A \subseteq B$ if and only if $\mu(x) \leq \mu'(x)$ and $\nu(x) \geq \nu'(x)$ $\forall x \in S$.

For any family $\{A_i = (\mu_i, \nu_i), i \in \Delta\}$ of IFS in any ordered semigroup S, we define:

(i) $\bigcup\limits_{i \in \Delta} A_i = \{(\bigvee\limits_{i \in \Delta} \mu_i, \bigwedge\limits_{i \in \Delta} \nu_i) | i \in \Delta\}$ and

(ii) $\bigcap\limits_{i \in \Delta} A_i = \{(\bigwedge\limits_{i \in \Delta} \mu_i, \bigvee\limits_{i \in \Delta} \nu_i) | i \in \Delta\}$;

where for any $\alpha_i : S \to [0, 1]$, $((\bigvee\limits_{i \in \Delta} \alpha_i)(x) = (\bigvee\limits_{i \in \Delta} \alpha_i(x))$ and $((\bigwedge\limits_{i \in \Delta} \alpha_i)(x) = (\bigwedge\limits_{i \in \Delta} \alpha_i(x))$ for all $x \in S$.

Remark When the index set Δ is finite, we shall be using maximum and minimum in place of $\bigvee$ and $\bigwedge$, standing for least upper bound and greatest lower bound respectively, in the sequel without further mention.

Let $(S, ., \leq)$ be an ordered semigroup and $I \subseteq S$. The intuitionistic characteristic function (see [23]) $\chi_I := \{x; \mu_{\chi_I}, \nu_{\chi_I} | x \in S\}$, where $(\mu_{\chi_I}, \nu_{\chi_I})$ is an intuitionistic fuzzy subset of S defined as:

$$\mu_{\chi_I} : S \to [0, 1] \text{ defined by } \qquad \mu_{\chi_I}(x) = \begin{cases} 1 & \text{if } x \in I; \\ 0 & \text{if } x \notin I. \end{cases}$$

$$\nu_{\chi_I} : S \to [0, 1] \text{ defined by } \qquad \nu_{\chi_I}(x) = \begin{cases} 0 & \text{if } x \in I; \\ 1 & \text{if } x \notin I. \end{cases}$$

Let $(S, ., \leq)$ be an ordered semigroup and let $A = (\mu(x), \nu(x))$ be an IFS of S. Then $A = (\mu, \nu)$ is called an intuitionistic fuzzy subsemigroup of S (see [23]) if $\mu(xy) \geq \min\{\mu(x), \mu(y)\}$ and $\nu(xy) \leq \max\{\nu(x), \nu(y)\}$ $(\forall x, y \in S)$.

An IFS A of S is called an intuitionistic fuzzy left (resp. right) ideal (see [24]) if (i) $x \leq y \Rightarrow \mu(x) \geq \mu(y)$ and $\nu(x) \leq \nu(y)$; (ii) $\mu(xy) \geq \mu(y) (resp.\ \mu(x))$ and $\nu(xy) \leq \nu(y) (resp.\ \nu(x))$ $\forall x, y \in S$.

An IFS $A = (\mu, \nu)$ of S is called an intuitionistic fuzzy ideal of S (see [23]) if

(i) $x \leq y \Rightarrow \mu(x) \geq \mu(y)$ and $\nu(x) \leq \nu(y)$;
(ii) $\mu(xy) \geq \min\{\mu(x), \mu(y)\}$ and $\nu(xy) \leq \max\{\nu(x), \nu(y)\}$ $\forall x, y \in S$.

Let $(S, ., \leq)$ be an ordered semigroup and $A = (\mu, \nu)$ be an intuitionistic fuzzy subsemigroup of S. Then $A = (\mu, \nu)$ is called an intuitionistic fuzzy bi-ideal of S if

(i) $x \leq y \Rightarrow \mu(x) \geq \mu(y)$ and $\nu(x) \leq \nu(y) \quad (\forall x, y \in S)$;
(ii) $\mu(xy) \geq \min\{\mu(x), \mu(y)\}$ and $\nu(xy) \leq \max\{\nu(x), \nu(y)\} \quad (\forall x, y \in S)$;
(iii) $\mu(xyz) \geq \min\{\mu(x), \mu(z)\}$ and $\nu(xyz) \leq \max\{\nu(x), \nu(z)\} \quad (\forall x, y, z \in S)$.

In whatever follows, we treat any ordered semigroup S itself as an IFS by defining $S(x) = (S, \widetilde{S})(x) \equiv [1, 0]$ for each $x \in S$, where $S(x) = 1$ and $\widetilde{S}(x) = 0, \forall x \in S$.

3 Characterizations of Various Intuitionistic Fuzzy Ideals

In this section, we characterize various intuitionistic fuzzy ideals of an ordered semigroup.

Let $(S, ., \leq)$ be an ordered semigroup and k denote an arbitrary element of $[0, 1)$ unless otherwise specified. For an IFP x_u^v and IFS $A = (\mu, \nu)$ of S, where $u \in (0, 1]$ *and* $v \in [0, 1)$, we say that

(i) $x_u^v q_k A$ if $\mu(x) + k + u > 1$ and $\nu(x) + k + v < 1$;
(ii) $x_u^v \in \vee q_k A$ if $x_u^v \in A$ or $x_u^v q_k A$;
(iii) $x_u^v \overline{\alpha} A$ if $x_u^v \alpha A$ does not hold for $\alpha \in \{q_k, \in \vee q_k\}$.

Definition An IFS A of an ordered semigroup S is called an $(\in, \in \vee q_k)$-intuitionistic fuzzy subsemigroup of S if $x_u^v \in A$ and $y_{u_1}^{v_1} \in A$ imply that $(xy)_{\min\{u,u_1\}}^{\max\{v,v_1\}} \in \vee q_k A$, for all $x, y \in S$ and $u, u_1 \in (0, 1]$ and $v, v_1 \in [0, 1)$.

Definition An IFS A of an ordered semigroup S is called an $(\in, \in \vee q_k)$-intuitionistic fuzzy left (resp. right) ideal of S if for all $u \in (0, 1]$ and $v \in [0, 1)$ and $x, y \in S$, the following conditions are satisfied:

(i) $x \leq y \Rightarrow \mu(x) \geq \mu(y)$ *and* $\nu(x) \leq \nu(y)$;
(ii) $x \in S, y_u^v \in A \Rightarrow (xy)_u^v \in \vee q_k A$ (resp. $(yx)_u^v \in \vee q_k A$).

Further A is called an $(\in, \in \vee q_k)$-intuitionistic fuzzy ideal if it is both an $(\in, \in \vee q_k)$-intuitionistic fuzzy left ideal and $(\in, \in \vee q_k)$-intuitionistic fuzzy right ideal of S.

Theorem 1 *Let A be any intuitionistic fuzzy subset of S. Then A is an* $(\in, \in \vee q_k)$-*intuitionistic fuzzy subsemigroup of S if and only if* $\mu(xy) \geq \min\{\mu(x), \mu(y), \frac{1-k}{2}\}$ *and* $\nu(xy) \leq \max\{\nu(x), \nu(y), \frac{1-k}{2}\}$ *for all* $x, y \in S$.

Proof Let A be any IFS of an ordered semigroup S. Suppose to the contrary that $\mu(xy) < \min\{\mu(x), \mu(y), \frac{1-k}{2}\}$ and $\nu(xy) > \max\{\nu(x), \nu(y), \frac{1-k}{2}\}$ for some $x, y \in S$, then there exists $u \in (0, \frac{1-k}{2})$ and $v \in (\frac{1-k}{2}, 1)$ such that $\mu(xy) < u \leq \min\{\mu(x), \mu(y), \frac{1-k}{2}\}$ and $\nu(xy) > v \geq \max\{\nu(x), \nu(y), \frac{1-k}{2}\}$. This implies that x_u^v, $y_u^v \in A$. As A is an $(\in, \in \vee q_k)$-intuitionistic fuzzy subsemigroup, $(xy)_u^v \in \vee q_k A$. Since $\mu(xy) < u$ and $\nu(xy) > v$. So $(xy)_u^v \notin A$. Therefore $(xy)_u^v \overline{\in \vee q_k} A$. This is a contradiction. Hence $\mu(xy) \geq \min\{\mu(x), \mu(y), \frac{1-k}{2}\}$ and $\nu(xy) \leq \max\{\nu(x), \nu(y), \frac{1-k}{2}\}$ for all $x, y \in S$.

Conversely, suppose that $x, y \in S$ and $u, u_1 \in (0, 1], v, v_1 \in [0, 1)$ be such that $x_u^v \in A$ and $y_{u_1}^{v_1} \in A$. Then $\mu(x) \geq u,\ \nu(x) \leq v$ and $\mu(y) \geq u_1,\ \nu(y) \leq v_1$. Now $\mu(xy) \geq \min\{\mu(x), \mu(y), \frac{1-k}{2}\} \geq \min\{u, u_1, \frac{1-k}{2}\}$ and $\nu(xy) \leq \max\{\nu(x), \nu(y), \frac{1-k}{2}\} \leq \max\{v, v_1, \frac{1-k}{2}\}$.

Case(i): If $\min\{u, u_1\} \leq \frac{1-k}{2}$ and $\max\{v, v_1\} \geq \frac{1-k}{2}$. So $\mu(xy) \geq \min\{u, u_1\}$ and $\nu(xy) \leq \max\{v, v_1\}$. Therefore $(xy)_{\min\{u,u_1\}}^{\max\{v,v_1\}} \in \vee q_k A$.
Case(ii): If $\min\{u, u_1\} \geq \frac{1-k}{2}$, then $\max\{v, v_1\} \leq \frac{1-k}{2}$. Therefore $\mu(xy) \geq \frac{1-k}{2}$ and $\nu(xy) \leq \frac{1-k}{2}$. Now it follows that $(xy)_{\min\{u,u_1\}}^{\max\{v,v_1\}} \in A$ or $\mu(xy) + \min\{u, u_1\} \geq \frac{1-k}{2} + \min\{u, u_1\} = 1 - k$ and $\nu(xy) + \max\{v, v_1\} \leq \frac{1-k}{2} + \max\{v, v_1\} = 1 - k$, i.e., $(xy)_{\min\{u,u_1\}}^{\max\{v,v_1\}} q_k A$. Therefore $(xy)_{\min\{u,u_1\}}^{\max\{v,v_1\}} \in \vee q_k A$.
Case(iii): If $\min\{u, u_1\} \leq \frac{1-k}{2}$ and $\max\{v, v_1\} \leq \frac{1-k}{2}$. We show that $(xy)_{\min\{u,u_1\}}^{\max\{v,v_1\}} \in \vee q_k A$. The Proof in this case is similar to the proof in above cases.
Case(iv): If $\min\{u, u_1\} \geq \frac{1-k}{2}$ and $\max\{v, v_1\} \geq \frac{1-k}{2}$. We show that $(xy)_{\min\{u,u_1\}}^{\max\{v,v_1\}} \in \vee q_k A$. The Proof in this case is similar to the proof in above cases.

Hence A is an $(\in, \in \vee q_k)$-intuitionistic fuzzy subsemigroup of S.

Theorem 2 *Let S be an ordered semigroup. Then an IFS A of S is an $(\in, \in \vee q_k)$-intuitionistic fuzzy left ideal of S if and only if A satisfies*

(i) $x \leq y \Rightarrow \mu(x) \geq \mu(y)$, *and* $\nu(x) \leq \nu(y)$;
(ii) $\mu(xy) \geq \min\{\mu(y), \frac{1-k}{2}\}$ *and* $\nu(xy) \leq \max\{\nu(y), \frac{1-k}{2}\}$ *for all* $x, y \in S$.

Proof Let A be an $(\in, \in \vee q_k)$-intuitionistic fuzzy left ideal of S. As A is an $(\in, \in \vee q_k)$-intuitionistic fuzzy left ideal, the condition (i) is satisfied.

Next we show that the condition (ii) also holds.

Case(i): If $\mu(y) \leq \frac{1-k}{2}$ and $\nu(y) \geq \frac{1-k}{2}$. We show that $\mu(xy) \geq \mu(y)$ *and* $\nu(xy) \leq \nu(y)$. Suppose to the contrary that $\mu(xy) < \mu(y)$ and $\nu(xy) > \nu(y)$, then there exist $u_1 \in (0, \frac{1-k}{2})$ and $v_1 \in (\frac{1-k}{2}, 1)$ such that $\mu(xy) < u_1 \leq \mu(y)$ and $\nu(xy) > v_1 \geq \nu(y)$. It follows that $y_{u_1}^{v_1} \in A$, but $(xy)_{u_1}^{v_1} \notin A$ and $\mu(xy) + u_1 < 2u_1 < 1 - k$ and $\nu(xy) + v_1 > 2v_1 > 1 - k$. Therefore $(xy)_{u_1}^{v_1} \overline{q_k} A$. So, $(xy)_{u_1}^{v_1} \overline{\in \vee q_k} A$, a contradiction. Hence $\mu(xy) \geq \mu(y)$ *and* $\nu(xy) \leq \nu(y)$.
Case(ii): Now we show that $\mu(xy) \geq \frac{1-k}{2}$ *and* $\nu(xy) \leq \frac{1-k}{2}$. If $\mu(y) \geq \frac{1-k}{2}$ *and* $\nu(y) \leq \frac{1-k}{2}$. Then $x \in S$ and $(y)_{\frac{1-k}{2}}^{\frac{1-k}{2}} \in A$. So we have $(xy)_{\frac{1-k}{2}}^{\frac{1-k}{2}} \in \vee q_k A$. This implies that $(xy)_{\frac{1-k}{2}}^{\frac{1-k}{2}} \in A$ or $(xy)_{\frac{1-k}{2}}^{\frac{1-k}{2}} q_k A$. Therefore $\mu(xy) \geq \frac{1-k}{2},\ \nu(xy) \leq \frac{1-k}{2}$ *or* $\mu(xy) + \frac{1-k}{2} + k > 1,\ \nu(xy) + \frac{1-k}{2} + k < 1$. Hence $\mu(xy) \geq \frac{1-k}{2},\ \nu(xy) \leq \frac{1-k}{2}$; otherwise $\mu(xy) + \frac{1-k}{2} + k < \frac{1-k}{2} + \frac{1-k}{2} + k = 1,\ \ \nu(xy) + \frac{1-k}{2} + k > \frac{1-k}{2} + \frac{1-k}{2} + k = 1$, a contradiction. Therefore $\mu(xy) \geq \frac{1-k}{2}$ and $\nu(xy) \leq \frac{1-k}{2}$.
Case(iii): If $\mu(y) \leq \frac{1-k}{2}$ and $\nu(y) \leq \frac{1-k}{2}$. We show that $\mu xy \geq \mu(y)$ and $\nu(xy) \leq \frac{1-k}{2}$. The Proof in this case is similar to the proof in above cases.
Case(iv): If $\mu(y) \geq \frac{1-k}{2}$ and $\nu(y) \geq \frac{1-k}{2}$. We show that $\mu xy \geq \frac{1-k}{2}$ and $\nu(xy) \leq \nu(y)$. The Proof in this case is similar to the proof in above cases. Therefore, $\mu(xy) \geq \min\{\mu(x), \mu(y), \frac{1-k}{2}\}$ and $\nu(xy) \leq \max\{\nu(x), \nu(y), \frac{1-k}{2}\}$ for all $x, y \in S$.

Conversely, suppose that $x \in S$, $y_u^v \in A$. Then $\mu(y) \geq u$, $\nu(y) \leq v$. Now $\mu(xy) \geq \min\{u, \frac{1-k}{2}\}$ and $\nu(xy) \leq \max\{v, \frac{1-k}{2}\}$.

Case(i): If $u > \frac{1-k}{2}$ and $v < \frac{1-k}{2}$. So $\mu(xy) \geq \frac{1-k}{2}$ and $\nu(xy) \leq \frac{1-k}{2}$. Now $\mu(y) + u > \frac{1-k}{2} + \frac{1-k}{2} = 1-k$ and $\nu(y) + v < \frac{1-k}{2} + \frac{1-k}{2} = 1-k$. This implies that $(xy)_u^v q_k A$. Therefore $(xy)_u^v \in \vee q_k A$

Case(ii): If $u \leq \frac{1-k}{2}$ and $v \geq \frac{1-k}{2}$. So $\mu(xy) \geq u$ and $\nu(xy) \leq v$. This implies that $(xy)_u^v \in A$. Therefore $(xy)_u^v \in \vee q_k A$

Case(iii): If $u \geq \frac{1-k}{2}$ and $v \geq \frac{1-k}{2}$. We show that $(xy)_u^v \in \vee q_k A$. The Proof in this case is similar to the proof in above cases.

Case(iv): If $u \leq \frac{1-k}{2}$ and $v \leq \frac{1-k}{2}$. We show that $(xy)_u^v \in \vee q_k A$. The Proof in this case is similar to the proof in above cases.

Hence IFS A is an $(\in, \in \vee q_k)$-intuitionistic fuzzy left ideal of S.

Dually we have the following:

Theorem 3 *Let S be an ordered semigroup and A be an IFS of S. Then A is an $(\in, \in \vee q_k)$-intuitionistic fuzzy right ideal of S if and only if A satisfies*

(i) $x \leq y \Rightarrow \mu(x) \geq \mu(y)$, *and* $\nu(x) \leq \nu(y)$;
(ii) $\mu(xy) \geq \min\{\mu(x), \frac{1-k}{2}\}$ *and* $\nu(xy) \leq \max\{\nu(x), \frac{1-k}{2}\}$ *for all* $x, y \in S$.

Definition An IFS A of S is called an $(\in, \in \vee q_k)$-intuitionistic fuzzy bi-ideal of S, if it satisfies the following conditions,

(i) $x \leq y \Rightarrow \mu(x) \geq \mu(y)$ *and* $\nu(x) \leq \nu(y)$;
(ii) $x_{u_1}^{v_1} \in A$ *and* $y_{u_2}^{v_2} \in A \Rightarrow (xy)_{\min\{u_1,u_2\}}^{\max\{v_1,v_2\}} \in \vee q_k A$;
(iii) $x_{u_1}^{v_1} \in A$ *and* $z_{u_2}^{v_2} \in A \Rightarrow (xyz)_{\min\{u_1,u_2\}}^{\max\{v_1,v_2\}} \in \vee q_k A$.

For all $u_1, u_2 \in (0, 1]$ *and* $v_1, v_2 \in [0, 1)$ and $x, y, z \in S$.

Clearly every $(\in, \in \vee q_k)$-intuitionistic fuzzy bi-ideal of S is an $(\in, \in \vee q_k)$-intuitionistic fuzzy subsemigroup of S.

Theorem 4 *Let S be an ordered semigroup. Then an IFS $A = (\mu, \nu)$ of S is an $(\in, \in \vee q_k)$-intuitionistic fuzzy bi-ideal of S if and only if A satisfies*

(i) $x \leq y \Rightarrow \mu(x) \geq \mu(y)$ *and* $\nu(x) \leq \nu(y)$;
(ii) $\mu(xy) \geq \min\{\mu(x), \mu(y), \frac{1-k}{2}\}$ *and* $\nu(xy) \leq \max\{\nu(x), \nu(y), \frac{1-k}{2}\}$;
(iii) $\mu(xyz) \geq \min\{\mu(x), \mu(z), \frac{1-k}{2}\}$ *and* $\nu(xyz) \leq \max\{\nu(x), \nu(z), \frac{1-k}{2}\}$ *for all* $x, y, z \in S$.

Proof Suppose A be an $(\in, \in \vee q_k)$-intuitionistic fuzzy bi-ideal of S. Then, by definition and Theorem 1, conditions (i) and (ii) are satisfied. Now, we show that condition (iii) also holds. Suppose to the contrary that, $\mu(xyz) < \min\{\mu(x), \mu(z), \frac{1-k}{2}\}$ and $\nu(xyz) > \max\{\nu(x), \nu(z), \frac{1-k}{2}\}$ for some $x, y, z \in S$, then $\exists\ u_1 \in (0, \frac{1-k}{2})$ and $v_1 \in (\frac{1-k}{2}, 1)$ such that $\mu(xyz) < u_1 \leq \min\{\mu(x), \mu(z), \frac{1-k}{2}\}$ and $\nu(xyz) > v_1 \geq \max\{\nu(x), \nu(z), \frac{1-k}{2}\}$. From this, we have $x_{u_1}^{v_1} \in A$ and $z_{u_1}^{v_1} \in A$. As A is an $(\in, \in \vee q_k)$-intuitionistic fuzzy bi-ideal, $(xyz)_{u_1}^{v_1} \in \vee q_k A$. Since $\mu(xyz) < u_1$ and $\nu(xyz) > v_1$. So, $(xyz)_{u_1}^{v_1} \overline{\in \vee q_k} A$, a contradiction. Hence $\mu(xyz) \geq \min\{\mu(x), \mu(z), \frac{1-k}{2}\}$ and $\nu(xyz) \leq \max\{\nu(x), \nu(z), \frac{1-k}{2}\}$ for all $x, y, z \in S$.

Conversely suppose that conditions (i), (ii) and (iii) are satisfied by A. Let $x, y \in S$ and $u_1, u_2 \in (0, 1]$ and $v_1, v_2 \in [0, 1)$ such that $x_{u_1}^{v_1} \in A$ and $y_{u_2}^{v_2} \in A$. Then $\mu(x) \geq u_1, \nu(x) \leq v_1$ and $\mu(y) \geq u_2, \nu(y) \leq v_2$. Now $\mu(xy) \geq \min\{\mu(x), \mu(y), \frac{1-k}{2}\} \geq \min\{u_1, u_2, \frac{1-k}{2}\}$ and $\nu(xy) \leq \max\{\nu(x), \nu(y), \frac{1-k}{2}\} \leq \max\{v_1, v_2, \frac{1-k}{2}\}$.

Case(i): If $\min\{u_1, u_2\} > \frac{1-k}{2}$ and $\max\{v_1, v_2\} < \frac{1-k}{2}$. So $\mu(xy) \geq \frac{1-k}{2}$ and $\nu(xy) \leq \frac{1-k}{2}$ which implies that $\mu(xy) + \min\{u_1, u_2\} + k > 1$ and $\nu(xy) + \max\{v_1, v_2\} + k < 1$, i.e., $(xy)_{\min\{u_1,u_2\}}^{\max\{v_1,v_2\}} q_k A$. Therefore $(xy)_{\min\{u_1,u_2\}}^{\max\{v_1,v_2\}} \in \vee q_k A$.
Case(ii): If $\min\{u_1, u_2\} \leq \frac{1-k}{2}$ and $\max\{v_1, v_2\} \geq \frac{1-k}{2}$, then $\mu(xy) \geq \min\{u_1, u_2\}$ and $\nu(xy) \leq \max\{v_1, v_2\}$, thus $(xy)_{\min\{u_1,u_2\}}^{\max\{v_1,v_2\}} \in A$. Therefore $(xy)_{\min\{u_1,u_2\}}^{\max\{v_1,v_2\}} \in \vee q_k A$.
Case(iii): If $\min\{u_1, u_2\} > \frac{1-k}{2}$ and $\max\{v_1, v_2\} > \frac{1-k}{2}$. We show that $(xy)_{\min\{u_1,u_2\}}^{\max\{v_1,v_2\}} \in \vee q_k A$. The Proof in this case is similar to the proof in above cases.
Case(iv): If $\min\{u_1, u_2\} < \frac{1-k}{2}$ and $\max\{v_1, v_2\} < \frac{1-k}{2}$. We show that $(xy)_{\min\{u_1,u_2\}}^{\max\{v_1,v_2\}} \in \vee q_k A$. The Proof in this case is similar to the proof in above cases.

Similarly, we may show that $(xyz)_{\min\{u_1,u_2\}}^{\max\{v_1,v_2\}} \in \vee q_k A$ for all $x, y, z \in S$. Hence IFS $A = (\mu, \nu)$ is an $(\in, \in \vee q_k)$-intuitionistic fuzzy bi-ideal of S.

Now, we show, by an example, that in general union of $(\in, \in \vee q_k)$-intuitionistic fuzzy bi-ideals need not be $(\in, \in \vee q_k)$-intuitionistic fuzzy bi-ideal.

Example Let $S := \{a, b, c, d\}$ be an ordered semigroup with respect to the order relation $a \leq d$ and the operation '. ' defined in the following cayley table

.	a	b	c	d
a	a	a	a	a
b	a	a	d	a
c	a	a	a	a
d	a	a	a	a

Let $A = (\mu, \nu)$ and $B = (\mu', \nu')$ be two IFS's of S such that $\mu(a) = 0.35$, $\mu(b) = 0.20$, $\mu(c) = 0.25$, $\mu(d) = 0.20$, $\mu'(a) = 0.40$, $\mu'(b) = 0.30$, $\mu'(c) = 0.20$, $\mu'(d) = 0.10$ and $\nu(a) = 0.20$, $\nu(b) = 0.30$, $\nu(c) = 0.40$, $\nu(d) = 0.40$, $\nu'(a) = 0.30$, $\nu'(b) = 0.50$, $\nu'(c) = 0.20$, $\nu'(d) = 0.40$. Then IFS's A and B both are $(\in, \in \vee q_k)$-intuitionistic fuzzy bi-ideals of S. $(\mu \cup \mu')(bc) = (\mu \cup \mu')(d) = \max\{\mu(d), \mu'(d)\} = \min\{0.20, 0.10\} = .20$, but $\min\{(\mu \cup \mu')(b), (\mu \cup \mu')(c), \frac{1-k}{2}\} = \min\{0.30, 0.25, \frac{1-k}{2}\} = 0.25$, i.e., $A \cup B$ is not an $(\in, \in \vee q_k)$-intuitionistic fuzzy subsemigroup of S for any $k \in [0, 1)$. Therefore $A \cup B$ is not an $(\in, \in \vee q_k)$-intuitionistic fuzzy bi-ideal of S. In the next Theorem, we show, under some conditions, that $\bigcup_{i \in \Delta} A_i$ of a family of $(\in, \in \vee q_k)$-intuitionistic fuzzy bi-ideals is again an $(\in, \in \vee q_k)$-intuitionistic fuzzy bi-ideal with the help of the following Lemma.

Lemma 1 *Let $\{A_i | i \in \Delta\}$, where $A_i = (\mu_i, \nu_i)$ for each $i \in \Delta$, be a family of $(\in, \in \vee q_k)$-intuitionistic fuzzy bi-ideals of S such that $A_i \subseteq A_j$ or $A_j \subseteq A_i$ for all $i, j \in \Delta$. Then $\bigvee_{i \in \Delta} \{\min\{\mu_i(x), \mu_i(y), \frac{1-k}{2}\}\} = \min\{\bigvee_{i \in \Delta} \mu_i(x), \bigvee_{i \in \Delta} \mu_i(y), \frac{1-k}{2}\}$ and $\bigwedge_{i \in \Delta} \{\max\{\nu_i(x), \nu_i(y), \frac{1-k}{2}\}\} = \max\{\bigwedge_{i \in \Delta} \nu_i(x), \bigwedge_{i \in \Delta} \nu_i(y), \frac{1-k}{2}\}$.*

Proof It is obvious that $\bigvee_{i\in\Delta}\{\min\{\mu_i(x), \mu_i(y), \frac{1-k}{2}\}\} \leq \min\{\bigvee_{i\in\Delta} \mu_i(x), \bigvee_{i\in\Delta} \mu_j(i), \frac{1-k}{2}\}$ and $\bigwedge_{i\in\Delta}\{\max\{\nu_i(x), \nu_i(y), \frac{1-k}{2}\}\} \geq \max\{\bigwedge_{i\in\Delta} \nu_i(x), \bigwedge_{i\in\Delta} \nu_i(y), \frac{1-k}{2}\}$.

Suppose that $\bigvee_{i\in\Delta}\{\min\{\mu_i(x), \mu_i(y), \frac{1-k}{2}\}\} \neq \min\{\bigvee_{i\in\Delta} \mu_i(x), \bigvee_{i\in\Delta} \mu_i(y), \frac{1-k}{2}\}$ and $\bigwedge_{i\in\Delta}\{\max\{\nu_i(x), \nu_i(y), \frac{1-k}{2}\}\} \neq \max\{\bigwedge_{i\in\Delta} \nu_i(x), \bigwedge_{i\in\Delta} \nu_i(y), \frac{1-k}{2}\}$.

Then $\exists\ u, v$ such that $\bigvee_{i\in\Delta}\{\min\{\mu_i(x), \mu_i(y), \frac{1-k}{2}\}\} < u < \min\{\bigvee_{i\in\Delta} \mu_i(x), \bigvee_{i\in\Delta} \mu_i(y), \frac{1-k}{2}\}$ and $\bigwedge_{i\in\Delta}\{\max\{\nu_i(x), \nu_i(y), \frac{1-k}{2}\}\} > v > \max\{\bigwedge_{i\in\Delta} \nu_i(x), \bigwedge_{i\in\Delta} \nu_i(y), \frac{1-k}{2}\}$. Since $A_i \subseteq A_j$ or $A_j \subseteq A_i$ for all $i, j \in \Delta$, we have $\mu_i \leq \mu_j$ *and* $\nu_i \leq \nu_j$ or $\mu_j \leq \mu_i$ *and* $\nu_j \leq \nu_i$ for all $i, j \in \Delta$. Then $\exists\ n_1, n_2 \in \Delta$ such that $u < \min\{\mu_{n_1}(x), \mu_{n_1}(y), \frac{1-k}{2}\}$ or $v > \max\{\nu_{n_2}(x), \nu_{n_2}(y), \frac{1-k}{2}\}$. On the other hand $\min\{\mu_i(x), \mu_i(y), \frac{1-k}{2}\} < u$ and $\max\{\nu_i(x), \nu_i(y), \frac{1-k}{2}\} > v$ for $i \in \Delta$. This is a contradiction.

Hence $\bigvee_{i\in\Delta}\{\min\{\mu_i(x), \mu_i(y), \frac{1-k}{2}\}\} = \min\{\bigvee_{i\in\Delta} \mu_i(x), \bigvee_{i\in\Delta} \mu_i(y), \frac{1-k}{2}\}$ and $\bigwedge_{i\in\Delta}\{\max\{\nu_i(x), \nu_i(y), \frac{1-k}{2}\}\} = \max\{\bigwedge_{i\in\Delta} \nu_i(x), \bigwedge_{i\in\Delta} \nu_i(y), \frac{1-k}{2}\}$.

Theorem 5 *Let $\{A_i | i \in \Delta\}$, where $A_i = (\mu_i, \nu_i)$ for each $i \in \Delta$, be a family of $(\in, \in \vee q_k)$-intuitionistic fuzzy bi-ideals of S such that $A_i \subseteq A_j$ or $A_j \subseteq A_i$ for all $i, j \in \Delta$. Then $A = (\mu, \nu) = \bigcup_{i\in\Delta}\{A_i | i \in \Delta\}$ is an $(\in, \in \vee q_k)$-intuitionistic fuzzy bi-ideal of S, where $\bigcup_{i\in\Delta} A_i = \{(\bigvee_{i\in\Delta} \mu_i, \bigwedge_{i\in\Delta} \nu_i) | i \in \Delta\}$.*

Proof By Theorem 4, we show that conditions (i)–(iii) are satisfied.

(i) Let $x, y \in S$ with $x \leq y$. Then, we have to show that $\mu(x) \geq \mu(y)$ and $\nu(x) \leq \nu(y)$. Since A_i is an $(\in, \in \vee q_k)$-intuitionistic fuzzy bi-ideal of S for each $i \in \Delta$, we have $\mu_i(x) \geq \mu_i(y)$ and $\nu_i(x) \leq \nu_i(y)$ for each $i \in \Delta$. Thus

$$\begin{aligned}\mu(x) &= \bigvee_{i\in\Delta}(\mu_i(x)) \\ &\geq \bigvee_{i\in\Delta}(\mu_i(y)) = \mu(y)\end{aligned}$$

and

$$\begin{aligned}\nu(x) = \left(\bigwedge_{i\in\Delta} \nu_i\right)(x) &= \bigwedge_{i\in\Delta}(\nu_i(x)) \\ &\leq \bigwedge_{i\in\Delta}(\nu_i(y)) = \nu(y).\end{aligned}$$

(ii) Let $x, y \in S$. As each A_i is an $(\in, \in \vee q_k)$ intuitionistic fuzzy bi-ideal,

$$\mu_i(xy) \geq \min\left\{\mu_i(x), \mu_i(y), \frac{1-k}{2}\right\} \text{ and } \nu_i(xy) \leq \max\left\{\nu_i(x), \nu_i(y), \frac{1-k}{2}\right\}.$$

Now

$$\begin{aligned}
\mu(xy) = \left(\bigvee_{i\in\Delta} \mu_i\right)(xy) &\geq \bigvee_{i\in\Delta} \min\left\{\mu_i(x), \mu_i(y), \frac{1-k}{2}\right\} && \text{(by definition)} \\
&= \min\left\{\bigvee_{i\in\Delta} \mu_i(x), \bigvee_{i\in\Delta} \mu_i(y), \frac{1-k}{2}\right\}, && \text{(by Lemma 1)} \\
&= \min\left\{\mu(x), \mu(y), \frac{1-k}{2}\right\}
\end{aligned}$$

and

$$\begin{aligned}
\nu(xy) = \left(\bigwedge_{i\in\Delta} \nu_i\right)(xy) &\leq \bigwedge_{i\in\Delta} \max\left\{\nu_i(x), \nu_i(y), \frac{1-k}{2}\right\} && \text{(by definition)} \\
&= \max\{\bigwedge_{i\in\Delta} \nu_i(x), \bigwedge_{i\in\Delta} \nu_i(y), \frac{1-k}{2}\}, && \text{(by Lemma 1)} \\
&= \max\{\nu(x), \nu(y), \frac{1-k}{2}\}.
\end{aligned}$$

(iii) Let $x, y, z \in S$. As each A_i is an $(\in, \in \vee q_k)$-intuitionistic fuzzy bi-ideal of S, $\mu_i(xyz) \geq \min\{\mu_i(x), \mu_i(z), \frac{1-k}{2}\}$ and $\nu_i(xyz) \leq \max\{\nu_i(x), \nu_i(z), \frac{1-k}{2}\}$.

Now

$$\begin{aligned}
\mu(xyz) = \left(\bigvee_{i\in\Delta} \mu_i\right)(xyz) &\geq \bigvee_{i\in\Delta} \min\left\{\mu_i(x), \mu_i(z), \frac{1-k}{2}\right\} && \text{(by definition)} \\
&= \min\left\{\bigvee_{i\in\Delta} \mu_i(x), \bigvee_{i\in\Delta} \mu_i(z), \frac{1-k}{2}\right\}, && \text{(by Lemma 1)} \\
&= \min\left\{\mu(x), \mu(z), \frac{1-k}{2}\right\}
\end{aligned}$$

and

$$\begin{aligned}
\nu(xyz) = \left(\bigwedge_{i\in\Delta} \nu_i\right)(xyz) &\leq \bigwedge_{i\in\Delta} \max\left\{\nu_i(x), \nu_i(z), \frac{1-k}{2}\right\} && \text{(by definition)} \\
&= \max\left\{\bigwedge_{i\in\Delta} \nu_i(x), \bigwedge_{i\in\Delta} \nu_i(z), \frac{1-k}{2}\right\}, && \text{(by Lemma 1)}
\end{aligned}$$

$$= \max\left\{\nu(x), \nu(z), \frac{1-k}{2}\right\}.$$

Therefore A is an $(\in, \in \vee q_k)$-intuitionistic fuzzy bi-ideal of S.

Definition An IFS $A = (\mu, \nu)$ of an ordered semigroup S is called strongly convex if $\mu(x) = \bigvee_{x \leq y} \mu(y)$ and $\nu(x) = \bigwedge_{x \leq y} \nu(y)$ respectively $\forall x \in S$.

Remark In [13, 25], an IFS $A = (\mu, \nu)$ of an ordered semigroup S has been defined strongly convex as follows: Let $(A] = ((\mu], (\nu])$, where $(\mu]$ and $(\nu]$ are defined by $(\mu](x) = \bigvee_{x \leq y} \mu(y)$ and $(\nu](x) = \bigwedge_{x \leq y} \nu(y)$ respectively for all $x \in S$. The IFS A of S is called strongly convex if $A = (A]$.

Proposition 1 *Let S be an ordered semigroup. Then an IFS $A = (\mu, \nu)$ of S is strongly convex intuitionistic fuzzy subset of S if and only if $x \leq y \Rightarrow \mu(x) \geq \mu(y)$ and $\nu(x) \leq \nu(y)$.*

Proof Let $x, y \in S$ and $x \leq y$. Since A is a strongly convex intuitionistic fuzzy subset of S, $\mu(x) = \bigvee_{x \leq z} \mu(z) \geq \mu(y)$ and $\nu(x) = \bigwedge_{x \leq z} \nu(z) \leq \nu(y)$.

Conversely, for any $x, \ y \in S$ and $x \leq y$, as $\mu(x) \geq \mu(y)$, $\nu(x) \leq \nu(y)$, we have $\bigvee_{x \leq y} \mu(y) \leq \mu(x)$ and $\bigwedge_{x \leq y} \nu(y) \geq \nu(x)$. As $x \leq x$, we have $\mu(x) \leq \bigvee_{x \leq y} \mu(y)$ and $\nu(x) \geq \bigwedge_{x \leq y} \nu(y)$. Therefore $\mu(x) = \bigvee_{x \leq y} \mu(y)$ and $\nu(x) = \bigwedge_{x \leq y} \nu(y)$. Hence A is a strongly convex intuitionistic fuzzy subset.

Definition For any IFS $A = (\mu, \nu)$ of an ordered semigroup S and $u \in (0, 1]$, $v \in [0, 1)$, $k \in [0, 1)$, the subset $[A]_u^v = \{x \in S \mid x_u^v \in \vee q_k A\}$ of S is called $(\in \vee q_k)$-level subset of A.

Theorem 6 *Let S be an ordered semigroup. Then*

(1) *The subset $[A]_u^v$ of S, for any $(\in, \in \vee q_k)$-intuitionistic fuzzy bi-ideal A of S, is a bi-ideal of S.*
(2) *If A is strongly convex intuitionistic fuzzy subset of S and the subset $[A]_u^v$ of S is a bi-ideal of S, then A is an $(\in, \in \vee q_k)$-intuitionistic fuzzy bi-ideal of S.*

Proof (1) Suppose A is an $(\in, \in \vee q_k)$-intuitionistic fuzzy bi-ideal of S. Let $y \in [A]_u^v$ and $x \leq y$. Then $y_u^v \in \vee q_k A$, that is $\mu(y) \geq u, \nu(y) \leq v$ *or* $\mu(y) + u + k > 1$, $\nu(y) + v + k < 1$. Now, by Theorem 4, we have $\mu(x) \geq \mu(y) \geq u$, $\nu(x) \leq \nu(y) \leq v$ *or* $\mu(x) + u + k \geq \mu(y) + u + k > 1$ and $\nu(x) + v + k \leq \nu(y) + v + k < 1$. Therefore $x_u^v \in \vee q_k A$. So $x \in [A]_u^v$.

Take any $x, y \in [A]_u^v$. Then $x_u^v \in \vee q_k A$ and $y_u^v \in \vee q_k A$, i.e., $\mu(x) \geq u, \nu(x) \leq v$ or $\mu(x) + u + k > 1, \nu(x) + v + k < 1$ and $\mu(y) \geq u, \nu(y) \leq v$ or $\mu(y) + u + k > 1, \nu(y) + v + k < 1$. Since IFS A is an $(\in, \in \vee q_k)$-intuitionistic fuzzy bi-ideal of S, we have $\mu(xy) \geq \min\{\mu(x), \mu(y), \frac{1-k}{2}\}$ and $\nu(xy) \leq \max\{\nu(x), \nu(y), \frac{1-k}{2}\}$.

Now four cases arise:

Case(i). $\mu(x) \geq u, \nu(x) \leq v$ and $\mu(y) \geq u, \nu(y) \leq v$;
Case(ii). $\mu(x) \geq u, \nu(x) \leq v$ and $\mu(y) + u + k > 1, \ \nu(y) + v + k < 1$;
Case(iii). $\mu(x) + u + k > 1, \ \nu(x) + v + k < 1$ and $\mu(y) \geq u, \nu(y) \leq v$;
Case(iv). $\mu(x) + u + k > 1, \ \nu(x) + v + k < 1$ and $\mu(y) + u + k > 1, \ \nu(y) + v + k < 1$.
Case(i): Suppose $\mu(x) \geq u, \nu(x) \leq v$ and $\mu(y) \geq u, \nu(y) \leq v$. If $u > \frac{1-k}{2}$, then $v < \frac{1-k}{2}$. Now $\mu(xy) \geq \min\{\mu(x), \mu(y), \frac{1-k}{2}\} \geq \min\{u, u, \frac{1-k}{2}\} = \frac{1-k}{2}$ and $\nu(xy) \leq \max\{\nu(x), \nu(y), \frac{1-k}{2}\} \leq \max\{v, v, \frac{1-k}{2}\} = \frac{1-k}{2}$. As $\mu(xy) + u + k > \frac{1-k}{2} + \frac{1-k}{2} + k = 1$ and $\nu(xy) + u + k < \frac{1-k}{2} + \frac{1-k}{2} + k = 1$, we have $(xy)_u^v q_k A$. If $u \leq \frac{1-k}{2}$, then $v \geq \frac{1-k}{2}$. Again $\mu(xy) \geq \min\{\mu(x), \mu(y), \frac{1-k}{2}\} = u$ and $\nu(xy) \leq \max\{\nu(x), \nu(y), \frac{1-k}{2}\} = v$. Therefore, $(xy)_u^v \in A$. Hence $(xy)_u^v \in \vee q_k A$.
Case(ii): Suppose that $u > \frac{1-k}{2}$ and $v < \frac{1-k}{2}$. Then $1 - u - k < \frac{1-k}{2}$ and $1 - v - k > \frac{1-k}{2}$. So, we have $\mu(xy) \geq \min\{\mu(x), \mu(y), \frac{1-k}{2}\} = \min\{\mu(y), \frac{1-k}{2}\} = 1 - u - k$ and $\nu(xy) \leq \max\{\nu(x), \nu(y), \frac{1-k}{2}\} = \max\{\nu(y), \frac{1-k}{2}\} = 1 - v - k$. This implies that $\mu(xy) + u + k > 1, \ \nu(xy) + v + k < 1$. Thus $(xy)_u^v q_k A$. If $u \leq \frac{1-k}{2}$ *and* $v \leq \frac{1-k}{2}$, then $1 - u - k \leq \frac{1-k}{2}$ and $1 - v - k \geq \frac{1-k}{2}$. So $\mu(xy) \geq \min\{\mu(x), \mu(y), \frac{1-k}{2}\} = u$ and $\nu(xy) \leq \max\{\nu(x), \nu(y), \frac{1-k}{2}\} = v$. Therefore $(xy)_u^v \in A$. Hence $(xy)_u^v \in \vee q_k A$.
Case(iii): The proof in this case is similar to the proof in Case (ii).
Case(iv): Suppose $\mu(x) + u + k > 1, \ \nu(x) + v + k < 1$ and $\mu(y) + u + k > 1, \ \nu(y) + v + k < 1$. If $u > \frac{1-k}{2}, v < \frac{1-k}{2}$, then $\mu(xy) \geq \min\{\mu(x), \mu(y), \frac{1-k}{2}\} > \min\{1 - u - k, \frac{1-k}{2}\} = 1 - u - k$, and $\nu(xy) \leq \max\{\nu(x), \nu(y), \frac{1-k}{2}\} < \max\{1 - v - k, \frac{1-k}{2}\} = 1 - v - k$. Thus $(xy)_u^v q_k A$. If $u \leq \frac{1-k}{2}, v \geq \frac{1-k}{2}$, then $\mu(xy) \geq \min\{\mu(x), \mu(y), \frac{1-k}{2}\} \geq \min\{1 - u - k, \frac{1-k}{2}\} = u$, and $\nu(xy) \leq \max\{\nu(x), \nu(y), \frac{1-k}{2}\} \leq \max\{1 - v - k, \frac{1-k}{2}\} = v$. Thus $(xy)_u^v \in A$. Hence $(xy)_u^v \in \vee q_k A$. Thus, in all cases, we have $(xy)_u^v \in \vee q_k A$. Therefore $xy \in [A]_u^v$. Finally take any $x, z \in [A]_u^v$ for $u \in (0, 1]$, $v \in [0, 1)$. Then $x_u^v \in \vee q_k A$ and $z_u^v \in \vee q_k A$; that is, $\mu(x) \geq u, \nu(x) \leq v$ or $\mu(x) + u + k > 1, \nu(x) + v + k < 1$ and $\mu(z) \geq u, \nu(z) \leq v, \mu(z) + u + k > 1, \nu(z) + v + k < 1$. Now for any $y \in S$, as IFS A is an $(\in, \in \vee q_k)$-intuitionistic fuzzy bi-ideal of S, $\mu(xyz) \geq \min\{\mu(x), \mu(z), \frac{1-k}{2}\}$ and $\nu(xyz) \leq \max\{\nu(x), \nu(z), \frac{1-k}{2}\}$. Now on the lines similar to the above proof, we may show that $(xyz)_u^v \in \vee q_k A$. Thus, $xyz \in [A]_u^v$.

(2) let IFS A be a strongly convex intuitionistic fuzzy subset of S and $u \in (0, 1]$, $v \in [0, 1)$ be such that $[A]_u^v$ is a bi-ideal of S. Suppose to the contrary that $\mu(xy) < \min\{\mu(x), \mu(y), \frac{1-k}{2}\}$ and $\nu(xy) > \max\{\nu(x), \nu(y), \frac{1-k}{2}\}$ for some $x, y \in S$, then there exists $u \in (0, \frac{1-k}{2}]$ and $v \in (\frac{1-k}{2}, 1)$ such that $\mu(xy) < u \leq \min\{\mu(x), \mu(y), \frac{1-k}{2}\}$ and $\nu(xy) > v \geq \max\{\nu(x), \nu(y), \frac{1-k}{2}\}$. This implies that $x, y \in [A]_u^v$. Therefore $xy \in [A]_u^v$ implying that $\mu(xy) \geq u, \nu(xy) \leq v$ *or* $\mu(xy) + u + k > 1, \nu(xy) + v + k < 1$, which is impossible. Therefore $\mu(xy) \geq \min\{\mu(x), \mu(y), \frac{1-k}{2}\}$ *and* $\nu(xy) \leq \max\{\nu(x), \nu(y), \frac{1-k}{2}\}$ for all $x, y \in S$. Similarly we may show that $\mu(xyz) \geq \min\{\mu(x), \mu(z), \frac{1-k}{2}\}$ and $\nu(xyz) \leq \max\{\nu(x), \nu(z), \frac{1-k}{2}\}$ for

all $x, y, z \in S$. Now A is an $(\in, \in \vee q_k)$-intuitionistic fuzzy bi-ideal of S follows by Proposition 1 and Theorem 4.

Definition An IFS A of an ordered semigroup S is called an $(\in, \in \vee q_k)$-intuitionistic fuzzy generalized bi-ideal of S if $\forall u, u_1 \in (0, 1], v, v_1 \in [0, 1)$ and $x, y, z \in S$, the following conditions are satisfied:

(i) $x \leq y \Rightarrow \mu(x) \geq \mu(y)$ *and* $\nu(x) \leq \nu(y)$;
(ii) $x_u^v \in A,\ z_{u_1}^{v_1} \in A \Rightarrow (xyz)_{\min\{u,u_1\}}^{\max\{v,v_1\}} \in \vee q_k A$.

Similar to the proof of Theorem 4, we may prove.

Theorem 7 *Let A be an intuitionistic fuzzy subset of an ordered semigroup S. Then A is an $(\in, \in \vee q_k)$-intuitionistic fuzzy generalized bi-ideal of S if and only if A satisfies:*

(i) $x \leq y \Rightarrow \mu(x) \geq \mu(y)$ *and* $\nu(x) \leq \nu(y)$*;*
(ii) $\mu(xyx) \geq \min\{\mu(x), \mu(z), \frac{1-k}{2}\}$ *and* $\mu(xyx) \leq \max\{\nu(x), \nu(z), \frac{1-k}{2}\}$.

Theorem 8 *Let S be an ordered semigroup. Then*

(1) *The subset $[A]_u^v$ of S, for any $(\in, \in \vee q_k)$-intuitionistic fuzzy generalized bi-ideal A of S, is a generalized bi-ideal of S.*
(2) *If A is strongly convex intuitionistic fuzzy subset of S and $[A]_u^v$ is a generalized bi-ideal of S, then A is an $(\in, \in \vee q_k)$-intuitionistic fuzzy generalized bi-ideal of S.*

Proof The proof follows on the lines similar to the proof of Theorem 6.

Definition Let $(S, ., \leq)$ be an ordered semigroup. For $x \in S$, let

$$I_x = \{(y, z) \in S \times S | x \leq yz\}.$$

For any IFS $A = (\mu, \nu)$ and $B = (\mu', \nu')$ of S, we define $A \circ B = (\mu \circ \mu', \nu \circ \nu')$, where

$$(\mu \circ \mu')(x) = \begin{cases} \bigvee_{(y,z)\in I_x} \min\{\mu(y), \mu'(z)\} & \text{if } I_x \neq \emptyset; \\ 0 & \text{if } I_x = \emptyset. \end{cases}$$

and

$$(\nu \circ \nu')(x) = \begin{cases} \bigwedge_{(y,z)\in I_x} \max\{\nu(y), \nu'(z)\} & \text{if } I_x \neq \emptyset; \\ 1 & \text{if } I_x = \emptyset. \end{cases}$$

for all $x \in S$.

Definition Let A be an IFS of an ordered semigroup S. For any $k \in [0, 1)$, let

$$\mu_k(x) = \min\left\{\mu(x), \frac{1-k}{2}\right\}$$

$$\nu_k(x) = \max\left\{\nu(x), \frac{1-k}{2}\right\} \qquad \text{for all } x \in S.$$

Then IFS (μ_k, ν_k) of S is denoted by A_k and is called the k-lower part of A.

Definition Let $A = (\mu, \nu)$ and $B = (\mu', \nu')$ be two IFS of an ordered semigroup S. For any $k \in [0, 1)$, define $A \cap_k B$ and $A \circ_k B$ of S by

$$(A \cap_k B)(x) = \left(\min\left\{(\mu \cap \mu')(x), \frac{1-k}{2}\right\}, \max\left\{(\nu \cap \nu')(x), \frac{1-k}{2}\right\}\right)$$

and $$(A \circ_k B)(x) = \left(\min\left\{(\mu \circ \mu')(x), \frac{1-k}{2}\right\}, \max\left\{(\nu \circ \nu')(x), \frac{1-k}{2}\right\}\right)$$

respectively for all $x \in .S$.

It is easy to see that $A \cap_k B$ and $A \circ_k B$ are IFS of S.

The following Lemma easily follows:

Lemma 2 *Let A and B be IFS of an ordered semigroup S and $k \in [0, 1)$. Then*

(i) $(A_k)_k = A_k, \quad A_k \subseteq A$;
(ii) *If* $A \subseteq B$ *and* $C \in I(S)$, *then* $A \circ_k C \subseteq B \circ_k C$, $C \circ_k A \subseteq C \circ_k B$, *where* $I(S)$ *be the set of all IFS of S;*
(iii) $A \cap_k B = A_k \cap B_k$;
(iv) $A \circ_k B = A_k \circ B_k$;
(v) $A \circ_k S = A_k \circ S$, $S \circ_k A = S \circ A_k$, $A \circ_k S \circ_k A = A_k \circ S \circ A_k$ *and* $S \circ_k A \circ_k S = S \circ A_k \circ S$.

Example Let $S = \{a, b, c\}$ be a ordered semigroup with respect to the order relation $a \leq b$ and the operation '. ' defined in the following cayley table

.	a	b	c
a	a	a	c
b	a	b	c
c	a	a	c

Define an IFS $A = (\mu, \nu)$ by $\mu(a) = 0.50,\ \mu(b) = 0.45,\ \mu(c) = 0.60$ and $\nu(a) = 0.30,\ \nu(b) = 0.35,\ \nu(c) = 0.20$. Then IFS $A = (\mu, \nu)$ is an $(\in, \in \vee q_k)$ intuitionistic fuzzy ideal of S for any $k \in [0, 1)$. But A is not an intuitionistic fuzzy ideal of S as $\mu(ca) \geq \mu(c)$ implies that $\mu(a) \geq \mu(c) \Rightarrow 0.50 \geq 0.60$ which is not possible.

The above example shows that an $(\in, \in \vee q_k)$-intuitionistic fuzzy ideal of an ordered semigroup S is not necessarily an intuitionistic fuzzy ideal of S. However, in the following Theorem, we show that if IFS A is an $(\in, \in \vee q_k)$-intuitionistic fuzzy ideal of an ordered semigroup S, then A_k is an intuitionistic fuzzy ideal of S.

Theorem 9 *Let A be an $(\in, \in \vee q_k)$-intuitionistic fuzzy ideal of an ordered semigroup S. Then A_k is an intuitionistic fuzzy ideal of S.*

Proof Let $A = (\mu, \nu)$ be an $(\in, \in \vee q_k)$-intuitionistic fuzzy ideal of an ordered semigroup S and $x, y \in S$. Then, by Theorem 3, $\mu(xy) \geq \min\{\mu(x), \frac{1-k}{2}\}$ and $\nu(xy) \leq \max\{\nu(x), \frac{1-k}{2}\}$. Now $\mu_k(xy) = \min\{\mu(xy), \frac{1-k}{2}\} \geq \min\{\mu(x), \frac{1-k}{2}\} = \mu_k(x)$, $\nu_k(xy) = \max\{\nu(xy), \frac{1-k}{2}\} \leq \max\{\nu(x), \frac{1-k}{2}\} = \nu_k(x)$. Also, if $x \leq y$, we need to show that $\mu_k(x) \geq \mu_k(y), \nu_k(x) \leq \nu_k(y)$. Since A is an $(\in, \in \vee q_k)$-intuitionistic fuzzy ideal, we have $\mu(x) \geq \mu(y), \nu(x) \leq \nu(y)$. Now $\mu_k(x) = \min\{\mu(x), \frac{1-k}{2}\} \geq \min\{\mu(y), \frac{1-k}{2}\} = \mu_k(y)$ and $\nu_k(x) = \max\{\nu(x), \frac{1-k}{2}\} \leq \max\{\nu(y), \frac{1-k}{2}\} = \nu_k(y)$. Therefore A_k is an intuitionistic fuzzy right ideal of S. Similarly, we may show that A_k is an intuitionistic fuzzy left ideal of S. Hence A_k is an intuitionistic fuzzy ideal of S.

The following Proposition may easily be proved.

Proposition 2 *Let A be an IFS of an ordered semigroup S and let I, J be any two nonempty subsets of S. Then for any $k \in [0, 1)$:*

(i) $\chi_I \cap_k \chi_J = (\chi_{I \cap J})_k$;
(ii) $\chi_I \circ_k \chi_J = (\chi_{(IJ]})_k$;
(iii) $S \circ_k \chi_I = (\chi_{(SI]})_k$, $\chi_I \circ_k S = (\chi_{(IS]})_k$, $\chi_I \circ_k S \circ_k \chi_I = (\chi_{(ISI]})_k$ *and* $S \circ_k \chi_I \circ_k S = (\chi_{(SIS]})_k$.

Theorem 10 *Let A be an IFS of an ordered semigroup S and let I be a nonempty subset of S. Then I is a left ideal of S if and only if the characteristic function χ_I of I is an $(\in, \in \vee q_k)$-intuitionistic fuzzy left ideal of S.*

Proof Let $x, y \in S$ be such that $x \leq y$. If $y \in I$, then $\mu_{\chi_I}(y) = 1,\ \nu_{\chi_I}(y) = 0$. As $x \leq y$ and $y \in I$, we have $x \in I$. Then $\mu_{\chi_I}(x) = 1,\ \nu_{\chi_I}(x) = 0$. Thus $\mu_{\chi_I}(x) \geq \mu_{\chi_I}(y)$ and $\nu_{\chi_I}(x) \leq \nu_{\chi_I}(y)$. If $y \notin I$, then $\mu_{\chi_I}(y) = 0,\ \nu_{\chi_I}(y) = 1$. Since $x \in S$, we have $\mu_{\chi_I}(x) \geq 0,\ \nu_{\chi_I}(x) \leq 1$. Thus $\mu_{\chi_I}(y) = 0 \leq \mu_{\chi_I}(x)$, and $\nu_{\chi_I}(y) = 1 \geq \nu_{\chi_I}(x)$. Next we show that $\mu_{\chi_I}(xy) \geq \min\{\mu_{\chi_I}(y), \frac{1-k}{2}\}$ and $\nu_{\chi_I}(xy) \leq \max\{\nu_{\chi_I}(y), \frac{1-k}{2}\}$.

Let $x \in S$ *and* $y \in I$. Since I is a left ideal of S, $xy \in I$. Then $\mu_{\chi_I}(xy) = 1 \geq \min\{\mu_{\chi_I}(y), \frac{1-k}{2}\}$ and $\nu_{\chi_I}(xy) = 0 \leq \max\{\nu_{\chi_I}(y), \frac{1-k}{2}\}$. If $y \notin I$, then $\mu_{\chi_I}(y) = 0,\ \nu_{\chi_I}(y) = 1$. Now $\min\{\mu_{\chi_I}(y), \frac{1-k}{2}\} = 0$ and $\max\{\nu_{\chi_I}(y), \frac{1-k}{2}\} = 1$. As $xy \in S$, $\mu_{\chi_I}(xy) \geq 0,\ \nu_{\chi_I}(xy) \leq 1$. Thus $\mu_{\chi_I}(xy) \geq \min\{\{\mu_{\chi_I}(y), \frac{1-k}{2}\}$ and $\nu_{\chi_I}(xy) \leq \max\{\{\nu_{\chi_I}(y), \frac{1-k}{2}\}$.

Conversely let $x, y \in I$. Now $\mu_{\chi_I}(xy) \geq \min\{\mu_{\chi_I}(y)\frac{1-k}{2}\} = \min\{1, \frac{1-k}{2}\} = \frac{1-k}{2} > 0$ and $\nu_{\chi_I}(xy) \leq \max\{\nu_{\chi_I}(y), \frac{1-k}{2}\} = \max\{0, \frac{1-k}{2}\} = \frac{1-k}{2} < 1$. This implies that $xy \in I$. Let $x \in S,\ y \in I$ and $x \leq y$. Since $y \in I$, $\mu_{\chi_I}(y) = 1,\ \nu_{\chi_I}(y) = 0$. As χ_I is an $(\in, \in \vee q_k)$-intuitionistic fuzzy left ideal of S and $x \leq y$, $\mu_{\chi_I}(x) \geq \mu_{\chi_I}(y) = 1,\ \nu_{\chi_I}(x) \leq \nu_{\chi_I}(y) = 0$. Since $x \in S$, $\mu_{\chi_I}(x) \leq 1$ and $\nu_{\chi_I}(x) \geq 0$. Therefore $\mu_{\chi_I}(x) = 1$ and $\nu_{\chi_I}(x) = 0$. This implies that $x \in I$.

Dually we may prove the following:

Theorem 11 *Let A be an IFS of an ordered semigroup S and let I be a nonempty subset of S. Then I is a right ideal of S if and only if the characteristic function* χ_I *of I is an* $(\in, \in \vee q_k)$*-intuitionistic fuzzy right ideal of S.*

Combining Theorems 10 and 11, we have the following:

Theorem 12 *Let A be an IFS of an ordered semigroup S and I be a nonempty subset of S. Then I is a ideal of S if and only if the characteristic function* χ_I *of I is an* $(\in, \in \vee q_k)$*-intuitionistic fuzzy ideal of S.*

Theorem 13 *Let IFS A of an ordered semigroup S and I be a nonempty subset of S. Then I is a bi-ideal of S if and only if the characteristic function* χ_I *of I is an* $(\in, \in \vee q_k)$*-intuitionistic fuzzy bi-ideal of S.*

Proof The proof follows on the similar to the proof of Theorem 10.

Lemma 3 *Let A be a strongly convex intuitionistic fuzzy subset of an ordered semigroup S. Then A is an* $(\in, \in \vee q_k)$*-intuitionistic fuzzy subsemigroup of S if and only if* $A \circ_k A \subseteq A_k$.

Proof Let $A = (\mu, \nu)$ be an $(\in, \in \vee q_k)$-intuitionistic fuzzy subsemigroup of S and $x \in S$.

If $I_x = \emptyset$, then $(\mu \circ_k \mu)(x) = (\mu_k \circ \mu_k)(x) = 0 \leq \mu_k(x)$,

$$(\nu \circ_k \nu)(x) = (\nu_k \circ \nu_k)(x) = 1 \geq \nu_k(x).$$

If $I_x \neq \emptyset$, then

$$\begin{aligned}
(\mu \circ_k \mu)(x) &= \min\left\{(\mu \circ \mu)(x), \frac{1-k}{2}\right\} = \bigvee_{(y,z)\in I_x} \min\left\{\mu(y), \mu(z), \frac{1-k}{2}\right\} \\
&\leq \bigvee_{(y,z)\in I_x} \min\left\{\mu(yz), \frac{1-k}{2}\right\} && \text{(by Theorem 1)} \\
&\leq \bigvee_{(y,z)\in I_x} \min\left\{\mu(x), \frac{1-k}{2}\right\} && \text{(since } A \text{ is strongly convex)} \\
&= \min\left\{\mu(x), \frac{1-k}{2}\right\} = \mu_k(x)
\end{aligned}$$

and

$$\begin{aligned}
(\nu \circ_k \nu)(x) &= \max\left\{(\nu \circ \nu)(x), \frac{1-k}{2}\right\} = \bigwedge_{(y,z)\in I_x} \max\left\{\nu(y), \nu(z), \frac{1-k}{2}\right\} \\
&\geq \bigwedge_{(y,z)\in I_x} \max\left\{\nu(yz), \frac{1-k}{2}\right\} && \text{(by Theorem 1)}
\end{aligned}$$

$$\geq \bigwedge_{(y,z)\in I_x} \max\left\{\nu(x), \frac{1-k}{2}\right\} \qquad \text{(since } A \text{ is strongly convex)}$$

$$= \max\left\{\nu(x), \frac{1-k}{2}\right\} = \nu_k(x).$$

Therefore $A \circ_k A \subseteq A_k$.

Conversely suppose that $A \circ_k A \subseteq A_k$. Now for any $x,\ y \in S$

$$\begin{aligned}\mu(xy) \geq \mu_k(xy) &\geq (\mu \circ_k \mu)(xy) = (\mu_k \circ \mu_k)(x) \qquad \text{(by Lemma 2(iv))}\\ &= \bigvee_{(b,c)\in I_{xy}} \min\{\mu_k(b), \mu_k(c)\}\\ &\geq \min\{\mu_k(x), \mu_k(y)\} = \min\left\{\mu(x), \mu(y), \frac{1-k}{2}\right\}\end{aligned}$$

and

$$\begin{aligned}\nu(xy) \leq \nu_k(xy) &\leq (\nu \circ_k \nu)(xy) = (\nu_k \circ \nu_k)(x) \qquad \text{(by Lemma 2(iv))}\\ &= \bigwedge_{(b,c)\in I_{xy}} \max\{\nu_k(b), \nu_k(c)\}\\ &\leq \max\{\nu_k(x), \nu_k(y)\} = \max\left\{\nu(x), \nu(y), \frac{1-k}{2}\right\}.\end{aligned}$$

Therefore, by Theorem 1, A is an $(\in, \in \vee q_k)$-intuitionistic fuzzy subsemigroup of S.

Lemma 4 *An IFS A of an ordered semigroups is an $(\in, \in \vee q_k)$-intuitionistic fuzzy left ideal of S if and only if*

(i) $x \leq y \Rightarrow \mu(x) \geq \mu(y)$ *and* $\nu(x) \leq \nu(y)\ \forall x, y \in S$ *and*
(ii) $S \circ_k A \subseteq A_k$.

Proof Let IFS A be an $(\in, \in \vee q_k)$-intuitionistic fuzzy left ideal of S and $x \in S$. If $I_x = \emptyset$, then $(S \circ_k \mu)(x) = (S \circ \mu_k)(x) = 0 \leq \mu_k(x)$ and $(\widetilde{S} \circ_k \nu)(x) = (\widetilde{S} \circ \nu_k)(x) = 1 \geq \nu_k(x)$.

If $I_x \neq \emptyset$, then by Theorem 2, we have

$$\begin{aligned}(S \circ_k \mu)(x) &= \min\left\{(S \circ \mu)(x), \frac{1-k}{2}\right\} = \bigvee_{(y,z)\in I_x} \min\left\{S(y), \mu(z), \frac{1-k}{2}\right\}\\ &= \bigvee_{(y,z)\in I_x} \min\left\{1, \mu(z), \frac{1-k}{2}\right\} \leq \bigvee_{(y,z)\in I_x} \min\left\{\mu(yz), \frac{1-k}{2}\right\}\\ &\leq \bigvee_{(y,z)\in I_x} \min\left\{\mu(x), \frac{1-k}{2}\right\} = \min\left\{\mu(x), \frac{1-k}{2}\right\} = \mu_k(x)\end{aligned}$$

$$\text{and } (\widetilde{S} \circ_k \nu)(x) = \max\left\{(\widetilde{S} \circ \nu)(x), \frac{1-k}{2}\right\} = \bigwedge_{(y,z)\in I_x} \max\left\{\widetilde{S}(y), \nu(z), \frac{1-k}{2}\right\}$$
$$= \bigwedge_{(y,z)\in I_x} \max\left\{0, \nu(z), \frac{1-k}{2}\right\} \geq \bigwedge_{(y,z)\in I_x} \max\left\{\nu(yz), \frac{1-k}{2}\right\}$$
$$\geq \bigwedge_{(y,z)\in I_x} \max\left\{\nu(x), \frac{1-k}{2}\right\} = \max\left\{\nu(x), \frac{1-k}{2}\right\} = \nu_k(x)$$

which implies that $S \circ_k A \subseteq A_k$.

Conversely suppose that $x \leq y$. Then $\mu(x) \geq \mu(y)$ and $\nu(x) \leq \nu(y)\ \forall\ x, y \in S$ and $S \circ_k A \subseteq A_k$.

Let $b, c \in S$ and $a = bc$. Then as $S \circ_k A \subseteq A_k$, we have

$$\mu(bc) = \mu(a) \geq \mu_k(a) \geq (S \circ_k \mu)(a) = (S \circ \mu_k)(a) \qquad \text{(by Lemma 2(v))}$$
$$= \bigvee_{(d,e)\in I_a} \min\{S(d), \mu_k(e)\}$$
$$\geq \min\{S(b), \mu_k(c)\} = \mu_k(c) = \min\left\{\mu(c), \frac{1-k}{2}\right\}$$
$$\text{and} \quad \nu(bc) = \nu(a) \leq \nu_k(a) \leq (\widetilde{S} \circ_k \nu)(a) = (S \circ \nu_k)(a) \qquad \text{(by Lemma 2(v))}$$
$$= \bigwedge_{(d,e)\in I_a} \max\{\widetilde{S}(d), \nu_k(e)\}$$
$$\leq \max\{\widetilde{S}(b), \nu_k(c)\} = \nu_k(c) = \max\left\{\nu(c), \frac{1-k}{2}\right\}.$$

Hence A, by Theorem 2, is an $(\in, \in \vee q_k)$-intuitionistic fuzzy left ideal of S.

The proof of the following follows on the lines similar to the proof of Lemma 4.

Lemma 5 *Let A be an IFS of an ordered semigroup S. Then A is an $(\in, \in \vee q_k)$-intuitionistic fuzzy right ideal of S if and only if*

(i) $x \leq y \Rightarrow \mu(x) \geq \mu(y)$ *and* $\nu(x) \leq \nu(y)\ \forall x, y \in S$;
(ii) $A \circ_k S \subseteq A_k$.

Combining Lemmas 4 and 5, we get the following:

Lemma 6 *Let A be an IFS of an ordered semigroup S. Then A is an $(\in, \in \vee q_k)$-intuitionistic fuzzy ideal of S if and only if*

(i) $x \leq y \Rightarrow \mu(x) \geq \mu(y)$ *and* $\nu(x) \leq \nu(y)\ \forall x, y \in S$;
(ii) $S \circ_k A \subseteq A_k$, $A \circ_k S \subseteq A_k$.

Theorem 14 *If IFS A is an $(\in, \in \vee q_k)$-intuitionistic fuzzy generalized bi-ideal of an ordered semigroup S, then $A \circ_k S \circ_k A \subseteq A_k$.*

Proof Let A be an $(\in, \in \vee q_k)$-intuitionistic fuzzy generalized bi-ideal of an ordered semigroup S. Then $(A \circ_k S \circ_k A)(a) \subseteq A_k(a)$, for all $a \in S$. Since A_k is a fuzzy subset of S, $\mu_k(a) \geq 0$ *and* $\nu_k(a) \leq 1$ for all $a \in S$. If $(A \circ_k S \circ_k A)(a) = 0$, then trivially $(\mu \circ_k S \circ_k \mu)(a) \leq \mu_k(a)$ and $(\nu \circ_k \widetilde{S} \circ_k \nu)(a) \geq \nu_k(a)$. If $(A \circ_k S \circ_k A)(a) \neq 0$, by Lemma 2(v), $(A_k \circ S \circ A_k)(a) = (A \circ_k S \circ_k A)(a) \neq 0$. Then $\exists\, x, y, x_1, y_1 \in S$ such that $(x, y) \in I_a$ *and* $(x_1, y_1) \in I_x$. Then $a \leq xy$ and $x \leq x_1 y_1$. Since A is an $(\in, \in \vee q_k)$-intuitionistic fuzzy generalized bi-ideal of S, $\mu(x_1 y_1 y) \geq \min\{\mu(x_1), \mu(y), \frac{1-k}{2}\}$ and $\nu(x_1 y_1 y) \leq \max\{\nu(x_1), \nu(y), \frac{1-k}{2}\}$. Now, by Lemma 2, we have

$$
\begin{aligned}
(\mu \circ_k S \circ_k \mu)(a) = (\mu_k \circ S \circ \mu_k)(a) &= \bigvee_{(x,y)\in I_a} \min\{(\mu_k \circ S)(x), \mu_k(y)\} \\
&= \bigvee_{(x,y)\in I_a} \left\{ \bigvee_{(x_1,y_1)\in I_x} \min\{\mu_k(x_1), S(y_1), \mu_k(y)\} \right\} \\
&= \bigvee_{(x,y)\in I_a} \left\{ \bigvee_{(x_1,y_1)\in I_x} \min\left\{\mu(x_1), \mu(y), \frac{1-k}{2}\right\} \right\} \\
&\leq \bigvee_{(x,y)\in I_a} \left\{ \bigvee_{(x_1,y_1)\in I_x} \min\left\{\mu(x_1 y_1 y), \frac{1-k}{2}\right\} \right\} \\
&\leq \bigvee_{(x,y)\in I_a} \min\left\{\mu(xy), \frac{1-k}{2}\right\} \\
&\leq \min\left\{\mu(a), \frac{1-k}{2}\right\} = \mu_k(a)
\end{aligned}
$$

and

$$
\begin{aligned}
(\nu \circ_k \widetilde{S} \circ_k \nu)(a) = (\nu_k \circ \widetilde{S} \circ \nu_k)(a) &= \bigwedge_{(x,y)\in I_a} \max\{(\nu_k \circ \widetilde{S})(x), \nu_k(y)\} \\
&= \bigwedge_{(x,y)\in I_a} \left\{ \bigwedge_{(x_1,y_1)\in I_x} \max\{\nu_k(x_1), \widetilde{S}(y_1), \nu_k(y)\} \right\} \\
&= \bigwedge_{(x,y)\in I_a} \left\{ \bigwedge_{(x_1,y_1)\in I_x} \max\left\{\nu(x_1), \nu(y), \frac{1-k}{2}\right\} \right\} \\
&\geq \bigwedge_{(x,y)\in I_a} \left\{ \bigwedge_{(x_1,y_1)\in I_x} \max\left\{\nu(x_1 y_1 y), \frac{1-k}{2}\right\} \right\} \\
&\geq \bigwedge_{(x,y)\in I_a} \max\left\{\nu(xy), \frac{1-k}{2}\right\}
\end{aligned}
$$

$$\geq \max\left\{\nu(a), \frac{1-k}{2}\right\} = \nu_k(a).$$

Therefore $A \circ_k S \circ_k A \subseteq A_k$.

The proof of the following theorem follows easily by Lemma 3.

Theorem 15 *If IFS A is an* $(\in, \in \vee q_k)$*-intuitionistic fuzzy bi-ideal of an ordered semigroup S, then* $A \circ_k A \subseteq A$ *and* $A \circ_k S \circ_k A \subseteq A_k$.

Lemma 7 *Let* $A = (\mu, \nu)$ *be an IFS of an ordered semigroup S and* $A \circ_k S \circ_k A \subseteq A_k$. *Then* $\mu(xyz) \geq \min\{\mu(x), \mu(z), \frac{1-k}{2}\}$ *and* $\nu(xyz) \leq \max\{\nu(x), \nu(z), \frac{1-k}{2}\}$ $\forall x, y, z \in S$.

Proof Let $a = xyz$ for all $x, y, z \in S$. Since $A \circ_k S \circ_k A \subseteq A_k$, we have

$$\begin{aligned}
\mu(xyz) &= \mu(a) \geq \mu_k(a) \geq (\mu \circ_k S \circ_k \mu)(a) = (\mu_k \circ S \circ \mu_k)(a) \\
&= \bigvee_{(b,c)\in I_a} \min\{(\mu_k \circ S)(b), \mu_k(c)\} \geq \min\{(\mu_k \circ S)(xy), \mu_k(z)\} \qquad (\text{as } (xy, z) \in I_a) \\
&= \bigvee_{(u,v)\in I_{xy}} \min\{\mu_k(u), S(v), \mu_k(z)\} \geq \min\{\mu_k(x), S(y), \mu_k(z)\} \\
&\geq \min\left\{\min\left\{\mu(x), \frac{1-k}{2}\right\}, 1, \min\left\{\mu(z), \frac{1-k}{2}\right\}\right\} = \min\left\{\mu(x), \mu(z), \frac{1-k}{2}\right\}
\end{aligned}$$

and

$$\begin{aligned}
\nu(xyz) &= \nu(a) \leq \nu_k(a) \leq (\nu \circ_k \widetilde{S} \circ_k \nu)(a) = (\nu_k \circ \widetilde{S} \circ \nu_k)(a) \\
&= \bigwedge_{(b,c)\in I_a} \max\{(\nu_k \circ \widetilde{S})(b), \nu_k(c)\} \leq \max\{(\nu_k \circ \widetilde{S})(xy), \nu_k(z)\} \qquad (\text{as } (xy, z) \in I_a) \\
&= \bigwedge_{(u,v)\in I_{xy}} \max\{\nu_k(u), \widetilde{S}(v), \nu_k(z)\} \leq \max\{\nu_k(x), \widetilde{S}(y), \nu_k(z)\} \\
&\leq \max\left\{\max\left\{\nu(x), \frac{1-k}{2}\right\}, 0, \max\left\{\nu(z), \frac{1-k}{2}\right\}\right\} = \max\left\{\nu(x), \nu(z), \frac{1-k}{2}\right\}.
\end{aligned}$$

This completes the proof.

Theorem 16 *Let IFS A and B be two* $(\in, \in \vee q_k)$*-intuitionistic fuzzy generalized bi-ideals of an ordered semigroup S. Then* $A \circ_k B$ *is an* $(\in, \in \vee q_k)$*-intuitionistic fuzzy bi-ideal of S.*

Proof As IFS $A = (\mu, \nu)$ and $B = (\mu', \nu')$ are $(\in, \in \vee q_k)$-intuitionistic fuzzy generalized bi-ideals of S, by Theorem 14, we have $\mu \circ_k S \circ_k \mu \leq \mu_k$, $\nu \circ_k \widetilde{S} \circ_k \nu \geq \nu_k$ and $\mu' \circ_k S \circ_k \mu' \leq \mu'_k$, $\nu' \circ_k \widetilde{S} \circ_k \nu' \geq \nu'_k$.

Now, by Lemma 2, we have $(\mu \circ_k \mu') \circ_k (\mu \circ_k \mu') \leq \mu \circ_k (\mu' \circ_k S \circ_k \mu') \leq \mu \circ_k \mu'$ and $(\nu \circ_k \nu') \circ_k (\nu \circ_k \nu') \geq \nu \circ_k (\nu' \circ_k \widetilde{S} \circ_k \nu') \geq \nu \circ_k \nu'$. Clearly $\mu \circ_k \mu'$ and $\nu \circ_k \nu'$ are strongly convex intuitionistic fuzzy subsets of S. Therefore, by Lemma 3, $\mu \circ_k \mu'$ and $\nu \circ_k \nu'$ are $(\in, \in \vee q_k)$-intuitionistic fuzzy subsemigroups of S. By

Lemma 2, we have $(\mu \circ_k \mu') \circ_k S \circ_k (\mu \circ_k \mu') = \mu \circ_k \mu' \circ_k (S \circ_k \mu) \circ_k \mu' \leq \mu \circ_k (\mu' \circ_k S \circ_k \mu') \leq \mu \circ_k \mu'$ and $(\nu \circ_k \nu') \circ_k \widetilde{S} \circ_k (\nu \circ_k \nu') = \nu \circ_k \nu' \circ_k (\widetilde{S} \circ_k \nu) \circ_k \nu' \geq \nu \circ_k (\nu' \circ_k \widetilde{S} \circ_k \nu') \geq \nu \circ_k \nu'$. By Lemma 7, $(\mu \circ_k \mu')(xyz) \geq \min\{(\mu \circ_k \mu')(x), (\mu \circ_k \mu')(y), \frac{1-k}{2}\}$ and $(\nu \circ_k \nu')(xyz) \leq \max\{(\nu \circ_k \nu')(x), (\nu \circ_k \nu')(y), \frac{1-k}{2}\}$ for all $x, y, z \in S$.

Let $x \leq y$, then $(\mu \circ_k \mu')(x) \geq (\mu \circ_k \mu')(y)$ and $(\nu \circ_k \nu')(x) \leq (\nu \circ_k \nu')(y)$. If $I_y = \emptyset$, then $(\mu \circ_k \mu')(y) = (\mu_k \circ \mu'_k)(y) = 0$ and $(\nu \circ_k \nu')(y) = (\nu_k \circ \nu'_k)(y) = 1$. Since $(\mu \circ_k \mu')$ and $(\nu \circ_k \nu')$ are intuitionistic fuzzy subsets of S, $(\mu \circ_k \mu')(x) \geq 0 = (\mu \circ_k \mu')(y)$ and $(\nu \circ_k \nu') \leq 1 = (\nu \circ_k \nu')(y)$. Again, if $I_y \neq \emptyset$, as $x \leq y$, we have $I_y \subseteq I_x$. Now, by Lemma 2, we have

$$\begin{aligned}(\mu \circ_k \mu')(y) = (\mu_k \circ \mu'_k)(y) &= \bigvee_{(b,c)\in I_y} \min\{\mu_k(b), \mu'_k(c)\} \\ &\leq \bigvee_{(b,c)\in I_x} \min\{\mu_k(b), \mu'_k(c)\} = (\mu_k \circ \mu'_k)(x) \\ &= (\mu \circ_k \mu')(x)\end{aligned}$$

and

$$\begin{aligned}(\nu \circ_k \nu')(y) = (\nu_k \circ \nu'_k)(y) &= \bigwedge_{(b,c)\in I_y} \max\{\nu_k(b), \nu'_k(c)\} \\ &\geq \bigwedge_{(b,c)\in I_x} \max\{\nu_k(b), \nu'_k(c)\} = (\nu_k \circ \nu'_k)(x) \\ &= (\nu \circ_k \nu')(x).\end{aligned}$$

Hence $A \circ_k B$ is an $(\in, \in \vee q_k)$-intuitionistic fuzzy bi-ideal of S.

4 Regular and Intra-regular Ordered Semigroups

Definition An ordered semigroup S is called regular if for each $a \in S$, there exits an element $x \in S$ such that $a \leq axa$.

Definition An ordered semigroup S is called intra-regular if for each $a \in S$, there exits $x, y \in S$ such that $a \leq xa^2y$.

In this section, we characterize regular and intra-regular ordered semigroups by $(\in, \in \vee q_k)$-intuitionistic fuzzy left ideals, $(\in, \in \vee q_k)$-intuitionistic fuzzy right ideals, $(\in, \in \vee q_k)$-intuitionistic fuzzy bi-ideals and $(\in, \in \vee q_k)$-intuitionistic fuzzy generalized bi-ideals.

Lemma 8 (25, Lemma 4.1) *An ordered semigroup S is regular if and only if $(ISI] = I$, where I is any bi-ideal of S.*

Lemma 9 (25, Lemma 4.6) *Let S be an ordered semigroup. Then the following statements are equivalent:*

(i) *S is regular;*
(ii) $I \cap L \subseteq (IL]$ *for each bi-ideal I and each left ideal L of S;*
(iii) $R \cap I \cap L \subseteq (RIL]$ *for each bi-ideal I, each right ideal R and each left ideal L of S.*

Lemma 10 (25, Lemma 5.1) *Let S be an ordered semigroup. Then the following statements are equivalent:*

(i) *S is intra-regular;*
(ii) $R \cap L \subseteq (LR]$ *for each left ideal L and each right ideal R of S.*

Theorem 17 *Let S be an ordered semigroup. Then the following statements are equivalent:*

(i) *S is regular;*
(ii) $A_k = A \circ_k S \circ_k A$ *for each* $k \in [0, 1)$ *and for each* $(\in, \in \vee q_k)$*-intuitionistic fuzzy bi-ideal A of S.*

Proof Suppose S is regular and let $A = (\mu, \nu)$ be any $(\in, \in \vee q_k)$ intuitionistic fuzzy bi-ideals of S and $a \in S$. Since S is regular, there exits $x \in S$ such that $a \leq axa$. Now

$$\begin{aligned}(\mu \circ_k S \circ_k \mu)(a) = (\mu_k \circ S \circ \mu_k)(a) &= \bigvee_{(y,z)\in I_a} \min\{(\mu_k \circ S)(y), \mu_k(z)\} \\ &\geq \min\{(\mu_k \circ S)(ax), \mu_k(a)\} = \bigvee_{(b,c)\in I_{ax}} \min\{\mu_k(b), S(c), \mu_k(a)\} \\ &\geq \min\{\mu_k(a), S(x), \mu_k(a)\} = \min\{\mu_k(a), 1, \mu_k(a)\} = \mu_k(a)\end{aligned}$$

and

$$\begin{aligned}(\nu \circ_k \widetilde{S} \circ_k \nu)(a) = (\nu_k \circ \widetilde{S} \circ \nu_k)(a) &= \bigwedge_{(y,z)\in I_a} \max\{(\nu_k \circ \widetilde{S})(y), \nu_k(z)\} \\ &\leq \max\{(\nu_k \circ \widetilde{S})(ax), \nu_k(a)\} = \bigwedge_{(b,c)\in I_{ax}} \max\{\nu_k(b), \widetilde{S}(c), \nu_k(a)\} \\ &\leq \max\{\nu_k(a), \widetilde{S}(x), \nu_k(a)\} = \max\{\nu_k(a), 0, \nu_k(a)\} = \nu_k(a).\end{aligned}$$

So, we have $A \circ_k S \circ_k A \supseteq A_k$. Since IFS A is an $(\in, \in \vee q_k)$intuitionistic fuzzy bi-ideal of S, by Theorem 15, we have $A \circ_k S \circ_k A \subseteq A_k$. Hence $A \circ_k S \circ_k A = A_k$.

Conversely suppose that I be any bi-ideal of S. Then, by Theorem 13, χ_I is an $(\in, \in \vee q_k)$-intuitionistic fuzzy bi-ideal of S. Now, by Proposition 2, we have $(\chi_{ISI})_k = \chi_I \circ_k S \circ_k \chi_I = (\chi_I)_k$ which implies that $(ISI] = I$. Hence, by Lemma 8, S is regular.

The proof of the following theorem follows on lines similar to the proof of Theorem 17.

Theorem 18 *Let S be any ordered semigroup. Then the following statements are equivalent:*

(i) *S is regular;*
(ii) $A_k = A \circ_k S \circ_k A$ *for each* $k \in [0, 1)$ *and for each* $(\in, \in \vee q_k)$*-intuitionistic fuzzy generalized bi-ideal* A *of* S.

Theorem 19 *Let* S *be any ordered semigroup. Then the following statements are equivalent:*

(i) *S is regular;*
(ii) $A \circ_k B \circ_k A = A \cap_k B$ *for each* $k \in [0, 1)$ *and for each* $(\in, \in \vee q_k)$*-intuitionistic fuzzy bi-ideal* A *and* $(\in, \in \vee q_k)$*-intuitionistic fuzzy ideal* B *of* S.

Proof Let $A = (\mu, \nu)$ and $B = (\mu', \nu')$ be any $(\in, \in \vee q_k)$-intuitionistic fuzzy bi-ideal and any $(\in, \in \vee q_k)$-intuitionistic fuzzy ideal of S respectively. Now, by Lemma 2 and Theorem 15, $A \circ_k B \circ_k A = A_k \circ B_k \circ A_k \subseteq S \circ B_k \circ S \subseteq S \circ B_k \subseteq B_k$.

Thus $A \circ_k B \circ_k A \subseteq A_k \cap B_k = A \cap_k B$. Let $a \in S$. Since S is regular, $\exists\, x \in S$ such that $a \leq axa \leq (axa)xa$. Since B is an $(\in, \in \vee q_k)$ intuitionistic fuzzy ideal of S, $\mu'(xax) \geq \min\{\mu'(ax), \frac{1-k}{2}\} \geq \min\{\mu'(a), \frac{1-k}{2}\}$ and $\nu'(xax) \leq \max\{\nu'(ax), \frac{1-k}{2}\} \leq \max\{\nu'(a), \frac{1-k}{2}\}$. Now $\mu'_k(xax) = \min\{\mu'(xax), \frac{1-k}{2}\} \geq \min\{\mu'(a), \frac{1-k}{2}, \frac{1-k}{2}\} = \mu'_k(a)$, $\nu'_k(xax) = \max\{\nu'(xax), \frac{1-k}{2}\} \leq \max\{\nu'(a), \frac{1-k}{2}, \frac{1-k}{2}\} = \nu'_k(a)$. Thus

$$\begin{aligned}(\mu \circ_k \mu' \circ_k \mu)(a) = (\mu_k \circ \mu'_k \circ \mu_k)(a) = &\bigvee_{(y,z)\in I_a} \min\{\mu_k(y), (\mu'_k \circ \mu_k)(z)\}\\ &\geq \min\{\mu_k(a), (\mu'_k \circ \mu_k)(xaxa)\} \geq \min\{\mu_k(a), \mu'_k(a), \mu_k(a)\}\\ &= (\mu \cap_k \mu')(a)\end{aligned}$$

and

$$\begin{aligned}(\nu \circ_k \nu' \circ_k \nu)(a) = (\nu_k \circ \nu'_k \circ \nu_k)(a) = &\bigwedge_{(y,z)\in I_a} \max\{\nu_k(y), (\nu'_k \circ \nu_k)(z)\}\\ &\leq \max\{\nu_k(a), (\nu'_k \circ \nu_k)(xaxa)\} \leq \max\{\nu_k(a), \nu'_k(a), \nu_k(a)\}\\ &= (\nu \cup_k \nu')(a).\end{aligned}$$

Therefore $A \circ_k B \circ_k A \supseteq A \cap_k B$. Hence $A \circ_k B \circ_k A = A \cap_k B$.

Conversely, as S itself is an $(\in, \in \vee q_k)$-intuitionistic fuzzy ideal of S, $A_k = A_k \cap S = A \cap_k S = A \circ_k S \circ_k A$. Therefore, by Theorem 17, S is regular.

On the lines similar to the proof of Theorem 19, we get the following:

Theorem 20 *Let* S *be an ordered semigroup. Then the following statements are equivalent:*

(i) *S is regular;*
(ii) $A \circ_k B \circ_k A = A \cap_k B$ *for each* $k \in [0, 1)$ *and for each* $(\in, \in \vee q_k)$*-intuitionistic fuzzy generalized bi-ideal* A *and* $(\in, \in \vee q_k)$*-intuitionistic fuzzy ideal* B *of* S.

Theorem 21 *Let S be an ordered semigroup. Then the following statements are equivalent:*

(i) *S is regular;*
(ii) $A \circ_k B = A \cap_k B$ *for each* $k \in [0, 1)$ *and for each* $(\in, \in \vee q_k)$*-intuitionistic fuzzy right ideal A and* $(\in, \in \vee q_k)$*-intuitionistic fuzzy left ideal B of S.*

Theorem 22 *Let S be an ordered semigroup. Then the following statements are equivalent:*

(i) *S is intra-regular;*
(ii) $A \cap_k B \subseteq B \circ_k A$ *for each* $k \in [0, 1)$ *and for each* $(\in, \in \vee q_k)$*-intuitionistic fuzzy right ideal A and* $(\in, \in \vee q_k)$*-intuitionistic fuzzy left ideal B of S.*

Proof Suppose that S is an intra-regular ordered semigroup and $A = (\mu, \nu)$ and $B = (\mu', \nu')$ be any $(\in, \in \vee q_k)$-intuitionistic fuzzy right and $(\in, \in \vee q_k)$-intuitionistic fuzzy left ideal of S respectively. Let $a \in S$. Then $\exists\, x, y \in S$ such that $a \leq xa^2y$.

Now $(\mu' \circ_k \mu)(a) = (\mu'_k \circ \mu_k)(a) = \bigvee\limits_{(y,z)\in I_a} \min\{\mu'_k(y), \mu_k(z)\}$

$$\geq \min\left\{\mu'_k(xa), \mu_k(ay)\right\} \geq \min\left\{\min\left\{\mu'(a), \frac{1-k}{2}\right\}, \min\left\{\mu(a), \frac{1-k}{2}\right\}\right\}$$
$$= \min\left\{\mu'_k(a), \mu_k(a)\right\} = (\mu_k \cap \mu'_k)(a) = (\mu \cap_k \mu')(a).$$

and $(\nu' \circ_k \nu)(a) = (\nu'_k \circ \nu_k)(a) = \bigwedge\limits_{(y,z)\in I_a} \max\{\nu'_k(y), \nu_k(z)\}$

$$\leq \max\left\{\nu'_k(xa), \nu_k(ay)\right\} \leq \max\left\{max\left\{\nu'(a), \frac{1-k}{2}\right\}, \max\left\{\nu(a), \frac{1-k}{2}\right\}\right\}$$
$$= \max\left\{\nu'_k(a), \nu_k(a)\right\} = (\nu_k \cup \nu'_k)(a) = (\nu \cup_k \nu')(a).$$

Therefore $A \cap_k B \subseteq B \circ_k A$.

Conversely, let R and L be any right and left ideal of S respectively and $a \in R \cap L$. By Theorem 10, χ_R is an $(\in, \in \vee q_k)$-intuitionistic fuzzy right ideal and χ_L is an $(\in, \in \vee q_k)$-intuitionistic fuzzy left ideal of S respectively. By Proposition 2, we have $(\chi_{R \cup L})_k = \chi_R \cup_k \chi_L \subseteq \chi_L \circ \chi_R = (\chi_{(LR]})_k$, which implies that $R \cap L \subseteq (LR]$. So, by Lemma 10, S is intra-regular.

Theorem 23 *Let IFS* $A = (\mu, \nu)$ *and* $B = (\mu', \nu')$ *of an ordered semigroup S, then the following statements are equivalent:*

(i) *S is regular and intra-regular.*
(ii) $A \circ_k A = A_k$*, for every* $(\in, \in \vee q_k)$*-intuitionistic fuzzy bi-ideals of S.*
(iii) $A \cap_k B \subseteq (A \circ_k B) \cap (B \circ_k A)$*, for any* $(\in, \in \vee q_k)$*-intuitionistic fuzzy bi-ideals A and B of S.*
(iv) $A \cap_k B \subseteq (A \circ_k B) \cap (B \circ_k A)$*, for any* $(\in, \in \vee q_k)$*-intuitionistic fuzzy bi-ideals A and every* $(\in, \in \vee q_k)$*-intuitionistic fuzzy left ideals B of S.*

(v) $A \cap_k B \subseteq (A \circ_k B) \cap (B \circ_k A)$, *for any* $(\in, \in \vee q_k)$-*intuitionistic fuzzy right ideals A and every* $(\in, \in \vee q_k)$-*intuitionistic fuzzy bi-ideals B of S.*

(vi) $A \cap_k B \subseteq (A \circ_k B) \cap (B \circ_k A)$, *for any* $(\in, \in \vee q_k)$-*intuitionistic fuzzy right ideals A and every* $(\in, \in \vee q_k)$-*intuitionistic fuzzy left ideals B of S.*

Proof The proof follows on the lines similar to the proof of Theorems 21 and 22.

References

1. Atanassov, K.T.: Intuitionistic fuzzy sets. Fuzzy Sets Syst. **20**(1), 87–96 (1986)
2. Atanassov, K.T.: New operations defined over the intuitionistic fuzzy sets. Fuzzy Sets Syst. **61**, 137–172 (1994)
3. Bhakat, S.K., Das, P.: $(\in, \in \vee q)$-fuzzy subgroups. Fuzzy Sets Syst. **80**, 359–368 (1996)
4. Bhakat, S.K.: $(\in, \in \vee q)$-fuzzy normal, quasinormal and maximal subgroups. Fuzzy Sets Syst. **112**, 299–312 (2000)
5. Bhakat, S.K.: $(\in \vee q)$-level subset. Fuzzy Sets Syst. **103**, 529–533 (1999)
6. Biswas, R.: On fuzzy sets and intuitionistic fuzzy sets. NIFS **3**, 3–11 (1997)
7. Davvaz, B., Dudek, W.A., Jun, Y.B.: Intuitionistic fuzzy Hv-submodules. Inform. Sci. **176**, 285–300 (2006)
8. Davvaz, B.: $(\in, \in \vee q)$-fuzzy subnearrings and ideals. Soft Comput. **10**, 206–211 (2006)
9. Hong, Y., Jiang, C.: Characterizing regular semigroups using intuitionistic fuzzy sets. Notes IFS **10**, 20–31 (2004)
10. Jun, Y.B., Khan, A., Shabir, M.: Ordered semigroups characterized by their $(\in, \in \vee q)$-fuzzy bi-ideals. Bull. Malays. Math. Soc. **32**(3), 391–408 (2009)
11. Kazanci, O., Yamak, S.: Generalized fuzzy bi-ideals of semigroup. Soft. Comput. **12**, 1119–1124 (2008)
12. Kehayopulu, N., Tsingelis, M.: Fuzzy bi-ideals in ordered semigroups. Inf. Sci. **171**, 13–28 (2005)
13. Kehayopulu, N., Tsingelis, M.: The embedding of an ordered groupoids into a poe-groupoid in terms of fuzzy sets. Inf. Sci. **152**, 231–236 (2003)
14. Kehayopulu, N.: On intra-regular ordered semigroups. Semigroup Forum **46**, 271–278 (1993)
15. Kehayopulu, N.: On regular ordered semigroups. Math. Jpn. **45**, 549–553 (1997)
16. Kim, K.H., Jun, Y.B.: Intuitionistic fuzzy ideals of semigroups. Indian J. Math. Sci. **33**, 443–449 (2002)
17. Kuroki, N.: Fuzzy bi-ideals in semigroups. Comment. Math. Univ. St. Pauli **28**, 17–21 (1979)
18. Kuroki, N.: On fuzzy semigroups. Inform. Sci. **53**, 203–236 (1991)
19. Mordeson, J.N., Malik, D.S., Kuroki, N: Fuzzy Semigroups. In: Studies in Fuzziness and Soft Computing, vol. 131. Springer, Berlin (2003)
20. Murali, V.: Fuzzy points of equivalent fuzzy subsets. Inf. Sci. **158**, 277–288 (2004)
21. Pu, P.M., Liu, Y.M.: Fuzzy topology I, neighborhood structure of a fuzzy point and Moore-Smith convergence. J. Math. Anal. Appl. **76**, 571–599 (1980)
22. Rosenfeld, A.: Fuzzy subgroups. J. Math. Anal. Appl. **35**, 512–517 (1971)
23. Shabir, M., Khan, A.: Intuitionistic fuzzy filters of ordered semigroups. J. Appl. Math. Inform. **26**, 1071–1084 (2008)
24. Xie, X.Y., Tang, J.: Regular ordered semigroups and intra regular ordered semigroups in terms of fuzzy subsets. Iran. J. Fuzzy Syst. **7**(2), 121–140 (2010)
25. Xie, X.Y., Tang, J.: Fuzzy radicals and prime fuzzy ideals of ordered semigroups. Inform. Sci. **178**, 4357–4374 (2008)
26. Zadeh, L.A.: Fuzzy sets. Inf. Control **8**, 338–353 (1965)

On a Problem of Satyanarayana Regarding the Recognizability of Codes

R.D. Giri

Abstract M. Satyanarayana posed a problem in 2001: which infinite codes are recognizable. Recognizable code means a code accepted by a finite automaton. Equivalently a code X is recognizable iff $u^{-1}X$ is finite for $u \in A^*$. It is well known that finite codes are recognizable. We partially answer the problem of Satyanarayana. We know that a right complete semaphore suffix code is recognizable. We study here recognizablity of right complete semaphore codes with conditions other than suffix and further dropping semaphore condition also. Barring right completeness, the problem for general infinite codes is still open.

Keywords Recognizability of codes · Right complete codes · Semaphore codes

AMS Mathematics subject classification (2010) 20M45

1 Introduction

An alphabet A is a non empty set of symbols, called letters. A word w over A is a finite concatenation of letters, a word without any letter is called the empty word denoted by 1.

The set of all words including 1 under concatenation of words forms the free monoid A^* generated by A. However non empty words over A forms a semigroup A^+. Thus $A^+ = A^* - \{1\}$.

Any proper subset X of A^+ is called a code if a message over X is unique namely $x_1 \dots x_r = y_1 \dots y_s$, with $x_i, y_j \in X$ implies $r = s$ and $x_i = y_i$. The set of messages $x_1 \dots x_r$ forms a monoid X^*, called information.

R.D. Giri (✉)
RTM Nagpur University, Nagpur, India
e-mail: rdgiri39@yahoo.in

S.T. Rizvi et al. (eds.), *Algebra and its Applications*, Springer Proceedings in Mathematics & Statistics 174, DOI 10.1007/978-981-10-1651-6_27

A word $u \in A^*$ is called a left (right) factor of w if $uv = w$ $(vu = w)$ for some $v \in A^*$. If $v = 1$, the factor is called improper otherwise if $v \neq 1$, it is called proper factor. Note that 1 is always a left (right) factor of any word $w \in A^+$.

2 Preliminaries

In this section we provide some notation, basic definitions, and results.

For a code X, the set of all left factors is denoted by P_X and the set of all proper left factors is denoted by L_X that is, $L_X = P_X \backslash (X \cup 1)$.

The set of all right factors is denoted by R_X.

Some basic definitions are provided with specific references.

Definition 2.1 ([1]) A code X is called right complete if for every $u \in A^*$, there exists a word $v \in A^*$ such that $uv \in X^*$. In other words, every word in A^* can be completed on the right to a message.

Remark 2.1 (Theorem 3.3. [1]) A maximal prefix code can be defined as a right complete prefix code.

Definition 2.2 ([1, 3]) A code X over an alphabet A is said to satisfy F-1 condition if $A^+XA^+ \cap X = \phi$.

Definition 2.3 ([1, 3]) Maximal prefix codes which satisfy the F-1 condition are called semaphore codes.

Definition 2.4 ([1]) A code X is said to be suffix P_X-closed when $uv \in P_X$ implies $v \in P_X$ where P_X is the set of all left factors of X-words.

For example the sets $X = b^*a$ or $X = \{a, b^2, ba\}$ are suffix P_X-closed over $A = \{a, b\}$.

Definition 2.5 ([1, 3]) A code X over an alphabet A is called a thin code if there exists a word $u \in A^*$ such that $A^*uA^* \cap X = \phi$.

Definition 2.6 ([1]) The codes of the form A^n are called fixed length code or uniform code because each X-word has length n, where n is a natural number.

Definition 2.7 ([1]) For two non-negative integers such that $m + n \neq 0$, a code X is called an (m, n)-limited code if for any sequence $\{u_0, u_1, \ldots, u_{m+n}\}$ of words in A^*, the consecutive pairs $u_0u_1, u_1u_2, \ldots, u_{m+n-1}u_{m+n} \in X^*$ imply $u_m \in X^*$.

The examples of the above definitions can be found in [1].

In order to study recognizability of codes, we need to know the notions of language and automaton [2].

Any proper subset L of A^+ is called language and code is a special language wherein every message has a unique factorization. On the other hand combinatorial properties of a language are recognized by finite automaton where finite automata are defined as follows.

Definition 2.8 ([2]) A sequential machine denoted by a pair (I, R) is called a finite automaton (FA), where I denotes a finite set of internal states and R denotes a set of rules governing the current state after an input symbol.

Definition 2.9 ([1]) A code X over an alphabet A is called recognizable if it is accepted by a finite automaton. Equivalently according to Berstel and Perrin [1] (p. 15) the family of the non empty sets $u^{-1}X, u \in A^*$ is finite where $u^{-1}X = \{v \in A^* | uv \in X\}$. In fact the non empty subsets $u^{-1}X$ are called states of the minimal automaton accepting the code X.

In the entire text, we presume that no code is a subset of its alphabet. For undefined terms, we refer to [1].

The following well known results are frequently used in proving our main results. Their proofs are in [4].

Lemma 2.1 ([4, Theorem 1]) *Let X be a right complete code over an alphabet A. Then $A^* = X^*P$.*

Lemma 2.2 *If X is a semaphore code over an alphabet A then $R_X \subseteq X \cup L_X$.*

Proof Let $u \in R_X$, if $u \notin X \cup L_X$ then $u \notin L_X$ yielding $vu \in X$ for $v \in A^+$ therefore $uv \in X \cap A^+XA^+$ contradicting F-1 condition, hence $u \in X$. □

Lemma 2.3 *Let X be a $(1, 1)$ limited code over an alphabet A satisfying the F-1 condition, then $L_X \cap R_X \subseteq X$. In addition if X is a right complete code then $R_X \subseteq X$.*

Proof If $u \in L_X \cap R_X$ then, $xu, uy \in X$ for some $x, y \in A^+$. Thus by $(1, 1)$-limited condition, $u \in X^+$. If $u \in X^n$ for $n > 1$ then $xu \in X \cap A^+X^n$, contradicting F-1 condition for X. Hence $u \in X$. If X is a right complete code then by Lemma 2.2, $R_X \subseteq X \cup L_X$ yielding $R_X \subseteq X$. □

Lemma 2.4 *Let X be a semaphore code over the alphabet A and $X \neq A$. If $R_X \subseteq X$ then $X = (B')^*B$ where B', B are non empty subsets of A such that $A = B \cup B'$, $B \cap B' = \phi$ and $B = X \cap A$.*

Proof Since $R_X \subseteq X$, so every right factor of an X-word is an X-word. Note a right factor is a letter or concatenation of more than one letter which is in fact an $X-$word ending with a letter. Set $B = X \cap A \neq \phi$, $X \neq A$, $A = B \cup B'$, $B \cap B' = \phi$, $B' \neq \phi$. By the F-1 condition $A^+XA^+ \cap X = \phi$, no letter of B occurs as an internal factor of any X-word, hence $X \subseteq (B')^*B$. But $(B')^*B$ is prefix and X is maximal prefix, so $X = (B')^*B$. □

Proposition 2.5 *Let X be a semaphore code, $X \neq A$. Then the following are equivalent:*

(i) $X = (B')^*B$ *where B, B' are non empty subsets of A such that* $A = B \cup B'$, $B \cap B' = \phi$ *and* $B = X \cap A$
(ii) *X is a $(1, 0)$—limited code.*
(iii) *X is a $(1, 1)$—limited code.*

(iv) $R_X \subseteq X$ *where* R_X *has its usual meaning.*
(v) $L_X \cap R_X = \phi$ *where* L_X *and* R_X *have their usual meanings.*

Proof $(i) \Rightarrow (ii)$, follows from Theorem 10 ([3], p. 368).
$(ii) \Rightarrow (iii)$ As is well known ([1], p. 329)
an (m, n) limited code is also an $(m+s, n+t)$ limited code for $s, t \geq 0$
$(iii) \Rightarrow (iv)$ It follows by Lemma 2.3
$(iv) \Rightarrow (v)$. On the contrary if $u \in L_X \cap R_X$ then by Lemma 2.2, $u \in X \cup L_X$ which yields $ux \in X$, a contradiction to prefix property. Hence $L_X \cap R_X = \phi$
$(v) \Rightarrow (iv)$ Let $u \in R_X, u \notin X$. By Lemma 2.2, $R_X \subseteq X \cup L_X$, so $u \in R_X \cap L_X \neq \phi$, a contradiction, hence $R_X \subseteq X$. □

Corollary 2.1 *Let* X *be a semaphore code,* $X \neq A$. *If* $R_X \subseteq X$ *then* $X = (B')^* B$.

Proof Let $uxv \in X$ for $x \in X,\ u, v \in A^+$ which implies $xv \in R_X \subseteq X$. Clearly $x \in X$ is a left factor of an X-word xv contradicting prefix property of X, so by Proposition 2.5 $X = (B')^* B$. □

3 A Key Result

Now we prove a key result to establish the recognizability of semaphore codes.

Theorem 3.1 *Let* $X \neq \{a, b\}$ *be a semaphore code. If* $L_X \cap R_X \subseteq A$. *Then* X *has one of the following forms:*
(1) $\{a, ba, b^2\}$ *(2)* b^*a *(3)* $\{a^2, b^2, ab, ba\}$ *(4)* $\{a^2 \cup ab \cup b^+a\}$
(5) $\{b^2 \cup ba \cup a^+b^2 \cup ab \cup a^+ba\}$ *(6)* $\{a^+b^+a \cup b^+a\}$
(7) $\{ab^2 \cup (ab)^2 \cup b^+a^+b \cup a^+b\}$ *(8)* $\{ab^+ab \cup a^+b\}$
(9) codes obtained from above on interchanging a and b. The total partition classes are sixteen.

Proof Let M be the family of all codes under various possible options that are mutually pairwise disjoint such that there union is M itself.

Case 1: $a \in X,\ b \in X$. It is vacuous as $X \neq \{a, b\}$

Case 2: $a \in X,\ b \notin X,\ X \neq \{a, b\}$

Subcase 2.1: X words end with a so $ba \in X$.

Subcase 2.2: $b^2 \in X$

Considering the two subcases together we have $X = \{a, ba, b^2\}$

Subcase 2.3: $b^2 \notin X$ then $b^n \notin X$. Claim $b^n a \in X$. If not then by Lemma 3.1 $b^n a = X^r P_X$ for $r \geq 1$ which forces some power of b is an X-word which is false as $b \notin X$ and $b^n \notin X$. Now $b^n a \in X$ for all n hence $X = b^* a$.

Case 3: $a \notin X,\ b \in X$. It is dual to case 2, and lies in class (9).

Case 4: $a \notin X, b \notin X$. We obtain all possible codes systematically.

Subcase 4.1: Under above hypothesis, X-words end with both a as well as b. therefore $ab, ba \in X$. Further there are three more subcase of this;

Subcase 4.1.1: $a^2 \in X, b^2 \in X$, then required maximal prefix code is $X = \{ab, ba, a^2, b^2\}$.

Subcase 4.1.2: $a^2 \in X$ but $b^2 \notin X$, X-word does not end with b then $b^n \notin X$ for all natural numbers n i.e. $b^n t \in X$ for some $t \in A^+$. By assumption $L_X \cap R_X \subseteq A$, $X \neq A$ so $R_X \subseteq X \cup A$. Suppose $t \notin A$ then $t \in X$. Since X-word does not end with b so $t \notin b^+$ so $t = a$ or a^2. The second option is ruled out as it contradicts (F-1)-conditions as $ba \in X$. In first option we get $b^n a \in X$ for all n. Hence $X = \{a^2 \cup ab \cup b^+ a\}$.

Subcase 4.1.3: $a^2 \notin X$, $b^2 \in X$ and X-word does not end with a. Clearly $a^n \notin X$ as $a^2 \notin X$ for all values of n. We claim $a^n b^2, a^n ba \in X$. By Lemma 3.1 words, $a^n b^2 \in X^r P_X$, $a^n ba \in X^s P_X$ for $r, s \geq 1$. This shows $x = a^r \in a^+$ and $y = a^s b \in a^+ b$ are X-words contradicting $a^+ \cap X = \phi$ and $a^+ b \cap X = \phi$. Hence the claim. Therefore required code $X = \{b^2 \cup ab \cup a^+ b^2 \cup a^+ ba\}$.

Subcase 4.2: $a^2 \notin X$, $b^2 \notin X$

Subcase 4.2.1: X-words end with a but does not end with b. We obtain the required codes. Since $a^2 \notin X, b^2 \notin X$ so $a^+ \cap X = \phi$, $b^+ \cap X = \phi$. Since X-word do not end with b so $a^n b \notin X$ for all n. We note that $a^n b^m \notin X$ for all natural numbers $n, m \geq 2$. In case otherwise $a^n b^m = a(a^{n-1} b^{m-1}) b \in A^+ X A^+ \cap X$ a contradiction to (F-1)-condition. Let $a^n b^m a \notin X$. Then by Lemma 3.1 $a^m b^m a = x_1.x_2 \ldots\ldots x_r q$, $x_i \in X, 1 \leq i \leq r, q \in P_X$. For $r \geq 1$, we obtain $x_1 \in \{a^+ \cup a^+ b^+\}$ which is false, since $ba \in X$, X is prefix and satisfies (F-1)-condition so $a^n b^m a \notin P_X$. Therefore $a^n b^m a \in X$ for all natural numbers n and m. Similarly we can show $b^n a \in X$. Hence maximal prefix code is $X = \{a^+ b^+ a \cup b^+ a\}$.

Subcase 4.2.2: X-words end with b but do not end with a. We further divide into subcases.

Subcase 4.2.2 (i): In addition to above, we have condition that $ab^2 \in X$. Since $aba \notin X$ so $aba \in P_X$ yielding $(ab)^2 \in X$ since X-words do not end with a so $b^n a \notin X$ for all n. Dual to Case 4.2.1 we obtain $b^n a^m \in X$ for all $n, m \geq 2$. Similarly we show that $a^n b \in X$. Hence maximal prefix code is

$$X = \{ab^2, (ab)^2 \cup b^+ a^+ b \cup a^+ b\}$$

Subcase 4.2.2 (ii): In addition to above $ab^2 \notin X$ then $ab^n \notin X$. Clearly $ab \in X$. As X-word does not end with a so $ab^n a \notin X$ and hence $ab^n a \in P_X$ for all n. Therefore $ab^n ab \in X$ for all $n \geq 1$. Further $a^2 \notin X$ implies $a^n \notin X$ so $a^n b \in X$. If not then $a^n b \in P_X$ for all n i.e. $b^n t \in X$ for some $t \in A^+$. By assumption $L_X \cap R_X \subseteq A$, $X \neq A$ so $R_X \subseteq X \cup A$. Suppose $t \notin A$ then $t \in X$. Since X-words do not end with a so $t \in b^+$ i.e. $t = b$ or $t = b^2$. Second option is ruled out as it leads to a contradiction to (F-1)-condition as $ab \in X$. For first option $t = b$; $a^n b \in X$. Therefore maximal prefix code $X = \{ab^+ab \cup a^+b\}$. **Subcase 4.3**: Neither is there any word ending with a not is there any word ending with b. This case is vacuous as X-word has to end with a or b.

This completes the partition of all possible mutually disjoint sets of codes and the proof. □

4 Recognizability

Now we are ready to prove some results on recognizability.

Theorem 4.1 *Let X be a semaphore code, $X \neq A$. Then X is recognizable if it satisfies anyone of the equivalent conditions of Proposition 2.5.*

Proof By virtue of Proposition 2.5, it is sufficient to prove that $X = (B')^*B$ is recognizable. Note that, $u^{-1}X \neq \phi$ iff $u \in X \cup P_X$. Let $u \in X \cup \{1\}$ and $u^{-1}X \neq \phi$ then $u \in L_X = (B')^+$. If $uy \in X$ with $y \in A^+$ then $y \in X$ by condition (iv) of Proposition 2.5 therefore $u^{-1}X \subseteq X$. If $x \in X$ then $ux \in (B')^+B \subseteq X$. Thus $X = u^{-1}X$. Thus X is recognizable. □

Corollary 4.1 *Let X be a maximal prefix code $X \neq A$ then X is recognizable code if $R_X \subseteq X$.*

Proof It follows from Corollary 2.1 and Theorem 4.1. □

Theorem 4.2 *Let X be a semaphore code over $A = \{a, b\}$; $X \neq A$. Then X is recognizable code if $L_X \cap R_X \subseteq A$.*

Proof By virtue of key Theorem 3.1, we prove that all subsets given in the theorem are recognizable. This follows from the fact that each set is given by a regular expression. □

Now we turn our attention to non-semaphore codes and non-recognizability.

It is well known (Theorem 5.5 [1]) that semaphore codes are suffix P_X-closed. It is obvious that suffix P_X-closed sets b^*a and $\{a, ba, b^2\}$ are recognizable. First we look at what happens when the maximal prefix property (right completeness) of semaphore code is weakened and suffix P_X-closed is retained.

Proposition 4.1 *Let X be a prefix code over an alphabet $A = \{a, b\}$ containing a. If X is suffix P_X-closed then $X \subseteq a \cup b^m \cup b^+a$ where m is some fixed natural number.*

Proof Clearly $b^m a^n \in X$ and $b^m a^n A^+ \nsubseteq X$ by prefix property of X. But if $b^m a^n \in X$ then $b^m a \in P_X$ which gives $a \in P_X$ by suffix P_X-condition of X. Since X is prefix so $a \notin P_X$, a contradiction. Hence $b^m a \in X$. Now factorizing $b^m a$ into X-words and $a \in X$ yields some power of b is an X-word by definition of code. Hence $X \subseteq a \cup b^m \cup b^+a$ for a fixed $m \geq 1$. □

Now we see the conditions of recognizability, right completeness and suffix ness give specific structure to the code.

Proposition 4.2 *Let X be a recognizable, right complete and suffix code over an alphabet A. Then for any $a \in A$, X contains a power of a.*

Proof Any $w \in X$ has to have all letters a only yielding the assertion. □

Now we give two results on non-recognizatibility.

Theorem 4.3 *Let X be a right complete suffix code over an alphabet $A = \{a, b\}$ and $X \nsubseteq A$. If X contains a letter a and $(ba)^n b \notin X$ for all natural numbers n. Then X is non recognizable.*

Proof Since $X \neq A, a \in X$ so $b \notin X$. By contrapositive definition of suffix code $(ba)^n b \notin X$ implies $(ba)^n \notin X$ for every natural number n. By Lemma 2.1 we have $(ba)^n b = x_1 \ldots x_r.q;\ x_i \in X, q \in P_X \subset A^+$. By observation $x_1 = (ba)^n t_1$ for some $t_1 \in A^+ \Rightarrow (ba)^n b = (x_1)x_2 \ldots x_r q = (ba)^n t_1 x_2 \ldots x_r q \Rightarrow b = t x_2 \ldots x_r q$ since $t, q \in A^+$ so length of LHS is 1 and RHS is greater than or equal to 2, a contradiction. Therefore $(ba)^n = x_1 t_1$ for $t \in A^+$ as X is prefix. If $n = 1,\ ba = x_1 t_1$ implying $b = x_1 \in X$ a contradiction to $b \notin X$. Similarly for $n > 1$ $(ba)(ba)^{n-1} = x_1 t_1 (ba)^{n-1} t_2$.

Now $(ba)^{n-1} \notin X$ so $(ba)^{n-1} t_2 \in X$ for $t_2 \in A^+$. Hence $ba = t_2 t_1$ yielding $b = t_2$. This shows $(ba)^n b \in X$, a contradiction to hypothesis. Therefore $(ba)^{n-1} = x_1 t_2 \in XA^+$. This contradiction proves $(ba)^n b \in P_X$ for all natural numbers n. Hence $\{[(ba)^n b]^{-1} X\}$ is an infinite family, proving X is not recognizable. □

If any one of the conditions in the hypothesis of the theorem fails then the theorem breaks down.

Counter Example 4.1 Let $X = a \cup b^3 \cup b^2 a \cup (ba)^+ b^2 \cup (ba)^+ a^+ b$ since $baab$, $(ba)^2 ab \in X$ so X is not suffix code. Further $a \in X$, X is maximal prefix so it is right complete. For every natural number n, $[(ba)^n]^{-1} X = (ba)^* b^2 \cup (ba)^* a^+ b$, $[(ba)^n b]^{-1} X = a(ba)^* b^2 \cup a(ba)^* a^+ b$ and $[(ba)^n a^m]^{-1} X = a^* b$ for all n and m so X is recognizable. The literal diagram is as follows.

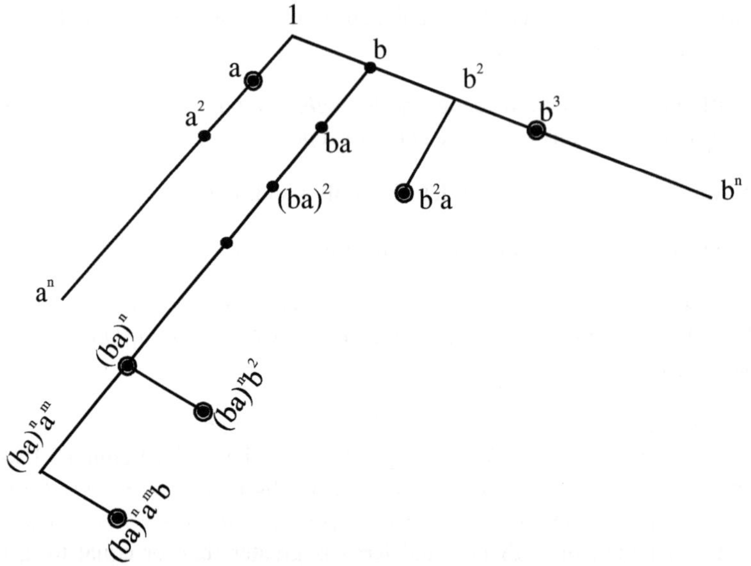

Counter Example 4.2 Let $X = \{a^2, b^2, ab, ba\}$. Clearly X satisfies all conditions except the letters are not X-words namely $A \cap X = \phi$. So X is recognizable being a finite code.

Counter Example 4.3 Let $X = a \cup b^2 \cup ba^+b$ whose literal diagram is as follows:

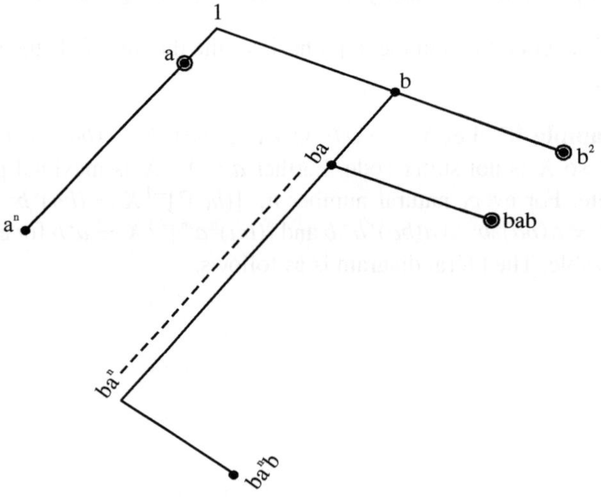

Clearly X is maximal prefix code hence right complete also. But $bab \in X$ is violation of the one of the hypotheses. Since $(ba^n)^{-1}X = A^*b$, for every natural number n. Hence X is a recognizable code.

Theorem 4.4 *Let X be a right complete suffix code over an alphabet $A = \{a, b\}$ and $X \not\subseteq A$. If no letter is an X-word, $(ba)^n b \notin X$ for any natural number n and $aba \in X$, then X is not a recognizable code.*

Proof Given that $(ba)^n = (ba)^{n-2}.b(aba) \notin X$ but $aba \in X$ so by the suffix property $(X \cap A^+X) = \phi$ we have $(ba)^n \in A^+X$. We claim that $(ba)^n \in P$. On the contrary let $(ba)^n \notin P$. By Lemma 2.1, $(ba)^n = x_1 \dots x_r.q$ with $x_i \in X, q \in P$. Since $(ba)^n = (ba)^{n-1}(ba) \notin X$ so by suffix property $(ba)^{n-1} \notin X$ i.e. $(ba)^{n-1}.t \in X$ for some $t \in A^+$. That is $ba = tx_1 \in tA^+$ implying $t = b$ so $(ba)^{n-1}b \in XA^+$ as $(ba)^{n-1} \notin X$. Using the fact $(ba)^n b \notin X$ for every n, on reduction, in a finite number of steps we have $ba \in XA^+$ implying $b \in X$, contrary to our hypothesis. This proves our claim $(ba)^n \in P$ for every n. Thus $\{[ba]^{-1}X\}$ is an infinite family of subsets of X which are mutually distinct by the suffix condition. Hence X is not recognizable.

Counter Example 4.4 The code $X = \{a^2, b^2, ab, ba\}$ given in counter Example 4.2 works for this theorem, because it satisfies all conditions except $aba \in X$.

Counter Example 4.5 The code $X = a^2 \cup b^2 \cup ab^+a \cup ba^+b$ is given by following literal diagram:

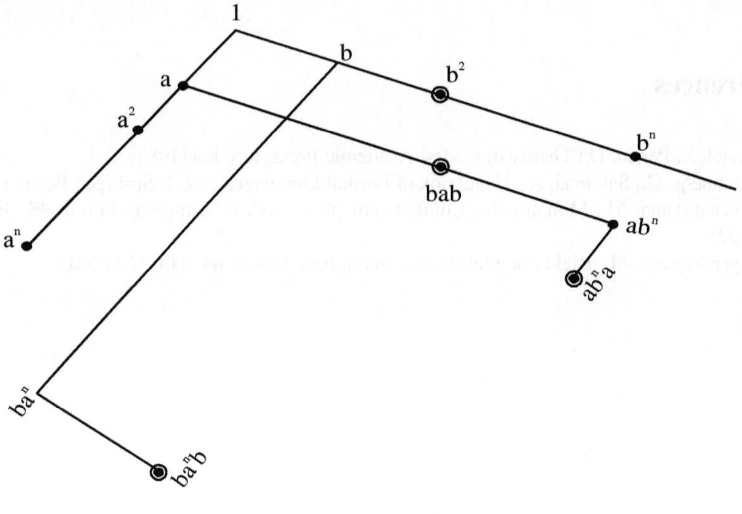

Clearly X is maximal prefix and suffix code. The condition $(ba)^n b \notin X$ of the Theorem 4.5 is violated as $bab \in X$. Hence X is recognizable because $(ab^n)^{-1}X = b^*a$ and $(ba^n)^{-1}X = a^*b$ for all natural numbers n.

Counter Example 4.6 Let $X = a^2 \cup b^2 \cup ba \cup ab^+a$. It is maximal prefix. But it is not suffix as $aba \in X$ and $ba \in X$. However condition $(ba)^n b \notin X$ holds because in particular if $bab \in X$ and $ba \in X$ will violate prefix nature of X. Further $(ab^n)^{-1}X = b^*a$ for every natural number n. Hence X is recognizable.

Problem 4.1 Prove Theorem 4.3 with the weaker condition $(ba)^n b \notin X$ for $n > 1$.

Remark 4.1 Since the recognizable codes are thin ([1], p. 69) and then right complete codes are maximal prefix codes ([1] Theorem 3.7, p. 101), the recognizability of right complete codes reduces to maximal prefix codes. Therefore we have the following problem:-

Problem 4.2 Find conditions under which suffix maximal prefix codes are recognizable.

This problem arises because we have an example of right complete code which is suffix over $A = \{a, b\}$ and $a \in X$ Further it violates condition $(ba)^n b \notin X$ for natural numbers n of the Theorem 4.3, yet it is not recognizable, contrary to counter Examples 4.1–4.3. This is because, the code is not thin ($X \cap A^+ X A^+ = \phi$). The example is as follows:-

Counter Example 4.7 $X = \{uab^{|u|} | u \in A^*\}$ over $A = \{a, b\}$. This is not thin because $bab \in X, b^2ab^2 \in X$ so $b^2ab^2 \in X \cap A^+ X A^+ \neq \phi$, contrary to the definition of thin.

Acknowledgments I am thankful to Professor M. Satyanarayana for his fruitful discussion during his visit to me in 2001. I am also grateful to Helmut Jurgensen for useful suggestions for revision.

References

1. Berstel, J., Perrin, D.: Theory of Codes. Academic Press, London (1985)
2. Rozenberg, G., Salomaa, A.: Handbook of Formal Languages, vol. 1. Springer, Berlin (1997)
3. Satyanarayana, M., Mohanty, S.: Limited semaphore codes. Semigroup Forum **45**, 364–371 (1992)
4. Satyanarayana, M.: Right complete codes. Semigroup Forum **64**, 443–452 (2002)

MIX
Papier aus verantwortungsvollen Quellen
Paper from responsible sources
FSC® C105338

If you have any concerns about our products,
you can contact us on
ProductSafety@springernature.com

In case Publisher is established outside the EU,
the EU authorized representative is:
Springer Nature Customer Service Center GmbH
Europaplatz 3, 69115 Heidelberg, Germany

Printed by Libri Plureos GmbH
in Hamburg, Germany